Fundamental Concepts in Electrical and Computer Engineering with Practical Design Problems

Second Edition

Reza Adhami
Peter M. Meenen, III
Dennis Hite

Universal Publishers
Boca Raton, Florida

Fundamental Concepts in Electrical and Computer Engineering with Practical Design Problems
Second Edition

Universal Publishers
Boca Raton, Florida • USA
2007

ISBN: 1-58112-971-8 (paperback)
13-ISBN: 978-1-58112-971-7 (paperback)

ISBN: 1-58112-973-4 (ebook)
13-ISBN: 978-1-58112-973-1 (ebook)

www.universal-publishers.com

Preface

Purpose

In many cases, the beginning engineering student is thrown into upper-level engineering courses without an adequate introduction to the basic material. This, at best, causes undue stress on the student as they feel unprepared when faced with unfamiliar material, and at worst, results in students dropping out of the program or changing majors when they discover that their chosen field of engineering is not what they thought it was. The purpose of this text is to introduce the student to a general cross-section of the field of electrical and computer engineering. The text is aimed at incoming freshmen, and as such, assumes that the reader has a limited to nonexistent background in electrical engineering and knowledge of no more than pre-calculus in the field of mathematics. By exposing students to these fields at an introductory level, early in their studies, they will have both a better idea of what to expect in later classes and a good foundation of knowledge upon which to build.

Organization

This text consists of twenty chapters organized into five distinct units. In the first unit, the student is presented with a review of some key mathematical concepts. This overview is designed to fill in any gaps in the student's previous mathematical background and to provide a foundation for their study of the remaining chapters of the text. The first unit consists of three chapters. Each of these chapters focuses on a different important area of mathematics with which the beginning engineering student should be acquainted. The topics, covered in the first three chapters of the text, include, numeric bases, complex numbers, and basic elements of vectors and matrices.

Chapter 1 introduces the student to the concept of numerical bases. They will learn what a base is, how to convert back and forth between a given base and decimal, and how to work with some of the most common numerical bases, including binary, octal, and hexadecimal.

Chapter 2 covers the basics of complex numbers. The student will learn what a complex number is, how to represent these numbers in both the Cartesian and polar coordinate systems, how to convert between these systems, and how to perform basic arithmetic operations using complex numbers.

Chapter 3, the final math review chapter, introduces the student to some basic concepts in linear algebra. The chapter begins with the introduction of vectors and some simple vector arithmetic. The focus is then expanded to include matrices. Basic matrix arithmetic, such as addition, subtraction, and multiplication, is covered. Other basic matrix operations are also introduced. These include the concepts of matrix transpose, minors, cofactors, determinants, and inverses. The chapter concludes with an application of the matrix inverse. The use of the matrix inverse to obtain the solution of systems of linear equations is covered and related to the practical engineering application of circuit analysis.

Once the basics in mathematics have been established, the student may move on to the other four units of the text. The second unit consists of three chapters that are designed to provide the student with a better understanding of basic concepts in the area of electrical engineering.

Chapter 4 focuses on providing the student with the basics of electricity. In addition, the electrical components that are used to build electronic devices are examined.

Chapter 5 provides the student with an overview of electrical power and the power industry. The student is exposed to topics including the generation and supply of electrical power to the end user.

Chapter 6 introduces the student to the concepts involved in basic circuit analysis. They are first presented with the basics of resistors in series and parallel and taught how to find the equivalent resistances of these configurations. They then learn about voltage and current dividers and Kirchoff's voltage and current laws. Finally, the student is presented with two approaches to more complicated circuit analysis. The basics of both mesh and nodal analysis are presented.

The third unit of the text is designed to provide the student with a background in important computer engineering concepts. Its two chapters provide the student with basic concepts in the areas of digital logic and computer organization.

Chapter 7 presents the basic concepts of digital logic. The student is introduced to the basic logic gates, truth tables and Boolean expressions. They are taught how to analyze a logic circuit to create truth tables and how to build a logic circuit that satisfies a given set of conditions. Finally, students are taught the basics of logic simplification and introduced to the practical considerations involved in the construction of a logic circuit from physical components.

Chapter 8 continues with the student's introduction to computer engineering topics by describing the organization of a typical computer. The various functional components that make up a computer are described and the student is introduced to the way in which these components work together to perform the computer's basic functions.

The fourth unit covers the areas of digital signal processing, image processing, and communication. The four chapters in this unit provide the student with an overview of DSP, sound and image processing, communication theory, and information coding.

Chapter 9 is designed to provide the student with a basic introduction to the area of digital signal processing. The student is presented with an introduction to signals and systems. In addition, some of the basic operations involved in basic signal processing are introduced on a limited scale.

Chapter 10 provides an application of digital signal processing to the area of sound. The concept of audio signals is introduced, and several applications of digital signal processing within this area are presented.

Chapter 11 continues the examination of digital signal processing applications by looking at two-dimensional signals or images. The basics of photography and human vision are presented to show the importance of analog signals. Digital equivalents are given and the concepts of filtering, edge detection, and noise removal are explored.

Chapter 12 deals with the concepts involved in communication. The five parts of a basic communication system are examined. In addition, this chapter also explores the concept of encoding information for ease of transmission and the detection or correction of errors. Several coding techniques are presented.

The fifth and final unit provides several laboratory tutorials (modules), exercises, and design problems written to provide students with an introduction to several concepts and tools they will be utilizing throughout their education in engineering.

Chapter 13 is a self contained Module that introduces students to the MATLAB software package. Topics covered include MATLAB's basic commands, functions, plotting, and MATLAB M-Files

Chapter 14 introduces students to the CAD software package Multisim. Several topics are included such as voltage, current, power, and resistance measurements. This material can be used as a precursor to having the students make real measurements in the laboratory.

Chapter 15 introduces students to serial communications and the RS232 standard utilizing a personal computer and the Windows communication software HyperTerminal.

Chapter 16 covers the topic of operational amplifiers. It provides students with enough information to use op amps as building blocks in simple circuits. Several basic op amp circuits are discussed including the inverting, non-inverting, current to voltage, and summing amplifier configurations.

Chapter 17 introduces basic diodes. It introduces the general purpose silicon diode, light emitting diode (LED), and photodiode. In section 2 (optional) the characterization of semiconductor materials and a more detailed explanation of the pn junction are provided.

Chapter 18 introduces the concepts needed to understand the basic behavior of light and how fiber optic cable acts as a waveguide. The topics of light reflection, refraction, and total internal reflection are discussed.

Chapter 19 introduces students to the concept of PCB (Printed Circuit Board) design utilizing the software package ExpressPCB, A package available for free download at the ExpressPCB website. The software is introduced by having the student enter a simple predefined board design.

Finally, Chapter 20 contains several exercises, and an example design exercise is also presented. The exercises have been designed to provide students with practice applying the material presented in Chapters 13 through 19.

The goal of each of the chapters in this text is not to give the student an in-depth understanding of the topics that they cover, but rather to expose the student to the basic ideas and concepts at a level that their background allows.

Usage

The organization of this text allows instructors to present the material in a variety of different ways. Each of the chapters is self-contained and can stand alone should an instructor desire to present it in such a way. The recommended approach to this text is to present the first three chapters (Unit 1) first. After the students have the required math background, the remaining chapters can be approached in virtually any order.

About the Authors

Dr. Reza Adhami

Dr. Reza Adhami serves as both a professor and the Chair of the Electrical and Computer Engineering Department at The University of Alabama in Huntsville (UAH). He has been a member of the electrical engineering faculty at UAH since 1983. During that time he has held many positions including Instructor, Associate Professor, Assistant Professor, Professor, Chair of the department, and Chair of the Inter-Campus Program Coordinating Board for the Shared Ph.D. degree in Computer Engineering at the University of Alabama in Birmingham and The University of Alabama in Huntsville. His research interests include communication systems, signal processing, image processing, biomedical systems, and VLSI. More specifically, his work has involved such topics as wavelet theory and its applications in signal processing, biomedicine and biometrics. As a faculty advisor, he has successfully supervised numerous Masters Theses and Ph.D. dissertations. He has authored/co-authored over 100 technical papers that have been presented and published in many technical journals and conferences. Dr. Adhami has multiple years of industrial experience during which he has designed and implemented numerous hardware/software systems.

Dr. Peter Meenen

Dr. Peter Meenen has more than seven years of experience teaching introductory engineering concepts to students, primarily at the University of Alabama in Huntsville. He obtained a Bachelor of Science degree from Berry College in the areas of Physics and Computer Science, and a Master of Science and Ph.D. from the University of Alabama in Huntsville in the area of Electrical Engineering with an emphasis in signal processing and pattern recognition. He currently works in industry as a senior engineer where he focuses on developing technology and products in the areas of medical imaging and biometric security.

Mr. Dennis Hite

Mr. Dennis Hite has more than 10 years of experience managing and teaching introductory physics and engineering laboratories. He obtained a Bachelor of Science degree from Purdue University in Physics, and a Master of Science in Engineering with an option in Electrical Engineering from the University of Alabama in Huntsville. He currently holds the position of Lecturer/Information Technology Specialist at the University of Alabama in Huntsville.

Table of Contents

Table of Contents

Table of Contents

Table of Contents

UNIT 1

Basic Mathematics

Chapter 1. Numeric Bases

Chapter Outline

1.1   Introduction to Numeric Bases

No matter what your occupation, you will encounter and work with numbers on a regular basis. Whether you are making change at a cash register, balancing your checkbook, or calculating power consumption of an electrical circuit, you must have a basic understanding of how numbers work. Take for example, the simple task of counting from 0 to 100. While this is a trivial task, you must know something about numbers to accomplish it. You begin counting using a single placeholder, the "ones" place. You know that when you reach the largest digit that can fit in this place (in this case 9) you must add an additional placeholder to continue. Thus, to proceed beyond 9, you add the "tens" place. One is added to this new place, and the original "ones" place is reset to zero. By doing so, you can proceed from 9 to 10. Likewise, each time you reach a 9 in the "ones" place, the next number is determined by incrementing the "tens" place and resetting the "ones" place. Following this procedure, you can count all the way up to 99. At this point, an additional placeholder, the "hundreds" place is used. By extending this procedure, it is possible to count to numbers far larger than 100.

While this example may seem obvious, it illustrates an important point about numbers. As you saw, there was a particular number that, to be properly represented required the addition of an additional placeholder. In this example we found that each time a placeholder reached the number 10, we had to add a new placeholder to our number in order to represent it. When the "ones" place reached 10 we had to add the "tens" place, and when the "tens" place reached 10, we had to add the "hundreds" place. The number that, once reached, requires this addition of a new placeholder is called the radix. So, from our example, we find that the numbers we are used to working with have a radix of 10. Knowing the radix gives us important information about the number system that we are using, thus, the radix number is often referred to as the base of the number system. In other words, we typically work with numbers using a radix of 10, so we can say we use a base 10 number system. This base 10 system is often referred to as the decimal number system (from the Latin word "Decima" which means a tenth).

The decimal number system is not the only one used, however. Many applications, especially in electrical engineering and other technical fields, require the use of number systems based on a different radix. One good example is the binary, or base 2, number system. It is commonly used in the areas of digital logic and computers. These areas deal with signals that can have only two possible states, on or off. Thus a number system with a radix of two is perfectly suited for describing events in these systems.

Before we look too closely at the use of other bases, let's get a better understanding of the mechanics of numbers in the decimal system. In order to see exactly how decimal numbers work, let's look at an example of a typical decimal number. In this case, let's use the decimal number $(1279.875)_{10}$. You may have noticed that this number is not written quite the way you might expect. The parentheses and subscript 10 is part of a notation method that is often used when dealing with numbers of different bases. It helps us avoid confusion by keeping track of what base each number is in. If we did not specify the base each number is in, it would be easy to get confused, and that could lead to unwanted errors. Now that we are comfortable with this notation, let's look at what this decimal number is really telling us. Figure 1.1, shown below, breaks the number down into its respective places.

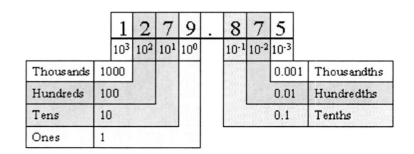

Figure 1.1. An example of a typical decimal (Base 10) number.

As you can see, by breaking the number down, we find that each of the digits gives us important information. The numeric value in each of the placeholders indicates the degree to which its respective power of the base is represented in the total number. In this example, it is easy to see that there are 1×10^3 or 1 "thousand," 2×10^2 or 2 "hundreds," 7×10^1 or 7 "tens," 9×10^0 or 9 "ones," 8×10^{-1} or 8 "tenths," 7×10^{-2} or 7 "hundredths," and 5×10^{-3} or 5 "thousandths" in the number. As a result the final number can be thought of as $(1 \times 10^3) + (2 \times 10^2) + (7 \times 10^1) + (9 \times 10^0) + (8 \times 10^{-1}) + (7 \times 10^{-2}) + (5 \times 10^{-3})$ or $1000 + 200 + 70 + 9 + 0.8 + 0.07 + 0.005 = 1279.875$. While this may seem obvious when working with decimal numbers, the same concept can be used to work with numbers of any base. To better understand this, let's first look at how this works for an arbitrary base b. We will then examine numbers in some of the more commonly used bases.

1.2   Working With Arbitrary Bases

While working with numbers in a specific base is useful, it is far more beneficial to be able to work with numbers in any base. Since you never know when you may encounter numbers in an unfamiliar base, it is important to have general tools that can be adapted to work with any base you choose. In this section, we will explore the basics of bases. We will learn how counting works in different bases, and we will learn how to convert numbers between an arbitrary base and decimal.

1.2.1   Counting in an Arbitrary Base

Once we have determined what base we are working with, let's call it b, we automatically know two important things about our number. First, we know that each digit or placeholder in the number is going to represent a particular power of the base, and second, there are exactly b values that can be used in each placeholder. These values will range from 0 to $(b - 1)$. As a result, when we count in a base b number system, each placeholder will begin at 0 and increment up to $b - 1$. Once a placeholder has reached $b - 1$, a new placeholder must be added to count beyond that point.

To help clarify this concept, let's look at an example.

Example 1.1
_____

Problem: Count to $(20)_{10}$ in base 5

Solution: Since we have been given a base to work in, we can assign a value to our arbitrary base b. In this case let $b = 5$. Since we now know that b is equal to 5, we know that each digit of any number represented in this base gives us an indication of the degree to which each power of five is represented in the number as a whole. We also know that each digit of the number will range from 0 to $(5 - 1)$ or 4. So, let's count to $(20)_{10}$ in base 5.

Decimal	Base 5
0	0
1	1
2	2
3	3
4	4

As you can see we have reached the value of $b - 1$, which in this case is 4. Since our radix is 5, we know that when a placeholder reaches 5 a new placeholder must be added to represent the number. Let's add a placeholder and continue.

Decimal	Base 5
5	10
6	11
7	12
8	13
9	14
10	20

At this point, we have reached $(20)_5$. However, this does not mean we have finished counting. If we look at the table, we will see that $(20)_5$ is equivalent to only $(10)_{10}$. We are only half - way there. Let's count the rest of the way now.

Decimal	Base 5
11	21
12	22
13	23
14	24
15	30
16	31
17	32
18	33
19	34
20	40

As you can see, we have now reached $(20)_{10}$ in the base five number system. From the chart we see that $(40)_5$ is equal to $(20)_{10}$.

This example gives us a good idea of how counting in a base other than decimal works. At this point you may ask, is there any way to tell what a number in some other base b will equal in decimal without building a chart and counting? The answer is yes.

1.2.2   Converting an Arbitrary Base to Decimal

The process used to determine the decimal equivalent of a number expressed in a different base b is actually very simple. The key is to remember the function of each of the placeholders in the number. Each placeholder determines to what degree a certain power of the base is represented in the final number. In addition, each placeholder to the left of the radix point (note that this is what a decimal point is called in number systems that are not base 10) represents an increasing positive power of the base, and that each placeholder to the right of the radix point represents an increasing negative power of the base. Let's take a look at what this would look like for a general base b.

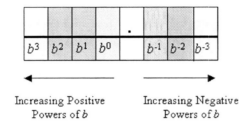

Increasing Positive
Powers of b Increasing Negative
                                Powers of b

Figure 1.2. An illustration of the relationship between each placeholder and its respective power of the base.

Since each placeholder represents a specific power of the base, we can obtain the decimal value represented by a given placeholder by simply multiplying the value in that placeholder by the value obtained by raising the base to the appropriate power. If we calculate this value for each placeholder and sum the results, we will arrive at the decimal equivalent of the number represented in base b. To illustrate this, let's look at base 5 once again. Figure 1.3 displays the powers associated with each placeholder in the base 5 system.

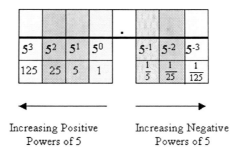

**Figure 1.3 A representation of the powers of 5 associated
with the placeholders in a base 5 number.**

Continuing with our base 5 example, let's find the equivalent decimal number for $(13412.213)_5$.

Example 1.2

Problem: Find the equivalent decimal number for $(13412.213)_5$.

Solution: Our first step in converting this number to its decimal equivalent is to multiply the value in each placeholder by its respective power of the base, in this case 5. The following table illustrates this process.

Placeholder Value	Associated Power of the Base	Resulting Product
1	5^4=625	1 x 625 = 625
3	5^3=125	3 x 125 = 375
4	5^2=25	4 x 25 = 100
1	5^1=5	1 x 5 = 5
2	5^0=1	2 x 1 = 2
.	Radix Pt.	
2	5^{-1}=0.2	2 x 0.2 = 0.4
1	5^{-2}=0.04	1 x 0.04 = 0.04
3	5^{-3}=0.008	3 x 0.008 = 0.024

Now that we know the decimal value associated with each placeholder, all we have to do is add these values together to arrive at our answer.

$$
\begin{array}{r}
625.000 \\
375.000 \\
100.000 \\
5.000 \\
2.000 \\
0.400 \\
0.040 \\
+\quad 0.024 \\
\hline
1107.464
\end{array}
$$

So, now we know that $(13412.213)_5$ is equivalent to $(1107.464)_{10}$.

In the preceding example, we found the decimal equivalent of a base 5 number. The method we used in this example will work to convert a number in any base to its decimal equivalent. Simply multiply the value in each placeholder by its respective power of the base and add the results.

> **MATLAB:** MATLAB has a function that allows for the conversion of an arbitrary base to its decimal equivalent number. This function has the following usage syntax.
>
> BASE2DEC(S,B)
>
> Where S is a string containing the number in base B, and B is the base that the number is in. Keep in mind that this function only converts positive integer values. It will not work on negative numbers or numbers containing fractional parts.
>
> **Example:** Base2Dec('13412',5) will return 1107 which is the decimal equivalent of the specified base 5 number.

1.2.3   Converting Decimal to an Arbitrary Base

Now that we know how to find the decimal equivalent of a number given in another base, our next goal is to learn how to work in the other direction. We want to be able to represent a given decimal number in another base. The procedure required to perform this conversion is slightly more complicated than the conversion to decimal, but it is still not too difficult. Unlike the process of converting a given base to decimal, the reverse process requires separate processing for the whole and fractional parts of the number.

Let's look at the whole number part first. The easiest way to determine the representation of the whole number portion of a decimal number in a different base is by successively dividing the decimal number by the radix of the new base and keeping track of the remainder. This sounds difficult at first, but is actually quite easy. If we wish to perform the conversion of a whole number from decimal to base b, we must divide our number by b. The remainder of this division will be the rightmost digit of our base b number. Next, we divide the quotient from the previous division by b. The remainder from this division is the next digit in our base b number. This division process is repeated until the quotient is zero. At this point we have successfully converted our decimal number to the new base. Figure 1.4 illustrates this process.

$$
\begin{array}{r}
0 \\
\overline{b\,)\,Q_3} \\
\overline{b\,)\,Q_2} \\
\overline{b\,)\,Q_1} \\
\overline{b\,)\,Decimal}
\end{array}
\quad
\begin{array}{l}
R_4 \\
R_3 \\
R_2 \\
R_1
\end{array}
$$

$$(\text{Decimal})_{10} \longrightarrow (R_4 R_3 R_2 R_1)_b$$

Figure 1.4. Division method for base conversion.

Having completed the conversion of the whole number portion of the decimal number, we can now focus on the fractional portion. While the whole number portion was converted using division, the fractional portion of the number is converted using multiplication. In this case, the fractional part of the number is multiplied by the radix of the new base. After the multiplication we record the whole number portion of the result. Then, if the fractional part of the number is different from any of the previously encountered fractional parts, we continue to multiply and track the whole number portion. It should be noted that the multiplication at each stage is applied to only the fractional part of the number. Any whole number portion from previous multiplications is ignored. The process comes to an end when either, the fractional portion of the number becomes zero or, a duplicate fractional portion is encountered. If the fractional portion of the number becomes zero, the conversion to the new base was successful and the resulting number can be represented with a finite number of placeholders. If a duplicate fractional portion is encountered, the representation in the new base requires a repeating number (Infinite Placeholders). In this case, all the numbers between the first and second occurrence of the duplicate fractional part will repeat. Figure 1.5 illustrates the procedure.

$$
\begin{array}{r}
0.dddd_1 \\
\underline{x \quad b} \\
w_1.dddd_2 \\
\underline{x \quad b} \\
w_2.dddd_3 \\
\underline{x \quad b} \\
w_3.0000
\end{array}
$$

$$(0.dddd)_{10} \longrightarrow (0.w_1 w_2 w_3)_b$$

Figure 1.5. An illustration of the conversion of fractional numbers to a new base.

The best way to truly understand the conversion from decimal to another base is by actually looking at an example. First, let's look at the number $(1107.464)_{10}$. This is the number that we arrived at after our conversion from base 5 to decimal in the previous example. If we use the preceding methods to convert the number to base 5, we should end up with the same base 5 number we had before it was converted to decimal.

Example 1.3

Problem: Convert the number $(1107.464)_{10}$ to its base 5 equivalent.

Solution: To begin we need to decide whether we want to process the whole or fractional part of the number first. For this example, we will start with the whole number portion. Starting with the number $(1107)_{10}$, we need to repeatedly divide it by the radix of our desired base and track the remainders. Since we are moving to base 5 we will divide the number by 5. By performing these divisions we will get the following.

$$
\begin{array}{ll}
0 & \\
5\,)\overline{1} & R=1 \\
5\,)\overline{8} & R=3 \\
5\,)\overline{44} & R=4 \\
5\,)\overline{221} & R=1 \\
5\,)\overline{1107} & R=2 \\
\end{array}
$$

$$(1107)_{10} \longrightarrow (13412)_5$$

Now it is time to process the fractional portion of the number. To complete this conversion we will take $(0.464)_{10}$, which is the fractional part of our number, and repeatedly multiply it by the radix for our desired base. Doing this yields the following.

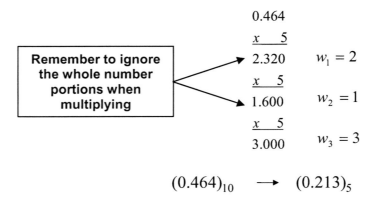

0.464	
$x\quad 5$	
2.320	$w_1 = 2$
$x\quad 5$	
1.600	$w_2 = 1$
$x\quad 5$	
3.000	$w_3 = 3$

Remember to ignore the whole number portions when multiplying

$$(0.464)_{10} \longrightarrow (0.213)_5$$

Putting everything back together we find that $(1107.464)_{10}$ is equivalent to $(13412.213)_5$, which is the number we previously converted to decimal.

Now that we understand the conversion will work, let's look at one more example. In this case, we will convert $(275.625)_{10}$ to base 7.

Example 1.4

Problem: Convert $(275.625)_{10}$ to base 7.

Solution: Once again, we will begin with the whole number portion. As we did in the previous example, we will determine the base 7 equivalent of this whole number portion by using division. Performing the division gives us the following.

$$
\begin{array}{ll}
\dfrac{0}{7\,)\,5} & R=5 \\[4pt]
\dfrac{}{7\,)\,39} & R=4 \\[4pt]
\dfrac{}{7\,)\,275} & R=2
\end{array}
\qquad \downarrow
$$

$$(275)_{10} \longrightarrow (542)_7$$

Now, let's calculate the fractional portion.

$$
\begin{array}{rl}
0.625 & \\
\underline{x \quad 7} & \\
4.375 & w_1 = 4 \\
\underline{x \quad 7} & \\
2.625 & w_2 = 2
\end{array}
$$

$$(0.625)_{10} \rightarrow (0.\overline{42})_7$$

As you can see, in this case the multiplication operation ended due to a repeated fractional portion in the calculation. Thus, the base 7 equivalent will contain a repeating value. The values between the first and second occurrence of the duplicate fractional portion will repeat. In this case, that includes the 4 from the first multiplication and the 2 from the second multiplication. You could continue to perform multiplications, but you would quickly find that the 4 and 2 would indeed continue to repeat.

Putting our two parts together, we can see that $(275.625)_{10}$ is equivalent to $(542.424242...)_7$.

By working through these two conversion examples, we have explored conversion from decimal to two different bases. We discovered what happens both when a fractional conversion results in a repeating fractional portion of the number and when it does not.

At this point you have all the tools you need to convert back and forth between decimal and any other base that you may choose. Thus far we have been discussing operations that will work with any base. The next section will discuss some of the more commonly used bases and will provide some examples of both basic and special conversions between these bases.

> **MATLAB:** MATLAB has a function that allows for the conversion of a decimal number to its equivalent in an arbitrary base. This function has the following usage syntax.
>
>
>
> DEC2BASE(D,B)
> or
> DEC2BASE(D,B,N)
>
>
>
> Where D is the decimal number, B is the desired base, and N is the desired number of placeholders to use when representing the result. If the result requires fewer placeholders, the number will be zero padded to reach the correct number of places. Keep in mind that this function only converts positive integer values. It will not work on negative numbers or numbers containing fractional parts. In addition, the returned value will be represented using a text string, not an actual number.
>
> **Example:** Dec2Base(1107,5) will return '13412' which is the base 5 equivalent of the specified decimal number. Dec2Base(1107,5,7) will return '0013413' which is the base 5 equivalent of the specified decimal number padded with zeros to reach 7 placeholders.

1.3   Operations in Common Bases

In the previous section, we looked at methods of dealing with numbers in different bases. We were not concerned about what specific base we were using. Instead, we wanted to be able to understand and work with any arbitrary base. In this section, we will look at some specific bases. These particular numeric bases are commonly used in engineering and computer applications. While it is important to be able to work with numbers regardless of their base, it is also beneficial to be comfortable with numbers that are presented in bases that you will deal with on a regular basis. The three bases that we will deal with here are binary (base 2), octal (base 8), and hexadecimal (base 16). You will encounter binary numbers at numerous points in your electrical engineering studies, particularly in the areas of digital logic, communications, and signal processing. Octal and hexadecimal will be seen to a lesser extent. They are primarily used in computer-based applications.

1.3.1   Binary (Base 2) Numbers

The binary, or base 2 number system has a radix of 2. This means that each placeholder in a binary number can hold only two possible values, 0 or 1. This type of a number system is ideally suited to the description of logic events. For example, if you have a series of 3 light bulbs, each of which will either be on or off at a given time, you can represent the state of those light bulbs at a given instant by representing a light that is on with a binary 1 and a light that is off with a binary 0. Since we have three light bulbs and each bulb can be either on or off (two possibilities), we have 2 x 2 x 2, or 8 possible states for our light bulb arrangement. These states range from all the bulbs being off to some of the bulbs being on to all of the bulbs being on. The following chart displays all of the eight possible states for our three light arrangements.

Arrangement Number	Light 1	Light 2	Light 3	Binary Representation
0	Off (0)	Off (0)	Off (0)	000
1	Off (0)	Off (0)	On (1)	001
2	Off (0)	On (1)	Off (0)	010
3	Off (0)	On (1)	On (1)	011
4	On (1)	Off (0)	Off (0)	100
5	On (1)	Off (0)	On (1)	101
6	On (1)	On (1)	Off (0)	110
7	On (1)	On (1)	On (1)	111

As you can see, all eight possible combinations have been represented above. Since each light has only two possible states, we can describe the state of the three lights using a string of three binary digits instead of describing the state of each bulb individually. While an example using only three lights is trivial, the idea can be expanded to handle any number of lights. In addition, lights are only one simple example of systems that can be described using binary. Any digital logic circuit, including the contents of computer memory can be expressed in this way. Thus, we can see the usefulness of binary numbers in electrical engineering applications.

Since binary numbers are commonly encountered and are useful in engineering applications, we will look specifically at them. As we noted before, binary is based on the number 2, and as such, allows only the digits 0 and 1. That being said, let's look at the basic structure of a binary number. Figure 1.6 shows the progression of the various powers of two associated with the placeholders of a binary number.

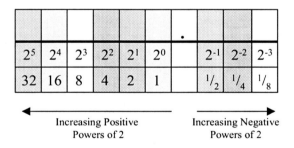

Figure 1.6. Placeholder values for a binary number.

This representation of a binary number is exactly what we would expect based upon our previous experience with numbers of different bases. Now that we know what a binary number will look like, let's look at some basic operations involving binary.

The first operation we will look at is counting in binary. Counting is a simple concept, but a familiarity with it will simplify later tasks. A good engineering student should be familiar with at least the first 16 binary numbers (0 to 15)$_{10}$. You should be able to recognize and convert between these numbers and decimal in your head. Since we want to get acquainted with these numbers, let's count to (15)$_{10}$ in binary.

Example 1.5

Problem: Count to $(15)_{10}$ in binary.

Solution: This is a simple operation that works in exactly the same way as example 1.1, except in this case we are working with base 2 instead of 5. This simply means that each time we reach a value of 2 in a given placeholder, we must add a new placeholder to represent it. If we look at figure 1.6 we will note that $(16)_{10}$ is equal to 2^4. Thus, we will need 4 placeholders to represent the numbers up to $(15)_{10}$. If we count using 4 placeholders we will get the following.

Decimal	Binary
0	0000
1	0001
2	0010
3	0011
4	0100
5	0101
6	0110
7	0111
8	1000
9	1001
10	1010
11	1011
12	1100
13	1101
14	1110
15	1111

So, as we can see in the table above, counting in binary is not a difficult activity. It is a good idea to make yourself familiar with these 16 binary numbers, as you will be seeing a lot more of them in the future.

Having looked at counting in binary, our next goal is to look at the process of converting a binary number to decimal. While, as we discovered, the conversion from any base to decimal is an easy task, it is particularly easy in binary. Since each placeholder can hold only a 0 or a 1, the decimal equivalent can be obtained by simply summing the powers of two associated with each placeholder that contains a one. Binary simplifies your task since you do not have to perform the multiplication between the contents of the placeholder and its power of the base. To see how this works, let's look at an example.

Example 1.6

Problem: Find the decimal equivalent of $(1101101.011)_2$.

Solution: Finding the decimal equivalent to a binary number, as mentioned, is a bit easier than finding the decimal equivalent of any other base. Since the power of two for each placeholder is either multiplied by a zero or a one, we can arrive at our answer simply by adding the powers of two associated with the placeholders containing a one. First, let's find out which placeholders we need to add. The following table displays each placeholder in our number and its power of two.

Placeholder Value	Associated Power of 2
1	$2^6=64$
1	$2^5=32$
0	$2^4=16$
1	$2^3=8$
1	$2^2=4$
0	$2^1=2$
1	$2^0=1$
.	Binary Pt.
0	$2^{-1}=0.500$
1	$2^{-2}=0.250$
1	$2^{-3}=0.125$

Now, if we add up each of the powers of two with a one in its respective placeholder, we will get the following.

```
     64.000
     32.000
      8.000
      4.000
      1.000
      0.250
 +    0.125
   _____
    109.375
```

So, we discover that $(1101101.011)_2$ is equivalent to $(109.375)_{10}$.

The conversion of binary to decimal is not a difficult task. Having explored this example, you should be able to easily convert any binary number that you come across to its decimal equivalent.

MATLAB: MATLAB has a function that allows for the conversion of a binary number to its decimal equivalent. This function has the following usage syntax.

BIN2DEC(B)

Where B is a string containing the binary number. Keep in mind that this function only converts positive integer values. It will not work on negative numbers or numbers containing fractional parts.

Example: Bin2Dec('1101101') will return 109 which is the decimal equivalent of the specified binary number.

The final example we will look at in this section dedicated to binary numbers involves the conversion of decimal numbers to their binary equivalents. As you would expect, this is just a special case of the arbitrary base conversions that were explored previously. To convert to binary, the whole number portion must be divided by two, and the fractional part must be multiplied by two. Let's take a look at an example of this.

Example 1.7

Problem: Convert $(145.625)_{10}$ to its binary equivalent.

Solution: Let's start by taking the whole number part, $(145)_{10}$, and repeatedly dividing by two while keeping track of our remainder.

$$
\begin{array}{r}
0 \\
\hline
2\overline{)1} \qquad R=1 \\
2\overline{)2} \qquad R=0 \\
2\overline{)4} \qquad R=0 \\
2\overline{)9} \qquad R=1 \\
2\overline{)18} \qquad R=0 \\
2\overline{)36} \qquad R=0 \\
2\overline{)72} \qquad R=0 \\
2\overline{)145} \qquad R=1
\end{array}
$$

$$(145)_{10} \longrightarrow (10010001)_2$$

Now, let's take care of the fractional part of the number by multiplying by two and tracking the whole number portion.

$$
\begin{array}{r}
0.625 \\
\underline{x \quad 2} \\
1.250 \qquad W=1 \\
\underline{x \quad 2} \\
0.500 \qquad W=0 \\
\underline{x \quad 2} \\
1.000 \qquad W=1
\end{array}
$$

$$(0.625)_{10} \longrightarrow (0.101)_2$$

Putting everything together, we find that $(145.625)_{10}$ is equivalent to $(10010001.101)_2$.

By now, you should be fairly familiar with binary numbers. You should be able to count in binary, as well as easily convert between binary and decimal. The need to work with binary numbers arises quite often in electrical engineering. As a result, it is always a good idea to be familiar with the operations and conversions associated with binary numbers.

MATLAB: MATLAB has a function that allows for the conversion of a decimal number to its binary equivalent. This function has the following usage syntax.

DEC2BIN(D)

or

DEC2BIN(D,N)

Where D is the decimal number and N is the number of placeholders to use when representing the resulting binary number. Keep in mind that this function only converts positive integer values. It will not work on negative numbers or numbers containing fractional parts.

Example: Dec2Bin(145) will return '10010001' which is the binary equivalent of the specified decimal number. Dec2Bin(145,10) will return '0010010001' which is the binary equivalent of the specified decimal number. The number has been zero padded to use 10 placeholders.

1.3.2   Octal (Base 8) Numbers

While binary numbers are very useful and, in fact, are widely used, they do have their drawbacks. The most notable is their length. When using binary numbers to describe computer memory contents, or the logic state of a large number of outputs, the length of the resulting binary representation of the system can become cumbersome. As a result, engineers started to use numbers in bases that were larger then base two. While the use of a larger base will result in smaller representations of system states, the numbers will not be useful unless they can be easily converted to binary. One of the first bases adopted for this purpose is octal, or base eight. If we look at base eight in a little detail, we will see why it was used. First, we know that since our radix is 8, each placeholder of our number can hold a value between 0 and $(8-1) = 7$. We know from our study of binary numbers that the number $(7)_{10}$ is represented as $(111)_2$. So we discover that we can represent any combination of the first three binary placeholders using only one octal placeholder. If we extend our experiment, we will find that for each three binary placeholders we add, we only need to add one octal placeholder to be able to express an equivalent number. Thus, we have discovered that an octal number will be three times shorter than its binary equivalent. This occurs because eight is a power of two. In a future section, we will explore bases that are powers of other bases in more detail. We will also look at some easy methods to convert between those bases. For now, however, let's look at the octal number system. Figure 1.7 shows some of the placeholders in an octal number along with their associated powers of eight.

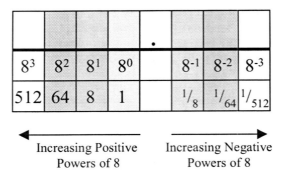

Increasing Positive   Increasing Negative
Powers of 8   Powers of 8

Figure 1.7. Placeholder values for an octal number.

As was the case when we explored the binary number system, we will begin by looking at the concept of counting. It is important to get used to what the first 15 numbers in a given base will look like. As a result, we will once again count to $(15)_{10}$. This time we will do it in octal.

Example 1.8

Problem: Count to $(15)_{10}$ in Octal.

Solution: This is a simple operation that works exactly the same way as our previous counting examples. In this case we are working with base 8, so each time we reach a value of 8 in a given placeholder, we must add a new placeholder to represent it. If we look at figure 1.7 we will note that $(16)_{10}$ is less than 8^2, but greater than 8. Thus, we will need 2 placeholders to represent the numbers up to $(15)_{10}$. If we count using 2 placeholders we will get the following. Note that we have left in the binary equivalent numbers from example 1.5 so that you can easily see the correspondence between all three bases.

Decimal	Binary	Octal
0	0000	00
1	0001	01
2	0010	02
3	0011	03
4	0100	04
5	0101	05
6	0110	06
7	0111	07
8	1000	10
9	1001	11
10	1010	12
11	1011	13
12	1100	14
13	1101	15
14	1110	16
15	1111	17

So, as we can see in the table above, counting in octal is not a difficult activity. It is a good idea to make yourself familiar with these 16 octal numbers, just as you did with the binary.

Now that we have seen the process of counting in octal, let's look at the conversion processes. First we will explore the conversion from octal to decimal, and then we will look at the process for going the other way. The process used to convert from octal to decimal is exactly the same as the one we used in example 1.2 except that this time the base is eight. Let's look at an example.

Example 1.9

Problem: Convert the octal number $(4715.72)_8$ to its decimal equivalent.

Solution: The procedure here is exactly the same as the one we used previously. We simply need to multiply the contents of each placeholder by its respective power of eight and then sum up the results. The following table shows the process of obtaining the products for our particular number.

Placeholder Value	Associated Power of 8	Resulting Product
4	$8^3=512$	4 x 512 = 2048
7	$8^2=64$	7 x 64 = 448
1	$8^1=8$	1 x 8 = 8
5	$8^0=1$	5 x 1 = 5
.	Octal Pt.	
7	$8^{-1}=0.125$	7 x 0.125 = 0.875
2	$8^{-2}=0.015625$	2 x 0.015625 = 0.03125

Now, if we add up all of the products obtained above, we will get the following.

```
      2048.00000
       448.00000
         8.00000
         5.00000
         0.87500
    +    0.03125
      2509.90625
```

So, we discover that $(4715.72)_8$ is equivalent to $(2509.90625)_{10}$.

As would be expected, the conversion from octal to decimal is a straightforward process.
    The last procedure to examine for octal numbers is the conversion from a decimal number to its octal equivalent. Once again, this process works just like many of the base to decimal conversions we have done in the past. The whole number portion of the number is determined by repeated division, and the fractional part is found by repeated multiplication. Let's look at an example.

MATLAB: MATLAB has a function that allows for the conversion of an octal number to its decimal equivalent. This function has the following usage syntax.

$$OCT2DEC(N)$$

Where N is the octal number. Keep in mind that this function only converts positive integer values. It will not work on negative numbers or numbers containing fractional parts.

Example: Oct2Dec(4715) will return 2509 which is the decimal equivalent of the specified octal number.

Example 1.10

Problem: Find the Octal equivalent of $(476.0625)_{10}$.

Solution: Let's start by finding the equivalent whole number portion of the number. To do so, we must divide $(476)_{10}$ by 8 and keep track of our remainder. This is done below.

$$
\begin{array}{ll}
\dfrac{0}{8\,\overline{)\,7}} & R=7 \\[6pt]
8\,\overline{)\,59} & R=3 \\[6pt]
8\,\overline{)\,476} & R=4
\end{array}
$$

$$(476)_{10} \longrightarrow (734)_8$$

Thus, our whole number part is $(764)_8$. Now let's find the equivalent fractional portion. We need to multiply $(0.0625)_{10}$ by 8 and keep track of the whole number portion.

$$
\begin{array}{ll}
0.0625 & \\
\underline{x \quad\; 8} & \\
0.5000 & W=0 \\
\underline{x \quad\; 8} & \\
4.0000 & W=4
\end{array}
$$

$$(0.0625)_{10} \longrightarrow (0.04)_8$$

So, the equivalent fractional portion is $(0.04)_8$. If we put both parts together we arrive at the final answer. The equivalent of $(476.0625)_{10}$ is $(734.04)_8$.

MATLAB: The easiest way to convert a decimal number to its octal equivalent using MATLAB is through the use of the DEC2BASE function. The syntax of this function as it would be used for octal conversion is shown below.

DEC2BASE(D,8)
or
DEC2BASE(D,8,N)

Where D is the decimal number and N is the desired number of placeholders to use when representing the result in octal. If the result requires fewer placeholders, the number will be zero padded to reach the correct number of places. Keep in mind that this function only converts positive integer values. It will not work on negative numbers or numbers containing fractional parts. In addition, the returned value will be represented using a text string, not an actual number.

Example: Dec2Base(476,8) will return '734' which is the octal equivalent of the specified decimal number. Dec2Base(1107,8,4) will return '0734' which is the octal equivalent of the specified decimal number padded with zeros to reach 4 placeholders.

Having seen examples of counting and conversions in octal, you should now be prepared to work with numbers in this base without difficulty. While octal is not seen as often as some of the other common bases, it is still used in some applications, and it is important to be aware of how to deal with it.

1.3.3   Hexadecimal (Base 16) Numbers

The final commonly used base that we will explore is hexadecimal, or base 16. The common use of this base is yet another product of the search for a shorter way to express lengthy binary representations of memory contents or system states. As computer memory got larger and system buses got wider, there was a need to reduce binary representations even more than the reduction that octal was achieving. The solution was to use a base with a radix of the next higher power of two. Where octal had a radix of $2^3 = 8$, hexadecimal has a radix of 2^4, or 16. Thus, based upon our results with the octal base, you would expect each hexadecimal placeholder to represent four binary placeholders. This is in fact the case. A hexadecimal number will typically be four times shorter than its binary equivalent, and because hexadecimal is based on a power of 2, it retains the added benefit of easy conversion back and forth to binary (see bases that are powers of other bases later in this chapter).

The one difficulty that arose when looking at hexadecimal as a base was that since we have always used decimal for our common calculations, we only have 10 placeholder values, 0 – 9. This left us with the need for 6 additional placeholder values. The solution to this problem was to use alphabetic characters, specifically A through F, to fill in for the needed values. Thus, we let A = 10, B = 11, C = 12, D = 13, E = 14, and F = 15. This gives us our 16 needed values for each placeholder and enables us to use base 16.

Now that know a little about the hexadecimal number system, let's explore it in a bit more detail. Figure 1.8 shows some of the powers of 16 associated with the placeholders of a typical hexadecimal number.

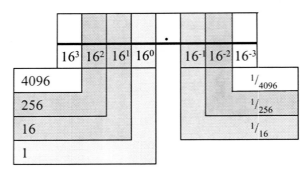

Figure 1.8. Placeholder values for a hexadecimal number.

As we can see from the figure, since each placeholder is based on a power of 16, the actual number multiplied by the value stored in each placeholder will grow quickly as you move to the left. By the time you reach the fifth placeholder to the left of the hexadecimal point, you are already multiplying the placeholder values by $16^4 = (65536)_{10}$.

Since we have begun our exploration of each of the common bases by looking at the process of counting in that base, we will continue that practice here. Once again we will count to $(15)_{10}$. This time we will do it in hexadecimal.

Example 1.11

Problem: Count to $(15)_{10}$ in Hexadecimal.

Solution: Once again, this simple operation works in exactly the same way as our previous counting examples. In this case we are working with base 16, so each time we want to represent a number larger than $(15)_{10}$ or $(F)_{16}$ in a given placeholder, we must add a new placeholder to represent it. Since we are counting up to exactly $(15)_{10}$ in this example, we will need only one hexadecimal placeholder. If we count using our single placeholder we will get the following. Note that we have left in the binary equivalent numbers from example 1.5 and the octal equivalent numbers from example 1.8 so that you can easily see the correspondence between all four bases.

Decimal	Binary	Octal	Hexadecimal
0	0000	00	0
1	0001	01	1
2	0010	02	2
3	0011	03	3
4	0100	04	4
5	0101	05	5
6	0110	06	6
7	0111	07	7
8	1000	10	8
9	1001	11	9
10	1010	12	A
11	1011	13	B
12	1100	14	C
13	1101	15	D
14	1110	16	E
15	1111	17	F

As you can see from this table, we have counted up to the maximum number that can be represented in a single hexadecimal placeholder. If we had counted any higher, an additional placeholder would have been needed. This table gives you an excellent summary of the first 15 numbers of all four of our commonly used bases. It will be beneficial to become familiar with this table.

Having seen how we count in hexadecimal, we will now turn our attention to the conversion processes. While conversions between hexadecimal and decimal follow the same methods we have been using up until now, they have the added complexity of having to deal with characters (A – F) as well as the numbers we are more accustomed to. In some cases, these characters can cause confusion and error. It is often easy to forget which number each of the characters represents. If you have not dealt with hexadecimal before, you may find it helpful to refer back to the table in example 1.11 to refresh your memory. That being said, let's look at the process of converting from a hexadecimal number to its decimal equivalent.

Example 1.12

Problem: Convert the hexadecimal number $(AB6.E)_{16}$ to its decimal equivalent.

Solution: The procedure is similar to the procedures we have previously used to convert from a given base to decimal. We simply need to multiply the contents of each placeholder by its respective power of sixteen and then sum up the results. The only difference here is that we must remember what decimal value each of the characters A – F represents so that we can multiply by the correct value. The following table shows the process of obtaining the products for our number.

Placeholder Value	Associated Power of 16	Resulting Product
A	$16^2=256$	$10 \times 256 = 2560$
B	$16^1=16$	$11 \times 16 = 176$
6	$16^0=1$	$6 \times 1 = 6$
.	Hexadecimal Pt.	
E	$16^{-1}=0.0625$	$14 \times 0.0625 = 0.875$

Now, if we add up all of the products obtained above, we will get the following.

```
      2560.000
       176.000
         6.000
   +     0.875
      ---------
      2742.875
```

So, we discover that $(AB6.E)_{16}$ is equivalent to $(2742.875)_{10}$.

As you can see, the conversion from hexadecimal to decimal is not too difficult as long as you can remember which numbers the letters A – F represent.

> **MATLAB:** MATLAB has a function that allows for the conversion of a hexadecimal number to its decimal equivalent. This function has the following usage syntax.
>
> $$\text{HEX2DEC(N)}$$
>
> Where N is a string containing the hexadecimal number. Keep in mind that this function only converts positive integer values. It will not work on negative numbers or numbers containing fractional parts.
>
> **Example:** Hex2Dec('AB6') will return 2742 which is the decimal equivalent of the specified hexadecimal number.

The final procedure we will look at relating to hexadecimal numbers is the conversion from a decimal number to its hexadecimal equivalent. Once again, this process works just like many of the base to decimal conversions we have done previously, but it is somewhat complicated by the usage of the characters A – F. As usual, the whole number portion of the number is determined by repeated division, and the fractional part is found by repeated multiplication. Let's look at an example.

Example 1.13

Problem: Convert $(1594.5625)_{10}$ to its hexadecimal equivalent.

Solution: As usual, let's start by finding the equivalent whole number portion of the number. To do so, we must divide $(1594)_{10}$ by 16 and keep track of our remainder. This is done below.

$$
\begin{array}{l}
\quad\quad\ \ 0 \\
16\overline{)6} \quad\quad R=6 \\
16\ \overline{)99} \quad\quad R=3 \\
16\ \overline{)1594} \quad\quad R=10=A
\end{array}
$$

$$(1594)_{10} \longrightarrow (63A)_{16}$$

Thus, our whole number part is $(63A)_{16}$. Now we need to find the equivalent fractional portion. So, let's multiply $(0.5625)_{10}$ by 16 and keep track of the whole number portion.

$$
\begin{array}{l}
0.5625 \\
\underline{x\quad 16} \\
9.0000 \quad\quad W=9
\end{array}
$$

$$(0.5625)_{10} \longrightarrow (0.9)_{16}$$

In this case, we found the equivalent fractional portion very quickly. In fact, it only took one multiplication. So, now we know that $(0.5625)_{10}$ is equivalent to $(0.9)_{16}$.

Putting our two parts together we find that $(1594.5625)_{10}$ is equivalent to $(63A.9)_{16}$.

This example has illustrated the process of converting a decimal number to its equivalent hexadecimal number. Once again, the process is straightforward as long as care is taken when dealing with the alphabetic characters.

> **MATLAB:** MATLAB has a function that allows for the conversion of a decimal number to its hexadecimal equivalent. This function has the following usage syntax.
>
>
>
> DEC2HEX(D)
> or
> DEC2HEX(D,N)
>
>
>
> Where D is the decimal number and N is the number of placeholders to use when representing the resulting hexadecimal number. Keep in mind that this function only converts positive integer values. It will not work on negative numbers or numbers containing fractional parts.
>
> **Example:** Dec2Hex(1594) will return '63A' which is the hexadecimal equivalent of the specified decimal number. Dec2Hex(1594,4) will return '063A' which is the binary equivalent of the specified decimal number. The number has been zero padded to use 4 placeholders.

At this point, we have seen examples of counting and conversions in all of the most commonly used bases. You should feel comfortable working in any of these bases and should have no trouble converting from any base to decimal or from decimal to any base. In the next section we will look at some special case conversions that can be done when the radix of one base is the radix of another base raised to a power. This is a trait possessed by our three common bases, Binary, Octal and Hexadecimal. The radix of each of these bases is a power of two. We will find that this makes conversion between these bases somewhat easier.

1.4   Bases that are Powers of Other Bases

As I am sure you have noticed, the base conversion techniques that we have covered up to now have all dealt with the conversion between decimal and some other base. There has been no discussion of the conversion between two bases where neither of those bases is decimal. The main reason for this is that the conversion is, in most cases, not worth the effort. It is usually easier, and far less confusing, to convert a given base to decimal and then convert the decimal number to the final desired base.

One notable exception to this is when the radix of one base is equal to the radix of the other base raised to a power. An example of such a pair of bases is binary and octal. In this case, octal has a radix of 8, which is 2^3. At this point you may be wondering why this relationship between the two bases makes it any easier to convert between them. A closer investigation will quickly reveal the reason. Let's take a look at the structure of these two bases. Figure 1.9 gives us a side-by-side comparison of the placeholders in the two bases.

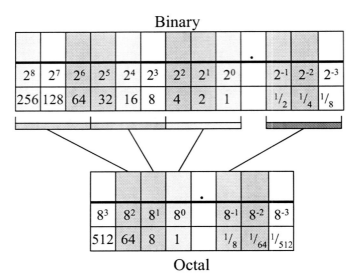

Binary

Octal

**Figure 1.9. Placeholder relationships between the
binary and octal bases.**

As we see in Figure 1.9, the largest number that can be represented with the first three placeholders to the left of the binary point in the binary number is $(7)_{10}$. This is also the largest number that can be represented with the first placeholder to the left of the octal point in the octal number. If we add three more binary placeholders, we can represent numbers up to 63, which is the largest number that can be represented by adding a second placeholder to our octal number. So, what we have discovered here is that each octal placeholder corresponds to three binary placeholders. We can extend this into more general terms. Let's call our smaller base b. Then our base that is a power of this base can be referred to as b^n, where n is the power to which the smaller base was raised. Based on this terminology, we will find that each placeholder of our base b^n will be equivalent to n placeholders of the smaller base b. So, from figure 9.1 we had b = 2, or binary and $b^n = 2^3 = 8$, or octal. Thus, based on our general relationship, we would expect there to be three binary placeholders equivalent to each octal placeholder. This relationship between the bases makes conversion between the two a simple task. To convert from the larger base to the smaller base, we simply convert the contents of each placeholder to its n placeholder equivalent in the smaller base and append them all together. Let's look at an example of how this will work.

Example 1.14

Problem: Convert the hexadecimal number $(AB7C.3D)_{16}$ to binary.

Solution: In this case we are dealing with a base that is the power of another base. Our smaller base is 2 and the larger base is $2^4 = 16$. Thus, we will expect each hexadecimal digit to be equal to four binary digits. Our first step is to find the four digit binary equivalents to each of the hexadecimal digits. If you do not remember the corresponding binary representations, you can refer back to the table in example 1.11 to refresh your memory. The following table displays the binary representations for each of the hexadecimal digits.

Hexadecimal	Binary
A	1010
B	1011
7	0111
C	1100
.	.
3	0011
D	1101

Now that we have all of the four digit binary equivalents defined, we just need to put them together to yield our desired result.

$$\underbrace{1010}_{A}\underbrace{1011}_{B}\underbrace{0111}_{7}\underbrace{1100}_{C}.\underbrace{0011}_{3}\underbrace{1101}_{D}$$

$$A\quad B\quad 7\quad C\ .\ 3\quad D$$

If we want to double-check our result, we can convert both of these numbers to decimal and see if they really are the same.

$(AB7C.3D)_{16}$ =(10x4096)+(11x256)+(7x16)+(12x1)+(3x0.0625)+(13x0.00390625)
=40960+2816+112+12+0.1875+0.05078125
=(43900.23828125)$_{10}$

$(1010101101111100.00111101)_2$=32768+8192+2048+512+256+64+32+16+8+4+0.125+
0.0625+0.03125+0.015625+0.00390625
=(43900.23828125)$_{10}$

So, we have found that $(AB7C.3D)_{16}$ is equivalent to $(1010101101111100.00111101)_2$.

Now that we have looked at the conversion from the larger base to the smaller one, let's look at what we need to do to convert in the other direction. Basically what we need to do is the exact opposite of what we did to convert from the larger base to the smaller base. To convert the smaller base to the larger one, you must group the placeholders of the number into sets of n digits, where n is the power to which the smaller base was raised to arrive at the larger base. These groupings start at the radix point and group to the left for the whole number portion and to the right for the fractional portion. If there are not enough digits to form a full group, the number can be zero-padded to create a full group. Once these groups are created, they can be treated as separate numbers and converted into the new base. These groups will always convert to represent a single digit in the new base. These single digits can then be grouped together to arrive at the equivalent number in the larger base. Let's look at an example to help clarify this process.

Example 1.15

Problem: Convert the number $(21202.212)_3$ to base 9.

Solution: In this case we find that we are dealing with base 3 and base 9. It is clear that nine is equal to three squared, so we know that the power we are dealing with is two. Thus we need to group our smaller base into two digit groups. We can then convert each of these two digit groups to single base 9 digits and combine these to generate our solution. First, let's generate our two digit groups.

$$0\underline{2}\ \underline{12}\ \underline{02}.\underline{21}\ \underline{2}0$$

As you can see we needed to pad both the whole number and fractional parts of the number to achieve full groups. Now that we have our two digit groups, we can convert these to base 9 and create our final result.

Base 3	Base 9
02	2
12	5
02	2
.	.
21	7
20	6

Now, if we put our base 9 digits together we arrive at our solution $(252.76)_9$.

Once again, we can check this solution by converting both numbers to decimal.

$(21202.212)_3$ $=(2\times81)+(1\times27)+(2\times9)+(2\times1)+(2\times\frac{1}{3})+(1\times\frac{1}{9})+(2\times\frac{1}{27})$
                       $=162+27+18+2+\frac{2}{3}+\frac{1}{9}+\frac{2}{27}$
                       $=\mathbf{209\ {}^{23}/_{27}}$

$(252.75)_9$ $=(2\times81)+(5\times9)+(2\times1)+(7\times\frac{1}{9})+(6\times\frac{1}{81})$
                       $=162+45+2+\frac{7}{9}+\frac{2}{27}$
                       $=\mathbf{209\ {}^{23}/_{27}}$

Thus, our solution is verified and we know that $(21202.212)_3$ is equivalent to $(252.75)_9$.

From these examples, it should be clear that when we work with bases that are powers of other bases, it is an easy task to convert between the two bases. As was mentioned before, the methods used here will only work on bases that have a power relationship. Conversion between bases that are not related in this way is easier done by first converting to decimal.

1.5   Simple Addition in Other Bases

Our final topic relating to numbers in different bases is addition. While it is possible to convert any number you may come across to decimal for the purposes of addition, this often adds some unnecessary steps. In most cases, performing addition in other bases is easier than performing all of the conversions just so that the work can be done in the decimal system. We will explore the concept of addition in other bases by first looking at the mechanics of addition in the decimal system and then expanding those mechanics to cover any base we may encounter. So, what happens when we add two numbers in decimal? Let's look at the sum of the numbers 1299.75 and 1307.42. When we add these two numbers, we start by adding the contents of the left-most right-most placeholder. We put the results of this addition into the left-most right-most position of our result. If the resulting sum is larger than 9, we record the "ones" place of the result as our left-most right-most digit and "carry" the remaining digits over into our sum for the placeholder to the left. This process is continued until all of the placeholders have been summed. Let's look at what happens when we add our two example numbers. Figure 1.10 displays the results of the addition.

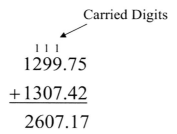

$$\begin{array}{r} \text{Carried Digits} \\ 1\ 1\ 1 \\ 1299.75 \\ +\ 1307.42 \\ \hline 2607.17 \end{array}$$

Figure 1.10. The results of adding two numbers in decimal.

As Figure 1.10 shows, when we arrived at numbers that were larger than 9 we carried the "tens" portion of our number over to the next placeholder. For example, when we added 7 and 4 we arrived at the number 11. Since this number was too large to hold in one placeholder, we stored the one from the "ones" place and "carried" the one from the "tens" place. This is the same method that most students have been using to add numbers since early in their education. The addition of numbers in a base other than decimal follows the same rules. The numbers in the corresponding placeholders are added beginning with the right-most digit. If any of these additions result in a sum that is larger than the radix of the base, carrying must take place. To be sure we understand what is going on when two numbers are added in a different base, let's look at a couple of examples. First we will look at a simple example using binary, and then we will look at an example in hexadecimal.

Example 1.16

Problem: Add the following binary numbers: $(11010101.101)_2$ and $(10111011.001)_2$.

Solution: Here we are dealing with binary. We will begin at the rightmost digit and begin adding the corresponding placeholders. Since this base has a radix of 2, we must perform carrying if a sum larger than one is encountered. The following table summarizes the operations that we must perform to sum the two numbers. Note that the sum of the placeholders is given in binary.

Carry	Number 1	Number 2	Sum	Store	Operation
	1	1	1 + 1 = 10	0	Store 0, Carry 1
1	0	0	1 + 0 + 0 = 01	1	Store 1
	1	0	1 + 0 = 01	1	Store 1
	.	.	.	.	
	1	1	1 + 1 = 10	0	Store 0, Carry 1
1	0	1	1 + 0 + 1 = 10	0	Store 0, Carry 1
1	1	0	1 + 1 + 0 = 10	0	Store 0, Carry 1
1	0	1	1 + 0 + 1 = 10	0	Store 0, Carry 1
1	1	1	1 + 1 + 1 = 11	1	Store 1, Carry 1
1	0	1	1 + 0 + 1 = 10	0	Store 0, Carry 1
1	1	0	1 + 1 + 0 = 10	0	Store 0, Carry 1
1	1	1	1 + 1 + 1 = 11	1	Store 1, Carry 1
1			1	1	Store 1

The addition of the numbers can also be done in the standard notation, as shown here.

$$\begin{array}{r} {\scriptstyle 1\ 1\ 1\ 1\ 1\ 1\ 1\ 1 \quad\ 1} \\ 11010101.101 \\ +\ 10111011.001 \\ \hline 110010000.110 \end{array}$$

Thus, we discover that the sum of our two numbers is $(110010000.110)_2$.

Our next example will look at addition in the hexadecimal number system.

Example 1.17

Problem: Add the hexadecimal numbers $(A7B2.34)_{16}$ and $(15C2.8A)_{16}$.

Solution: In this example, we are working with hexadecimal. Thus any placeholder sum that is larger than $(15)_{10}$ or $(F)_{16}$ will require us to "carry" to the next placeholder. As always we start adding in the rightmost placeholder. The following table summarizes the operation.

Carry	Number 1	Number 2	Sum	Store	Operation
	4	A	4 + A = E	E	Store E
	3	8	3 + 8 = B	B	Store B
	.	.	.	.	
	2	2	2 + 2 = 4	4	Store 4
	B	C	B + C = 17	7	Store 7, Carry 1
1	7	5	1 + 7 + 5 = D	D	Store D
	A	1	A + 1 = B	B	Store B

This can also be expressed in conventional notation as follows.

$$\begin{array}{r} 1 \\ A7B2.34 \\ +\ 15C2.8A \\ \hline BD74.BE \end{array}$$

So, our two numbers add up to $(BD74.BE)_{16}$.

These examples show the basic processes that should be undertaken when adding numbers, regardless of the base. From the examples, we can see that addition works essentially the same no matter what base the numbers are in. The only difference is the value that must be reached within a given placeholder summation before carrying occurs. Thus, we have discovered that addition in bases other than decimal is a simple activity.

1.6 Chapter Summary

The contents of this chapter have dealt with the concept of numbers defined by bases other than 10 (decimal). In this chapter we have learned the following:

- We discovered how counting operates in various bases and looked at examples in both arbitrary and common bases. We found that in order to count, we simply continue to add one to the "ones" place of our number. When the value in this placeholder reaches the radix for our base, we increment the placeholder to the immediate left, adding a new placeholder if needed.

- We learned the process involved in converting a number in a different base back to our familiar decimal system. We simply multiplied the value held in the placeholder by its respective power of the base and added up the results.

- Likewise, the process for converting a decimal number to its representation in a different base was examined. It was found that the whole number portion of a decimal number must simply be divided by the radix of the desired base while keeping track of the remainders. We continue to divide the quotient until we arrive at a quotient of zero. The resulting series of remainders give us the digits of our new whole number representation. In a similar manner, we found that the new fractional part of the number is found by multiplying the fractional portion of the original number by the radix of the desired base and keeping track of the whole number portion. The process terminates when either the fractional portion reaches zero or repeats.

- We examined the three most commonly used non-decimal bases. We found the binary, octal, and hexadecimal are very useful bases, especially when dealing with logic and computers. Several examples of the basic conversions between numbers in these bases and decimal numbers were explored.

- The special relationship held between a given base, b, and another base that has a radix that is a power of b was examined. It was found that conversions between bases with such a relationship could be made easily without having to first convert to decimal.

- Finally, the concept of addition was explored. It was found that addition was basically the same in all bases. The contents of all corresponding placeholders are added beginning at the rightmost position. If the results of the addition exceed or equal the radix of the base, carrying occurs. Once all corresponding groups of placeholders have been summed, the operation is complete.

1.7   Review Exercises

Section 1.2 – Working with Arbitrary Bases
_____

1. Count to $(20)_{10}$ in the following bases.

   a. Base 3   b. Base 5
   c. Base 9   d. Base 4

2. Count to $(25)_{10}$ in the following bases.

   a. Base 7   b. Base 3
   c. Base 9   d. Base 6

3. Convert the following numbers to decimal.

   a. $(135)_6$ b. $(2121)_3$
   c. $(4121)_5$ d. $(872)_9$

4. Convert the following numbers to decimal. (Round your result to 4 decimal places.)

   a. $(0.361)_6$ b. $(183.462)_9$
   c. $(122.211)_3$ d. $(312.131)_4$

5. Convert the decimal number $(234)_{10}$ to the following bases.

   a. Base 3   b. Base 5
   c. Base 7   d. Base 9

6. Convert the decimal number $(136.372)_{10}$ to the following bases. (Report up to four places beyond the radix point)

   a. Base 4   b. Base 6
   c. Base 5   d. Base 9

7. Perform the following conversions. (Report up to four places beyond the radix point)

    a. $(176.29)_{10} \rightarrow (?)_{3}$ b. $(113.23)_{4} \rightarrow (?)_{10}$

    c. $(1762.45)_{10} \rightarrow (?)_{9}$ d. $(2122.221)_{3} \rightarrow (?)_{10}$

    e. $(32231.233)_{4} \rightarrow (?)_{10}$ f. $(722.5)_{10} \rightarrow (?)_{6}$

Section 1.3 – Operations in Common Bases

8. Count to $(25)_{10}$ in the following bases.

    a. Binary
    b. Hexadecimal
    c. Octal

9. Count to $(45)_{10}$ in hexadecimal.

10. Count to $(31)_{10}$ in binary.

11. Count to $(40)_{10}$ in octal.

12. Convert the following numbers to decimal.

    a. $(110111)_{2}$ b. $(762)_{8}$

    c. $(1001101)_{2}$ d. $(A7B)_{16}$

    e. $(276)_{8}$ f. $(3BC)_{16}$

13. Convert the following numbers to decimal. (Round your result to four decimal places.)

    a. $(372.43)_{8}$ b. $(769.B)_{16}$

    c. $(100111.101)_{2}$ d. $(1100110.00101)_{2}$

    e. $(277.34)_{8}$ f. $(AB.3)_{16}$

14. Perform the following conversions. (Report up to four places beyond the radix point.)

    a. $(1100110)_{2} \rightarrow (?)_{10}$ b. $(377.63)_{10} \rightarrow (?)_{8}$

    c. $(ABB)_{16} \rightarrow (?)_{10}$ d. $(972.36)_{10} \rightarrow (?)_{16}$

    e. $(A37B.C)_{16} \rightarrow (?)_{10}$ f. $(101.1101)_{2} \rightarrow (?)_{10}$

    g. $(372.33)_{8} \rightarrow (?)_{10}$ h. $(107.63)_{10} \rightarrow (?)_{2}$

15. Perform the following conversions. (Report up to four places beyond the radix point.)

     a. $(732.5)_8 \rightarrow (?)_{10}$ b. $(1023.25)_{10} \rightarrow (?)_2$

     c. $(A9B.3)_{16} \rightarrow (?)_{10}$ d. $(1772.4)_{10} \rightarrow (?)_8$

     e. $(11101011.11)_2 \rightarrow (?)_{10}$ f. $(2763.0625)_{10} \rightarrow (?)_{16}$

     g. $(256.5625)_{10} \rightarrow (?)_2$ h. $(FA.7)_{16} \rightarrow (?)_{10}$

16. Perform the following conversions.

     a. Convert the current date to binary.
     b. Convert your zip code to hexadecimal.
     c. Convert your telephone area code to octal.

Section 1.4 – Bases that are Powers of Other Bases

17. Convert the following binary numbers to octal.

     a. $(1010011101.100101)_2$ b. $(1101111001001.11101)_2$

     c. $(1011111011.111011)_2$ d. $(1111110111.0001)_2$

18. Convert the following binary numbers to hexadecimal.

     a. $(1111111011001.11101)_2$ b. $(1111100001010111.11)_2$

     c. $(100000011.10001)_2$ d. $(11111111111.11111)_2$

19. Convert the following octal numbers to binary.

     a. $(3732.35)_8$ b. $(747.321)_8$

     c. $(11223.145)_8$ d. $(2761.103)_8$

20. Convert the following hexadecimal numbers to binary.

     a. $(7F42.3B)_{16}$ b. $(E3B5.221)_{16}$

     c. $(C00B.17)_{16}$ d. $(4CAB.73)_{16}$

21. Perform the following conversions.

    a. $(1011101110.110101)_2 \rightarrow (?)_{16}$ b. $(A7B6.D1)_{16} \rightarrow (?)_2$

    c. $(1011101101.1011)_2 \rightarrow (?)_8$ d. $(1732.61)_8 \rightarrow (?)_2$

    e. $(6D4.BC)_{16} \rightarrow (?)_2$ f. $(1100001010.110101)_2 \rightarrow (?)_8$

    g. $(101101011101010.0101011)_2 \rightarrow (?)_{16}$ h. $(762.14)_8 \rightarrow (?)_2$

22. Perform the following conversions.

    a. $(21221.22)_3 \rightarrow (?)_9$ b. $(100111011.1101)_2 \rightarrow (?)_4$

    c. $(F32E.C1)_{16} \rightarrow (?)_4$ d. $(3122.23)_4 \rightarrow (?)_2$

    e. $(8776.311)_9 \rightarrow (?)_3$ f. $(13221.103)_4 \rightarrow (?)_{16}$

    g. $(3110.021)_4 \rightarrow (?)_2$ h. $(72FE.0B)_{16} \rightarrow (?)_4$

Section 1.5 – Simple Addition in Other Bases

23. Perform the following binary additions.

    a.
$$\begin{array}{r} 1100110.101 \\ +\ \ 101110.1001 \\ \hline \end{array}$$
    b.
$$\begin{array}{r} 1101110.111 \\ +\ 1011001.001 \\ \hline \end{array}$$
    c.
$$\begin{array}{r} 100111.0101 \\ +\ 1111001.1101 \\ \hline \end{array}$$

24. Perform the following octal additions.

    a.
$$\begin{array}{r} 271.34 \\ +\,622.12 \\ \hline \end{array}$$
    b.
$$\begin{array}{r} 137.621 \\ +\,223.410 \\ \hline \end{array}$$
    c.
$$\begin{array}{r} 365.4 \\ +\ \ 77.76 \\ \hline \end{array}$$

25. Perform the following hexadecimal additions.

    a.
$$\begin{array}{r} 6CB.72 \\ +\ \ AB \\ \hline \end{array}$$
    b.
$$\begin{array}{r} 17D.FE \\ +\ CC3.BA \\ \hline \end{array}$$
    c.
$$\begin{array}{r} ABCD.EF \\ +\ FEDC.BA \\ \hline \end{array}$$

26. Add the following pairs of numbers.

    a. $(372.17)_9 + (822.76)_9$ b. $(122.201)_3 + (123.21)_3$

    c. $(312.211)_4 + (332.11)_4$ d. $(455.321)_6 + (2133.45)_6$

Additional Exercises

27. Perform the following conversions. Report four places beyond the radix point. (Hint: You may need to convert to another base before the requested one.)

a. $(1221.12)_3 \rightarrow (?)_5$

b. $(733.21)_8 \rightarrow (?)_4$

c. $(877.63)_9 \rightarrow (?)_2$

d. $(10110110.10111)_2 \rightarrow (?)_6$

e. $(521.33)_6 \rightarrow (?)_{16}$

f. $(DBF3.CC)_{16} \rightarrow (?)_9$

g. $(178.31)_9 \rightarrow (?)_7$

h. $(2211.2)_3 \rightarrow (?)_2$

28. Add the following sets of numbers. Express your answers in the base that is indicated. Report your results four places past the radix point. (Hint: You need to convert the numbers to a common base before adding them.)

a. $(11011.0011)_2 + (107.3)_{10} + (722.B)_{16} = (?)_{10}$

b. $(1100110110.1011)_2 + (762.0625)_{10} + (ACD.3)_{16} = (?)_{16}$

c. $(10110110.101)_2 + (10110.001)_2 + (11101.01)_2 = (?)_2$

d. $(372.12)_8 + (10110.01)_2 + (B7.2)_{16} = (?)_8$

Chapter 2. Complex Numbers

Chapter Outline

2.1   Introduction to Complex Numbers

So far, we have been looking at different ways that numbers can be represented. More specifically, we have examined the concept of numeric bases. We saw that numbers can be represented based on a radix other than ten, and we discovered techniques to use when we encounter these numeric representations. In this chapter we will, once again, explore a different way of looking at numbers. In this case, we will discover that the numbers we are used to dealing with on a daily basis are only a small subset of a larger group of numbers known as complex numbers. At this point, one may ask, what exactly is a complex number?

2.1.1   Background

To fully understand the concept of a complex number, it is necessary to explore the general classifications into which numbers are divided. Figure 2.1 contains a tree diagram illustrating the hierarchy of the different numerical classifications.

As we can see, numbers can be broken down into several groupings. To better understand complex numbers, which is the top set of numbers in our tree, we need to understand the sets of numbers that make them up. The simplest set of numbers is composed of the natural or counting numbers. These are the numbers that most children learn first in their school career. Children are taught to count by beginning at one and proceeding upward from there. Thus the set of natural numbers contains the values 1, 2, 3, 4, 5, etc. It is important to note that this set does not include the number zero. The addition of zero comes with the next set which is the set

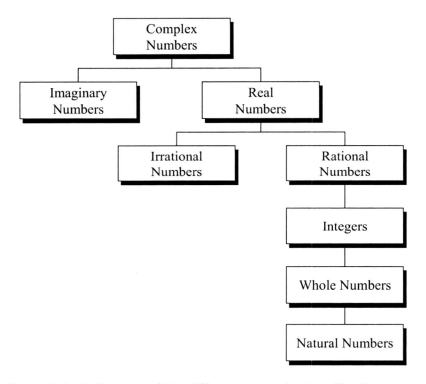

Figure 2.1. A diagram of the different numeric classification groups and their relationships.

of whole numbers. Whole numbers are composed of the natural numbers with the addition of zero (0, 1, 2, 3, 4, 5, etc.). Next, we add the negatives of the whole numbers to the mix. By doing so, we arrive at the group of numbers known as integers (..., -3, -2, -1, 0, 1, 2, 3, ...). These integer values can then be combined to form ratios, or fractions. The combination of integers to form the numerator and denominator of a fraction leads to the set of rational numbers ($\frac{1}{2}$, $\frac{3}{4}$, $\frac{-17}{629}$, etc.).

For a period of time in the history of mathematics, it was thought that the solution to any problem could be expressed as a ratio, and thus, rational numbers were the top level of the hierarchy. In fact, most problems encountered up to that point had a solution from the set of rational numbers. Later, numbers such as the square root of two, pi, and e came along. These numbers cannot be represented as a simple ratio of integers. They consist of a series of digits to the right of the decimal point that do not end or repeat. In fact the value for pi (3.141592 …) has been calculated to millions of decimal places and has not been found to end or repeat. So, what does one call a number that is not a rational number? It was decided to call numbers such as these irrational numbers. By combining these new irrational numbers with the existing rational set, we arrive at the set of real numbers.

As was the case before, it was thought that real numbers were all that would ever be needed to solve any math problem. This belief ended, however, when it became necessary to solve problems like $x^2 = -1$. Clearly no value from our set of real numbers could equal a negative value when squared. Since no real number could satisfy the equation, a new type of number was needed. In this case, we need a number that is negative when squared. A value that satisfies this condition is the square root of negative one. Such a value is clearly not a real number. So, in keeping with our naming convention, non-real numbers are called imaginary numbers. This name is somewhat unfortunate. The name imaginary implies something that does not exist. These numbers do exist, however. They simply received their name because they did not fit into the "real" numbers group. Finally, if we take our newly discovered imaginary numbers and combine them with the group of real numbers, we will arrive at our final destination, the set of complex numbers.

DEFINITION: j

    1.   j (also known as i) is equal to the square root of negative one.
         In other words, $j = \sqrt{-1}$ and $j^2 = -1$.

    2.   j is an operator that provides a 90 degree counter-clockwise rotation, or phase shift, to a quantity.

2.1.2   What is a Complex Number?

Now that we have explored the background, we can finally properly describe what a complex number is. As was mentioned before, a complex number consists of both real and imaginary parts. Thus, a typical complex number is written as $Z = a + jb$, where Z is the complex number, a is a real number, or the real part of the number, and jb is the imaginary part of the number. Looking at $Z = a + jb$, it is easy to comprehend the fact that a is simply a quantity taken from our set of real numbers, but what is the quantity jb? If we break this quantity down, we discover that b is simply another number from our set of real numbers that is being multiplied by the quantity j. So, what we really want to know is, what is j? In the field of electrical engineering, there are two commonly used definitions for j (Note: some texts, particularly math texts, will use i instead of j). The first, and most common, definition of j is that it is equal to the square root of negative one. It becomes tiresome to write the square root of negative one out in mathematical equations, so it was assigned a universally recognized abbreviation, much like pi and e. The second definition of j is a more practical, engineering definition of the quantity. This definition states that j is an operator that provides a 90° counter-clockwise rotation to a given quantity. This rotation is commonly referred to as a phase shift. Thus, multiplying a quantity by j will shift the phase of that quantity by 90 degrees in the counter-clockwise direction.

The two definitions of j raise an interesting question. What happens when j is multiplied by itself or is raised to a power? To answer this question we will look at what happens when j is raised to nine different powers, 0 to 8. The following chart displays what happens to the resulting quantity based upon both definitions of j.

Quantity	Numerical Result	Phase Result
j^0	1	0 Degree Phase Shift
j^1	j	90 Degree Phase Shift
j^2	-1	180 Degree Phase Shift
j^3	-j	270 Degree Phase Shift
j^4	1	0 Degree Phase Shift (360 Degrees)
j^5	j	90 Degree Phase Shift (450 Degrees)
j^6	-1	180 Degree Phase Shift (540 Degrees)
j^7	-j	270 Degree Phase Shift (630 Degrees)
j^8	1	0 Degree Phase Shift (720 Degrees)

By examining this table, we find an interesting result. When j is raised to successive powers, the result is periodic with a period of four. In other words, there are only four possible values that can arise from raising j to any integer power, and these values continuously repeat every four values. This fact can be extremely useful when one encounters a problem that involves j raised to a large power, for example j^{73}. Normally, it would be a time consuming task to create a table or try to calculate out what this quantity would equal. However, thanks to the periodicity of j when raised to a power the result can be found quickly by dividing the power by four and raising j to the power of the remainder. The remainder will always be a value between zero and three. Thus, if one knows the results of the first four powers of j, 0 through 3, they can easily determine which of these four values the large power of j is equivalent to. An example should help clarify this concept.

Example 2.1

Problem: Determine the simplified equivalent value of j^{237}.

Solution: As we mentioned above, since j^n is periodic, its value repeats every four integer powers. Thus, if we divide the specified power by four and find j to the power of the remainder, we will have our simplified form. By doing this we find:

$$4\overline{)237} \quad \begin{array}{c}59\end{array} \quad R = 1$$

Thus, taking j to the power of our remainder we find that

$$j^{237} = j^1 = j$$

So, in this case, our solution was j. Another way we can look at this simplification is to use the fact that j^4 is equal to 1. Thus if we factor out as many instances of j^4 as possible, we will be left with the simplest form. If we look at the problem that way, we get:

$$j^{237} = (j^4)^{59} * j = (1)^{59} * j = 1 * j = j$$

Whether we think of the operation as a division or as factoring, we do essentially the same thing and arrive at the correctly simplified value of j.

___

So, from our discussion here, we now understand that j is simply an imaginary number. More specifically we know that j is the square root of negative one. Even though b is a "real" number by itself, the process of multiplying it by j transforms the product into an "imaginary" number, which forms the "imaginary" portion of our complex value. Thus, we should now understand that a complex number is basically any number that consists of a "real" and "imaginary" part. There is no rule that requires either the "real" or "imaginary" parts of a complex number to be non-zero. If a = 0, the real portion of the complex number can be ignored, and the number can be considered as a purely "imaginary" value. Likewise if b = 0, the "imaginary" portion of the number disappears, and we are left with a "real" number. So both "real" and "imaginary" numbers are just special cases of complex numbers.

DEFINITION: Complex Number

A complex number can be defined as any number that consists of "real" and "imaginary" parts and can be expressed in the form $Z = a + jb$, where a and b are "real" numbers and j is the square root of negative one (an "imaginary" number). Thus, a is the "real" part of the number and the product jb forms the "imaginary" part. Note that zero is a valid value for both a and b. Therefore, both purely "real" and "imaginary" numbers are simple special cases of complex numbers.

2.1.3 Why Worry About Complex Numbers?

So, now that we know what a complex number is and how it is related to many of the other numbers that we are familiar with, the next thing we would want to know is why we are so concerned with complex numbers. Are they really that important or useful in engineering? The answer to that question is absolutely. An understanding of complex numbers is essential to any engineering, science, or math student. They are of particular importance within our field of interest, electrical engineering. Practically every area of electrical engineering relies on the use of complex numbers. Communications engineers require complex numbers to study radio carrier signals and to model modulation and demodulation techniques. This is particularly true for frequency and phase modulation techniques, which are commonly used. Electrical engineers specializing in image and signal processing require complex numbers to aid in the design and modeling of effective filters and to analyze the frequency content of signals. In control theory, complex numbers aid in the construction of control systems and help in system stability modeling. The study of electromagnetic fields requires complex numbers, as does electric power. Just understanding the way in which household power works requires knowledge of complex numbers. These are only a few examples of the vast contribution that the understanding of complex numbers has made to electrical engineering. In fact, it would be very nearly impossible to make an effective study of electrical engineering without a firm understanding of complex numbers.

2.2 Representing Complex Numbers

Now that we know what a complex number is, and why we want to use them, our next goal is to learn how these numbers are represented and how we work with them. We will begin by looking at how complex numbers are represented in both a written and graphic format. However, before we do that, let's review the methods we use to represent non-complex numbers.

The written expression of non-complex numbers is fairly trivial. We simply express the number by writing it out as we have always done, for example 146.45, -924, or j4. What we want to take a closer look at is the way in which we graphically represent our regular numeric values. When we are dealing with a single number in one dimension, we represent that number as a point on a number line. So, if we wanted to represent the number 4.5 graphically we would simply draw a number line and place a point at 4.5. The result would look something like figure 2.2.

Figure 2.2. A visual representation of the value 4.5 on a number line.

If we extend our representation to two dimensions, we will have two numbers, *x* and *y* representing horizontal distance and vertical distance respectively. These numbers are usually presented as an ordered pair, for example (x, y). The graphical representation of such a pair of numbers is a point on a plane. This point is positioned x units to the right (or left if *x* is negative) and y units up (or down if *y* is negative) from some central point, or origin. The plane itself is constructed with a pair of perpendicular axes. The horizontal axis, called the abscissa, is usually referred to as the *x*-axis, and the vertical axis, called the ordinate, is usually referred to as the *y*-axis. These axes meet at a point called the origin and divide the plane into four sections, called quadrants.   Figure 2.3 displays the 2-dimensional coordinate plane and illustrates its four quadrants.

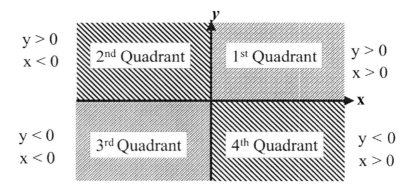

Figure 2.3. An illustration of the 2-Dimensional coordinate plane and its four quadrants.

This basic concept of graphical number representation is usually learned by students early in their mathematics education. Many students may be wondering how all of this relates to complex numbers. The answer is quite simple. When we look at an ordered pair, we realize that it simply represents a specific point on the x-y plane. Similarly, when we look at a complex number, we can think of it as representing a specific point in complex space, also known as the complex plane. Obviously, our next question should be, what does the complex plane look like and how do we represent a number on it?

2.2.1   The Complex Plane

The complex plane is actually very similar to the two dimensional coordinate system that we are used to using. Since a complex number consists of two parts, the real and imaginary parts, we can simply create a two dimensional coordinate system in which one dimension represents the real part of the number and the other dimension represents the imaginary part. So, if we call the horizontal axis the real axis and the vertical axis the imaginary axis, we can represent a complex number just as we would an ordered pair. Figure 2.4 illustrates the complex plane.

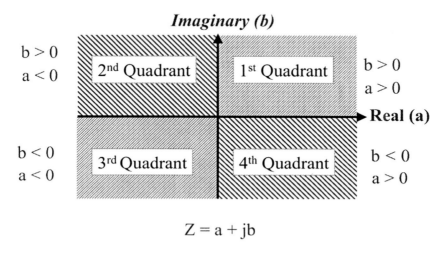

$$Z = a + jb$$

Figure 2.4. A representation of the complex number plane

By looking at figure 2.4 we see that we can represent any complex number simply by placing a point on the plane at the proper coordinates. One interesting aspect of the complex number plane that can be found by examining it closely is that, any point that falls on one of the axes will be a special case. If a point falls on the real axis, it will be a purely real number (the imaginary component is zero), and likewise, any point that falls on the imaginary axis will be purely imaginary (the real component is zero).

When representing a given point on the complex plane, we are not locked into one method of coordinate representation. There are many different ways of describing where a given point lies in space. The two coordinate systems that we will examine here are the rectangular, or Cartesian coordinate system and the polar coordinate system.

2.2.2   Complex Numbers in Rectangular Coordinates

The first coordinate system that we will examine is the rectangular or Cartesian coordinate system. This is the coordinate system that most people are accustomed to using when graphically representing numbers. The rectangular coordinate system, as the name suggests, is based on a grid approach. A specific point on the plane is located by moving a specific distance along the two axes. So, to plot a complex number in rectangular coordinates, we simply locate the position on the real axis that corresponds to the real portion of the complex number and the position on the imaginary axis that corresponds to the imaginary part. If we then imagine lines drawn perpendicular to these two points, we can easily find the location of our desired point where the two lines cross. The simplest way to illustrate this process is with an example.

Example 2.2

Problem: Plot the complex number Z = 4 - j3 using the rectangular coordinate system. In which quadrant does the point lie?

Solution: Plotting a point in rectangular coordinates is a simple task. First, we take the real portion of the number, in this case 4, and locate that position on the real axis. Then we find the value corresponding to the imaginary part of the number, in this case –3, on the imaginary axis. Now, if we imagine lines passing through these two points that are perpendicular to the respective axis, we can find the position of our desired point where the lines intersect. Performing this operation will yield the following:

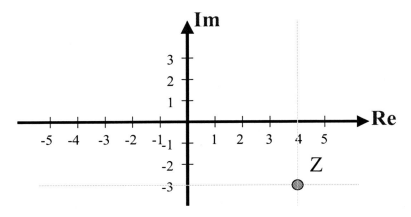

From this plot, we easily determine that the point Z representing our complex number lies within the fourth quadrant.

As we found in the example, the process of graphically representing a complex number using the rectangular coordinate system is a straightforward and simple task.

2.2.3   Complex Numbers in Polar Coordinates

While the rectangular coordinate system is based upon a rectangular grid, polar coordinates are based upon circles. Specifically, each point in space can be represented in polar coordinates by using a distance from the origin, called a magnitude, or radius and an angle from a given reference, in this case the positive real axis. So, instead of providing coordinates in the form of a distance along the real and imaginary axes, a point is specified in polar coordinates by giving a magnitude from the origin and an angle. In addition, a number expressed in polar form is written differently that one in rectangular. Since in polar form we are dealing with a magnitude and angle and not distance on the real and imaginary axes, we must represent our complex number in terms of that magnitude and angle. A typical complex number is formatted as follows:

$$Z = Me^{j\theta}$$

In this expression, M is our magnitude, theta is our angle, and j is the square root of negative one. But what is e? We briefly mentioned e when we referred to irrational numbers earlier in this chapter. Basically e is a constant that is used to represent a specific irrational number that serves as the base of the natural logarithm function. An approximate value of e is 2.718281828459 ... So, how is this number calculated? To see that, we must look at the following expression:

$$x = \left(1 + \frac{1}{n}\right)^n$$

Now, if we examine what happens to this expression as we increase the value of n, we will find that the larger the value we use for n, the closer the result is to the value we just gave for e. The following table gives the results for a few values of n.

n	result
1	2.00000000000000
2	2.25000000000000
3	2.37037037037037
4	2.44140625000000
100	2.70481382942153
1000	2.71692393223559
10000	2.71814592682493
100000	2.71826823719230

As we can see in the table, the larger the value of n the closer we get to e. Thus, we can say that our result approaches e as n approaches infinity and we can write:

$$e = \lim_{n \to \infty} \left(1 + \frac{1}{n}\right)^n$$

DEFINITION: e

The quantity e is an irrational number that forms the base of the natural logarithm function. It is defined by the following expression.

$$e = \lim_{n \to \infty} \left(1 + \frac{1}{n}\right)^n \approx 2.718281828459$$

Even though we always use the letter e to represent the constant value obtained through the above expression, it is important to remember that e is just like any other number. It can be added, subtracted, multiplied, raised to a power, etc. For example, $e^2 = 7.389056099 \ldots$

Now that we know what each part of our polar expression represents, we can look at how to plot such an expression. Plotting a polar number is not much more difficult than plotting a number in rectangular coordinates. The first thing we look at is the magnitude. This magnitude defines the radius of a circle centered at the origin. The point that we wish to plot must be located on this circle. It is the angle that tells us exactly where on this circle the point lies. So if we then measure the required angle from the positive real axis, we can precisely locate our desired point. An example will serve to clarify this process.

Example 2.3

Problem: Plot the complex number $Z = 3e^{j135°}$ on the complex plane using polar coordinates. In which quadrant does the point lie?

Solution: Here we can plainly see that we are given a number in polar form. From this value, we can extract our magnitude, or distance from the origin. In this case, our magnitude is 3. We can also find the angle that we need to measure from the positive real axis, in this case 135°. From this information, we know that our point lies on a circle centered at the origin with a radius of 3. In addition, we know that the position on that circle is determined by measuring 135° from the positive real axis. Using this information, we can plot our value as follows:

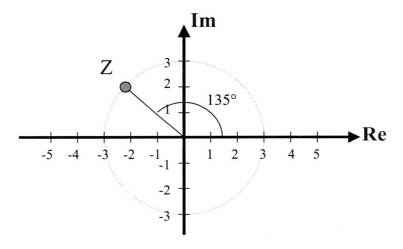

Now, we plainly see from our plot that the point lies in the second quadrant.

After working an example, we see that the process of plotting complex values in polar form is not much more difficult than plotting in rectangular coordinates.

2.2.4   Converting Between Coordinate Representations

We have now examined two different ways in which we can represent complex numbers. Both rectangular and polar coordinates provide an efficient method of representing a point in complex space. So why do we need to represent a complex number in more than one coordinate system? The reason is that some mathematical operations are easier to perform in one coordinate system than the other. We will study this in much more detail in section 2.3. We may also wonder whether there is any relationship between the coordinate systems. After all, it would be fairly useless to have our complex numbers in two different coordinate systems if we could not convert back and forth between representations. Luckily, there is a set of simple relationships between the two coordinate systems that allow us to easily convert between the two. We will examine the relationships for converting both to and from each system, beginning with how we convert from rectangular coordinates to polar.

To better understand the relationship between the rectangular and polar coordinate systems, we need to refer to a little bit of basic trigonometry. Let us start by considering the triangle in figure 2.5.

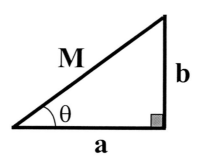

Figure 2.5. A representation of the trigonometric relationship between rectangular and polar coordinates.

If we assume that *a* and *b* are the rectangular coordinate values associated with our complex number along the real and imaginary axes, we can see that the line associated with M, or our polar magnitude, forms the hypotenuse of a right triangle. We know from our basic trigonometry that the square of the hypotenuse is equal to the sum of the squares of the other two sides. Thus, we can express our polar magnitude M in terms of our rectangular coordinates *a* and *b* as:

$$M = \sqrt{a^2 + b^2}$$

Next, we need to find an expression to determine the angle theta. Referring to basic trigonometry once again, we remember that the tangent of an angle is equal to the opposite side over the adjacent side. Thus, we can write:

$$\tan(\theta) = \frac{b}{a}$$

Now, if we take the inverse tangent of both sides of our expression, we will arrive at:

$$\theta = \tan^{-1}\left(\frac{b}{a}\right)$$

This expression gives us a method for calculating the angle theta from our rectangular coordinates, *a* and *b*.

When converting between rectangular and polar coordinate systems, there are two important things to keep in mind. First, magnitudes are always positive. If a negative magnitude is found, an error has occurred somewhere in the calculation. Second, it is important to be careful when calculating the angle theta. Most calculators will return an angle in the first or fourth quadrant, between 90 and –90 degrees. If the point lies in the second or third quadrant, 180 degrees needs to be added to the result given by the calculator or the answer will be incorrect. That being said, let us look at an example of the conversion between rectangular and polar coordinates.

DEFINITION: Rectangular to Polar Conversion

Rectangular to polar conversion is the process of converting a number represented with x and y coordinates (Rectangular) to one represented by a magnitude, M, and angle, theta (Polar). This is done using the following relationships.

$$M = \sqrt{a^2 + b^2} \qquad\qquad \theta = \tan^{-1}\left(\frac{b}{a}\right)$$

Where a is the x coordinate, or real part of the number and b is the y coordinate, or imaginary part of the number.

Example 2.4

Problem: Convert the complex number –4 + j3 to its equivalent polar representation.

Solution: We can see that the complex number that we have been given is represented in the standard Z = a + jb rectangular format. Thus, to arrive at our polar representation, we simply need to apply our two relationships to determine the values of the magnitude, M, and the angle, theta. Starting with the magnitude, we know that:

$$M = \sqrt{a^2 + b^2}$$

Thus, our magnitude will be:

$$M = \sqrt{(-4)^2 + 3^2} = \sqrt{16 + 9} = \sqrt{25} = 5$$

Next, we will determine the angle. We previously found that:

$$\theta = \tan^{-1}\left(\frac{b}{a}\right)$$

Using this we find that our angle will be:

$$\theta = \tan^{-1}\left(\frac{b}{a}\right) = \tan^{-1}\left(\frac{3}{-4}\right) \approx -36.87°$$

at least this is what our calculator tells us. However, -36.87 degrees puts our point in the fourth quadrant while an inspection of the original number in rectangular coordinates reveals that the point is in the second quadrant. Thus, as previously noted, we must add 180 degrees to put our point in the correct quadrant. So, we obtain the correct angle as follows:

$$\theta = -36.87° + 180° = 143.13°$$

Our final answer can then be represented in the polar complex form.

$$Z = 5e^{j143.13°}$$

If we plot the results, we can easily see that these two representations are equivalent.

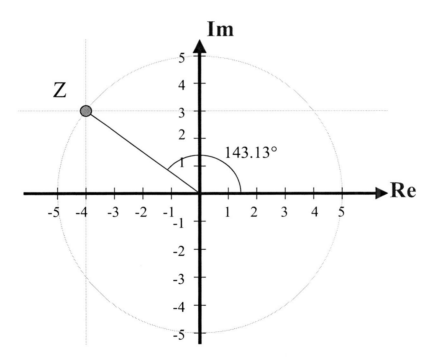

Next, we will look at the process of converting in the other direction, from polar to rectangular. In this case, we know the values for our magnitude, M, and angle, theta. We want to determine an expression for *a* and *b* in terms on M and theta. This is done, once again, by relying on basic trigonometry. If we examine Figure 2.6, we will see how magnitude and angle can be used to calculate the rectangular coordinates.

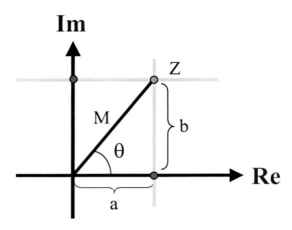

Figure 2.6. An illustration of the relationship between polar and rectangular coordinates.

First, we will find an expression for the real portion of the number, *a*. To do this, we make use of the fact that the cosine of theta is equal to the length of the adjacent side over the hypotenuse. So we can write:

$$\cos(\theta) = \frac{a}{M}$$

Then, if we multiply both sides of this equation by M, we will find:

$$a = M\cos(\theta)$$

Next, we find *b* by using the fact that the sine of theta is equal to the length of the opposite side over the hypotenuse. Therefore, we can write:

$$\sin(\theta) = \frac{b}{M}$$

MATLAB: MATLAB has a function that allows for the conversion from rectangular coordinates to polar coordinates. The syntax is as follows:

[theta, M] = CART2POL(a,b)

Where theta is the angle, *M* is the magnitude, *a* is the *x* or real component, and *b* is the *y* or imaginary component. It is important to note that MATLAB always works in radians. Thus the angle it provides will be in radians. This can be converted to degrees by multiplying by 180 and dividing by pi.

Example: [theta, M] = cart2pol(-4, 3) will return theta = 2.4981 radians, which can be converted to 143.13 degrees, and *M* = 5 which is the polar equivalent just calculated in the preceding example.

DEFINITION: Polar to Rectangular Conversion

Polar to rectangular conversion is the process of converting a number represented with magnitude and angle (Polar) to one represented by x and y coordinates (Rectangular). This is done using the following relationships.

$$a = M \cos(\theta) \qquad\qquad b = M \sin(\theta)$$

or

$$Z = a + jb = M \cos(\theta) + jM \sin(\theta)$$

Where *a* is the *x* coordinate, or real part of the number and *b* is the *y* coordinate, or imaginary part of the number, *M* is the polar magnitude, and θ is the polar angle.

Once again, we multiply both sides of the equation by M to yield:

$$b = M \sin(\theta)$$

Plugging these into our rectangular coordinate complex number form gives us:

$$Z = a + jb = M \cos(\theta) + jM \sin(\theta)$$

So, we discover that to convert from polar to rectangular form, all we must do is calculate the sine and cosine of the angle, theta, and multiply by the magnitude, M. Let us now look at an example of the process.

Example 2.5

Problem: Convert the complex number $4e^{j240°}$ to its rectangular coordinate representation.

Solution: This time we are clearly working with a polar number that has a magnitude of 4 and an angle of 240 degrees. So, to obtain our solution, we need to plug this magnitude and angle into the relationship that we previously developed. In this case, our relationship is as follows:

$$Z = a + jb = M \cos(\theta) + jM \sin(\theta)$$

Plugging in to this expression, we find:

$$Z = a + jb = 4\cos(240) + j4\sin(240)$$

$$Z = a + jb = 4\left(-\frac{1}{2}\right) + j4(-.866)$$

$$Z = a + jb = -2 - j3.464$$

So, our answer is simply:

$$Z = -2 - j3.464$$

MATLAB: MATLAB has a function that allows for the conversion from polar coordinates to rectangular coordinates. The syntax is as follows:

$$[a, b] = POL2CART(theta, M)$$

Where theta is the angle, M is the magnitude, a is the x or real component, and b is the y or imaginary component. It is important to note that MATLAB always works in radians. Thus it is important to supply the angle, theta, in radians. An angle in degrees can be converted to radians by multiplying by pi and dividing by 180.

Example: $[a, b] = pol2cart(4.1888, 4)$ will return $a = -2$ and $b = -3.46$ which is the rectangular equivalent that we just calculated in the preceding example.

2.2.5   Some Related Useful Equations

Now that we have successfully discovered how to convert between rectangular and polar coordinates, let us take a moment to reflect and take a closer look at some of the relationships we have developed. There are some interesting and useful equations that can be derived using what we have just learned.

Let us first look at the equations we developed to convert from polar to rectangular form. We know that a complex number represented in polar form is formatted as follows:

$$Z = Me^{j\theta}$$

We also know, from our conversions that:

$$Z = M\cos(\theta) + jM\sin(\theta)$$

If we combine these two expressions we will obtain:

$$Me^{j\theta} = M\cos(\theta) + jM\sin(\theta)$$

Now, dividing both sides of the equation by M yields:

$$e^{j\theta} = \cos(\theta) + j\sin(\theta)$$

DEFINITION: Euler's Equation

$$e^{j\theta} = \cos(\theta) + j\sin(\theta)$$

This equation also may be encountered in an alternate form such as:

$$\cos(\theta) = \frac{e^{j\theta} + e^{-j\theta}}{2} \qquad \text{or} \qquad \sin(\theta) = \frac{e^{j\theta} - e^{-j\theta}}{j2}$$

This expression is known as Euler's equation. With a little effort, we can use this equation to find expressions for the sine and cosine of an angle in terms of e. First, look at what we get if we use a negative angle for theta in the above equation.

$$e^{-j\theta} = \cos(-\theta) + j\sin(-\theta)$$

Now, from basic trigonometry, we know that the cosine of a negative angle is the same as the cosine of the same positive angle, and that the sine of a negative angle is equal to the negative of the sine of the same positive angle. In other words:

$$\cos(-\theta) = \cos(\theta)$$
$$\sin(-\theta) = -\sin(\theta)$$

Using this, we find:

$$e^{-j\theta} = \cos(\theta) - j\sin(\theta)$$

Now, if we add this equation and the original form of Euler's equation we will find:

$$e^{j\theta} + e^{-j\theta} = \left(\cos(\theta) + j\sin(\theta)\right) + \left(\cos(\theta) - j\sin(\theta)\right)$$

$$e^{j\theta} + e^{-j\theta} = 2\cos(\theta)$$

Dividing by two yields:

$$\cos(\theta) = \frac{e^{j\theta} + e^{-j\theta}}{2}$$

This result expresses the cosine in terms of e. It is considered to be an alternate form of Euler's equation.

By using a similar approach we can develop an expression for the sine. In this case, we want to subtract the two equations instead of adding them. Doing this gives us:

$$e^{j\theta} - e^{-j\theta} = \left(\cos(\theta) + j\sin(\theta)\right) - \left(\cos(\theta) - j\sin(\theta)\right)$$

$$e^{j\theta} - e^{-j\theta} = 2j\sin(\theta)$$

Thus, dividing by j2 will give us:

$$\sin(\theta) = \frac{e^{j\theta} - e^{-j\theta}}{j2}$$

This is the expression for sine in terms of e. It is also considered to be an alternate form of Euler's equation.

Before we move on, let us look at one final relationship. What will happen if Euler's equation is raised to a power? Let us try it and find out. Starting with Euler's equation:

$$e^{j\theta} = \cos(\theta) + j\sin(\theta)$$

We can raise it to the nth power:

$$\left(e^{j\theta}\right)^{n} = \left(\cos(\theta) + j\sin(\theta)\right)^{n}$$

Simplifying we get:

$$e^{j(n\theta)} = \left(\cos(\theta) + j\sin(\theta)\right)^{n}$$

Now, applying Euler's equation on the left gives us:

$$\cos(n\theta) + j\sin(n\theta) = \left(\cos(\theta) + j\sin(\theta)\right)^{n}$$

So we now have another useful relationship. This particular equation is known as De Moivre's Identity.

DEFINITION: De Moivre's Identity

$$\left(\cos(\theta) + j\sin(\theta)\right)^{n} = \cos(n\theta) + j\sin(n\theta)$$

2.3   Working With Complex Numbers

Now that we have completed our study of the conversions between different complex number representations, we can begin to look at how complex numbers operate in basic mathematical expressions. We will look at how addition, subtraction, multiplication, and division work in the world of complex numbers. We will then wrap up the chapter by looking at complex expression simplification, which will tie together everything we have learned so far. Before we jump into basic mathematical operations, however, we must look at one special relationship, the complex conjugate.

2.3.1   The Complex Conjugate

In the field of complex numbers, there exists a special relationship that sometimes proves useful. This relationship is known as the complex conjugate. Essentially, the complex conjugate of a number is simply the same number as before, but with j replaced by –j. The complex conjugate operation is usually denoted by an asterisk. For example, Z^* is read as the complex conjugate of Z. The importance of the complex conjugate, as we will find later in this chapter, is that a complex number multiplied by its complex conjugate will result in a purely real number.

Before we move on, however, let us look at an example of finding the complex conjugate of a number to be sure we all understand the concept.

DEFINITION: Complex Conjugate

> A complex number in which the real portion of the number remains the same, but j is replaced by –j. The complex Conjugate of Z is denoted Z^*.

Example 2.6

Problem: Find the complex conjugates of the complex numbers 4-j7 and $5e^{j35°}$.

Solution: The process of obtaining a complex conjugate is very simple. We must replace any instance of j in the number with –j.

Thus, for our first number we find:

$$(4 - j7)^* = 4 - (-j)7 = 4 + j7$$

And for our second number we get:

$$\left(5e^{j35°}\right)^* = 5e^{-j35°}$$

Now that we understand the basics, let us look at some of the typical mathematical operators applied to complex numbers. Where applicable we will look at the operations in both the rectangular and polar coordinate systems.

2.3.2   Addition

The first operation that we will examine is addition.   It is one of the easiest arithmetic operations to use with complex numbers.   To examine this operation, let us first define two complex numbers, call them x and y. Now, let $x = a + jb$ and $y = c + jd$. In order to find our sum, x + y, we need to add the two complex numbers, but how is it done? Actually, adding two complex numbers is very easy. We simply need to add the like parts. In other words we add the two real numbers together and the two imaginary numbers together. The result looks like this:

$$x + y = (a + jb) + (c + jd)$$

Combining like terms gives us:

$$x + y = (a + c) + j(b + d)$$

An example should help to clarify the procedure.

Example 2.7

Problem: Find the sum of the complex numbers x = 3 + j6 and y = 1 - j2.

Solution: The solution to this problem follows the procedure outlined above. We simply need to add the real and imaginary parts of the two numbers. Our addition should look like this.

$$x + y = (3 + j6) + (1 - j2)$$

Combining like terms gives us:

$$x + y = (3 + 1) + j(6 + (-2))$$

And simplifying yields:

$$x + y = 4 + j4$$

Complex addition is performed only in the rectangular coordinate system.

> **DEFINITION: Complex Addition**
>
> Complex addition is the process of adding two complex numbers in which the real and imaginary parts of the number are added separately.
>
> Let $x = a + jb$ and $y = c + jd$ then
>
> $$x + y = (a + c) + j(b + d)$$

2.3.3   Subtraction

Much like addition, complex subtraction operates on the real and imaginary parts of the number separately. Let us, once again, look at our two complex numbers, $x = a + jb$ and $y = c + jd$. This time we are concerned with what happens when the two are subtracted. We want to find x – y, so we write:

$$x - y = (a + jb) - (c + jd)$$

or

$$x - y = a + jb - c - jd$$

Then, by combining like terms we get:

$$x - y = (a - c) + j(b - d)$$

Let us look at an example of this before continuing.

Example 2.8

Problem: Find x - y given $x = 4 - j2$ and $y = 2 - j5$.

Solution: The solution to this problem follows the procedure outlined above. We simply need to perform the subtraction separately for the real and imaginary parts of the number. Our subtraction should look like this.

$$x - y = (4 - j2) - (2 - j5)$$

Expanding this yields:

$$x - y = 4 - j2 - 2 + j5$$

Combining like terms gives us:

$$x - y = (4 - 2) + j(-2 + 5)$$

And simplifying yields:

$$x - y = 2 + j3$$

As was the case with addition, subtraction is only performed in the rectangular coordinate system.

DEFINITION: Complex Subtraction

Complex subtraction is the process of subtracting two complex numbers in which the real and imaginary parts of the number are subtracted separately.

Let $x = a + jb$ and $y = c + jd$ then

$$x - y = (a - c) + j(b - d)$$

2.3.4   Multiplication

With the basics of addition and subtraction behind us, we can turn to the more complicated operations of multiplication and division. We will begin with multiplication, the simpler of the two operations. Unlike addition and subtraction, complex multiplication can be done in either the rectangular or polar coordinate systems. We will start by explaining the approach taken in the rectangular system and then move on to the polar method.

To determine how to multiply two complex numbers, let us look at our generic complex numbers x and y once again. Since a complex number is composed of two parts, we must treat it much the same way we would treat the polynomial x+1 when we perform multiplication. Thus to multiply two complex numbers in the rectangular coordinate field we must take the cross products into account in the same way we would when multiplying two polynomials. So, a typical complex multiplication would look like this:

$$xy = (a + jb)(c + jd)$$

Now, if we multiply the two numbers in the same way we would a polynomial, we will get:

$$xy = ac + jad + jbc + j^2 bd$$

We know from our previous study of j that j^2 is equal to –1, so we can write:

$$xy = ac + jad + jbc - bd$$

Finally, by combining like terms we arrive at:

$$xy = (ac - bd) + j(ad + bc)$$

Before we look at multiplication in the polar coordinate system, let us look at a couple of examples of multiplication in the rectangular coordinate system.

DEFINITION: Complex Multiplication (Rectangular Coordinates)

Complex multiplication in the rectangular coordinate system works much in the same way as the multiplication of polynomials. While the two respective parts of the number are multiplied, the cross terms must also be considered.

Let $x = a + jb$ and $y = c + jd$ then

$$xy = (ac - bd) + j(ad + bc)$$

Example 2.9

Problem: Find the product of X and Y where $X = 3 - j6$ and $Y = 4 - j3$

Solution: Here we can plainly see that a = 3, b = -6, c = 4, and d = -3. We could plug these values directly into our expression above yielding:

$$xy = \big((3)(4) - (-6)(-3)\big) + j\big((3)(-3) + (-6)(4)\big)$$

Simplifying this gives us:

$$xy = (12 - 18) + j(-9 - 24)$$
$$xy = -6 - j33$$

While this gives us the correct answer, it is always better to remember how the operation works, rather than memorizing the final formula. So let us work through this problem from the beginning. First, we start with:

$$xy = (3 - j6)(4 - j3)$$

Performing the multiplication gives us:

$$xy = 12 - j9 - j24 + j^2 18$$

Using the fact that j^2 is equal to -1 we can write:

$$xy = 12 - j9 - j24 - 18$$

If we combine like terms we get:

$$xy = (12 - 18) + j(-9 - 24)$$
$$xy = -6 - j33$$

This is the same answer we arrived at before, but we did it without having to remember the formula.

___

Next, let us look at what happens when a number is multiplied by its complex conjugate. We know from our discussion of the complex conjugate that we should expect a purely real number from such a multiplication. Let us find out if this is true.

Example 2.10

___

Problem: Find the product of x and y where x = 3 − j4 and y = 3 + j4.

Solution: From a quick examination of x and y, we see that y is the complex conjugate of x. Thus, we can expect a purely real number as a result of the multiplication. Once again, we begin by setting up the multiplication as follows:

$$xy = (3 - j4)(3 + j4)$$

Now, if we perform the multiplication, we get:

$$xy = 9 + j12 - j12 - j^2 16$$

Since j^2 is equal to −1 we can write:

$$xy = 9 + j12 - j12 + 16$$

Here we immediately see that the imaginary component of the expression will go to zero and we are left with only a real number as follows:

$$xy = 9 + 16 = 25$$

Thus we have verified the statement that we encountered earlier regarding complex conjugates. Multiplying a complex number by its complex conjugate does indeed result in a purely real number.

___

Now that we have encountered complex multiplication in the rectangular coordinate system, we can move on to the polar version. This is where we begin to see the benefits of representing complex numbers in polar form, and why we bothered to learn about the polar coordinate system in the first place. Working in polar coordinates greatly simplifies complex multiplication. To see this, let us take two complex numbers in polar form, call them x and y. If we let $x = Me^{j\theta}$ and $y = Ne^{j\vartheta}$ our multiplication will look like this:

$$xy = \left(Me^{j\theta}\right)\left(Ne^{j\vartheta}\right)$$

We can rearrange this expression to look like this:

$$xy = MN\left(e^{j\theta}e^{j\vartheta}\right)$$

We know that when exponential values are multiplied we simply add the exponents. Doing this gives us:

$$xy = MN\left(e^{j(\theta+\vartheta)}\right)$$

So, what this expression tells us is that when we multiply two complex numbers in polar form, we simply multiply the magnitudes and add the angles. This is a great simplification over the rectangular method and is typically preferred to performing the operation in rectangular coordinates since it is less prone to careless arithmetic errors. Let us work through an example of multiplication in polar form to wrap up our discussion of complex multiplication.

DEFINITION: Complex Multiplication (Polar Coordinates)

Complex multiplication in the polar coordinate system is far easier than multiplication in the rectangular coordinate system. The two magnitudes are multiplied and the two angles are added, resulting in a simple way to determine result.

Let $x = Me^{j\theta}$ and $y = Ne^{j\vartheta}$ then:

$$xy = MN\left(e^{j(\theta+\vartheta)}\right)$$

Example 2.11
_____

Problem: Multiply the following complex values: $x = 4e^{j45°}$ and $y = 7e^{j32°}$

Solution: Since we are dealing with the multiplication of complex numbers in their polar form, we will proceed according to the method we just outlined. We will simply multiply the magnitudes and add the angles. Thus, we begin by writing:

$$xy = (4e^{j45°})(7e^{j32°})$$

Next, we can rearrange the expression and write:

$$xy = (4)(7)e^{j(45+32)°}$$

And, finally, we can simplify this down to:

$$xy = 28e^{j77°}$$
_____

2.3.5   Division

The final operation relating to complex arithmetic that we will examine is complex division. As was the case with multiplication, division can be performed in either rectangular or polar coordinates. We will begin by examining the rectangular case and then look at the operation in polar coordinates.

The division of two complex numbers in the rectangular coordinate system can seem a little complicated at first, but we will take it one step at a time. Let us begin by defining our two complex numbers, $x = a + jb$ and $y = c + jd$. Next, we will write out what the division of x by y will look like.

$$\frac{x}{y} = \frac{a + jb}{c + jd}$$

Right now we have a ratio with complex numbers as both the numerator and the denominator. We want our result to be separable into its real and imaginary parts, or in other words, our answer must be in the form a + jb. In order to separate the numerator into real and imaginary parts, the denominator must be a real number. As we recall from our discussion of the complex conjugate, we can be assured a real number if we multiply a complex number by its complex conjugate. Thus, if we multiply the numerator and denominator by the complex conjugate of the denominator, we will obtain a real quantity for our denominator.

$$\frac{x}{y} = \left(\frac{a + jb}{c + jd}\right)\left(\frac{c - jd}{c - jd}\right) = \frac{(a + jb)(c - jd)}{(c + jd)(c - jd)}$$

Now, if we use what we learned while studying complex multiplication, we can perform the multiplications in the numerator and denominator.

$$\frac{x}{y} = \frac{ac - jad + jbc - j^2 bd}{c^2 - jcd + jcd - j^2 d^2} = \frac{(ac + bd) + j(bc - ad)}{(c^2 + d^2)}$$

This quantity can then be separated into its real and imaginary parts.

$$\frac{x}{y} = \frac{(ac + bd) + j(bc - ad)}{(c^2 + d^2)} = \frac{(ac + bd)}{(c^2 + d^2)} + j\frac{(bc - ad)}{(c^2 + d^2)}$$

Thus, we have arrived at an expression describing the results of the division of two complex numbers in rectangular form. It is not recommended that this final form be memorized, but rather that the entire process be followed since it leads to a better understanding of complex division and since it is never a good idea to just memorize formulas. Let us look at an example to help clarify the procedure.

Example 2.12
_____

Problem: Divide x by y given: $x = 4 + j6$ and $y = 2 - j3$.

Solution: Let us begin by writing out the equation for this complex division.

$$\frac{x}{y} = \frac{4+j6}{2-j3}$$

Now, we need to get a real number for the denominator. Thus, we must multiply the numerator and denominator by the complex conjugate of 2 – j3, which is 2 + j3. Doing this yields:

$$\frac{x}{y} = \left(\frac{4+j6}{2-j3}\right)\left(\frac{2+j3}{2+j3}\right) = \frac{(4+j6)(2+j3)}{(2-j3)(2+j3)}$$

Completing the multiplications gives us:

$$\frac{x}{y} = \frac{8+j12+j12+j^2 18}{4+j6-j6-j^2 9} = \frac{(8-18)+j(12+12)}{(4+9)}$$

Finally, simplifying and separating the numerator we get:

$$\frac{x}{y} = \frac{-10}{13} + j\frac{24}{13} \approx -0.7692 + j1.8462$$

DEFINITION: Complex Division (Rectangular Coordinates)

Complex division in the rectangular coordinate system is a fairly complex task. In order to separate the numerator into its real and complex parts, the numerator and denominator must be multiplied by the complex conjugate of the denominator.

Let $x = a + jb$ and $y = c + jd$ then

$$\frac{x}{y} = \frac{(ac+bd)}{(c^2+d^2)} + j\frac{(bc-ad)}{(c^2+d^2)}$$

So, as we have seen, division is a fairly complicated operation when working in rectangular coordinates. Fortunately, as was the case with multiplication, division is much easier in the polar coordinate system. To examine this, let us once again define two complex numbers. Let $x = Me^{j\theta}$ and $y = Ne^{j\vartheta}$. Using these two numbers we can write out what the division of x by y would look like.

$$\frac{x}{y} = \frac{Me^{j\theta}}{Ne^{j\vartheta}}$$

Rearranging we can write:

$$\frac{x}{y} = \left(\frac{M}{N}\right)\left(\frac{e^{j\theta}}{e^{j\vartheta}}\right)$$

Now, using the fact that when we divide two exponentials we subtract their exponents, we get:

$$\frac{x}{y} = \left(\frac{M}{N}\right)e^{j(\theta-\vartheta)}$$

DEFINITION: Complex Division (Polar Coordinates)

Complex division in the polar coordinate system is far easier than division in the rectangular coordinate system. The two magnitudes are simply divided and the two angles are subtracted, resulting in a simple to determine result.

Let $x = Me^{j\theta}$ and $y = Ne^{j\vartheta}$ then:

$$\frac{x}{y} = \left(\frac{M}{N}\right)e^{j(\theta-\vartheta)}$$

Thus, when we divide two complex numbers in the polar coordinate system, we simply divide their magnitudes and subtract their angles. This is much simpler than trying to perform the operation in rectangular coordinates, and thus, is the preferred method of performing complex division. Let us look at an example of complex division in polar coordinates to get a full understanding of the operation.

At this point, we have examined the basic arithmetic operations involving complex numbers. However, knowing how these operations work is only part of what is needed to successfully work with complex numbers in real applications. To succeed in the use of complex numbers, one must not only know how to perform the basic arithmetic operations, but also when they should be applied and in which coordinate system. In real applications, complex expressions can be encountered that are far more cumbersome than the simple examples we have encountered so far. As a result, the final section of this chapter will sum up all we have learned so far by looking at how to simplify these more complicated complex number expressions.

2.3.6   Expression Simplification

In this section, we will look at what happens when we have more than a simple two complex number problem to work with. In real applications we can see expressions that involve all of the basic arithmetic operations and include complex numbers represented in both rectangular and polar coordinates. In order to deal with expressions like these, it is important not only to know how to perform the necessary arithmetic operations, but also in what order to apply them and which coordinate system to use. The following guidelines should help in this process.

- Addition and subtraction can only be performed in rectangular coordinates. As a result, if an addition is encountered in which there is a mixture of rectangular and polar numbers, the polar numbers must be converted to rectangular in order to proceed.

- Multiplication and division can be performed in either coordinate system. However, the polar system is easier to work with, and thus, is the preferred system. If multiplication or division involves both complex numbers in both rectangular and polar form, it is usually easier to convert all of the terms to polar form before performing the operation. If all the terms are in rectangular form, it may not be worth the effort to convert them all to polar.

- Always simplify powers of j. If a power of j is encountered it should always be simplified to one of the four base representations (j, -1, -j, 1).

- Keep the final goal in mind. If the problem being solved requires magnitude and phase angle, then we want our final number in polar form. If the problem needs real and imaginary parts, then we need the number in rectangular format.

If these basic guidelines are followed, most complicated complex number expressions can be easily simplified down to more useful expressions. To wrap things up, let us look at one final example. In this example, we will simplify a complicated complex expression down to a more manageable size.

Example 2.13
_____

Problem: Simplify the following complex number expression. What is the magnitude and phase angle of the result?

$$x = 2 - j4 + \frac{1 + j2}{4 - j5} - 12 + 4e^{j30°}$$

Solution: Here we are dealing with a complicated expression of complex numbers. The key to simplifying it is to break it down into manageable parts and to keep the guidelines we mentioned before in mind. First, let us look at the expression. Most of it is composed of numbers in rectangular coordinates. There is only one number represented in polar. We should also note that there is a lot of addition in the expression, thus we know that we will be working primarily in rectangular coordinates. Most of these values can be added easily. The exceptions are the complex number represented in polar coordinates and the complex division. Thus, our goal is to get these two parts into a form that will allow them to be added along with the others. Let us start with the division. We could convert both of the numbers to polar coordinates. This would make the division easier, but then we would have to convert the result back to rectangular coordinates again. Thus, we will stay in the rectangular system for this division. Performing the division we get:

$$\frac{1 + j2}{4 - j5} = \frac{(1 + j2)(4 + j5)}{(4 - j5)(4 + j5)} = \frac{-6 + j13}{41} \approx -0.1463 + j0.3171$$

So, now that we know what the result of the division is we can place it in the original equation. This gives us:

$$2 - j4 - 0.1463 + j0.3171 - 12 + 4e^{j30°}$$

We can now simplify this by combining like parts and summing everything but the one number in polar form. Doing this yields:

$$-10.1463 - j3.6829 + 4e^{j30°}$$

Now, we must convert the number in polar form to rectangular form so that it can be added to the other number. Using our conversion equation, we get:

$$4e^{j30°} = 4\cos(30°) + j4\sin(30°) \approx 3.4641 + j2$$

Placing this in our equation yields:

$$-10.1463 - j3.6829 + 3.4641 + j2$$

This can then be simplified to:

$$-6.6822 - j1.6829$$

Now, since the problem asked for magnitude and phase, we must convert our result to polar form. Using our conversion equations, we get:

$$M = \sqrt{(-6.6822)^2 + (-1.6829)^2} \approx 6.8909$$

$$\theta = \tan^{-1}\left(\frac{-1.6829}{-6.6822}\right) \approx 14.1359°$$

However, since the real and imaginary parts are negative, we know our point is in the third quadrant. Thus we must add 180 degrees to the result returned by the calculator. So, our true phase angle is:

$$\theta = 14.1359° + 180° = 194.1359°$$

Thus, we know that the magnitude of the result is 6.8431 and the phase angle is approximately 194.1359 degrees. The final simplified form can be expressed as:

$$6.8909e^{j194.1359°}$$

2.4   Chapter Summary

In this chapter, we examined the concept of complex numbers.   We learned what a complex number is, why complex numbers are important, and how to work with these numbers.

- We began with a basic review of how numbers are categorized. We found that complex numbers are simply numbers that are made up of a combination of real and imaginary numbers. Thus, they often take the form

$$z = a + jb$$

  where a represents the real part of the number and jb represents the imaginary part. The imaginary term j can be defined as $\sqrt{-1}$ or a 90 degree counter-clockwise phase shift in the orientation angle when operating in polar coordinates.

- We also learned that complex numbers are widely used in the field of engineering and, as a result, a knowledge of the concepts related to complex numbers is essential for a good engineer.

- The concept of the complex number plane was introduced. It was much like a standard two-dimensional plane except the horizontal axis represents the real part of the number and the vertical axis represents the imaginary part of the number.   Thus, complex numbers can be represented as a given distance in the real and imaginary directions, much like any two dimensional point on a plane.

- Next, we saw that complex numbers can be represented in two different ways. First using Cartesian or rectangular coordinates in the form

$$z = a + jb,$$

  and secondly in polar form by using a magnitude from the origin and an angle from the positive real axis. A complex number in polar form is written in the form

$$z = Me^{j\theta}$$

  where M is the magnitude from the origin, θ is the angle from the positive real axis, and e is a special constant defined as

$$\lim_{n \to \infty}\left(1 + \frac{1}{n}\right)^n \approx 2.718281828459\ldots$$

- It was discovered that it was possible to convert a complex number between rectangular and polar representations. We learned that a complex number could be converted into a polar representation by finding its magnitude and angle with the equations

$$M = \sqrt{a^2 + b^2} \qquad \text{and} \qquad \theta = \tan^{-1}\left(\frac{b}{a}\right).$$

- Likewise, we found that we could convert a polar representation of a complex number into its rectangular equivalent by using the expression

$$M\cos(\theta) + jM\sin(\theta).$$

- In addition, we found some useful equations related to the equations that we used to convert between rectangular and polar form. Euler's equation can be derived from the polar to rectangular conversion expression. It allows us to express the sine and cosine functions in terms of exponentials.

$$\cos(\theta) = \frac{e^{j\theta} + e^{-j\theta}}{2} \qquad\qquad \sin(\theta) = \frac{e^{j\theta} - e^{-j\theta}}{j2}$$

- DeMoivre's can also be derived from the polar to rectangular relationship. It states that

$$(\cos(\theta) + j\sin(\theta))^n = \cos(n\theta) + j\sin(n\theta).$$

- Next, we looked at how basic arithmetic is performed using complex numbers. The first concept we examined in this area was the complex conjugate. If a typical complex number if represented as z, its complex conjugate is denoted as z^* and is found by replacing all occurrences of j in the complex number with $-j$.

- We found that addition between two complex numbers can only be performed in rectangular coordinates and consists of the simple independent additions of the real and imaginary parts of the numbers.

- Likewise, subtraction can only be performed in rectangular coordinates and the result is determined by independently subtracting the real and imaginary parts of the two numbers.

- Multiplication can be performed in both rectangular and polar coordinate systems, but it is much easier in polar form. Given two complex numbers in rectangular form, $x = a + jb$ and $y = c + jd$, the product in rectangular coordinates can be expressed as

$$xy = (ac - bd) + j(ad + bc).$$

Given two complex numbers in polar form $x = Me^{j\theta}$ and $y = Ne^{j\varphi}$, the product can be found in polar form by computing

$$xy = MNe^{j(\theta+\varphi)}$$

- Division can also be performed in either coordinate system, but, once again, it is much easier in polar form. Given two complex numbers in rectangular form, $x = a + jb$ and $y = c + jd$, the result of performing the division in rectangular coordinates can be expressed as

$$\frac{x}{y} = \frac{(ac + bd)}{(c^2 + d^2)} + j\frac{(bc - ad)}{(c^2 + d^2)}.$$

Given two complex numbers in polar form $x = Me^{j\theta}$ and $y = Ne^{j\varphi}$, the result of division be found in polar form by computing

$$\frac{x}{y} = \frac{M}{N}e^{j(\theta - \varphi)}$$

- Finally, we pulled all of our previous knowledge together and looked at the process involved in the simplification of complicated expressions using complex numbers. As long as care is take to follow the proper procedure, the simplification of complex expressions is not a difficult task.

2.5   Review Exercises

Section 2.1 – Background
_____

1. What is a complex number?

2. What are the two ways in which j can be described?

3. Using the periodic nature of j, reduce the following powers of j.

   a. j^{200} b. j^{123} c. j^{25}

   d. j^{14} e. j^{1463} f. j^{501}

4. Using the periodic nature of j, simplify the following powers of j to their simplest form.

   a. j^{629} b. j^{443} c. j^{100}

   d. j^{504} e. j^{7} f. j^{-15}

Section 2.2 – Representing Complex Numbers
_____

5. Which quadrant of the complex number plane contains the following complex numbers?

   a. $4-j3$ b. $-2+j6$ c. $-2-j4$

   d. $3+j7$ e. $14-j6$ f. $-3+j$

6. Describe where each of the following complex numbers can be found in the complex number plane.

   a. $4+j7$ b. $-3-j$ c. $4+j6$

   d. -17 e. $j32$ f. $4-j6$

7. Plot the following complex numbers.

   a. $4+j2$ b. $-3+j4$ c. $4e^{j30°}$

   d. $3-j3$ e. $3e^{-j180°}$ f. $2e^{j270°}$

8. Plot the following complex numbers.

   a. $-3 + j2$ b. $5e^{j45°}$ c. $2 + j4$

   d. $2e^{j90°}$ e. $5 - 3j$ f. $-j3$

9. Convert the following complex numbers from Cartesian to polar form.

   a. $4 + j3$ b. $3 - j4$ c. $-j$

   d. -17 e. $-2 - j4$ f. $-3 + j2$

10. Convert the following complex numbers from Cartesian to polar form.

   a. $2 - j5$ b. $4 + j7$ c. $-3 + j2$

   d. $j4$ e. 5 f. $-4 - j3$

11. Convert the following polar complex numbers into their rectangular representation.

   a. $5e^{j75°}$ b. $3e^{-j30°}$ c. $e^{j45°}$

   d. $7e^{j180°}$ e. $4e^{j90°}$ f. $7e^{j200°}$

12. Convert the following polar complex numbers into their rectangular representation.

   a. $5e^{j270°}$ b. $3e^{-j45°}$ c. $6e^{j15°}$

   d. $3e^{j135°}$ e. $10e^{j50°}$ f. $3e^{-j250°}$

Section 2.3 – Working With Complex Numbers

13. Find the complex conjugate of the following numbers.

   a. $3 + j7$ b. $4 - j3$ c. $-j2$

   d. $4e^{j30°}$ e. -2 f. $7e^{-j45°}$

14. Find the complex conjugate of the following numbers.

   a. $4 + 2j^3$ b. $17e^{-j47°}$ c. $7j^2$

   d. $3e^{j125°}$ e. $1 + 16j^{35}$ f. $-2 + j7$

15. Perform the following complex additions.

    a. $(3 - j2) + (4 + j3)$
    b. $7 + (3 + j2)$
    c. $j2 + (3 - j4)$
    d. $(6 + j3) + (2 - j4)$
    e. $(16 - j) + (-4 - j2)$
    f. $(3 + j4) + (7 - j4)$
    g. $(7 - j2) + (-5 - j)$
    h. $3e^{j60°} + 2e^{-j30°}$

16. Perform the following complex subtractions.

    a. $(6 - j4) - (2 + j3)$
    b. $(4 + j3) - (6 - j2)$
    c. $5 - (3 + j4)$
    d. $j7 - (3 + j2)$
    e. $(3 - j) - (6 + j2)$
    f. $(3 + j2) - 4e^{j90°}$
    g. $7e^{j45°} - 10$
    h. $3e^{j60°} - 4e^{j30°}$

17. Perform the following complex multiplications. Express your result in rectangular coordinates.

    a. $(j3) * (2 + j4)$
    b. $(6 - j4) * (3 + j6)$
    c. $(4 - j3) * (4 + j3)$
    d. $(2 - j) * (3 + j6)$
    e. $4 * (2 - j)$
    f. $5j * (1 + j7)$
    g. $(2 - j5) * 3e^{j60°}$
    h. $4e^{j30°} * 2e^{-j60°}$

18. Perform the following complex multiplications. Express the result in polar coordinates.

    a. $3e^{j25°} * 2e^{-j75°}$
    b. $6e^{j45°} * 2e^{j90°}$
    c. $e^{j45°} * 7e^{-j45°}$
    d. $2e^{j75°} * 4e^{-j15°}$
    e. $5 * 3e^{j45°}$
    f. $(4 - j5) * 3e^{j10°}$
    g. $5e^{-j30°} * (- j4)$
    h. $(3 + j7) * (-2 - j5)$

19. Perform the following complex divisions. Express your result in rectangular coordinates.

    a. $\dfrac{(3 + j6)}{(4 + j2)}$
    b. $\dfrac{(7 - j3)}{(2 + j4)}$
    c. $\dfrac{(- j4)}{(3 - j6)}$
    d. $\dfrac{(4 - j3)}{(16 - j12)}$

    e. $\dfrac{(5 + j3)}{(1 - j)}$
    f. $\dfrac{(2 - j3)}{4e^{j90°}}$
    g. $\dfrac{(- 5)}{3e^{j60°}}$
    h. $\dfrac{e^{j45°}}{(3 - j9)}$

20. Perform the following complex divisions. Express your result in polar coordinates.

a. $\dfrac{3e^{j45°}}{2e^{j15°}}$

b. $\dfrac{4e^{-j20°}}{2e^{j35°}}$

c. $\dfrac{2e^{j75°}}{e^{j15°}}$

d. $\dfrac{e^{-j30°}}{3e^{j90°}}$

e. $\dfrac{7e^{j25°}}{4e^{-j70°}}$

f. $\dfrac{5e^{j35°}}{(2-j6)}$

g. $\dfrac{(-4+j3)}{7e^{j35°}}$

h. $\dfrac{(2-j3)}{(7+j6)}$

21. Simplify the following expressions. Express the results in polar form.

a. $(3-j6)+j7*(-2+j7)-\dfrac{3e^{j60°}}{(4+j5)}$

b. $4e^{j25°}-(3+j2)+\dfrac{(8-j6)}{(-4+j3)}*2e^{j45°}$

22. Simplify the following expressions. Express the results in rectangular form.

a. $3e^{j45°}-(6+j2)^2+3j^{75}-\dfrac{(3+j7)}{2e^{j60°}}$

b. $7j^{215}+e^{j35°}*(6-j2)-\dfrac{3e^{j35°}}{(2-j4)}$

Chapter 3. Vectors and Matrices

Chapter Outline

3.1 Introduction

So far in this unit, we have been focusing on numbers as entities that convey a given quantity. For example, when we think of a specific number we typically think in terms of quantities, or magnitudes, for example, 4 meters, 12 kilograms, or 25 seconds. Numbers that convey only a quantity or magnitude, such as these, are known as **scalar** numbers. In the fields of engineering and physics, however, we often need to represent more than just a simple magnitude. While scalar numbers are fine for representing quantities such as mass, time, distance, and quantity of electric charge, concepts such as force, velocity, acceleration, and electric field require not only a magnitude, but a direction of application as well. Knowing that we are going to drive 100 kilometers per hour to reach a destination, for example, does not allow us to fully determine when we are going to arrive. To know this, we must also know in what direction we will travel. Without the direction of travel, we do not know if we are traveling toward or away from the planned destination. However, if we say we will be traveling due west at 100 kilometers per hour, we are in a much better position to determine an estimated time of arrival. An entity that conveys both magnitude and direction is called a **vector**. Vectors are extremely useful in both science and engineering. As a result, we will take a closer look at vectors and examine the basics of working with them in the following sections. Looking beyond vectors, there are even more complex arrangements of numbers that can be used to represent, for example, a system of linear equations. These structures are called **matrices**, and they are also very useful tools in the study of engineering. Thus, we will take some time to investigate the basic operations of matrices as well.

One should note that the concepts related to vectors and matrices can become very complex at times. We have chosen to present only a basic overview of some of the most important aspects of these structures. We will focus primarily on the application of vectors and matrices and, as a result, we will, at times, minimize our treatment of mathematical theory and proof. That is not to say that theory and proof are not important, but rather that these topics are left to an in-depth study of linear algebra where there is sufficient time to dedicate to their study. Students are encouraged to consult a text on linear algebra if they wish to obtain a more complete understanding of the concepts addressed in the following sections.

3.2 Introduction to Vectors

Let us begin our study by looking at the concept of a vector. As we mentioned before, there are basically two major ways of interpreting numbers. The first is to interpret them as a simple magnitude. Numbers interpreted in this way are called scalars. These numbers work well for describing physical quantities, such as, speed, time, temperature, distance, mass, charge, resistance, and capacitance. The second way of looking at numbers is as a magnitude oriented in a particular direction. Numbers interpreted in this way are called vectors. For the purposes of distinguishing vectors from scalars in this text, we will always print vector names in bold. In addition, quantities that refer to scalar ordered pairs will be enclosed in parentheses, for example (1, 1) and vector quantities will be enclosed in square brackets, for example [1, 1]. Since vectors have proven to be very useful in describing certain phenomena such as force, momentum, velocity, acceleration, and electric field intensity, they are worthy of our attention.

DEFINITION: Scalar Number

The term **Scalar Number** is essentially just another name for a real number. It refers to a quantity that expresses only magnitude.

> **DEFINITION: Vector**
>
> A **Vector** is a mathematical structure that, unlike a scalar, expresses both a magnitude and a direction. Vectors consist of a series of components that contain the portion of the overall vector in their respective direction or dimension.

As we know, a typical number cannot represent more than just a magnitude. As a result, vectors are usually composed of two or more numerical values. Each of these values represents a component of the overall vector oriented in a particular direction. By looking at the collective information provided by each of the components, it is possible determine both a magnitude and an angular direction for a given vector. To clarify this, let's look at a simple case. If we are given an ordered pair of scalar values, for example (2, 3), we know that we are talking about a point that is located two units to the right of the origin and three units up. Figure 3.1 shows this point.

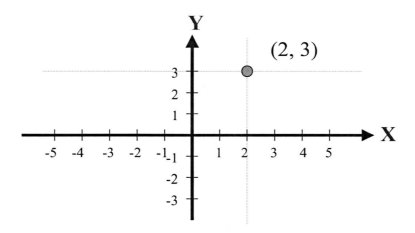

Figure 3.1. An illustration of the scalar point (2, 3).

Now, if we instead interpret our value as [2, 3], a vector, we no longer envision the quantity as a point in space. Let's say for example, this new vector represents a force being applied to an object. Knowing just the magnitude of this force does us little good. We must also know the direction in which this force is being applied to the object. Otherwise it would be impossible to know how the object will react. Luckily, a vector provides us with exactly this information. Each number in our vector is interpreted as a component force directed in a specific dimension. Thus, our example quantity would be interpreted as a two unit force in the x direction and a three unit force in the y direction. Figure 3.2 illustrates our quantity interpreted as a vector.

From the figure, we see that when a vector is represented graphically, it is drawn as an arrow with a length equal to the vector magnitude. With some closer inspection, we will also find that the magnitude and angular direction of the force that this vector describes can easily be obtained from its components through basic trigonometric relationships. This particular example illustrates a vector in two dimensions. However, vectors may exist in any number of dimensions. An n dimensional vector will simply contain n elements, one element to represent the vector's directional component in each of the n dimensions. In most cases vectors are represented as a

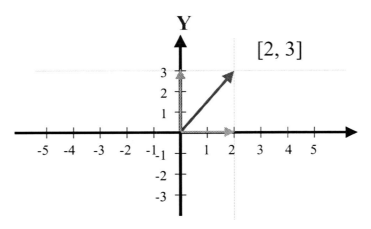

Figure 3.2. An illustration of the vector [2, 3].

series of directional components, as we have illustrated. However, in some cases, vectors will be written as a magnitude and an angle. So do not be surprised if you see them represented this way.

3.3 Working With Vectors

Since we use vectors to represent a wide variety of different phenomenon, it is obvious that we will desire to do more than just look at these quantities. A representation is useless unless it allows us to perform practical work. Thus, we should expect to be able to perform many of the basic arithmetic functions with vectors. This is indeed the case. In the following sections we will explore some of the typical arithmetic operations that can be performed with vectors, as well as some special operations that are specific to vectors. However, before that can be done we must get some basic vector terminology out of the way first.

3.3.1   Basic Concepts

The Zero Vector

As we know, when we work with regular scalar values, we have a special number to indicate the total absence of magnitude. This is the number 0. In order for vectors to operate correctly, there must be a vector with a magnitude of zero. This vector does exist. It is called the zero vector, and it is commonly written as **0**. This vector can contain as many elements as needed, as long as all of those elements are zeros. Thus, the zero vector for a three-dimensional coordinate system would be **0** = [0, 0, 0].

Vector Equality

Another important concept to keep in mind is vector equality. How do we determine that two vectors are exactly equal to each other? For this to occur, both vectors must have exactly the same magnitude and point in exactly the same direction. Obviously, for this to be true, the corresponding components of both vectors must be identical. Thus, if we have two vectors, **a** = [a_1, a_2, ..., a_n] and **b** = [b_1, b_2, ..., b_n], they can only be equal if all of the components are equal. In other words, **a** = **b**, if and only if $a_1 = b_1$, $a_2 = b_2$, ..., $a_n = b_n$.

Negative Vectors

Next let us look at the concept of negative vectors. As would be expected, for a vector to be the negative of another, it must have exactly the same magnitude and point in exactly the opposite direction. For this to occur, all of the corresponding components of the two vectors must have opposite signs. So, returning to our two vectors **a** and **b**, **a** = -**b** if and only if $a_1 = -b_1$, $a_2 = -b_2$, ..., $a_n = -b_n$.

Unit Vectors

A very important concept in the field of vectors is that of the unit vector. As its name suggests, a unit vector is simply a vector that has a magnitude of 1. In addition, unit vectors are directed in a single dimension. For example, within the three-dimensional coordinate system, there exist three unit vectors. The first of these vectors, often denoted as **i**, points along the x-axis and can be written **i** = [1, 0, 0]. The second vector, usually denoted as **j**, points along the y-axis, and can be written **j** = [0, 1, 0]. The final vector, which points along the z-axis, is denoted as **k** and can be written **k** = [0, 0, 1]. Since all vectors are composed of a series of components indicating magnitude in a given dimension, a vector can be written as a weighted sum of unit vectors. For example, the vector [5, 3, -1] can be written as 5**i** + 3**j** + -1**k**. It is not uncommon to see vectors expressed in this way. In fact, many physics and engineering texts make extensive use of this notation.

Vector Magnitude (Norm)

One final concept to consider before we look at vector arithmetic is the concept of the length of a vector. In some cases, it is useful to know the length or magnitude of the overall vector, rather than just the magnitude of its components. This overall vector magnitude, sometimes referred to as the vector norm, is determined using the formula for Euclidean distance. This formula is an extension of the basic trigonometric identity, the Pythagorean Theorem, which tells us that the length of the hypotenuse of a right triangle is equal to the square root of the sum of the squares of the other two sides, or $c = \sqrt{a^2 + b^2}$ where c is the length of the hypotenuse and a and b are the lengths of the other two sides. The formula for Euclidean distance extends this to an n dimensional case by stating that the distance from the origin can be calculated as the square root of the sum of the squares of the offsets in each dimension. Thus, if we have a vector, **v**, its magnitude, or norm, often written $\|\mathbf{v}\|$, is expressed as: $\|\mathbf{v}\| = \sqrt{v_1^2 + v_2^2 + ... + v_n^2}$. Let us look at an example to clarify this concept.

Example 3.1

Problem: Find the vector norm, $\|\mathbf{v}\|$, where **v** = [-4, 2, 4].

Solution: In this problem, we are asked to find the vector norm, or magnitude of a vector, **v**. To do this, we simply use the formula for Euclidean distance to yield:

$$\|\mathbf{v}\| = \sqrt{(-4)^2 + (2)^2 + (4)^2}$$

DEFINITION: Basic Vector Concepts

Let $\mathbf{a} = [a_1, a_2, ..., a_n]$ and $\mathbf{b} = [b_1, b_2, ..., b_n]$ then we can define the following:

Zero Vector: A special vector with a magnitude of zero. All components of this vector are zero. $\mathbf{0} = [0, 0, ..., 0]$.

Vector Equality: $\mathbf{a} = \mathbf{b}$, if and only if $a_1 = b_1$, $a_2 = b_2$, ..., $a_n = b_n$.

Negative Vectors: $\mathbf{a} = -\mathbf{b}$ if and only if $a_1 = -b_1$, $a_2 = -b_2$, ..., $a_n = -b_n$.

Unit Vectors: Special vectors of length 1 that point in a given dimension. For example, in three dimensions, $\mathbf{i}$ points in the x direction, $\mathbf{j}$ points in the y direction, and $\mathbf{k}$ points in the z direction.

Vector Magnitude (Norm): $\left\|\mathbf{a}\right\| = \sqrt{a_1^2 + a_2^2 + ... + a_n^2}$

$$\left\|\mathbf{v}\right\| = \sqrt{16 + 4 + 16}$$

$$\left\|\mathbf{v}\right\| = \sqrt{36}$$

$$\left\|\mathbf{v}\right\| = 6$$

So, here we have found the magnitude, or norm, of the vector $\mathbf{v} = [-4, 2, 4]$ is equal to 6.

Now that we have covered the preliminary information, we are ready to explore the concepts of vector arithmetic.

3.3.2   Vector Addition

First, let us examine vector addition. The addition of two or more vectors can be approached either from a graphical or algebraic perspective. The graphical approach is easier to visualize, so we will begin by looking at it. To add vectors graphically, we simply start by drawing our first vector starting from the origin. Next, we use the end point of this vector as a starting point from which to draw the second vector. We continue to use the endpoint of each vector as the starting point for the next until we have drawn all of our vectors. At this point, we draw a vector from the origin to the end point of the final vector. This new vector will represent the sum of all the others. Figure 3.3 illustrates the process with the vectors $\mathbf{a} = [1, 2]$, $\mathbf{b} = [2, 2]$, and $\mathbf{c} = [-1, -2]$, where $\mathbf{v} = \mathbf{a} + \mathbf{b} + \mathbf{c}$.

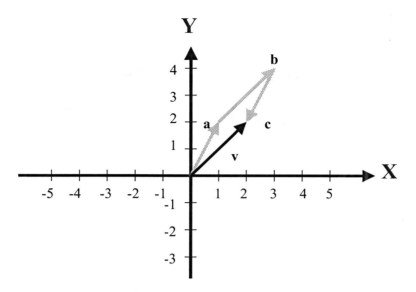

**Figure 3.3. An illustration of vector addition using the vectors
a = [1, 2], b = [2, 2], and c = [-1,-2].**

As the figure shows, we simply place our vectors end to end and our result is the vector that points from the origin to the end of our vector chain. We can verify this result by performing the same addition algebraically. Algebraic vector addition is very simple. To add multiple vectors we simply add their corresponding components. So for our example, we add all of the x components to arrive at our final x component and all the y components to arrive at our final y component. Thus, v = [1 + 2 + -1, 2 + 2 + -2]. Therefore, we find algebraically that our final vector should be v = [2, 2]. If we now refer back to figure 3.3, we find that this is exactly the same answer that we arrived at by using the graphical approach.   Obviously, the graphical addition of vectors is impractical for most applications, especially once vectors extend beyond three elements. In most cases, the algebraic approach to addition is the best method to use. Before we move on, we will look at a couple of examples.

Example 3.2

Problem: Add the vectors **a** = [3, 2], **b** = [1, -3], and **c** = [-2, 1] graphically. Verify your result by repeating the process algebraically.

Solution: Here we have been asked to add three vectors using the graphical method. Therefore, if we apply what we just learned, we will draw the first vector starting at the origin. From the end of this vector, we will draw the second vector, and from its end we will draw the third. Our final result should then be the vector that runs from the origin to the end of the third vector. If we perform this graphical representation we will find the following.

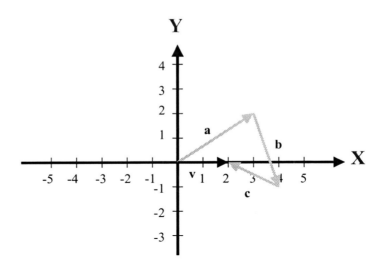

So, from our graph, we discover that the result of adding these three vectors in the vector:

v = [2, 0]

Now, we need to verify this result by using the algebraic approach to vector addition. Thus we will add the corresponding components of the three vectors as follows:

v = **a** + **b** + **c**

v = [3, 2] + [1, -3] + [-2, 1]

v = [3 + 1 + -2, 2 + -3 + 1]

v = [2, 0]

So, as we can clearly see, both of our results are the same and we have verified our graphical solution. Thus, we can be sure that the solution to this vector addition is **v** = [2, 0].

___

Example 3.3
___

Problem: Add the following Vectors: **a** = [2, 3, 7, 9], **b** = [5, -2, 1, 6], and **c** = [-1, 4, -2, -3].

Solution: Here we encounter three vectors that would be impractical to try to add graphically. As a result, we will stick to algebraic addition for these vectors. So, we will simply add the corresponding components of the three vectors as follows:

v = **a** + **b** + **c**

v = [2, 3, 7, 9] + [5, -2, 1, 6] + [-1, 4, -2, -3]

v = [2 + 5 + -1, 3 + -2 + 4, 7 + 1 + -2, 9 + 6 + -3]

v = [6, 5, 6, 12]

Thus, we arrive at our solution.

┌───┐
│ **DEFINITION: Vector Addition**   │
│   │
│ The addition of two or more vectors can be approached in two ways.   │
│   │
│ 1. Graphically: The first vector is drawn from the origin. Each successive vector is drawn from the end point of the vector before it, creating a chain of vectors. The solution will be a vector drawn from the origin to the final end point in the vector chain. This approach is only practical for vectors in two or three dimensions. │
│   │
│ 2. Algebraically: All of the corresponding components of the vectors are added together to form the result. Let **a** = $[a_1, a_2, ..., a_n]$ and **b** = $[b_1, b_2, ..., b_n]$. Then **a** + **b** = $[a_1 + b_1, a_2 + b_2, ..., a_n + b_n]$. This approach will work with vectors of any length. │
└───┘

3.3.3 Vector Subtraction

Our next area of attention is a vector subtraction. As one would expect, vector subtraction is very similar to vector addition. As with addition, it is easily performed both graphically and algebraically. The graphical approach proceeds much like it did for vector addition. We start by drawing the first vector beginning at the origin. From this point, we slightly modify the approach that we used before. Instead of drawing the next vector from the end of the first, as we would do for vector addition, we draw a vector that has the same magnitude but opposite direction. In other words we draw the negative of the original vector. So if we have the vector **v** = [1, -2] we would draw the vector −**v** = [-1, 2] from the end of our first vector. We then continue to chain on more of these negative vectors until we have represented all of the vectors we wish to subtract. Once all the vectors have been represented, we draw a new vector from the origin to the end of our final vector. This new vector is the solution to our subtraction problem. Figure 3.4 displays the graphical subtraction **v** = **a** − **b** − **c**, where **a** = [4, 2], **b** = [2, 2], and **c** = [-1, 2].

As we can see from the figure, the graphical approach to subtraction is the same as addition except that we are adding the negatives of the vectors. As was the case with addition, we can also perform our vector subtraction algebraically. To subtract two or more vectors algebraically, we simply perform the subtraction between the corresponding elements of each vector. So if we let **a** = $[a_1, a_2, ..., a_n]$, **b** = $[b_1, b_2, ..., b_n]$, and **c** = $[c_1, c_2, ..., c_n]$, we can define the algebraic subtraction of these vectors as **v** = **a** − **b** − **c** = $[a_1 - b_1 - c_1, a_2 - b_2 - c_2, ..., a_n - b_n - c_n]$. Some examples will help us better understand vector subtraction before we move on.

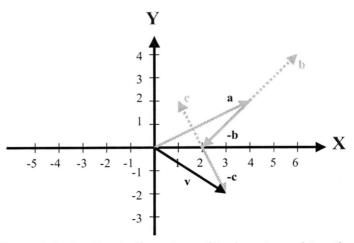

Figure 3.4. An illustration of graphical vector subtraction.

Example 3.4

Problem: Let **a** = [4, 2], **b** = [1, 2], and **c** = [2, -3]. Find the solution to **v** = **a** – **b** – **c** graphically. Verify your solution by performing the same subtraction algebraically.

Solution: In this case, we are given three vectors and asked to perform a graphical subtraction. Therefore, we will draw vector **a** from the origin. Then we will draw the vector –**b** from the end of vector **a** and vector –**c** from the end of –**b**. We can then draw a vector from the origin to the end of vector –**c** to obtain our solution. Doing this we get:

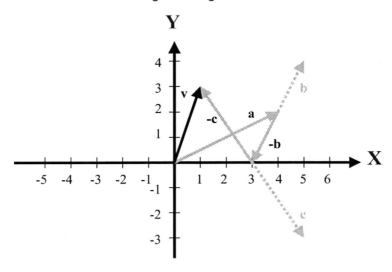

From our graph, we can see the solution to this graphical subtraction is [1, 3]. Next, we need to verify this by performing the same subtraction algebraically. Doing so gives us:

     v = **a** – **b** – **c**

$v = [4, 2] - [1, 2] - [2, -3]$

$v = [4 - 1 - 2, 2 - 2 - -3]$

$v = [1, 3]$

Thus, the algebraic and graphical solutions are the same and our answer is verified.

Example 3.5

Problem: Let $a = [4, 7, 9, 3]$, $b = [1, 4, 3, 2]$, and $c = [-1, 4, -3, 6]$. Perform the subtraction $v = a - b - c$.

Solution: Once again we are asked to perform a subtraction. In this case, we are dealing with vectors with four components. Thus, the most practical approach is to use algebraic vector subtraction. Doing so yields:

$v = a - b - c$

$v = [4, 7, 9, 3] - [1, 4, 3, 2] - [-1, 4, -3, 6]$

$v = [4 - 1 - -1, 7 - 4 - 4, 9 - 3 - -3, 3 - 2 - 6]$

$v = [4, -1, 9, -5]$

Thus, we have found the solution to our subtraction.

```
DEFINITION: Vector Subtraction

    The subtraction of two or more vectors can be approached in two ways.

    1.  Graphically:  The first vector is drawn from the origin.  The negative of each
        successive vector is drawn from the end point of the vector before it, creating a
        chain of vectors.  The solution will be a vector drawn from the origin to the final
        end point in the vector chain.  This approach is only practical for vectors in two or
        three dimensions.

    2.  Algebraically:  All of the corresponding components of the vectors are subtracted
        to form the result.  Let a = [a₁, a₂, ..., aₙ] and b = [b₁, b₂, ..., bₙ].   Then
        a - b = [a₁ - b₁, a₂ - b₂, ..., aₙ - bₙ].  This approach will work with vectors of any
        length.
```

As we have found, vector subtraction works much like addition and can be performed either graphically, using the head-to-tail method, or algebraically using component subtraction.

3.3.4   Scalar Multiplication

Now that we have examined addition and subtraction, one might ask: what about multiplication? Well, vector multiplication is an interesting topic. This is because there are several forms of multiplication that can be performed involving vectors. We will look at these forms in the following sections, but for now, let us look at the simplest case. What happens when we want to multiply a scalar and a vector together? Let s be a scalar and $v = [v_1, v_2, ..., v_n]$ be a vector. Using these terms, our product would be written as sv or $s[v_1, v_2, ..., v_n]$. So, as you would expect, the scalar multiplication distributes across our vector and each term is multiplied by the scalar value. Thus, we have: $sv = [sv_1, sv_2, ..., sv_n]$. A quick numeric example should serve to further illustrate the concept.

DEFINITION: Scalar Multiplication

Let s be a scalar and $\mathbf{v} = [v_1, v_2, ..., v_n]$ be a vector. Then multiplication of a scalar and a vector is defined as:

$$sv = [sv_1, sv_2, ..., sv_n]$$

Example 3.6

Problem: Multiply the vector $\mathbf{v} = [7, 3, 1, 6]$ by the scalar value $s = 5$.

Solution: To solve this problem, we must perform a simple multiplication of a vector and a scalar. Thus, we must multiply each term of our vector by the scalar value to reach our solution. Doing this gives us:

$sv = 5[7, 3, 1, 6]$

$sv = [5(7), 5(3), 5(1), 5(6)]$

$sv = [35, 15, 5, 30]$

So, as we can see, the result of multiplying our vector by the scalar value of 5 was that each component of the vector was multiplied by 5.

Now that we have covered the scalar case, let us see what happens when we multiply two vectors together. The result of a vector multiplication can be either a scalar or another vector, based on the type of vector multiplication that is used. Thus, there are two types of vector multiplication. Let us first look at the form that yields a scalar result. This type of multiplication is called the scalar product.

3.3.5   The Scalar (Dot) Product

The scalar product is one of the two forms of vector multiplication. As was mentioned before, there are two ways to multiply vectors, one resulting in a scalar and the other in a vector. As the name suggests, the scalar product produces a scalar result. To distinguish between the two types

---

DEFINITION: Some Properties of Vector Addition and Scalar Multiplication

Let **a**, **b**, and **c** be vectors and r and s be scalars. Then we can define the following properties of addition and scalar multiplication:

Addition:

$$\mathbf{a} + \mathbf{b} = \mathbf{b} + \mathbf{a} \qquad\qquad \text{Commutative Law}$$
$$\mathbf{a} + (\mathbf{b} + \mathbf{c}) = (\mathbf{a} + \mathbf{b}) + \mathbf{c} \qquad \text{Associative Law}$$
$$\mathbf{a} + \text{-}\mathbf{a} = \mathbf{0} \qquad\qquad \text{Additive Inverse}$$
$$\mathbf{a} + \mathbf{0} = \mathbf{a} \qquad\qquad \text{Additive Identity}$$

Scalar Multiplication:

$$r\mathbf{a} = \mathbf{a}r \qquad\qquad \text{Commutative Law}$$
$$r(s\mathbf{a}) = (rs)\mathbf{a} \qquad\qquad \text{Associative Law}$$
$$r(\mathbf{a} + \mathbf{b}) = r\mathbf{a} + r\mathbf{b} \qquad \text{Distributive Law}$$
$$(r + s)\mathbf{a} = r\mathbf{a} + s\mathbf{a} \qquad \text{Distributive Law}$$

---

of vector multiplication, two different operators are used. In the case of the scalar product, a dot (•) is used. So, if we have two vectors, **a** and **b**, the scalar product is written as **a** • **b**. Because of this operator, the scalar product has become known as the dot product, and the terms dot and scalar are now used interchangeably when referring to this product. So, how does the dot product work? Well, a dot product is defined in the geometrical sense as the product of the magnitude of two vectors multiplied by the cosine of the angle between them. Thus we can write:

$$\mathbf{a} \bullet \mathbf{b} = \|\mathbf{a}\| \|\mathbf{b}\| \cos(\theta_{ab})$$

where θ is the angle between the two vectors. Basically, what we're doing here is multiplying the magnitude of **a** by the component of **b** that lies in the direction of **a**, or if you would rather think that the other way around, we are multiplying the magnitude of **b** by the component of **a** that is in the direction of **b**. Figure 3.5 illustrates this concept.

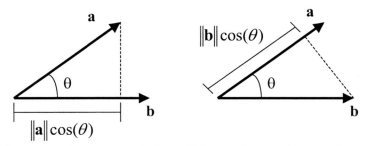

Figure 3.5. A geometric representation of the scalar, or dot, product of two vectors.

While we could use this method to calculate the dot product, our vectors are usually represented as components and not in terms of magnitude and angle. Thus, it would be beneficial if we could perform this calculation using our component-based vector representations. Luckily for us, there is a simple way to do this. We simply multiply the values of the corresponding terms and then sum the resulting products together. So, for our vectors $\mathbf{a} = [a_1, a_2, ..., a_n]$ and $\mathbf{b} = [b_1, b_2, ..., b_n]$, we can write:

$$\mathbf{a} \bullet \mathbf{b} = a_1 b_1 + a_2 b_2 + ... + a_n b_n$$

The dot product has many uses and is encountered often in vector mathematics, physics, and engineering. One of the most useful abilities of the dot product is that it can be used to find the angle between two vectors. As we recall from its definition:

$$\mathbf{a} \bullet \mathbf{b} = \|\mathbf{a}\| \|\mathbf{b}\| \cos(\theta_{ab})$$

If we divide both sides of this equation by $\|\mathbf{a}\| \|\mathbf{b}\|$, we get

$$\cos(\theta_{ab}) = \frac{\mathbf{a} \bullet \mathbf{b}}{\|\mathbf{a}\| \|\mathbf{b}\|}$$

From which we can write:

$$\theta_{ab} = \cos^{-1}\left(\frac{\mathbf{a} \bullet \mathbf{b}}{\|\mathbf{a}\| \|\mathbf{b}\|}\right)$$

So, the angle between any two vectors is simply the inverse cosine of the dot product of the vectors divided by the product of their magnitudes. Thus, we know that if our dot product is 0, the angle between the two vectors is 90 degrees. In other words, if the dot product is zero, the two vectors are perpendicular. Before we move on, let us look at a couple of example problems.

DEFINITION: The Scalar (Dot) Product

The Scalar product, commonly called the dot product is defined as:

$$\mathbf{a} \bullet \mathbf{b} = \|\mathbf{a}\| \|\mathbf{b}\| \cos(\theta_{ab})$$

Where $\mathbf{a}$ and $\mathbf{b}$ are vectors and θ is the angle between them.

If we let $\mathbf{a} = [a_1, a_2, ..., a_n]$ and $\mathbf{b} = [b_1, b_2, ..., b_n]$, then we can calculate the dot product of two vectors as follows:

$$\mathbf{a} \bullet \mathbf{b} = a_1 b_1 + a_2 b_2 + ... + a_n b_n$$

Example 3.7

Problem: Let **a** = [4, 2, -1, 7] and **b** = [3, 2, 4, -3]. Find the dot product of **a** and **b**.

Solution: here we have been given two vectors and asked to find their dot product. To obtain our solution, we must simply multiply the corresponding component values and sum the results. Thus, we write:

$$\mathbf{a} \bullet \mathbf{b} = [4, 2, -1, 7] \bullet [3, 2, 4, -3]$$

$$\mathbf{a} \bullet \mathbf{b} = (4)(3) + (2)(2) + (-1)(4) + (7)(-3)$$

$$\mathbf{a} \bullet \mathbf{b} = 12 + 4 - 4 - 21$$

$$\mathbf{a} \bullet \mathbf{b} = -9$$

Thus, the result of the dot product of **a** and **b** is –9.

DEFINITION: Finding the Angle Between Two Vectors

The angle between any two vectors can be found using the dot product as follows:

$$\theta_{ab} = \cos^{-1}\left(\frac{\mathbf{a} \bullet \mathbf{b}}{\|\mathbf{a}\|\|\mathbf{b}\|}\right)$$

Example 3.8

Problem: Find the angle between the vectors **a** = [1, 2, -1, 5] and **b** = [-2, 3, 4, 1].

Solution: In this case, we are looking for the angle between two vectors. We know from our discussion that this angle is simply the inverse cosine of the dot product of the two vectors divided by their magnitudes. So, we will first calculate the dot product a • b.

$$\mathbf{a} \bullet \mathbf{b} = (1)(-2) + (2)(3) + (-1)(4) + (5)(1)$$

$$\mathbf{a} \bullet \mathbf{b} = -2 + 6 - 4 + 5$$

$$\mathbf{a} \bullet \mathbf{b} = 5$$

Now, we must find the two magnitudes. Using our Euclidean distance equation we get:

$$\|\mathbf{a}\| = \sqrt{1^2 + 2^2 + (-1)^2 + 5^2} = \sqrt{1+4+1+25} = \sqrt{31}$$

$$\|\mathbf{b}\| = \sqrt{(-2)^2 + 3^3 + 4^2 + 1^2} = \sqrt{4+9+16+1} = \sqrt{30}$$

DEFINITION: Some Properties of the Scalar (Dot) Product

Let **a**, **b**, and **c** be vectors and r be a scalar. Then we can define the following properties of the scalar product:

a • **b** = **b** • **a**	Commutative Law
a • (**b** + **c**) = **a** • **b** + **a** • **c**	Distributive Law
r(**a** • **b**) = (r**a**) • **b** = **a** • (r**b**)	Homogeneity
a • **a** ≥ 0, and **a** • **a** = 0 if and only if **a** = **0**	Positivity

---

MATLAB: MATLAB has a function that allows for the calculation of the dot product of two vectors of the same length.

$$s = \text{DOT}(\mathbf{a}, \mathbf{b})$$

Where s is the scalar product of the vectors **a** and **b**.

Example: s = dot([4, 2, -1, 7], [3, 2, 4, -3]) will return -9 which is the same as the dot product calculated in example 3.7.

So, now we can find our angle by writing:

$$\theta_{ab} = \cos^{-1}\left(\frac{\mathbf{a} \bullet \mathbf{b}}{\|\mathbf{a}\|\|\mathbf{b}\|}\right) = \cos^{-1}\left(\frac{5}{\sqrt{31}\sqrt{30}}\right) \approx \cos^{-1}(0.1640) \approx 80.56°$$

So, the angle between our two vectors is 80.56 degrees.

3.3.6   The Cross Product

The final vector arithmetic operation that we will look at is the cross product. The cross product is the second form of vector multiplication in which the multiplication of two vectors yields another vector. To distinguish this form of multiplication for the scalar or dot product, we use a cross symbol (X) to represent it. Thus, if we have two vectors, **a** and **b**, we represent their cross product as: **v** = **a** x **b**. So far we know that the cross product is different from the dot product because it produces a vector, but how does the cross product work? To answer the question, we must once again look at the geometrical representation of the operation. The cross product can be broken down into two parts, a magnitude and a direction. The magnitude is defined by taking the product of the magnitudes of the two vectors, **a** and **b**, and multiplying by the sine of the angle between them. Here we are essentially finding the area of a parallelogram formed by the two vectors. This makes sense if we think of the magnitude of **a** as the width of the parallelogram. The magnitude of **b** multiplied by the sine of the angle determines the height of the parallelogram. Figure 3.6 illustrates this concept.

Once we have our magnitude, it must be multiplied by a unit vector pointing in the proper direction. The direction of the cross product is defined by a unit vector that is drawn normal (perpendicular) to the plane formed by the two vectors. This direction can be determined by the right-hand rule. This means that if we orient our right hand in such a way as our fingers curve from vector **a** toward vector **b**, the normal vector will point in the direction of our extended thumb. Figure 3.7 illustrates the use of the right-hand rule.

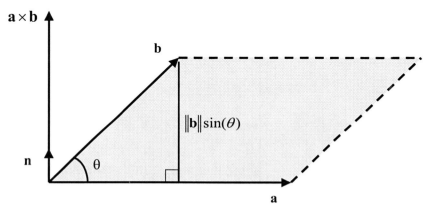

Figure 3.6. An illustration of the parallelogram formed by the two vectors in the cross product. The magnitude of the cross product is the area of this parallelogram.

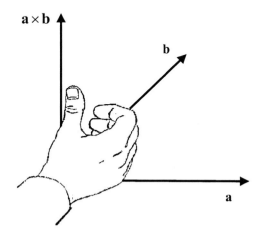

Figure 3.7. An illustration of the use of the right-hand rule to determine the direction of the cross product.

Now that we know how both the magnitude and direction of the cross product of two vectors are determined, we can put it all together and write:

$$\mathbf{a} \times \mathbf{b} = \|\mathbf{a}\| \|\mathbf{b}\| \sin(\theta_{ab})\mathbf{n}$$

where **a** and **b** are vectors, θ is the angle between them, and **n** is a unit vector drawn normal to the plane in which **a** and **b** are located. Because of the way in which the direction of the cross product is calculated, it is important to note that the order of the two vectors makes a difference. If the order of the vectors is reversed, so is the direction of the cross product. Thus:

a x b = -(b x a)

DEFINITION: The Cross Product

The cross product of two vectors is defined as:

$$\mathbf{a} \times \mathbf{b} = \|\mathbf{a}\| \|\mathbf{b}\| \sin(\theta_{ab})\mathbf{n}$$

Where **a** and **b** are vectors, θ is the angle between them, and **n** is a unit vector drawn normal to the plane formed by the vectors **a** and **b**.

If we let **a** = [a_1, a_2, a_3] and **b** = [b_1, b_2, b_3], then we can calculate the cross product of the two vectors as follows:

$$\mathbf{a} \times \mathbf{b} = [(a_2b_3 - a_3b_2), (a_3b_1 - a_1b_3), (a_1b_2 - a_2b_1)]$$

In addition, if we take the cross product of a vector with itself, the angle between the two vectors will be zero. The sine of zero is zero. Therefore, the cross product of a vector with itself is always equal to zero.

The calculation of a cross product using our standard component-based vectors is a bit more complicated than computation of the dot product. To explain the computation, we will need to make use of the unit vectors, **i**, **j**, and **k**. If we take two vectors **a** = [a_1, a_2, a_3] and **b** = [b_1, b_2, b_3], we can rewrite **a** and **b** as a sum of the unit vector components as follows:

a = a_1i + a_2j + a_3k

b = b_1i + b_2j + b_3k

Now, we can find an expression for the cross product of two vectors by multiplying **a** and **b** as if they were polynomial expressions. Thus, we write:

a x b = (a_1i + a_2j + a_3k) x (b_1i + b_2j + b_3k)

a x b = a_1b_1(i x i) + a_1b_2(i x j) + a_1b_3(i x k) + a_2b_1(j x i) + a_2b_2(j x j) + a_2b_3(j x k) + a_3b_1(k x i) + a_3b_2(k x j) + a_3b_3(k x k)

At this point, we have encountered a series of cross products involving the unit vectors. If we take a moment to apply the right hand rule to the three dimensional coordinate systems we will find the following:

$$i \times i = 0 \qquad i \times j = k \qquad i \times k = -j$$
$$j \times i = -k \qquad j \times j = 0 \qquad j \times k = i$$
$$k \times i = j \qquad k \times j = -i \qquad k \times k = 0$$

Using these relationships, we can simplify our expressions down as follows:

$$a \times b = a_1b_1(0) + a_1b_2(k) + a_1b_3(-j) + a_2b_1(-k) + a_2b_2(0) + a_2b_3(i) + a_3b_1(j) + a_3b_2(-i) + a_3b_3(0)$$

Combining like terms, we get:

$$a \times b = (a_2b_3 - a_3b_2)i + (a_3b_1 - a_1b_3)j + (a_1b_2 - a_2b_1)k$$

Thus, we have obtained an expression for the cross product in terms of the components of the vectors **a** and **b**. The cross product is a very useful vector operation that is used often in engineering.

Let us work an example of the cross product to wrap up our section on vector arithmetic.

Example 3.9

Problem: Find the cross product of the vectors **a** = [1, 3, 5] and **b** = [4, 2, -1].

Solution: This example can be solved either by plugging the respective values into the equation that we just found above, or by working back through the calculations. We will work through all of the calculations here. First, we need to rewrite our vectors as a sum of unit vectors. Thus, we write:

$$a = i + 3j + 5k$$

$$b = 4i + 2j - k$$

Taking the cross product yields:

$$a \times b = (i + 3j + 5k) \times (4i + 2j - k)$$

$$a \times b = 4(i \times i) + 2(i \times j) - (i \times k) + 12(j \times i) + 6(j \times j) - 3(j \times k) + 20(k \times i) + 10(k \times j) - 5(k \times k)$$

Simplifying the unit vector cross products gives us:

$$a \times b = 4(0) + 2(k) - (-j) + 12(-k) + 6(0) - 3(i) + 20(j) + 10(-i) - 5(0)$$

$$a \times b = 2k + j - 12k - 3i + 20j - 10i$$

Combining Like terms results in:

a x **b** = -13**i** + 21**j** – 10**k**

or, written as a standard vector:

a x **b** = [-13, 21, -10]

DEFINITION: Some Properties of the Cross Product

Let **a**, **b**, and **c** be vectors. Then we can define the following properties of the cross product:

Relationships:

a x **b** = -(**b** x **a**)	Anti-commutative
a x (**b** x **c**) ≠ (**a** x **b**) x **c**	Non- Associative
a x (**b** + **c**) = **a** x **b** + **a** x **c**	Distributive Law

Unit Vector Products:

i x **i** = 0	**i** x **j** = **k**	**i** x **k** = -**j**
j x **i** = -**k**	**j** x **j** = 0	**j** x **k** = **i**
k x **i** = **j**	**k** x **j** = -**i**	**k** x **k** = 0

MATLAB: MATLAB has a function that allows for the calculation of the cross product of two vectors of length three.

v=CROSS(a, b)

Where v is the cross product of the vectors **a** and **b**.

Example: v = cross([1, 3, 5], [4, 2, -1]) will return [-13, 21, -10], which is the same as the cross product calculated in example 3.9.

3.4 Introduction to Matrices

So far in this chapter, we have been discussing the concept of vectors. We discovered that vectors typically consist of an array of scalar values, or components, that when taken together allow us to represent both a magnitude and direction. We found that in many cases knowing the direction that is associated with a given magnitude is essential to properly interpreting how the magnitude will be applied. Now, we are going to broaden our scope and look at an even larger structure, the matrix. A matrix is defined as a rectangular, or two-dimensional, array of numbers consisting of rows and columns. This arrangement of values is usually enclosed in square brackets. Figure 3.8 illustrates the arrangement of a typical matrix.

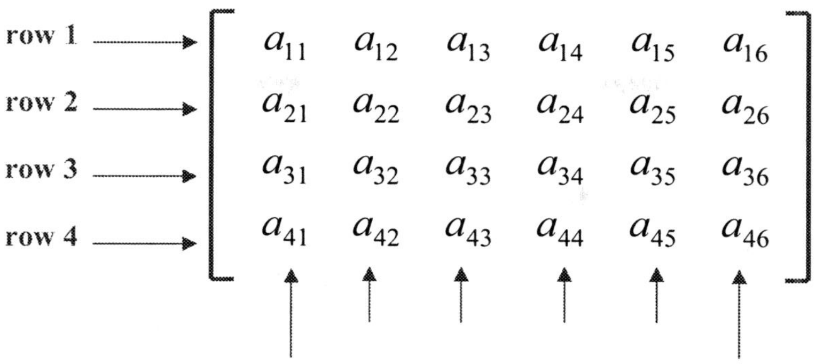

$$
\begin{array}{l}
\text{row 1} \longrightarrow \\
\text{row 2} \longrightarrow \\
\text{row 3} \longrightarrow \\
\text{row 4} \longrightarrow
\end{array}
\begin{bmatrix}
a_{11} & a_{12} & a_{13} & a_{14} & a_{15} & a_{16} \\
a_{21} & a_{22} & a_{23} & a_{24} & a_{25} & a_{26} \\
a_{31} & a_{32} & a_{33} & a_{34} & a_{35} & a_{36} \\
a_{41} & a_{42} & a_{43} & a_{44} & a_{45} & a_{46}
\end{bmatrix}
$$

Figure 3.8. An illustration of the structure of a matrix.

DEFINITION: Matrix

A matrix is defined as a rectangular, or two-dimensional, array of numbers consisting of rows and columns. This arrangement of values is usually enclosed in square brackets.

Let us take a moment to look at Figure 3.8 and interpret it more fully. One of the first things that we notice is the size of the matrix. A quick glance at the figure tells us that there are four rows and six columns. Thus, we would refer to this as a 4 x 6 matrix. When giving the size of a matrix, the number of rows is always given first. So, if we had a matrix with m rows and n columns, we would say that it was an m x n matrix. Next, notice the elements of the array (a_{11}, a_{12}, etc). Each of these elements is referenced by a subscript value. This subscript consists of the row and column that the element resides in. Thus, the element in row 3 and column 4 would be a_{34}. To avoid confusion between representations of the elements of matrices and the matrix as a whole, elements will always be printed in lowercase letters and matrix names will always be uppercase and bold. Now, if we want to give a more general representation of what a matrix looks like, we could write the following:

$$
\mathbf{A} = \begin{bmatrix}
a_{11} & a_{12} & \cdots & a_{1n} \\
a_{21} & a_{22} & \cdots & a_{2n} \\
\vdots & \vdots & \ddots & \vdots \\
a_{m1} & a_{m2} & \cdots & a_{mn}
\end{bmatrix}
$$

We could also describe the ith row of this matrix as:

$$
\begin{bmatrix} a_{i1} & a_{i2} & \cdots & a_{in} \end{bmatrix}, \quad for\ i = 1, 2, \cdots, m
$$

Likewise, the jth column could be referred to as:

$$\begin{bmatrix} a_{1j} \\ a_{2j} \\ \vdots \\ a_{mj} \end{bmatrix}, \quad for \; j = 1, 2, \cdots, n$$

Since each row and column is essentially a vector when looked at independently, they are sometimes referred to as row and column vectors.

Matrices are often used as a tool to solve problems in electrical engineering. In the following sections, we will explore some of the basic arithmetic operations that can be performed using matrices. In addition, we will look at some matrix-specific operations, and will wrap up the chapter with an engineering application of matrices in the area of circuit analysis.

3.5 Matrix Arithmetic

As was the case with vectors, we do not use matrices as simply a way of formatting numbers so that they look nice or are easier to comprehend. While this is often an added benefit of using matrices, the truth is that matrices are a powerful tool that can be applied to a wide range of science and engineering problems. Before we can see the true usefulness of matrices, however, it is necessary to understand the basics of working with them. Thus, in this section, we will first look at some basic concepts and terminology that is encountered when working with matrices. We will then examine the basic arithmetic operations as they relate to matrices.

3.5.1   Basic Concepts

Before we begin working with matrices, it is important to be aware of some basic concepts and terminology.

Matrix Equality

First, let us look at the concept of matrix equality. How do we know if we have two matrices that are equal? Well, just as was the case with vectors, we know that two matrices are equal when all of their corresponding elements are equal. For example, let us say we have been given the following two matrices, **A** and **B**. What must w, x, y, and z be equal to in order to make the two matrices equal?

$$\mathbf{A} = \begin{bmatrix} 1 & -5 & 0 \\ 3 & -2 & 2 \\ 4 & 0 & 4 \end{bmatrix} \qquad \mathbf{B} = \begin{bmatrix} 1 & w & x \\ 3 & y & z \\ 4 & 0 & 4 \end{bmatrix}$$

The answer is simple. For **A** to be equal to **B**, all of the corresponding elements must be the same. Thus, $w = -5$, $x = 0$, $y = -2$, and $z = 2$. If we used any other values, the matrices would not be equal. So, to generalize this, if **A** and **B** are m x n matrices, we can say:

$$\mathbf{A} = \mathbf{B} \text{ if and only if } a_{i,j} = b_{i,j}$$
for i = 1, 2, ..., m and j = 1, 2, ... n.

The Zero Matrix

As was the case with vectors, there exists a matrix equivalent to the scalar value of zero. In the case of vectors, it was a vector with all the components equal to zero. Likewise, in the case of matrices, the zero matrix, usually denoted as **0**, is a matrix with every element equal to zero. Thus, we can write:

$$\mathbf{0} = \begin{bmatrix} 0 & \cdots & 0 \\ \vdots & \ddots & \vdots \\ 0 & \cdots & 0 \end{bmatrix}$$

Square Matrices

Earlier, we talked about the size of a matrix. The size is always written as the number of rows in the matrix by the number of columns, for example, 3 x 4. This tells us that the matrix has 3 rows and 4 columns. When both the row and column sizes are the same, we have a special type of matrix called a square matrix. Thus, if we have an arbitrary m x n matrix, it can be called a square matrix if and only if $m = n$.

The Identity Matrix

One special type of square matrix is the identity matrix. As we know, when performing standard algebra, multiplying a quantity by one does not change the original value. The identity matrix, usually written as **I**, is simply the matrix version of the number one. Any matrix that is multiplied by an identity matrix is left unchanged. So, what does this matrix look like? It is simply a matrix with ones along its diagonal and zeros everywhere else. For example, a 4x4 identity matrix would look like this:

$$\mathbf{I}_4 = \begin{bmatrix} 1 & 0 & 0 & 0 \\ 0 & 1 & 0 & 0 \\ 0 & 0 & 1 & 0 \\ 0 & 0 & 0 & 1 \end{bmatrix}$$

The identity matrix can be as large or as small as needed, thus, we can write a general definition as:

$$\mathbf{I}_n = \begin{bmatrix} 1 & 0 & 0 & \cdots & 0 \\ 0 & 1 & 0 & \cdots & 0 \\ 0 & 0 & \ddots & & \vdots \\ \vdots & \vdots & & \ddots & 0 \\ 0 & 0 & \cdots & 0 & 1 \end{bmatrix}$$

or, we can define its components with:

$$i_{j,k} = \begin{pmatrix} 1 & \text{if } j = k \\ 0 & \text{otherwise} \end{pmatrix}$$

Linear Independence

An important concept we will encounter later in our study of matrices is the concept of linear independence. Earlier, we mentioned that a matrix can be thought of as a collection of row or column vectors. The concept of linear independence tells us whether any of these row or column vectors point along the same line, or in other words, lie directly on top of one another. Take for example, the two vectors $\mathbf{a} = [1, 2, 3]$ and $\mathbf{b} = [4, 8, 12]$. A quick inspection of these two vectors shows us that $\mathbf{b} = 4\mathbf{a}$. Thus, $\mathbf{a}$ and $\mathbf{b}$ both point in exactly the same direction. The only difference between $\mathbf{a}$ and $\mathbf{b}$ is that $\mathbf{b}$ has four times the magnitude of $\mathbf{a}$ ($\mathbf{b}$ is a scalar multiple of $\mathbf{a}$). When two row or column vectors of a matrix overlap in this way, they are said to be linearly dependent (they point along the same line). To put this more mathematically, let $\mathbf{v}$ and $\mathbf{w}$ be vectors and r be a scalar. Then $\mathbf{v}$ and $\mathbf{w}$ are said to be linearly dependent if $\mathbf{v} = r\mathbf{w}$ for some r not equal to zero. So, if no value other than zero exits for r in the previous relationship, the vectors are said to be linearly independent. Extending this concept to more than two vectors, we let $\mathbf{v_1, v_2}, \ldots, \mathbf{v_k}$ be a set of vectors of length n and $s_1, s_2, \ldots, s_k$ be a set of scalar values. Then we can write a dependence relationship for these vectors as follows:

$$s_1\mathbf{v_1} + s_2\mathbf{v_2} + \ldots + s_k\mathbf{v_k} = 0$$

where at least one s_j is not equal to zero. If nonzero values of s can be found such that this relationship holds, some of the vectors in the set are linearly dependent, and thus, the whole set is said to be linearly dependent. On the other hand, if no values other than zero can be found, the vectors are linearly independent. It is important to note that if there are more vectors in the set than there are elements in the vectors, the set will always be linearly dependent. For example, we cannot have a set containing more than three vectors that is linearly independent in three-dimensional space. This concept of linear independence is of great importance when discussing our next topic, matrix rank.

Matrix Rank

The last basic concept we will discuss is matrix rank. The rank of a matrix is determined by linear independence. In a given m x n matrix, we call the number of linearly independent column vectors the column rank and the number of linearly independent row vectors the row rank. The overall rank of the matrix is determined to be the smaller of these two ranks. Thus, if we have a matrix with a row rank of 4 and a column rank of 3, the overall rank of the matrix is 3. If our row rank is equal to the total number of rows, we say that the matrix has full row rank. Likewise if the column rank is equal to the total number of columns in the matrix, we say the matrix has full column rank. If the matrix has both full row and full column rank, the matrix is said to be a full rank matrix. It is important to note that for a matrix to have both full row and column rank, it must be a square matrix.

DEFINITION: Basic Matrix Concepts

Matrix Equality: Let **A** and **B** be $m \times n$ matrices, then we can write: **A = B**, if and only if $a_{i,j} = b_{i,j}$, for i = 1, 2, ... m and j = 1, 2, ... n

Zero Matrix: A special matrix. All the elements of this matrix are zero.

$$\mathbf{0} = \begin{bmatrix} 0 & \cdots & 0 \\ \vdots & \ddots & \vdots \\ 0 & \cdots & 0 \end{bmatrix}$$

Square Matrix: A matrix in which the number of columns is equal to the number of rows.

Identity Matrix: A special square matrix. Any matrix multiplied by the identity matrix remains unchanged. The identity matrix has ones on its diagonal and zeros everywhere else. These elements can be defined as:

$$i_{j,k} = \begin{pmatrix} 1 & \text{if } j = k \\ 0 & \text{otherwise} \end{pmatrix}$$

Linear Independence: Let $\mathbf{v_1}, \mathbf{v_2}, \ldots, \mathbf{v_k}$ be a set of vectors of length n and $s_1, s_2, \ldots, s_k$ be a set of scalar values. Then we can write:

$$s_1\mathbf{v_1} + s_2\mathbf{v_2} + \ldots + s_k\mathbf{v_k} = \mathbf{0}$$

If no scalar values other than zero satisfy this relationship, the set of vectors is linearly independent.

Row Rank: The number of linearly independent rows in a given matrix

Column Rank: The number of linearly independent columns in a given matrix

Rank: The rank of a matrix is the minimum of the row and column rank.

Now that we have looked at the basic background concepts, we can move on to the arithmetic operations. We will start by examining matrix addition.

3.5.2   Matrix Addition

The concept of matrix addition is not difficult to understand. Just as we did when we added two vectors by adding their corresponding components, we add two matrices by adding their corresponding elements. Thus, if we take two $m \times n$ matrices **A** and **B**, the addition would be written as follows:

$$\mathbf{A} + \mathbf{B} = \begin{bmatrix} a_{1,1} & a_{1,2} & \cdots & a_{1,n} \\ a_{2,1} & \ddots & & a_{2,n} \\ \vdots & & \ddots & \vdots \\ a_{m,1} & a_{m,2} & \cdots & a_{m,n} \end{bmatrix} + \begin{bmatrix} b_{1,1} & b_{1,2} & \cdots & b_{1,n} \\ b_{2,1} & \ddots & & b_{2,n} \\ \vdots & & \ddots & \vdots \\ b_{m,1} & b_{m,2} & \cdots & b_{m,n} \end{bmatrix}$$

And the solution would be:

$$\mathbf{A} + \mathbf{B} = \begin{bmatrix} a_{1,1} + b_{1,1} & a_{1,2} + b_{1,2} & \cdots & a_{1,n} + b_{1,n} \\ a_{2,1} + b_{2,1} & & & a_{2,n} + b_{2,n} \\ \vdots & & \ddots & \vdots \\ a_{m,1} + b_{m,1} & a_{m,2} + b_{m,2} & \cdots & a_{m,n} + b_{m,n} \end{bmatrix}$$

So, as we can see, the addition of two matrices simply requires an element-by-element addition. It is important to note, that only matrices of the same size can be added together. Let us look at an example before we move on.

DEFINITION: Matrix Addition

Let **A** and **B** be $m \times n$ matrices, then the operation **A + B** is defined as:

$$\mathbf{A} + \mathbf{B} = \begin{bmatrix} a_{1,1} + b_{1,1} & a_{1,2} + b_{1,2} & \cdots & a_{1,n} + b_{1,n} \\ a_{2,1} + b_{2,1} & & & a_{2,n} + b_{2,n} \\ \vdots & & \ddots & \vdots \\ a_{m,1} + b_{m,1} & a_{m,2} + b_{m,2} & \cdots & a_{m,n} + b_{m,n} \end{bmatrix}$$

Example 3.10

Problem: Add the following two matrices, **A** and **B**.

$$\mathbf{A} = \begin{bmatrix} 1 & -5 & 0 \\ 3 & -2 & 2 \\ 4 & 0 & 4 \end{bmatrix} \qquad \mathbf{B} = \begin{bmatrix} 2 & -2 & 1 \\ -1 & 0 & 4 \\ 0 & 3 & 2 \end{bmatrix}$$

Solution: To perform this addition, we must simply add the corresponding elements of the two matrices. Thus, we get:

$$\mathbf{A} + \mathbf{B} = \begin{bmatrix} 1+2 & -5-2 & 0+1 \\ 3-1 & -2+0 & 2+4 \\ 4+0 & 0+3 & 4+2 \end{bmatrix}$$

$$\mathbf{A} + \mathbf{B} = \begin{bmatrix} 3 & -7 & 1 \\ 2 & -2 & 6 \\ 4 & 3 & 6 \end{bmatrix}$$

So, as we can see, we arrived at our solution by simply adding the corresponding elements of the two matrices.

3.5.3   Matrix Subtraction

As we would expect, matrix subtraction works in much the same way as matrix addition. In both cases we operate on corresponding elements. The only difference is that we now want to subtract these elements instead of adding them. Therefore, if we take our matrices **A** and **B** we can define the operation of **A** − **B** as:

$$\mathbf{A} - \mathbf{B} = \begin{bmatrix} a_{1,1} & a_{1,2} & \cdots & a_{1,n} \\ a_{2,1} & \ddots & & a_{2,n} \\ \vdots & & \ddots & \vdots \\ a_{m,1} & a_{m,2} & \cdots & a_{m,n} \end{bmatrix} - \begin{bmatrix} b_{1,1} & b_{1,2} & \cdots & b_{1,n} \\ b_{2,1} & \ddots & & b_{2,n} \\ \vdots & & \ddots & \vdots \\ b_{m,1} & b_{m,2} & \cdots & b_{m,n} \end{bmatrix}$$

And, thus we get:

$$\mathbf{A} - \mathbf{B} = \begin{bmatrix} a_{1,1} - b_{1,1} & a_{1,2} - b_{1,2} & \cdots & a_{1,n} - b_{1,n} \\ a_{2,1} - b_{2,1} & & & a_{2,n} - b_{2,n} \\ \vdots & & & \vdots \\ a_{m,1} - b_{m,1} & a_{m,2} - b_{m,2} & \cdots & a_{m,n} - b_{m,n} \end{bmatrix}$$

Now, let us look at what happens when we subtract two actual matrices.

DEFINITION: Matrix Subtraction

Let **A** and **B** be $m \times n$ matrices, then the operation **A** - **B** is defined as:

$$A - B = \begin{bmatrix} a_{1,1} - b_{1,1} & a_{1,2} - b_{1,2} & \cdots & a_{1,n} - b_{1,n} \\ a_{2,1} - b_{2,1} & & & a_{2,n} - b_{2,n} \\ \vdots & & \ddots & \vdots \\ a_{m,1} - b_{m,1} & a_{m,2} - b_{m,2} & \cdots & a_{m,n} - b_{m,n} \end{bmatrix}$$

Example 3.11

Problem: Given the following matrices, **A** and **B**, perform the subtraction **A** − **B**.

$$A = \begin{bmatrix} 1 & -5 & 0 \\ 3 & -2 & 2 \\ 4 & 0 & 4 \end{bmatrix} \qquad B = \begin{bmatrix} 2 & -2 & 1 \\ -1 & 0 & 4 \\ 0 & 3 & 2 \end{bmatrix}$$

Solution: To perform this subtraction, we must simply subtract each element in **B** from its corresponding element in **A**. Thus, we get:

$$A - B = \begin{bmatrix} 1-2 & -5+2 & 0-1 \\ 3+1 & -2-0 & 2-4 \\ 4-0 & 0-3 & 4-2 \end{bmatrix}$$

$$A + B = \begin{bmatrix} -1 & -3 & -1 \\ 4 & -2 & -2 \\ 4 & -3 & 2 \end{bmatrix}$$

So, as we can see, we arrived at our solution by simply subtracting the corresponding elements of the two matrices.

3.5.4   Scalar Multiplication

Now that we have examined addition and subtraction, it is time to look at how multiplication is handled when dealing with matrices. There are two types of multiplication that we must take into account. The first is the multiplication of a matrix and a simple scalar number, and the second is the multiplication of two matrices. We will begin by looking at the case of scalar multiplication.

As was the case with vectors, scalar multiplication is the simplest type of multiplication. When performing a scalar-to-matrix multiplication, each element of the matrix is simply multiplied by the scalar value. Thus, if we have a matrix, **A**, and a scalar r, the resulting product would be:

$$r\mathbf{A} = \begin{bmatrix} ra_{1,1} & ra_{1,2} & \cdots & ra_{1,n} \\ ra_{2,1} & \ddots & & ra_{2,n} \\ \vdots & & \ddots & \vdots \\ ra_{m,1} & ra_{m,2} & \cdots & ra_{m,n} \end{bmatrix}$$

DEFINITION: Scalar Multiplication

Let **A** be an m x n matrix and r be a scalar, then the operation $r\mathbf{A}$ is defined as:

$$r\mathbf{A} = \begin{bmatrix} ra_{1,1} & ra_{1,2} & \cdots & ra_{1,n} \\ ra_{2,1} & \ddots & & ra_{2,n} \\ \vdots & & \ddots & \vdots \\ ra_{m,1} & ra_{m,2} & \cdots & ra_{m,n} \end{bmatrix}$$

Let us look at an example so that we can see how this works on an actual matrix.

Example 3. 12

Problem: Multiply the following matrix by the scalar value 3.

$$\mathbf{A} = \begin{bmatrix} 1 & -5 & 0 \\ 3 & -2 & 2 \\ 4 & 0 & 4 \end{bmatrix}$$

Solution: Here we are given a scalar-to-matrix multiplication to perform. As we mentioned above, all that we must do is multiply each matrix element by the scalar value, in this case 3. Doing this yields:

$$3\mathbf{A} = 3\begin{bmatrix} 1 & -5 & 0 \\ 3 & -2 & 2 \\ 4 & 0 & 4 \end{bmatrix} = \begin{bmatrix} 3(1) & 3(-5) & 3(0) \\ 3(3) & 3(-2) & 3(2) \\ 3(4) & 3(0) & 3(4) \end{bmatrix}$$

Therefore, our result is:

$$3\mathbf{A} = \begin{bmatrix} 3 & -15 & 0 \\ 9 & -6 & 6 \\ 12 & 0 & 12 \end{bmatrix}$$

So, we found our solution simply by multiplying each element by the scalar value 3.

3.5.5   Matrix Multiplication

As we have just seen, the process involved in multiplying a matrix and a scalar is very straightforward. The procedure used to multiply two matrices is a bit more complicated. First, the matrices being multiplied must be of compatible sizes. This is determined by looking at the number of columns in the fist matrix, and the number of rows in the second. These two numbers must be equal. In other words to multiply two matrices together, the first matrix must have as many columns as the second matrix has rows. The size of the matrix resulting from the multiplication is also determined by looking at the sizes of the two matrices being multiplied. The resulting matrix will have the same number of rows as the first matrix and the same number of columns as the second. Thus, let us look at two matrices: **A**, an m x p matrix, and **B**, a p x n matrix. If we want to find the product, **AB**, we must first check to see if the two matrices can be multiplied. A quick examination shows us that the first matrix, **A**, has p columns and the second matrix, **B**, has p rows. So, the number of columns of the first matrix is equal to the number of rows in the second. Therefore, we can perform the multiplication. In addition, we know that the result will have the same number of rows as **A** and the same number of columns as **B**. Thus, the resulting matrix will be of the size m x n.

Now that we know how to determine if two matrices can be multiplied and, if so, what the size of the resulting matrix will be, our next objective is to look at how the actual multiplication is performed. To do this, let us define the matrices. Let **A** be an m x p matrix, **B** be a p x n matrix, and **C** be an m x n matrix. So, our product can be written as **AB** = **C**. Now, let us say that we want to determine the value of the result located in the ith row and j^{th} column, in other words $\mathbf{C}_{i,j}$. To calculate this value, we look at the ith row of **A** and the jth column of **B**. We then multiply the elements in the ith row with their corresponding elements in the jth column. In other words, we multiply the first element in the column by the first element in the row and so on. When we have multiplied all the corresponding elements, we sum these products together to arrive at our result. So, each element in the resulting matrix **C** is determined by:

$$\mathbf{C}_{i,j} = \sum_{k=1}^{p} a_{i,k} b_{k,j}$$

Figure 3.9 illustrates this concept by showing the row and column used to obtain a given element of the result.

$$\begin{bmatrix} a_{11} & a_{12} & \cdots & a_{1p} \\ a_{21} & a_{22} & \cdots & a_{2p} \\ \vdots & \vdots & & \vdots \\ a_{i1} & a_{i2} & \cdots & a_{ip} \\ \vdots & \vdots & & \vdots \\ a_{m1} & a_{m2} & \cdots & a_{mp} \end{bmatrix} \begin{bmatrix} b_{11} & b_{12} & \cdots & b_{1j} & \cdots & b_{1n} \\ b_{21} & b_{22} & \cdots & b_{2j} & \cdots & b_{2n} \\ \vdots & \vdots & & \vdots & & \vdots \\ b_{p1} & b_{p2} & \cdots & b_{pj} & \cdots & b_{pn} \end{bmatrix} = \begin{bmatrix} c_{11} & c_{12} & \cdots & c_{1n} \\ c_{21} & c_{22} & \cdots & c_{2n} \\ \vdots & \vdots & c_{ij} & \vdots \\ c_{m1} & c_{m2} & \cdots & c_{mn} \end{bmatrix}$$

Figure 3.9. An illustration of the row of A and column of B used to obtain a given element of the result matrix, C.

In order to get a better idea of what is happening here, let us look at a simple example.

Example 3.13

Problem: Perform the following matrix multiplication.

$$C = \begin{bmatrix} 8 & -4 & 5 \end{bmatrix} \begin{bmatrix} 3 \\ 2 \\ -1 \end{bmatrix}$$

Solution: Here we are given two simple matrices and asked to find the product. A simple inspection shows us that the first matrix is a 1 x 3 matrix and the second is a 3 x 1 matrix. Thus, the number of columns in the first is equal to the number of rows in the second, and we can perform the multiplication. In addition, we see that the resulting matrix will be of the size 1 x 1, or basically a scalar value. Now, based upon what we discussed above, we know that we must multiply the elements in the rows of the first matrix by their counterparts in the columns of the second. We must then sum these products to arrive at our answer. Since we only have one row and one column, the operation is trivial. We can simply write:

C = [(8)(3) + (-4)(2) + (5)(-1)]

C = [24 – 8 – 5]

C = [11]

Thus we arrive at our result, the 1 x 1 matrix [11].

This example gives us a good idea of what is happening in a matrix multiplication. If we look back to our discussion on vector multiplication, we will discover that what we just calculated in the previous example is the dot product of two vectors. So, if we would like to think of matrix multiplication in a slightly different way, we can picture it as the process of finding the dot products of all the combinations that can be formed between the row vectors of the first matrix and the column vectors of the second. Thus to find $C_{i,j}$ we would simply find the dot product of the ith row vector of A and the jth column vector of B.

Lets look at one final example before moving on.

Example 3.14

Problem: Perform the following matrix multiplication.

$$C = \begin{bmatrix} 1 & -2 & 5 \\ 3 & 5 & -1 \\ 4 & 8 & -5 \end{bmatrix} \begin{bmatrix} 2 & 3 \\ 4 & 1 \\ -3 & 7 \end{bmatrix}$$

Solution: This time we have a 3 x 3 and a 3 x 2 matrix to multiply. First we check to see if we can multiply these two matrices. We see that the first matrix has 3 columns and the seconds has 3 rows. Thus, we can multiply these two matrices. Second we see that the resulting matrix will be a 3 x 2 matrix. Now, for each element, $C_{i,j}$ in the result we multiply the elements of the ith row of the first matrix and the jth column of the second and then sum the results. Doing so will yield the following:

$$C_{1,1} = (1)(2) + (-2)(4) + (5)(-3) = 2 - 8 - 15 = -21$$
$$C_{1,2} = (1)(3) + (-2)(1) + (5)(7) = 3 - 2 + 35 = 36$$
$$C_{2,1} = (3)(2) + (5)(4) + (-1)(-3) = 6 + 20 + 3 = 29$$
$$C_{2,2} = (3)(3) + (5)(1) + (-1)(7) = 9 + 5 - 7 = 7$$
$$C_{3,1} = (4)(2) + (8)(4) + (-5)(-3) = 8 + 32 + 15 = 55$$
$$C_{3,2} = (4)(3) + (8)(1) + (-5)(7) = 12 + 8 - 35 = -15$$

Putting this back into matrix form will give us:

$$C = \begin{bmatrix} -21 & 36 \\ 29 & 7 \\ 55 & -15 \end{bmatrix}$$

One final aspect of matrix multiplication that is important to note is that, unlike algebraic multiplication, matrix multiplication is not commutative. In other words, given two matrices, **A** and **B**, the product **AB** is not the same as **BA**. Let us show this with an example.

DEFINITION: Matrix Multiplication

Let **A** be an $m \times p$ matrix and **B** be a $p \times n$ matrix. Then, the elements of the resulting $m \times n$ matrix are defined as:

$$\mathbf{C}_{i,j} = \sum_{k=1}^{p} a_{i,k} b_{k,j}$$

Example 3.15

Problem: Given the following matrices, **A** and **B**, show that $\mathbf{AB} \neq \mathbf{BA}$.

$$\mathbf{A} = \begin{bmatrix} 2 & 3 \\ 1 & 8 \end{bmatrix} \qquad \mathbf{B} = \begin{bmatrix} -1 & 1 \\ 0 & 4 \end{bmatrix}$$

Solution: To show that matrix multiplication is not commutative, we will look at a proof by example. We will calculate the product **AB** and compare that to the product **BA**. If the two products are not equal, it proves by example that matrix multiplication is not commutative. Let us begin by calculating the product **AB**:

$$\mathbf{AB} = \begin{bmatrix} 2 & 3 \\ 1 & 8 \end{bmatrix} \begin{bmatrix} -1 & 1 \\ 0 & 4 \end{bmatrix} = \begin{bmatrix} (2)(-1)+(3)(0) & (2)(1)+(3)(4) \\ (1)(-1)+(8)(0) & (1)(1)+(8)(4) \end{bmatrix} = \begin{bmatrix} -2 & 14 \\ -1 & 33 \end{bmatrix}$$

Now, let us look at what we get by multiplying these two matrices in the opposite order.

$$\mathbf{BA} = \begin{bmatrix} -1 & 1 \\ 0 & 4 \end{bmatrix} \begin{bmatrix} 2 & 3 \\ 1 & 8 \end{bmatrix} = \begin{bmatrix} (-1)(2)+(1)(1) & (-1)(3)+(1)(8) \\ (0)(2)+(4)(1) & (0)(3)+(4)(8) \end{bmatrix} = \begin{bmatrix} -1 & 5 \\ 4 & 32 \end{bmatrix}$$

Here we can easily see that:

$$\begin{bmatrix} -2 & 14 \\ -1 & 33 \end{bmatrix} \neq \begin{bmatrix} -1 & 5 \\ 4 & 32 \end{bmatrix}$$

So, we have found that $\mathbf{AB} \neq \mathbf{BA}$

DEFINITION: Laws of Matrix Algebra

Let **A**, **B**, and **C** be matrices, **I** be the identity matrix, and r and s be scalars. Then we can define the following:

Matrix Addition:

$A + B = B + A$	Commutative Law
$(A + B) + C = A + (B + C)$	Associative Law
$A + 0 = 0 + A = A$	Additive Identity

Scalar Multiplication:

$r(sA) = (rs)A$	Associative Law
$(r + s)A = rA + sA$	Right Distributive Law
$r(A + B) = rA + rB$	Left Distributive Law
$A(rB) = r(AB) = (rA)B$	Scalars Pull Through

Matrix Multiplication:

$(AB)C = A(BC)$	Associative Law
$A(B + C) = AB + AC$	Left Distributive Law
$(B + C)A = BA + CA$	Right Distributive Law
$IA = A$ and $BI = B$	Matrix Multiplicative Identity
$AB \neq BA$	Non-Commutative

3.6 Other Matrix Operations

Now that we have established a working background in matrix algebra, we can turn our attention toward some other matrix operations. In the following sections, we will look at some of these special matrix operations and get a better idea of how they work and how we can use them to solve problems. Let us begin by taking a look at the transpose operation.

3.6.1   Transpose

The transpose of a given matrix is easy to obtain. We simply interchange the rows and columns of the matrix. In other words, we take the first row of the original matrix and make it the first column of the new matrix. The second row becomes the second column and so on. The transpose operation is usually denoted as a superscript T following the matrix name. Thus, if we have an m x n matrix **A** the transpose of that matrix would be written as A^T and the elements of the matrix would look something like this:

$$\mathbf{A}^T = \begin{bmatrix} a_{1,1} & a_{1,2} & \cdots & a_{1,n} \\ a_{2,1} & \ddots & & a_{2,n} \\ \vdots & & \ddots & \vdots \\ a_{m,1} & a_{m,2} & \cdots & a_{m,n} \end{bmatrix}^T = \begin{bmatrix} a_{1,1} & a_{2,1} & \cdots & a_{m,1} \\ a_{1,2} & \ddots & & a_{m,2} \\ \vdots & & \ddots & \vdots \\ a_{1,n} & a_{2,n} & \cdots & a_{m,n} \end{bmatrix}$$

or, we could describe the elements by letting $\mathbf{A} = [a_{i,j}]$. Then we can write:

$$\mathbf{A}^T = \left[a_{i,j}\right]^T = \left[a_{j,i}\right] \qquad \text{For i = 1, 2, ..., } m \text{ and j = 1, 2, ..., } n$$

DEFINITION: Transpose

Let **A** be an $m \times n$ matrix. Let us describe the elements of **A** as $\mathbf{A} = [a_{i,j}]$. Then, we can define the transpose operation as:

$$\mathbf{A}^T = \left[a_{i,j}\right]^T = \left[a_{j,i}\right]$$

For i = 1, 2, ..., m and j = 1, 2, ..., n

Now let us look at the transpose of some actual matrices.

Example 3.16

Problem: Find the transpose of the following matrices.

$$\mathbf{A} = \begin{bmatrix} 1 & 4 & -3 \\ 2 & 8 & 1 \\ -4 & -1 & 7 \\ 9 & 2 & 3 \end{bmatrix} \qquad\qquad \mathbf{B} = \begin{bmatrix} 5 & -2 & 7 \\ -2 & 8 & 3 \\ 7 & 3 & 4 \end{bmatrix}$$

Solution: We are given two matrices and asked to find their transposes. We know that all we need to do is take each row of the original matrix and rewrite it as a column. Let us do that for matrix **A**.

$$\mathbf{A}^T = \begin{bmatrix} 1 & 4 & -3 \\ 2 & 8 & 1 \\ -4 & -1 & 7 \\ 9 & 2 & 3 \end{bmatrix}^T = \begin{bmatrix} 1 & 2 & -4 & 9 \\ 4 & 8 & -1 & 2 \\ -3 & 1 & 7 & 3 \end{bmatrix}$$

So, our 4 x 3 matrix **A** becomes a 3 x 4 matrix when transposed. Now let us see what happens when we transpose matrix **B**.

$$\mathbf{B}^T = \begin{bmatrix} 5 & -2 & 7 \\ -2 & 8 & 3 \\ 7 & 3 & 4 \end{bmatrix}^T = \begin{bmatrix} 5 & -2 & 7 \\ -2 & 8 & 3 \\ 7 & 3 & 4 \end{bmatrix}$$

What happened here? The transposed matrix is the same as the original. This is because **B** is a special type of square matrix commonly called a symmetric matrix. Symmetric matrices have the special property that taking their transpose does not change the matrix. In other words, for a symmetric matrix **B**, $\mathbf{B}^T = \mathbf{B}$.

> **DEFINITION: Properties of the Transpose Operation**
>
> Let **A** and **B** be matrices, then we can define the following:
>
> $(\mathbf{A}^T)^T = \mathbf{A}$ Transpose of the Transpose
> $(\mathbf{A} + \mathbf{B})^T = \mathbf{A}^T + \mathbf{B}^T$ Transpose of a Sum
> $(\mathbf{AB})^T = \mathbf{B}^T \mathbf{A}^T$ Transpose of a Product

3.6.2   Determinant

The next important concept we will discuss is the determinant of a matrix. All square matrices have associated with them, a special scalar number called a determinant. This value is useful in a wide range of matrix algebra operations. In fact, we will use it later to help us find the inverse of a square matrix. The determinant of a matrix can be written in several different ways, and the method used will vary from text to text. The most common methods for denoting the determinant of a square matrix **A** are |**A**| and det(**A**). In this text, we will use these two notations interchangeably.

The calculation of the determinant value can be a complicated task. However, the process is fairly simple if it is restricted to 1 x 1, 2 x 2, and 3 x 3 matrices. We will look specifically at these cases and leave the discussion of the calculation of determinants for larger matrices for a class in linear algebra.

Let us begin with the trivial 1 x 1 case. A 1 x 1 matrix is essentially a scalar value, and the determinant of a 1 x 1 matrix is simply the value itself. Thus, if **A** is a 1 x 1 matrix, we can write:

$$\mathbf{A} = \begin{bmatrix} a_{1,1} \end{bmatrix}$$

And, thus:

$$|\mathbf{A}| = a_{1,1}$$

Next, we can turn our attention to the 2 x 2 case. For a 2 x 2 matrix, **A**, we find the determinant by subtracting the product of the upward diagonal terms from that of the downward diagonal terms. Figure 3.10 illustrates this process.

$$\mathbf{A} = \begin{bmatrix} a_{1,1} & a_{1,2} \\ a_{2,1} & a_{2,2} \end{bmatrix}$$

$$|\mathbf{A}| = a_{1,1}a_{2,2} - a_{1,2}a_{2,1}$$

**Figure 3.10. An illustration of the method for calculating
the determinant of a 2 x 2 matrix.**

Finally, let us look at the 3 x 3 case. There are two different ways to calculate a determinant for matrices of this size. The first method involves the calculation of three 2 x 2 determinants. To illustrate this method, we start with a 3 x 3 matrix, **A**:

$$\mathbf{A} = \begin{bmatrix} a_{1,1} & a_{1,2} & a_{1,3} \\ a_{2,1} & a_{2,2} & a_{2,3} \\ a_{3,1} & a_{3,2} & a_{3,3} \end{bmatrix}$$

Now, we cross out the first row and column of this matrix and take the determinant of the remaining 2 x 2 matrix. This determinant is multiplied by the value in position 1,1. Doing this gives us the first part of our answer. We will call it p_1:

$$p_1 = a_{1,1} \begin{vmatrix} a_{2,2} & a_{2,3} \\ a_{3,2} & a_{3,3} \end{vmatrix}$$

Next, we cross out the first row and second column of the matrix and take the determinant of the remaining 2 x 2 matrix. This determinant is multiplied by the negative of the value in position 1,2. We will call this part of the calculation p_2:

$$p_2 = -a_{1,2} \begin{vmatrix} a_{2,1} & a_{2,3} \\ a_{3,1} & a_{3,3} \end{vmatrix}$$

Finally, we cross out the first row and third column of the matrix and take the determinant of the remaining 2 x 2 matrix. We multiply this determinant by the value in position 1,3 of the matrix. We will call this result p_3.

$$p_3 = a_{1,3} \begin{vmatrix} a_{2,1} & a_{2,2} \\ a_{3,1} & a_{3,2} \end{vmatrix}$$

Now, if we add all three parts together, we will arrive at our final determinant for the 3 x 3 matrix. Doing so yields:

$$|\mathbf{A}| = p_1 + p_2 + p_3 = a_{1,1} \begin{vmatrix} a_{2,2} & a_{2,3} \\ a_{3,2} & a_{3,3} \end{vmatrix} - a_{1,2} \begin{vmatrix} a_{2,1} & a_{2,3} \\ a_{3,1} & a_{3,3} \end{vmatrix} + a_{1,3} \begin{vmatrix} a_{2,1} & a_{2,2} \\ a_{3,1} & a_{3,2} \end{vmatrix}$$

This can be simplified even further if we apply our method for calculating 2 x 2 determinants. Doing this will give us:

$$|\mathbf{A}| = a_{1,1}(a_{2,2}a_{3,3} - a_{3,2}a_{2,3}) - a_{1,2}(a_{2,1}a_{3,3} - a_{3,1}a_{2,3}) + a_{1,3}(a_{2,1}a_{3,2} - a_{3,1}a_{2,2})$$

$$|\mathbf{A}| = a_{1,1}a_{2,2}a_{3,3} - a_{1,1}a_{3,2}a_{2,3} - a_{1,2}a_{2,1}a_{3,3} + a_{1,2}a_{3,1}a_{2,3}$$
$$+ a_{1,3}a_{2,1}a_{3,2} - a_{1,3}a_{3,1}a_{2,2}$$

Thus, we have arrived at an equation for the determinant of a 3 x 3 matrix in terms of its elements. This first method of calculating the determinant of a 3 x 3 matrix is fairly involved. Luckily, there is a simpler way to calculate this determinant. We will begin by writing out the elements of our 3 x 3 matrix twice as follows:

$$
\begin{array}{cccccc}
a_{1,1} & a_{1,2} & a_{1,3} & a_{1,1} & a_{1,2} & a_{1,3} \\
a_{2,1} & a_{2,2} & a_{2,3} & a_{2,1} & a_{2,2} & a_{2,3} \\
a_{3,1} & a_{3,2} & a_{3,3} & a_{3,1} & a_{3,2} & a_{3,3}
\end{array}
$$

Now, we simply add the products of the first three downward diagonals and subtract the products of the first three upward diagonals, as shown below.

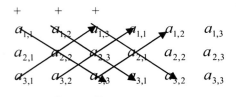

Doing this gives us:

$$|\mathbf{A}| = a_{1,1}a_{2,2}a_{3,3} + a_{1,2}a_{2,3}a_{3,1} + a_{1,3}a_{2,1}a_{3,2} - a_{3,1}a_{2,2}a_{1,3} - a_{3,2}a_{2,3}a_{1,1} - a_{3,3}a_{2,1}a_{1,2}$$

This equation is a slightly rearranged, but numerically identical version of the equation that we calculated using the first method.

DEFINITION: Determinant

The Determinant of an $n \times n$ matrix **A**, written |A| or det(**A**), is a scalar-valued function of **A**.

For n = 1: $\mathbf{A} = \begin{bmatrix} a_{1,1} \end{bmatrix}$ and $|\mathbf{A}| = a_{1,1}$

For n = 2: $\mathbf{A} = \begin{bmatrix} a_{1,1} & a_{1,2} \\ a_{2,1} & a_{2,2} \end{bmatrix}$ and $|\mathbf{A}| = a_{1,1}a_{2,2} - a_{1,2}a_{2,1}$

For n = 3: $\mathbf{A} = \begin{bmatrix} a_{1,1} & a_{1,2} & a_{1,3} \\ a_{2,1} & a_{2,2} & a_{2,3} \\ a_{3,1} & a_{3,2} & a_{3,3} \end{bmatrix}$ and

$$|\mathbf{A}| = a_{1,1}a_{2,2}a_{3,3} + a_{1,2}a_{2,3}a_{3,1} + a_{1,3}a_{2,1}a_{3,2} - a_{3,1}a_{2,2}a_{1,3} - a_{3,2}a_{2,3}a_{1,1} - a_{3,3}a_{2,1}a_{1,2}$$

Now let us look at some example problems.

Example 3.17

Problem: Find the determinant of the following 2 x 2 matrix.

$$\mathbf{A} = \begin{bmatrix} 3 & -2 \\ 2 & 1 \end{bmatrix}$$

Solution: We have been given a 2 x 2 matrix and we wish to find the determinant. In this case, we simply use our formula for the determinant of a 2 x 2 matrix. We subtract the product of the upward diagonal terms from the product of the downward diagonal terms. Doing this gives us:

$$|\mathbf{A}| = (3)(1) - (2)(-2)$$
$$|\mathbf{A}| = 3 + 4$$
$$|\mathbf{A}| = 7$$

So, in this case, we found that our determinant was equal to 7.

Example 3.18

___

Problem: Find the determinant of the following 3 x 3 matrix.

$$\mathbf{A} = \begin{bmatrix} 4 & 2 & 1 \\ 2 & 6 & 3 \\ 1 & 3 & 5 \end{bmatrix}$$

Solution: This time we have a 3 x 3 matrix. So, we could choose to use the first complex method to calculate our determinant, or we could opt for the simpler second method. In this example, we will use the second method. Thus, we start by writing out our matrix elements twice:

$$\begin{array}{cccccc} 4 & 2 & 1 & 4 & 2 & 1 \\ 2 & 6 & 3 & 2 & 6 & 3 \\ 1 & 3 & 5 & 1 & 3 & 5 \end{array}$$

Now, we will add the first three downward diagonals and subtract the first three upward diagonals. This will give us:

$$|\mathbf{A}| = (4)(6)(5) + (2)(3)(1) + (1)(2)(3) - (1)(6)(1) - (3)(3)(4) - (5)(2)(2)$$
$$|\mathbf{A}| = 120 + 6 + 6 - 6 - 36 - 20$$
$$|\mathbf{A}| = 70$$

Thus the determinant of our 3 x 3 matrix is 70,

___

DEFINITION: Some Properties of the Determinant

Let **A** and **B** be matrices, then we can define the following:

$\|\mathbf{A}^T\| = \|\mathbf{A}\|$	Determinant of the Transpose
$\|\mathbf{AB}\| = \|\mathbf{A}\| \|\mathbf{B}\|$	Determinant of a Product
If any 2 rows of a **A** are linearly dependent, $\|\mathbf{A}\| = 0$	Dependent Row Property

---

MATLAB: MATLAB has a function that allows for the calculation of determinants of matrices.

$$d = DET(A)$$

Where d is the determinant of matrix **A**.

Example: d = det([3, −2;2, 1]) will return 7, which is the same as the determinant calculated in example 3.17.

3.6.3   Minors

Now that we can calculate simple determinants, we can turn our attention to the concept of matrix minors. If we are given an $n \times n$ matrix **A**, we know that this matrix possesses n^2 elements, $a_{i,j}$ where $i = 1, 2, ..., n$ and $j = 1, 2, ..., n$. Each of these elements has associated with it a unique scalar called a minor, $m_{i,j}$. So, how do we find these unique scalar values? We determine a minor for a given element $m_{i,j}$ by crossing out the ith row and j^{th} column of the matrix and taking the determinant of the remaining values. In other words, if we wanted to find the minor associated with the element in the 2nd row and 3rd column, we would cross out the 2nd row and 3rd column of our matrix and find the determinant of the remaining rows and columns. Let us see how this works for a given 3 x 3 matrix.

DEFINITION: Matrix Minor

Let **A** be an $n \times n$ matrix with elements $a_{i,j}$. Each of these elements has associated with it a unique scalar, $m_{i,j}$. These values are determined by removing the ith row and j^{th} column and finding the determinant of the remaining elements.

Example 3.19

Problem: Find the minors for the following 3 x 3 matrix.

$$A = \begin{bmatrix} 4 & 2 & 1 \\ 2 & 6 & 3 \\ 1 & 3 & 5 \end{bmatrix}$$

Solution: Here we are dealing with a 3 x 3 matrix. Thus, we have 9 elements in our matrix and 9 minors to calculate. As we know, the minors are determined by crossing out their associated row and column and calculating the determinant of the remaining values. We will begin with the minor for the first row and first column and work across the rows.

$$m_{1,1} = \begin{vmatrix} 6 & 3 \\ 3 & 5 \end{vmatrix} = (6)(5) - (3)(3) = 30 - 9 = 21$$

$$m_{1,2} = \begin{vmatrix} 2 & 3 \\ 1 & 5 \end{vmatrix} = (2)(5) - (1)(3) = 10 - 3 = 7$$

$$m_{1,3} = \begin{vmatrix} 2 & 6 \\ 1 & 3 \end{vmatrix} = (2)(3) - (1)(6) = 6 - 6 = 0$$

$$m_{2,1} = \begin{vmatrix} 2 & 1 \\ 3 & 5 \end{vmatrix} = (2)(5) - (3)(1) = 10 - 3 = 7$$

$$m_{2,2} = \begin{vmatrix} 4 & 1 \\ 1 & 5 \end{vmatrix} = (4)(5) - (1)(1) = 20 - 1 = 19$$

$$m_{2,3} = \begin{vmatrix} 4 & 2 \\ 1 & 3 \end{vmatrix} = (4)(3) - (1)(2) = 12 - 2 = 10$$

$$m_{3,1} = \begin{vmatrix} 2 & 1 \\ 6 & 3 \end{vmatrix} = (2)(3) - (6)(1) = 6 - 6 = 0$$

$$m_{3,2} = \begin{vmatrix} 4 & 1 \\ 2 & 3 \end{vmatrix} = (4)(3) - (2)(1) = 12 - 2 = 10$$

$$m_{3,3} = \begin{vmatrix} 4 & 2 \\ 2 & 6 \end{vmatrix} = (4)(6) - (2)(2) = 24 - 4 = 20$$

Now, we can rewrite our results in matrix form.

$$\mathbf{M} = \begin{bmatrix} 21 & 7 & 0 \\ 7 & 19 & 10 \\ 0 & 10 & 20 \end{bmatrix}$$

Thus, we have determined the minors of our 3 x 3 matrix.

3.6.4   Cofactors

We can build on our idea of matrix minors to develop the concept of matrix cofactors. As was the case with minors, each element in the matrix has associated with it a value called a cofactor. The method used to calculate the cofactor for a given element is very similar to that used to calculate the minor. In fact, the only difference between the two is that the cofactor can have a different sign. The cofactor, $c_{i,j}$, for a given element, $a_{i,j}$, is determined as follows:

$$c_{i,j} = (-1)^{i+j} m_{i,j}$$

where $m_{i,j}$ is the minor associated with element $a_{i,j}$. Thus, we see that the cofactor of the element of a matrix is simply the minor for that element with its sign changed based upon the position of its associated element in the matrix. The minor will be multiplied by a positive one when the associated element is in a row and column such that the sum of the subscripts is equal to an even number and, it will be multiplied by a negative one when the sum of the subscripts of the associated element is odd. In other words:

$$c_{i,j} = m_{i,j} \quad \text{when i + j is even.}$$

And,

$$c_{i,j} = -m_{i,j} \quad \text{when i + j is odd.}$$

DEFINITION: Matrix Cofactor

Let **A** be an $n \times n$ matrix with elements $a_{i,j}$ and $m_{i,j}$ be the minor associated with element $a_{i,j}$. Then, the cofactor, $c_{i,j}$, associated with each element $a_{i,j}$ can be defined as:

$$c_{i,j} = (-1)^{i+j} m_{i,j}$$

Let us look at an example to see how cofactors are found.

Example 3.20

Problem: Find the cofactors of the following matrix.

$$\mathbf{A} = \begin{bmatrix} 4 & 2 & 1 \\ 2 & 6 & 3 \\ 1 & 3 & 5 \end{bmatrix}$$

Solution: In this problem, we are being asked to find the cofactors of a matrix. Since the definition of a cofactor required us to first know the matrix minors, they must be calculated first. The minors of this matrix were found in example 3.19. If we look back at that example, we find that the minors of the matrix were:

$$m_{1,1} = 21 \quad m_{1,2} = 7 \quad m_{1,3} = 0$$
$$m_{2,1} = 7 \quad m_{2,2} = 19 \quad m_{2,3} = 10$$
$$m_{3,1} = 0 \quad m_{3,2} = 10 \quad m_{3,3} = 20$$

Using these values, we can now easily calculate the cofactors.

$$c_{1,1} = (-1)^{1+1} m_{1,1} = (-1)^2 (21) = 21$$

$$c_{1,2} = (-1)^{1+2} m_{1,2} = (-1)^3 (7) = -7$$

$$c_{1,3} = (-1)^{1+3} m_{1,3} = (-1)^4 (0) = 0$$

$$c_{2,1} = (-1)^{2+1} m_{2,1} = (-1)^3 (7) = -7$$

$$c_{2,2} = (-1)^{2+2} m_{2,2} = (-1)^4 (19) = 19$$

$$c_{2,3} = (-1)^{2+3} m_{2,3} = (-1)^5 (10) = -10$$

$$c_{3,1} = (-1)^{3+1} m_{3,1} = (-1)^4 (0) = 0$$

$$c_{3,2} = (-1)^{3+2} m_{3,2} = (-1)^5 (10) = -10$$

$$c_{3,3} = (-1)^{3+3} m_{3,3} = (-1)^6 (20) = 20$$

Placing these cofactor values into a matrix form we arrive at:

$$\mathbf{C} = \begin{bmatrix} 21 & -7 & 0 \\ -7 & 19 & -10 \\ 0 & -10 & 20 \end{bmatrix}$$

So, now we have the cofactors for our 3 x 3 matrix.

On a side-note, if we look back at our first method for finding the determinant of a 3 x 3 matrix, we would discover that we actually used cofactors. Basically, to find the determinant of the matrix, we multiplied each element in the first row by its cofactor and summed the results. If we were to perform this operation on our example matrix we would get:

$$|\mathbf{A}| = (4)(21) + (2)(-7) + (1)(0)$$

$$|\mathbf{A}| = 84 - 14 + 0$$

$$|\mathbf{A}| = 70$$

This is the same value that we obtained for the determinant of the matrix in example 3.18. The cofactor method of finding the determinant can be used to find the determinant of larger matrices, by breaking them down into a series of smaller determinant operations.

3.6.5   Matrix Inverse

The matrix inverse is a very important and useful tool in the field of matrix algebra. To understand what it is and how it works, let us first look at the concept of an inverse in the realm of scalar numbers. If we have a scalar value a, we refer to its inverse, a^{-1}, as the scalar value that satisfies the following relationship.

$$\alpha\alpha^{-1} = \alpha^{-1}\alpha = 1$$

In other words, the inverse of a scalar value is simply the value that it must be multiplied by to get a result of 1. Thus, the scalar inverse of 2 would be ½. The matrix inverse operates in much the same way. If we are given a square matrix **A**, then its inverse is the matrix, **A**$^{-1}$, that satisfies the relationship:

$$\mathbf{A}\mathbf{A}^{-1} = \mathbf{A}^{-1}\mathbf{A} = \mathbf{I}$$

DEFINITION: Matrix Inverse

Let **A** be an n x n matrix. Then, the inverse of **A**, written as **A**$^{-1}$, is defined as a matrix that satisfies the following relationship:

$$\mathbf{A}\mathbf{A}^{-1} = \mathbf{A}^{-1}\mathbf{A} = \mathbf{I}$$

Where **I** is the identity matrix.

Thus, the inverse of a matrix, **A**, is a special matrix. When **A** is multiplied by this matrix from either side, the result will be the identity matrix. Let us take a look at such a matrix.

Example 3.21

Problem: Show that matrix **B** is the inverse of matrix **A**.

$$\mathbf{A} = \begin{bmatrix} 1 & 2 \\ 3 & 4 \end{bmatrix} \qquad \mathbf{B} = \begin{bmatrix} -2 & 1 \\ 1.5 & -0.5 \end{bmatrix}$$

Solution: To show that **B** is the inverse of **A**, all that we need to do is show that the product of the two matrices in either order will yield the identity matrix. First, we will find the product **AB**.

$$\mathbf{AB} = \begin{bmatrix} 1 & 2 \\ 3 & 4 \end{bmatrix}\begin{bmatrix} -2 & 1 \\ 1.5 & -0.5 \end{bmatrix} = \begin{bmatrix} -2+3 & 1-1 \\ -6+6 & 3-2 \end{bmatrix}$$

$$\mathbf{AB} = \begin{bmatrix} 1 & 0 \\ 0 & 1 \end{bmatrix}$$

Thus, the product **AB** results in an identity matrix. Next we will look at the product **BA**.

$$\mathbf{BA} = \begin{bmatrix} -2 & 1 \\ 1.5 & -0.5 \end{bmatrix} \begin{bmatrix} 1 & 2 \\ 3 & 4 \end{bmatrix} = \begin{bmatrix} -2+3 & -4+4 \\ 1.5-1.5 & 3-2 \end{bmatrix}$$

$$\mathbf{BA} = \begin{bmatrix} 1 & 0 \\ 0 & 1 \end{bmatrix}$$

So, the product **BA** also results in an identity matrix. Thus, **B** satisfies the relationship for an inverse, and we can say that $\mathbf{B} = \mathbf{A}^{-1}$.

Now that we know what a matrix inverse looks like, how do we find it? We will look at two methods that can be used to find the inverse of a square matrix. The first method will make use of much of what we have already discussed regarding determinants and cofactors, and the second will use elementary row operations. Let us begin by looking at what we will call the adjoint method of inverse calculation.

Finding the Matrix Inverse Using the Adjoint Method

The first method we will look at for finding the inverse of a matrix makes use of our knowledge of the determinant and the matrix cofactors. This method states that we can find the inverse of a given matrix by applying the following equation:

$$\mathbf{A}^{-1} = \frac{\mathbf{C}^T}{|\mathbf{A}|}$$

In other words, the inverse of the matrix is equal to the transpose of the cofactor matrix, **C**, divided by the determinant of **A**. The transpose of the cofactor matrix is often referred to as the adjoint matrix, thus the name of the method. So, if we call the transpose of the cofactors the adjoint, we can rewrite the equation as:

$$\mathbf{A}^{-1} = \frac{Adj(\mathbf{A})}{|\mathbf{A}|}$$

Now, let us look at the other method for finding the matrix inverse.

Finding the Matrix Inverse Using Elementary Row Operations

Our second method for finding the inverse makes use of what are called elementary row operations. Basically, if we are given a matrix, **A**, an elementary row operation on this matrix involves one of the following:

1. Swapping two rows: $\mathbf{A}_i \leftrightarrow \mathbf{A}_j$

2. Multiplying a row by a non-zero scalar: $\mathbf{A}_i \rightarrow \alpha\mathbf{A}_i$

3. Add the scalar multiple of one row to another: $\mathbf{A}_i \rightarrow \mathbf{A}_i + \alpha\mathbf{A}_j$

Now, to find the inverse of a matrix, we create a special, or augmented, matrix. This matrix consists of the original matrix, **A**, on the left, and an identity matrix, **I** of equal size on the right.

[A|I]

We then perform elementary row operations on this matrix to convert the left side to an identity matrix. Each operation applied to the left side is also applied to the right side of our augmented matrix. When our left side becomes an identity matrix, the right side, which started as an identity matrix, will now be the inverse of the original matrix **A**.

DEFINITION: Finding the Inverse

The Inverse of a matrix, **A,** can be found in one of two basic ways:

1. Create an augmented matrix consisting of the matrix, **A,** on the left and an identity matrix of equal size on the right. Reduce the matrix on the left to an identity matrix using elementary row operations. The matrix on the right will then be the inverse of **A**.

$$[\mathbf{A} \mid \mathbf{I}] \rightarrow [\mathbf{I} \mid \mathbf{A}^{-1}]$$

2. Use the determinant and cofactors of the matrix in the following relationship:

$$\mathbf{A}^{-1} = \frac{\mathbf{C}^{T}}{|\mathbf{A}|} = \frac{adj(\mathbf{A})}{|\mathbf{A}|}$$

Let us take a look at both of our methods in action.

Example 3.22

Problem: Use both elementary row operation and the adjoint method to find the inverse of the following matrix.

$$\mathbf{A} = \begin{bmatrix} 4 & 2 & 1 \\ 2 & 6 & 3 \\ 1 & 3 & 5 \end{bmatrix}$$

Solution: In this example, we must find the inverse of our matrix using both the elementary row operation and adjoint methods. We will begin by using the elementary row operation method. So, we will write out our augmented matrix:

$$\begin{bmatrix} 4 & 2 & 1 & 1 & 0 & 0 \\ 2 & 6 & 3 & 0 & 1 & 0 \\ 1 & 3 & 5 & 0 & 0 & 1 \end{bmatrix}$$

Now we perform our row operations to convert the left side into an identity matrix. First, we will multiply the first row by ¼.

$$\begin{bmatrix} 1 & 1/2 & 1/4 & 1/4 & 0 & 0 \\ 2 & 6 & 3 & 0 & 1 & 0 \\ 1 & 3 & 5 & 0 & 0 & 1 \end{bmatrix}$$

Next, we add –2 times our first row to the second row. This gives us:

$$\begin{bmatrix} 1 & 1/2 & 1/4 & 1/4 & 0 & 0 \\ 0 & 5 & 5/2 & -1/2 & 1 & 0 \\ 1 & 3 & 5 & 0 & 0 & 1 \end{bmatrix}$$

Now, add –1 times our first row to the third row giving us:

$$\begin{bmatrix} 1 & 1/2 & 1/4 & 1/4 & 0 & 0 \\ 0 & 5 & 5/2 & -1/2 & 1 & 0 \\ 0 & 5/2 & 19/4 & -1/4 & 0 & 1 \end{bmatrix}$$

Multiplying row two by $\dfrac{1}{5}$ gives us:

$$\begin{bmatrix} 1 & 1/2 & 1/4 & 1/4 & 0 & 0 \\ 0 & 1 & 1/2 & -1/10 & 1/5 & 0 \\ 0 & 5/2 & 19/4 & -1/4 & 0 & 1 \end{bmatrix}$$

Next, we add $\dfrac{-5}{2}$ times row two to row three we arrive at:

$$\begin{bmatrix} 1 & 1/2 & 1/4 & 1/4 & 0 & 0 \\ 0 & 1 & 1/2 & -1/10 & 1/5 & 0 \\ 0 & 0 & 7/2 & 0 & -1/2 & 1 \end{bmatrix}$$

Now, we multiply the third row by $\dfrac{2}{7}$ yielding:

$$\begin{bmatrix} 1 & 1/2 & 1/4 & 1/4 & 0 & 0 \\ 0 & 1 & 1/2 & -1/10 & 1/5 & 0 \\ 0 & 0 & 1 & 0 & -2/18 & 2/9 \end{bmatrix}$$

We have converted the lower half of the left part of the matrix, now we must work on the upper half. First, let us add -½ times the second row to the first row. This gives us:

$$\begin{bmatrix} 1 & 0 & 0 & 3/10 & -1/10 & 0 \\ 0 & 1 & 1/2 & -1/10 & 1/5 & 0 \\ 0 & 0 & 1 & 0 & -1/7 & 2/7 \end{bmatrix}$$

Finally, we add −½ times the third row to the second yielding:

$$\begin{bmatrix} 1 & 0 & 0 & 3/10 & -1/10 & 0 \\ 0 & 1 & 0 & -1/10 & 19/70 & -1/7 \\ 0 & 0 & 1 & 0 & -1/7 & 2/7 \end{bmatrix}$$

So, based on the elementary row operation method, we arrive at an inverse of:

$$\mathbf{A}^{-1} = \begin{bmatrix} 3/10 & -1/10 & 0 \\ -1/10 & 19/70 & -1/7 \\ 0 & -1/7 & 2/7 \end{bmatrix}$$

Now, let us find the inverse through the adjoint method. From example 3.20, we know that the cofactors of the matrix are:

$$\mathbf{C} = \begin{bmatrix} 21 & -7 & 0 \\ -7 & 19 & -10 \\ 0 & -10 & 20 \end{bmatrix}$$

and, from example 3.18, we know that the determinant if **A** is 70.

Using the fact that:

$$\mathbf{A}^{-1} = \frac{\mathbf{C}^T}{|\mathbf{A}|} = \frac{adj(\mathbf{A})}{|\mathbf{A}|}$$

we can write:

$$\mathbf{A}^{-1} = \frac{1}{70}\begin{bmatrix} 21 & -7 & 0 \\ -7 & 19 & -10 \\ 0 & -10 & 20 \end{bmatrix}^T = \frac{1}{70}\begin{bmatrix} 21 & -7 & 0 \\ -7 & 19 & -10 \\ 0 & -10 & 20 \end{bmatrix}$$

$$\mathbf{A}^{-1} = \begin{bmatrix} 3/10 & -1/10 & 0 \\ -1/10 & 19/70 & -1/7 \\ 0 & -1/7 & 2/7 \end{bmatrix}$$

Which is the same result we obtained through the first method.

Let us look at one final example before moving on.

Example 3.23

Problem: Find the inverse of the following matrix.

$$\mathbf{A} = \begin{bmatrix} -2 & 0 & 4 \\ 1 & 3 & -7 \\ 4 & 0 & -8 \end{bmatrix}$$

Solution: Let us begin by using elementary row operations to get the inverse:

$$\begin{bmatrix} -2 & 0 & 4 & 1 & 0 & 0 \\ 1 & 3 & -7 & 0 & 1 & 0 \\ 4 & 0 & -8 & 0 & 0 & 1 \end{bmatrix}$$

Multiply the first row by -½.

$$\begin{bmatrix} 1 & 0 & -2 & -1/2 & 0 & 0 \\ 1 & 3 & -7 & 0 & 1 & 0 \\ 4 & 0 & -8 & 0 & 0 & 1 \end{bmatrix}$$

Add −1 times the first row to the second row.

$$\begin{bmatrix} 1 & 0 & -2 & -1/2 & 0 & 0 \\ 0 & 3 & -5 & 1/2 & 1 & 0 \\ 4 & 0 & -8 & 0 & 0 & 1 \end{bmatrix}$$

Add −4 times the first row to the third row.

$$\begin{bmatrix} 1 & 0 & -2 & -1/2 & 0 & 0 \\ 0 & 3 & -5 & 1/2 & 1 & 0 \\ 0 & 0 & 0 & 2 & 0 & 1 \end{bmatrix}$$

At this point, we see that the third row of the left part of the matrix is all zeros. Thus, there is no way to convert the left portion of the augmented matrix to an identity matrix. So, we cannot find the inverse of the matrix using this method. Let us try the adjoint method and see if we can use it. We begin by finding the determinant of **A**.

$$|\mathbf{A}| = \begin{vmatrix} -2 & 0 & 4 \\ 1 & 3 & -7 \\ 4 & 0 & -8 \end{vmatrix}$$

$$|\mathbf{A}| = (-2)(3)(-8) + (0)(-7)(4) + (4)(1)(0) - (4)(3)(4) - (0)(-7)(-2) - (-8)(1)(0)$$

$$|\mathbf{A}| = 48 + 0 + 0 - 48 - 0 - 0$$

$$|\mathbf{A}| = 0$$

We have a problem here as well. If we plug a determinant of zero into the equation for the inverse, we find:

$$\mathbf{A}^{-1} = \frac{adj(\mathbf{A})}{|\mathbf{A}|} = \frac{adj(\mathbf{A})}{0}$$

This division by zero results in an undefined result. Thus, we cannot use this method to find the inverse either. In reality, this particular matrix does not have an inverse. It is what is known as a singular matrix. We will discuss singular matrices in the next section.

MATLAB: MATLAB has a function for finding the inverse of a matrix. Given a matrix, **A**:

$$\mathbf{B} = \text{INV}(\mathbf{A})$$

Example: B = inv([4, 2, 1; 2, 6, 3;1, 3, 5]) will return:

```
B =

        3/10          -1/10            0
       -1/10          19/70          -1/7
          0           -1/7           2/7
```

Which is the same as the inverse we calculated in example 3.22.

3.6.6   Singular and Non-Singular Matrices

As we discovered in the last example, not every matrix has an inverse. This brings up the concept of singular and non-singular matrices. Basically, if a given matrix has an inverse, it is said to be non-singular. However, if the matrix does not have an inverse, it is called a singular matrix. So, how do we determine if a matrix is singular or not? The reason that a given matrix is singular is because it does not have full rank. If we think back to our discussion of matrix rank, we will remember that a matrix has full rank only if all of its rows and columns are linearly independent. This means that if our matrix has any two rows or columns that are linearly dependent, we will be unable to find an inverse. Thus, matrices with linearly dependent rows and columns are always singular matrices. Therefore, if we inspect our matrix and find that two rows or columns are exactly equal or are scalar multiples of each other, we will know that the matrix is singular and we cannot compute an inverse. In some cases, however, the linear dependence of the rows in a matrix is not easily seen. In cases like this, we will know that we have a singular matrix when we get a row of all zeros when performing elementary row operations, or we get a determinant of zero. When either of these cases occurs, we know we have a singular matrix and the inverse cannot be found.

DEFINITION: Singular and Non-Singular Matrices

A **Singular** matrix is a matrix that contains linearly dependent rows or columns. An inverse cannot be calculated for such a matrix.

A **Non-Singular** matrix is a matrix that possesses full row and column rank. All of its rows and columns are linearly independent. Thus, an inverse can be calculated for a non-singular matrix.

3.7 Application of the Matrix Inverse – Systems of Linear Equations

After spending all this time discussing basic matrix operations and the matrix inverse, one might wonder if there is any actual practical application for all of this. There are many practical applications for matrices. One of the most important is the ability to work with systems of linear equations. In the following sections we will look at what is meant by a system of linear equations and how we can use matrices to solve these systems. We will wrap things up with a practical electrical engineering example that employs matrices to solve an electrical circuit analysis problem.

3.7.1   Introduction to Systems of Linear Equations

We have mentioned that matrices are very useful when working with systems of linear equations, before we can explain these benefits, we must first explain what a system of linear equations is. An equation of the form:

$$y = ax + b$$

is called a "linear equation." Equations such as this are seen often in the study of electrical engineering. In this equation, a and b are constants and x and y are variables. Equations of this type are called linear equations because they form a straight line when plotted. Figure 3.11 illustrates the plot of a linear equation.

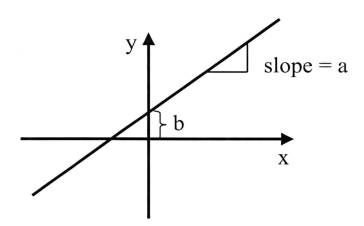

Figure 3.11. The plot of a typical linear equation.

As we can see from the figure, this form of the equation specifies a line by providing a slope, a, and an intercept point on the y-axis, b. Thus, this form is often referred to as the slope-intercept form of the linear equation. In many cases, however, we find it more beneficial to use a more general form of the linear equation. To arrive at this form, we will take the slope-intercept form and rearrange it slightly. Let's start by renaming the variables. In this case, we will replace y with x_1 and x with x_2. So, our slope-intercept form can now be written as

$$x_1 = ax_2 + b .$$

Next, we will rearrange the expression to get both variables onto the left side of the equal sign. Doing this yields

$$x_1 - ax_2 = b.$$

Now, we will label the constant values in front of each of the variables as a_1 and a_2 respectively, and the constant to the right of the equal sign will be renamed to b_1.

$$a_1 x_1 + a_2 x_2 = b_1$$

This general form makes the linear equations easier to work with in matrix form, but the equation can be easily converted back into the slope-intercept notation $(y = ax + b)$ using the following relationships:

$$y = x_1, \quad x = x_2, \quad a = \frac{-a_2}{a_1}, \quad \text{and } b = \frac{b_1}{a_1}$$

One of the main advantages of using the more general representation for a liner equation is that we can represent a series, or system, of these equations in a special shorthand using matrices. For example, let us assume that we have two linear equations. We could express these two equations normally:

$$a_{1,1} x_1 + a_{1,2} x_2 = b_1$$
$$a_{2,1} x_1 + a_{2,2} x_2 = b_2$$

Or, using what we know about matrix multiplication, we can rewrite these equations in a matrix form:

$$\mathbf{Ax = b}$$

Where:

$$\mathbf{A} = \begin{bmatrix} a_{1,1} & a_{1,2} \\ a_{2,1} & a_{2,2} \end{bmatrix}, \quad \mathbf{x} = \begin{bmatrix} x_1 \\ x_2 \end{bmatrix}, \quad \text{and } \mathbf{b} = \begin{bmatrix} b_1 \\ b_2 \end{bmatrix}.$$

Thus, we can write our two equations as follows:

$$\begin{bmatrix} a_{1,1} & a_{1,2} \\ a_{2,1} & a_{2,2} \end{bmatrix} \begin{bmatrix} x_1 \\ x_2 \end{bmatrix} = \begin{bmatrix} b_1 \\ b_2 \end{bmatrix}$$

Obviously, we can extend this form to represent any number of linear equations by adding additional elements to our matrices. Thus, for the more general case of *n* equations and *n* variables, we would have the following matrix representation:

$$\begin{bmatrix} a_{1,1} & a_{1,2} & \cdots & a_{1,n} \\ a_{2,1} & a_{2,2} & \cdots & a_{2,n} \\ \vdots & \vdots & \ddots & \vdots \\ a_{n,1} & a_{n,2} & \cdots & a_{n,n} \end{bmatrix} \begin{bmatrix} x_1 \\ x_2 \\ \vdots \\ x_n \end{bmatrix} = \begin{bmatrix} b_1 \\ b_2 \\ \vdots \\ b_n \end{bmatrix}$$

Assuming we are given a system of equations of this form, there are three general statements that we can make regarding the possible solutions to the system, regardless of the size of the system. These three conclusions are:

1. The system has one unique solution

2. The system has no solution

3. The system has an infinite number of solutions

Let us see how each of these possibilities can occur. To illustrate these cases, we will examine a simple case of two equations. The same idea can also be applied to larger systems.

Case 1: One Unique Solution

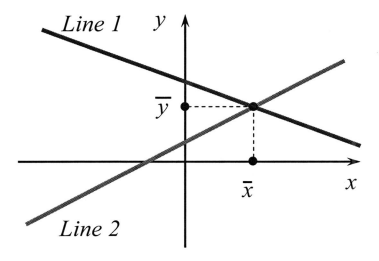

Figure 3.12. An illustration of a system of equations with a single solution.

In the case of a single solution, we have one point in space that all of our lines intersect. Thus, if we are dealing with two lines, there will be one point where these two lines cross. This point of intersection is the solution to the system of equations. Figure 3.12 illustrates this case. In this figure, we let Line 1 represent the first equation and Line 2 represent the second equation. The solution to the system is shown as the point, $(\bar{x}, \bar{y})$, where the two lines cross.

Case 2: No Solution

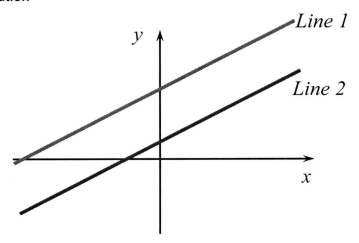

Figure 3.13. A graphical illustration of a system of linear equations with no solution.

The fact that there are no solutions to the system indicates that there is no point in space where all of the lines come into contact. In the case of two lines of infinite length, the only way that this can happen is if the two lines lie parallel to each other. Figure 3.13 shows what this would look like. In the figure, Line 1, representing the first linear equation, and Line 2, representing the second linear equation, are drawn parallel to each other. Since they have no point in common, there is no solution to the system.

Case 3: Infinite Solutions

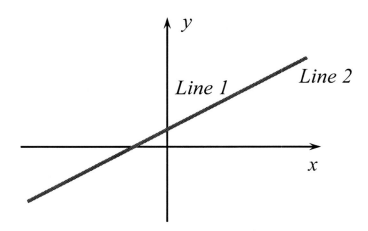

Figure 3.14. A graphical illustration of a system of linear equations with an infinite number of solutions.

For a system of equations to have an infinite number of solutions, the lines involved must have an infinite number of points in common. Therefore, the lines must lie directly on top of each other. Since the lines extend to infinity in both directions, if they lie on top of each other, there will be an infinite number of points in common between the two. Figure 3.14 illustrates this situation. In the figure Line 1, representing the first equation, lies directly on top of Line 2, representing the second equation. Thus the system has an infinite number of solutions.

Let us take a look at an example of actual systems of equations that fall into these three cases.

Example 3.24

Problem: Determine whether the following two-dimensional systems of equations have a single solution, no solution, or infinite solutions.

A) $\begin{bmatrix} 2 & -1 \\ -8 & 4 \end{bmatrix} \begin{bmatrix} x_1 \\ x_2 \end{bmatrix} = \begin{bmatrix} 2 \\ -6 \end{bmatrix}$

B) $\begin{bmatrix} 2 & -1 \\ 0 & 3 \end{bmatrix} \begin{bmatrix} x_1 \\ x_2 \end{bmatrix} = \begin{bmatrix} 2 \\ -6 \end{bmatrix}$

C) $\begin{bmatrix} -4 & -2 \\ 2 & 1 \end{bmatrix} \begin{bmatrix} x_1 \\ x_2 \end{bmatrix} = \begin{bmatrix} 2 \\ -1 \end{bmatrix}$

Solution: Since we are dealing with two-dimensional systems of equations, the easiest way to determine which of the three cases each system falls under is to look at the interaction of the two lines described. This can be done either by directly examining the equations for the two lines or by actually plotting the two lines to see how they interact. We will show both methods here. Let us begin with the first system.

A) First, we will convert the equations from their general form to the slope intercept form. This will allow us to get a better picture of what these lines look like. Performing this conversions gives us:

$$2x_1 - x_2 = 2$$
$$-8x_1 + 4x_2 = -6$$

$\rightarrow$

$$y = \frac{1}{2}x + 1$$
$$y = \frac{1}{2}x + \frac{3}{4}$$

If we examine these two equations, we can clearly see that they have identical slopes but different y-axis intercept points. Thus, these two lines are parallel to each other and there is no solution to the system of equations. If we plot these two lines we would get the following:

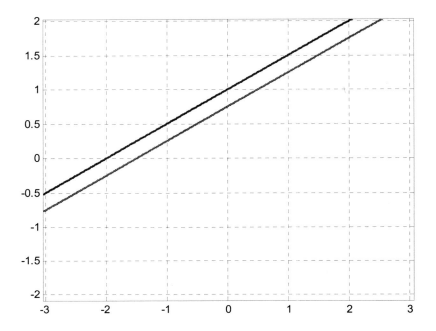

So, equation system A) has no solution.

B) As we did before, we will convert our equations to slope intercept form. This conversion yields:

$$2x_1 - x_2 = 2$$
$$3x_2 = -6$$

$\rightarrow$

$$y = \frac{1}{2}x + 1$$
$$x = -2$$

Here we have a line with a slope of ½ and a vertical line at x = -2. Obviously these lines will intersect. Thus we will have a single solution to our system of equations. The plot below verifies this conclusion.

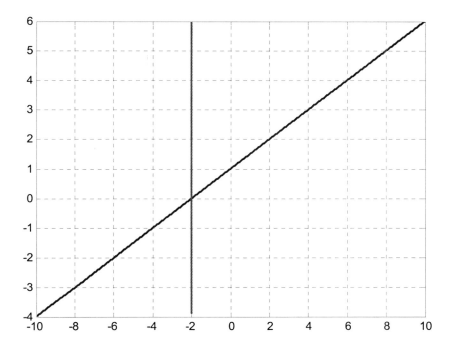

Thus, equation system B) has a single solution at the point (-2,0).

C) Now we will look at the final system. As usual, we begin by converting the equations from general to slope-intercept form:

$$-4x_1 - 2x_2 = 2$$
$$2x_1 + x_2 = -1$$

$\rightarrow$

$$y = -\frac{1}{2}x - \frac{1}{2}$$
$$y = -\frac{1}{2}x - \frac{1}{2}$$

As we can immediately see, the slope-intercept forms of these two equations are identical. Thus the two lines that they represent lie on top of each other. Therefore, we can say that equation system C) has an infinite number of solutions. Since the line equations are identical, it is not necessary to plot these two equations to see the solution.

___

Now that we know what a system of equations is and the three possible types of solutions that can be found, we can look at how to obtain a solution for systems that have a single solution by using matrices.

3.7.2   Solving Systems of Equations with the Matrix Inverse

So far we have established what a system of equations is. Now we want to figure out how we can solve these systems by using matrices. To do this, let us start by looking at the matrix representation of a system.

$$\mathbf{Ax} = \mathbf{b}$$

In this representation we have a coefficient matrix, **A**, that contains the coefficient values for each of the variables in the system, an unknown vector, **x** that contains the unknown values, and a result vector, **b**, that contains the results of the equations in the system. Now, if we find the inverse of the matrix, **A**, and multiply both sides of our equation by this inverse, we will get:

$$\mathbf{Ax} = \mathbf{b}$$

$$\mathbf{A}^{-1}\mathbf{Ax} = \mathbf{A}^{-1}\mathbf{b}$$

Since a matrix multiplied by its inverse yields the identity matrix, we can write:

$$\mathbf{Ix} = \mathbf{A}^{-1}\mathbf{b}$$

And since multiplying a matrix or vector by an identity matrix leaves the matrix unchanged we can write:

$$\mathbf{x} = \mathbf{A}^{-1}\mathbf{b}$$

Thus, our unknown values can be obtained by multiplying the result vector and the inverse of the coefficient matrix. It is important to note that if the matrix, **A**, does not have an inverse we cannot find a unique solution to the system. In other words, if our matrix of coefficient values is non-singular, we can take the inverse and find a single unique solution to the system. On the other hand, if the coefficient matrix is singular, the system of equation will fall under one of the other two solution cases. It will either have no solution, or infinite solutions. That being said, let us look at an example solution to a system of equations.

Example 3.25

Problem: Given the following system of equations, find the values of x, y, and z.

$$4x - 3y - 2z = -8$$
$$-3x + 5y + 6z = 25$$
$$-2x - y + 4z = 8$$

Solution: Here we have been given a system of three equations and three unknowns. Our first step in this situation is to convert our equations to their matrix shorthand notation. Doing this yields:

$$\begin{bmatrix} 4 & -3 & -2 \\ -3 & 5 & 6 \\ -2 & -1 & 4 \end{bmatrix} \begin{bmatrix} x \\ y \\ z \end{bmatrix} = \begin{bmatrix} -8 \\ 25 \\ 8 \end{bmatrix}$$

$$\quad\quad \mathbf{A} \quad\quad\quad \mathbf{x} \;=\; \mathbf{b}$$

Now, we know from our previous discussion that the solution to the system can be found by taking the inverse of the coefficient matrix and multiplying that inverse by the result vector. Therefore, we need to find the inverse of our coefficient matrix first. This can be done using either of the methods that we discussed for obtaining the inverse of a matrix. In this case, however, we will use the adjoint method to obtain our inverse. So, first we will find the determinant of **A**.

$$|\mathbf{A}| = (4)(5)(4) + (-3)(6)(-2) + (-2)(-3)(-1) - (-2)(5)(-2) - (-1)(6)(4) - (4)(-3)(-3)$$

$$|\mathbf{A}| = 80 + 36 - 6 - 20 + 24 - 36$$

$$|\mathbf{A}| = 78$$

Since our determinant is not equal to zero, we know we can find an inverse to this matrix. Thus, we can continue on to the next step and find the minors.

$$m_{1,1} = \begin{vmatrix} 5 & 6 \\ -1 & 4 \end{vmatrix} = 20 + 6 = 26$$

$$m_{1,2} = \begin{vmatrix} -3 & 6 \\ -2 & 4 \end{vmatrix} = -12 + 12 = 0$$

$$m_{1,3} = \begin{vmatrix} -3 & 5 \\ -2 & -1 \end{vmatrix} = 3 + 10 = 13$$

$$m_{2,1} = \begin{vmatrix} -3 & -2 \\ -1 & 4 \end{vmatrix} = -12 - 2 = -14$$

$$m_{2,2} = \begin{vmatrix} 4 & -2 \\ -2 & 4 \end{vmatrix} = 16 - 4 = 12$$

$$m_{2,3} = \begin{vmatrix} 4 & -3 \\ -2 & -1 \end{vmatrix} = -4 - 6 = -10$$

$$m_{3,1} = \begin{vmatrix} -3 & -2 \\ 5 & 6 \end{vmatrix} = -18 + 10 = -8$$

$$m_{3,2} = \begin{vmatrix} 4 & -2 \\ -3 & 6 \end{vmatrix} = 24 - 6 = 18$$

$$m_{3,3} = \begin{vmatrix} 4 & -3 \\ -3 & 5 \end{vmatrix} = 20 - 9 = 11$$

So, our minor matrix will be:

$$\mathbf{M} = \begin{bmatrix} 26 & 0 & 13 \\ -14 & 12 & -10 \\ -8 & 18 & 11 \end{bmatrix}$$

Now we can find our cofactor matrix by changing the sign of the minors that have subscripts which sum to an odd number. Doing this yields:

$$\mathbf{C} = \begin{bmatrix} 26 & 0 & 13 \\ 14 & 12 & 10 \\ -8 & -18 & 11 \end{bmatrix}$$

The cofactor matrix can then be transposed to find the adjoint.

$$adj(\mathbf{A}) = \begin{bmatrix} 26 & 14 & -8 \\ 0 & 12 & -18 \\ 13 & 10 & 11 \end{bmatrix}$$

The inverse of **A** can now be calculated by using the equation:

$$\mathbf{A}^{-1} = \frac{adj(\mathbf{A})}{|\mathbf{A}|}$$

Thus we find:

$$\mathbf{A}^{-1} = \frac{1}{78} \begin{bmatrix} 26 & 14 & -8 \\ 0 & 12 & -18 \\ 13 & 10 & 11 \end{bmatrix}$$

Now that we have the inverse of the coefficient matrix, we simply need to multiply it with the result vector to find our solution.

$$\begin{bmatrix} x \\ y \\ z \end{bmatrix} = \frac{1}{78} \begin{bmatrix} 26 & 14 & -8 \\ 0 & 12 & -18 \\ 13 & 10 & 11 \end{bmatrix} \begin{bmatrix} -8 \\ 25 \\ 8 \end{bmatrix}$$

$$\begin{bmatrix} x \\ y \\ z \end{bmatrix} = \frac{1}{78} \begin{bmatrix} -208 + 350 - 64 \\ 300 - 144 \\ -104 + 250 + 88 \end{bmatrix} = \frac{1}{78} \begin{bmatrix} 78 \\ 156 \\ 234 \end{bmatrix}$$

$$\begin{bmatrix} x \\ y \\ z \end{bmatrix} = \begin{bmatrix} 1 \\ 2 \\ 3 \end{bmatrix}$$

Thus, we have found that $x = 1$, $y = 2$, and $z = 3$.

To verify our solution, we can plug these values back into our original equations and make sure that they work.

$$\begin{aligned} 4x - 3y - 2z &= -8 && \rightarrow && 4(1) - 3(2) - 2(3) = 4 - 6 - 6 = -8 \\ -3x + 5y - 6z &= 25 && \rightarrow && -3(1) + 5(2) + 6(3) = 3 + 10 + 18 = 25 \\ -2x - 1y + 4z &= 8 && \rightarrow && -2(1) - 1(2) + 4(3) = -2 - 2 + 12 = 8 \end{aligned}$$

Thus, we have indeed found a solution to our system of equations.

3.7.3   Electrical Engineering Application – Circuit Analysis

We will wrap up this chapter with an example of an electrical engineering application for the matrix inverse. There are many cases in electrical engineering when we encounter a system of linear equations that must be solved. One notable example is electrical circuit analysis. If we are given a circuit such as the circuit in Figure 3.15, we can determine, through various circuit analysis techniques, a series of equations to describe various aspects of that circuit.

If we solve this system of equations, we can learn about various aspects of the circuit's operation. For example, we can use known values such as applied voltages and the resistances in various branches of the circuit to learn about the current flow in the circuit. We will discuss the methods for doing this in more detail in chapter 6. For now, we will focus on how we can solve the system of equations generated from the circuit.

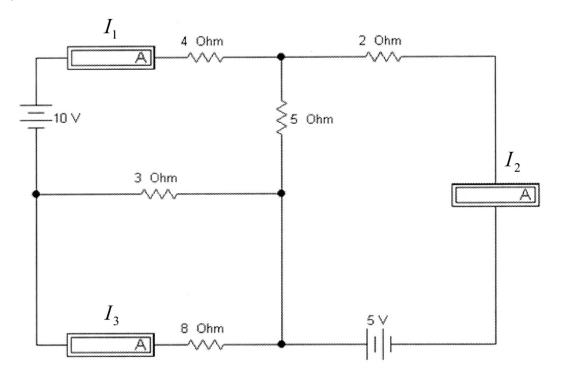

Figure 3.15. A circuit that can be analyzed using systems of equations and matrices.

Example 3.26

Problem: Find the electric current values for I_1, I_2, and I_3 in the circuit shown in figure 3.15 given the following equations:

$$12I_1 - 5I_2 - 3I_3 = 10$$
$$-5I_1 + 7I_2 = 5$$
$$-3I_1 + 11I_3 = 0$$

Solution: Here we have been given three equations that describe the circuit above. All that we need to do to find the three currents is to solve the system of equations. Let us first write the system in matrix notation.

$$\begin{bmatrix} 12 & -5 & -3 \\ -5 & 7 & 0 \\ -3 & 0 & 11 \end{bmatrix} \begin{bmatrix} I_1 \\ I_2 \\ I_3 \end{bmatrix} = \begin{bmatrix} 10 \\ 5 \\ 0 \end{bmatrix}$$

We now begin our solution by finding the inverse of the coefficient matrix, **A**. First we find the determinant.

$$|\mathbf{A}| = (12)(7)(11) + (-5)(0)(-3) + (-3)(-5)(0) - (-3)(7)(-3) - (0)(0)(12) - (11)(-5)(-5)$$
$$|\mathbf{A}| = 586$$

Next we find our minors:

$$m_{1,1} = \begin{vmatrix} 7 & 0 \\ 0 & 11 \end{vmatrix} = 77$$

$$m_{1,2} = \begin{vmatrix} -5 & 0 \\ -3 & 11 \end{vmatrix} = -55$$

$$m_{1,3} = \begin{vmatrix} -5 & 7 \\ -3 & 0 \end{vmatrix} = 21$$

$$m_{2,1} = \begin{vmatrix} -5 & -3 \\ 0 & 11 \end{vmatrix} = -55$$

$$m_{2,2} = \begin{vmatrix} 12 & -3 \\ -3 & 11 \end{vmatrix} = 123$$

$$m_{2,3} = \begin{vmatrix} 12 & -5 \\ -3 & 0 \end{vmatrix} = -15$$

$$m_{3,1} = \begin{vmatrix} -5 & -3 \\ 7 & 0 \end{vmatrix} = 21$$

$$m_{3,2} = \begin{vmatrix} 12 & -3 \\ -5 & 0 \end{vmatrix} = -15$$

$$m_{3,3} = \begin{vmatrix} 12 & -5 \\ -5 & 7 \end{vmatrix} = 59$$

So, our minor matrix will be:

$$\mathbf{M} = \begin{bmatrix} 77 & -55 & 21 \\ -55 & 123 & -15 \\ 21 & -15 & 59 \end{bmatrix}$$

Now changing the sign on elements whose index values sum to an odd number we get the cofactor matrix:

$$\mathbf{C} = \begin{bmatrix} 77 & 55 & 21 \\ 55 & 123 & 15 \\ 21 & 15 & 59 \end{bmatrix}$$

Now, we can write our inverse as:

$$\mathbf{A}^{-1} = \frac{\mathbf{C}^T}{|\mathbf{A}|} = \frac{1}{586} \begin{bmatrix} 77 & 55 & 21 \\ 55 & 123 & 15 \\ 21 & 15 & 59 \end{bmatrix}$$

Finally, we can multiply by our result vector to get the three desired current values:

$$\begin{bmatrix} I_1 \\ I_2 \\ I_3 \end{bmatrix} = \frac{1}{586} \begin{bmatrix} 77 & 55 & 21 \\ 55 & 123 & 15 \\ 21 & 15 & 59 \end{bmatrix} \begin{bmatrix} 10 \\ 5 \\ 0 \end{bmatrix} = \frac{1}{586} \begin{bmatrix} 770 + 275 \\ 550 + 615 \\ 210 + 75 \end{bmatrix}$$

$$\begin{bmatrix} I_1 \\ I_2 \\ I_3 \end{bmatrix} \approx \begin{bmatrix} 1.7833 \\ 1.9881 \\ 0.4863 \end{bmatrix}$$

Thus, if we built this circuit and physically measured the three currents, we would expect to find I_1 = 1.7833 Amps, I_2 = 1.9881 Amps, and I_3 = 0.4863 Amps. If we simulate this circuit using a circuit simulation tool, we will find the following:

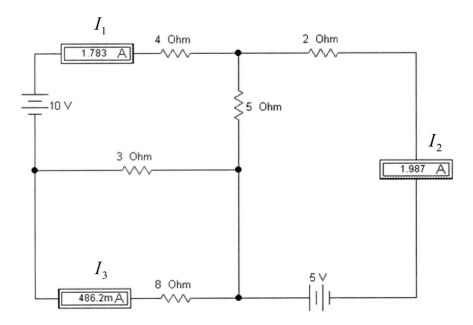

Thus, the simulated results agree with our solution. We have used a system of linear equations to solve an actual electrical engineering problem.

This example shows only one of the many possible uses of matrices in electrical engineering. So, from this we can see that a good understanding of matrices is essential in the study of electrical engineering.

3.8 Chapter Summary

In this chapter, we discussed the concepts of vectors and matrices. We discovered what these important structures are and how they are used. We also learned how basic arithmetic is performed and how these structures can be applied to engineering problems.

- We began our study by looking at the difference between a regular number that indicates only a quantity or magnitude, a scalar, and a value that provides both a magnitude and a direction, a vector.

- We learned how vectors are plotted, what a zero vector looks like, the concept of vector equality, and what the negative of a vector looks like. We also were introduced to the concept of vector magnitude, also known as the vector norm. We found that we can find the magnitude of a vector by simply taking the square root of the sum of the vector's elements squared. We saw that vectors can be expressed using a sum of unit vectors, or vectors that have a norm of 1. The most common unit vectors, $\mathbf{i}$, $\mathbf{j}$, and $\mathbf{k}$, point along the x, y, and z axes respectively.

- Next, we looked at how vectors can be added together. We saw that we could either add each of the vector's components algebraically or use the head to tail method to perform a graphical addition.

- We found that subtraction worked in much the same way as addition. The vectors could be subtracted element by element in an algebraic approach or head to tail through a graphical method.

- When it comes to multiplication, there are three approaches. A vector can be multiplied by a scalar in which case, each element of the vector is multiplied by the scalar value.

- A vector can also be multiplied by another vector to yield a scalar value. This approach, known as the dot product, requires that each pair of corresponding vector elements be multiplied together and then all of the products are summed up to yield a final scalar value.

- We saw that since the dot product between two vectors $\mathbf{a}$ and $\mathbf{b}$ is defined as

$$\mathbf{a} \bullet \mathbf{b} = \|\mathbf{a}\| \|\mathbf{b}\| \cos(\theta_{\mathbf{ab}})$$

    where θ is defined as the angle between the two vectors, we can use the cross product to find the angle between any two vectors. Thus, we find

$$\theta_{ab} = \cos^{-1}\left(\frac{\mathbf{a} \bullet \mathbf{b}}{\|\mathbf{a}\|\|\mathbf{b}\|}\right).$$

- Two vectors can also be multiplied together to produce a vector. This approach, known as the cross product is defined as

$$\mathbf{a} \times \mathbf{b} = \|\mathbf{a}\|\|\mathbf{b}\|\sin(\theta_{ab})\hat{\mathbf{n}}$$

where $\hat{\mathbf{n}}$ is a unit vector that points in a direction perpendicular to the plane that the two vectors lie within. The direction of this vector can be determined using the right-hand-rule.

- Next, we discovered that vectors can be grouped together to form rows or columns of larger structures known as matrices. Matrices are essentially a collection of numbers arranged in a square or rectangular array. These numbers are referenced based upon which row and column they are found in within the larger matrix structure.

- We examined the concepts of the zero matrix, the square matrix, and matrix equality. We looked at a special matrix called the identity matrix and briefly discussed the topics of linear dependence and matrix rank. These basic ideas gave us a starting point for working with matrices.

- With a good understanding of the basics, we next learned about basic matrix arithmetic. We found that as was the case with vectors, matrices are added by summing the corresponding elements. It is important that the matrices be the same size in order for them to be added together.

- Matrix subtraction was a very similar concept. Corresponding elements within the matrices were subtracted. Once again two matrices must be the same size for subtraction to occur.

- Multiplying a matrix by a scalar value works the same way that multiplying a vector by a scalar did. We simply multiply each element in the matrix by the scalar value.

- Multiplying two matrices together is slightly more complicated. First, the number of columns of the first matrix must equal the number of rows of the second matrix for multiplication to be possible. The resulting matrix will have the same number of rows as the first matrix and the same number of columns as the second. The value stored within an element of the resulting matrix is simply the dot product of the vector formed by the row of the first matrix equal to the row of the element and the column of the second matrix equal to the column of the element. For the product of two matrices $\mathbf{A}$ and $\mathbf{B}$, the element in the result matrix $\mathbf{C}$ in the ith row and j^{th} column can be expressed as

$$c_{i,j} = \sum_{k=1}^{p} a_{i,k} b_{k,j},$$

where p is the number of elements in the row and column.

- It is important to note that order is important in matrix multiplication. $\mathbf{AB} \neq \mathbf{BA}$.

  After looking at basic arithmetic using vectors, we turned out attention to other operations specific to matrices. The first of these was the transpose. When a matrix is transposed, the rows become columns. So, if we let $\mathbf{A}$ be an m x n matrix and describe the elements of $\mathbf{A}$ as $\mathbf{A} = [a_{i,j}]$. Then, we can define the transpose operation as:

  $$\mathbf{A}^T = \left[a_{i,j}\right]^T = \left[a_{j,i}\right]$$

  For i = 1, 2, ..., m and j = 1, 2, ..., n

- Next, we looked at the determinant which is a scalar value that is a function of a given square matrix. The determinant of a matrix $\mathbf{A}$ is typically written as $\mathrm{Det}(\mathbf{A})$ or $|\mathbf{A}|$. We looked at methods of finding the determinant for 1 x 1, 2 x 2, and 3 x 3 cases.

- We also learned that each element in a square matrix has a unique scalar value called a minor associated with it. The minor for the element of a matrix in a specific row and column is found by crossing out the row and column in the matrix and taking the determinant of the remaining values.

- From the minors matrix cofactors can be found. These values are obtained by changing the sign of minors that lie in a row and column that sum to an odd number. If $m_{i,j}$ is the minor associated with the ith row and j^{th} column of a matrix, then the associated cofactor $c_{i,j}$ can be described mathematically by writing

  $$c_{i,j} = (-1)^{i+j} m_{i,j} .$$

- Next, we looked at the matrix inverse. We learned that the inverse of a matrix is a matrix that when multiplied with the original matrix in any order results in an identity matrix. Two methods for finding the inverse of a matrix were discussed. The first method involved creating an augmented matrix containing the original matrix and an identity matrix. Elementary row operations are then employed to reduce the original matrix to an identity matrix. When this has been completed, the augmented portion of the matrix that originally contained the identity matrix will contain the inverse of the matrix.

- The second approach that we looked at was known as the adjoint method. In this method the adjoint, or transpose of the cofactor matrix, is divided by the determinant to find the matrix inverse.

- We also learned that some matrices do not have inverses. These are called singular matrices. They can be quickly identified by taking a determinant. All singular matrices will have a determinant of zero. Matrices that have an inverse are called non-singular matrices.

- Finally, we looked at an application of the matrix inverse. We found that a system of linear equations could be solved by taking the inverse of a matrix containing the coefficients from the equations and multiplying that with a vector containing the value that the equations were set equal to. This resulted in a vector containing the solution to the system of equations. We finished up the chapter by looking at how this technique could be used to solve a circuit analysis problem.

3.9 Review Exercises

Section 3.3 – Working With Vectors

1. Given the vectors $\mathbf{a} = \begin{bmatrix} 1 & -2 & 7 \end{bmatrix}$, $\mathbf{b} = \begin{bmatrix} -7 & 3 & -2 \end{bmatrix}$, $\mathbf{c} = \begin{bmatrix} 3 & 4 & -6 \end{bmatrix}$, and $\mathbf{d} = \begin{bmatrix} 4 & -2 & -3 \end{bmatrix}$, determine the following.

   a. $\|\mathbf{a}\|$ b. $\|\mathbf{b}\|$ c. $\|\mathbf{c}\|$ d. $\|\mathbf{d}\|$

2. Write the vectors in problem 1 in terms of the unit vectors $\mathbf{i}$, $\mathbf{j}$, and $\mathbf{k}$.

3. Perform the following vector additions and subtractions.

   a. $\begin{bmatrix} 1 & -2 & 3 \end{bmatrix} + \begin{bmatrix} 4 & 3 & -1 \end{bmatrix}$ b. $\begin{bmatrix} 3 & 7 & -5 \end{bmatrix} - \begin{bmatrix} 2 & -1 & -7 \end{bmatrix}$

   c. $\begin{bmatrix} 5 & -6 & -1 & 5 \end{bmatrix} + \begin{bmatrix} -6 & 3 & -2 & 1 \end{bmatrix}$ d. $\begin{bmatrix} 4 & -2 & 7 \end{bmatrix} + \begin{bmatrix} 15 & -3 & 2 \end{bmatrix}$

   e. $5\mathbf{i} + 3\mathbf{j} - 7\mathbf{k} - \begin{bmatrix} 3 & 7 & 10 \end{bmatrix}$ f. $\begin{bmatrix} 1 & 3 & -9 \end{bmatrix} + \begin{bmatrix} 2 & -7 & -8 \end{bmatrix}$

4. Given the vectors $\mathbf{a} = \begin{bmatrix} 4 & -2 & 7 \end{bmatrix}$, $\mathbf{b} = \begin{bmatrix} 3 & 1 & -6 \end{bmatrix}$, and $\mathbf{c} = \begin{bmatrix} -4 & 2 & 3 \end{bmatrix}$, determine the following

   a. $\mathbf{a} + \mathbf{b}$ b. $\mathbf{b} + \mathbf{c}$ c. $\mathbf{a} + \mathbf{c}$ d. $\mathbf{c} - \mathbf{a}$

   e. $\mathbf{b} - \mathbf{a}$ f. $\mathbf{a} + \mathbf{b} - \mathbf{c}$ g. $\|\mathbf{a} + \mathbf{b} + \mathbf{c}\|$ h. $\|\mathbf{a} - \mathbf{b} - \mathbf{c}\|$

5. Given the vectors $\mathbf{a} = \begin{bmatrix} 1 & 2 & 3 \end{bmatrix}$, $\mathbf{b} = \begin{bmatrix} -7 & 3 & -1 \end{bmatrix}$, and $\mathbf{c} = \begin{bmatrix} 8 & -2 & 1 \end{bmatrix}$, determine the following

   a. $6\mathbf{a} - 3\mathbf{b}$ b. $5(\mathbf{a} - \mathbf{c})$ c. $2\mathbf{a} + 3\mathbf{b} - 2\mathbf{c}$ d. $8\mathbf{a} + 3\mathbf{c}$

   e. $\mathbf{a} + 2\mathbf{c} - \mathbf{b}$ f. $3(2\mathbf{a} - 4\mathbf{b}) + 7\mathbf{c}$ g. $4\mathbf{a} - 3\mathbf{b} + \mathbf{c}$ h. $-2\mathbf{a} + 7\mathbf{c}$

6. Given the vectors $\mathbf{a} = \begin{bmatrix} 1 & 2 & -3 \end{bmatrix}$, $\mathbf{b} = \begin{bmatrix} 2 & -1 & 7 \end{bmatrix}$, and $\mathbf{c} = \begin{bmatrix} -3 & 6 & 9 \end{bmatrix}$, determine the following

   a. $\|\mathbf{a}\|$ b. $\|\mathbf{b}\|$ c. $\mathbf{a} \bullet \mathbf{b}$ d. The angle between $\mathbf{a}$ and $\mathbf{b}$

   e. $\|\mathbf{c}\|$ f. $\mathbf{a} \bullet \mathbf{c}$ g. $\mathbf{b} \bullet \mathbf{c}$ h. The angle between $\mathbf{a}$ and $\mathbf{c}$

   i. The angle between $\mathbf{b}$ and $\mathbf{c}$

7. Given the vectors $\mathbf{a} = \begin{bmatrix} 1 & 4 & -3 \end{bmatrix}$, $\mathbf{b} = \begin{bmatrix} 6 & -2 & -4 \end{bmatrix}$, and $\mathbf{c} = \begin{bmatrix} -6 & 2 & -9 \end{bmatrix}$

    a. Calculate $\mathbf{a} \bullet \mathbf{b}$
    b. Calculate $\mathbf{b} \bullet \mathbf{a}$
    c. Is $\mathbf{a} \bullet \mathbf{b} = \mathbf{b} \bullet \mathbf{a}$? Which property of the dot product does this illustrate?
    d. Calculate $\mathbf{a} \bullet (\mathbf{b} \bullet \mathbf{c})$
    e. Calculate $(\mathbf{a} \bullet \mathbf{b}) \bullet \mathbf{c}$
    f. Does $\mathbf{a} \bullet (\mathbf{b} \bullet \mathbf{c}) = (\mathbf{a} \bullet \mathbf{b}) \bullet \mathbf{c}$? Which property of the dot product does this illustrate?

8. Determine the cross product of the following vectors.

    a. $\begin{bmatrix} 1 & 2 & 3 \end{bmatrix} \times \begin{bmatrix} 3 & 2 & 1 \end{bmatrix}$
    b. $(3\mathbf{i} + 2\mathbf{j} - \mathbf{k}) \times (-2\mathbf{i} + 6\mathbf{j} - \mathbf{k})$
    c. $\begin{bmatrix} 2 & -1 & 3 \end{bmatrix} \times \begin{bmatrix} 4 & -1 & -1 \end{bmatrix}$
    d. $\begin{bmatrix} -3 & -1 & -1 \end{bmatrix} \times (4\mathbf{i} + 6\mathbf{j} - \mathbf{k})$

9. Given the vectors $\mathbf{a} = \begin{bmatrix} 1 & 1 & 3 \end{bmatrix}$, $\mathbf{b} = \begin{bmatrix} -1 & 1 & 4 \end{bmatrix}$, and $\mathbf{c} = \begin{bmatrix} 3 & 4 & -1 \end{bmatrix}$, answer the following

    a. $\mathbf{a} \times \mathbf{b}$
    b. $\mathbf{b} \times \mathbf{a}$
    c. Does $\mathbf{a} \times \mathbf{b} = \mathbf{b} \times \mathbf{a}$? What property of the cross product does this illustrate?
    d. $\mathbf{a} \times \mathbf{b} + \mathbf{a} \times \mathbf{c}$
    e. $\mathbf{a} \times \mathbf{b} \times \mathbf{c}$

Section 3.5 – Matrix Arithmetic

---

10. Let $A = \begin{bmatrix} 1 & -1 & 2 \\ 3 & 4 & -7 \\ -6 & 9 & 2 \end{bmatrix}$, $B = \begin{bmatrix} 4 & 3 & -7 \\ -6 & 2 & 1 \\ 0 & 3 & -2 \end{bmatrix}$, and $C = \begin{bmatrix} -1 & -1 & 0 \\ 6 & -7 & 1 \\ 3 & 1 & -7 \end{bmatrix}$

a. Find $A + B$	b. Find $B + C$	c. Find $A + B + C$
d. Find $A - B$	e. Find $B - C$	f. Find $A + B - C$
g. Find $C - A + B$	h. Find $A - B + C$	i. Find $3A - 4C$
j. Find $2C + 7A$	k. Find $3A - 4B + 2C$	l. Find $-2A + 3B - 4C$

11. If the following matrix operations can be performed, determine the result. If they cannot be performed, state why this is the case.

a. $\begin{bmatrix} 1 & -2 & 4 \end{bmatrix} \begin{bmatrix} 3 \\ 7 \\ -1 \end{bmatrix}$

b. $\begin{bmatrix} 1 & 4 \\ -2 & 3 \end{bmatrix} \begin{bmatrix} 7 \\ -3 \end{bmatrix}$

c. $\begin{bmatrix} 1 & -2 \\ 4 & 6 \\ -8 & 0 \end{bmatrix} \begin{bmatrix} 3 \\ -1 \end{bmatrix}$

d. $\begin{bmatrix} 1 & -2 & 7 \\ 3 & 6 & -1 \end{bmatrix} \begin{bmatrix} 1 \\ 4 \end{bmatrix}$

e. $\begin{bmatrix} 1 & 6 & -3 \\ 4 & -1 & 0 \\ -2 & 7 & 4 \end{bmatrix} \begin{bmatrix} 3 & 7 \\ -4 & 2 \\ 6 & -3 \end{bmatrix}$

f. $\begin{bmatrix} 3 & 2 \\ 0 & -1 \end{bmatrix} \begin{bmatrix} 4 & -6 \\ -3 & 7 \end{bmatrix}$

g. $\left(\begin{bmatrix} 3 & 1 & 0 \\ -2 & 0 & 1 \end{bmatrix} + \begin{bmatrix} 4 & -6 & 3 \\ 0 & 1 & 5 \end{bmatrix} \right) \begin{bmatrix} 4 \\ -2 \\ 1 \end{bmatrix}$

h. $\begin{bmatrix} 4 & 0 & 2 \\ 3 & -1 & 5 \\ 0 & -7 & 8 \end{bmatrix} \left(\begin{bmatrix} 1 & 2 \\ -2 & 0 \\ 3 & 5 \end{bmatrix} + \begin{bmatrix} 4 & 0 \\ 2 & 0 \\ 1 & 5 \end{bmatrix} \right)$

i. $\begin{bmatrix} 4 & -1 \\ 2 & 6 \\ 0 & -2 \end{bmatrix} \begin{bmatrix} 3 & -5 \\ -7 & 0 \end{bmatrix} \begin{bmatrix} 4 \\ -2 \end{bmatrix}$

j. $\begin{bmatrix} 1 & 4 & -2 \\ 6 & -7 & 8 \\ 9 & 0 & -1 \end{bmatrix} \begin{bmatrix} 2 & -1 \\ 0 & 4 \\ 5 & 0 \end{bmatrix} \begin{bmatrix} 4 & -1 \\ 0 & 6 \end{bmatrix}$

Section 3.6 – Other Matrix Operations

12. Find the transpose of the following matrices.

a. $\begin{bmatrix} 1 & 4 \\ 2 & 6 \\ -1 & 0 \end{bmatrix}$

b. $\begin{bmatrix} 1 & -1 & 1 \\ 0 & 4 & 2 \\ 1 & 2 & 3 \end{bmatrix}$

c. $\begin{bmatrix} 1 & -4 & 0 \\ 3 & 7 & 1 \end{bmatrix}$

d. $\begin{bmatrix} 1 & 2 & 3 \\ 2 & 7 & -3 \\ 3 & -3 & 6 \end{bmatrix}$

e. $\begin{bmatrix} 1 & 4 \\ 3 & 7 \\ 6 & 0 \end{bmatrix} + \begin{bmatrix} 5 & -2 \\ 1 & 6 \\ 0 & 3 \end{bmatrix}$

f. $\begin{bmatrix} 1 & 2 \\ -5 & 7 \end{bmatrix} \begin{bmatrix} 6 & 3 \\ -2 & 1 \end{bmatrix}$

13. Find the determinant of the following matrices.

a. $\begin{bmatrix} 2 & -7 \\ 0 & 3 \end{bmatrix}$
b. $\begin{bmatrix} 7 & 10 \\ -4 & 6 \end{bmatrix}$
c. $\begin{bmatrix} -2 \end{bmatrix}$
d. $\begin{bmatrix} -4 & 2 \\ -2 & 1 \end{bmatrix}$

e. $\begin{bmatrix} 4 & -3 & 2 \\ -1 & 7 & 0 \\ 3 & 1 & -4 \end{bmatrix}$
f. $\begin{bmatrix} 2 & 1 & 0 \\ -1 & 0 & 7 \\ 0 & 2 & -4 \end{bmatrix}$
g. $\begin{bmatrix} 1 & 2 & 3 \\ 4 & 5 & 6 \\ 7 & 8 & 9 \end{bmatrix}$
h. $\begin{bmatrix} 2 & 4 & 0 \\ 0 & 1 & 3 \\ 7 & 0 & 0 \end{bmatrix}$

14. Given the 3 x 3 matrix $\mathbf{A} = \begin{bmatrix} 1 & -5 & 7 \\ -2 & 6 & 3 \\ 0 & 1 & -2 \end{bmatrix}$, show that $|\mathbf{A}| = |\mathbf{A}^T|$.

15. Using the 3 x 3 matrix $\mathbf{A} = \begin{bmatrix} a_{1,1} & a_{1,2} & a_{1,3} \\ a_{2,1} & a_{2,2} & a_{2,3} \\ a_{3.1} & a_{3,2} & a_{3,3} \end{bmatrix}$, show that $|\mathbf{A}| = |\mathbf{A}^T|$.

16. Find the minors for the following matrices.

a. $\begin{bmatrix} 2 & -1 \\ 4 & 7 \end{bmatrix}$
b. $\begin{bmatrix} 3 & -6 \\ -1 & 2 \end{bmatrix}$
c. $\begin{bmatrix} 0 & -1 \\ 4 & 6 \end{bmatrix}$
d. $\begin{bmatrix} 3 & -1 & 2 \\ 0 & 4 & -7 \\ 0 & 3 & 8 \end{bmatrix}$

e. $\begin{bmatrix} 4 & -3 & 2 \\ -1 & 7 & 0 \\ 3 & 1 & -4 \end{bmatrix}$
f. $\begin{bmatrix} 2 & 0 & 1 \\ 1 & 4 & 7 \\ 0 & 3 & 8 \end{bmatrix}$
g. $\begin{bmatrix} -1 & 0 & 0 \\ 2 & 4 & -3 \\ 7 & -1 & 5 \end{bmatrix}$
h. $\begin{bmatrix} 1 & 2 & 3 \\ 4 & 5 & 6 \\ 7 & 8 & 9 \end{bmatrix}$

17. Determine the cofactors for each matrix in problem 16.

18. Find the inverse for each matrix in problem 16 using the adjoint method. If an inverse does not exist, explain why and provide the adjoint of the matrix.

19. Find the inverse of each matrix in problem 16 using elementary row operations. If an inverse does not exist, explain why.

Section 3.7 – Systems of Linear Equations

20. Determine if the following 2 x 2 systems of equations have one solution, no solution, or infinite solutions.

a. $\begin{bmatrix} 3 & -2 \\ 2 & -2 \end{bmatrix} \begin{bmatrix} x_1 \\ x_2 \end{bmatrix} = \begin{bmatrix} 15 \\ 4 \end{bmatrix}$
b. $\begin{bmatrix} 2 & -4 \\ 8 & 16 \end{bmatrix} \begin{bmatrix} x_1 \\ x_2 \end{bmatrix} = \begin{bmatrix} 2 \\ 8 \end{bmatrix}$
c. $\begin{bmatrix} 3 & -9 \\ 6 & -18 \end{bmatrix} \begin{bmatrix} x_1 \\ x_2 \end{bmatrix} = \begin{bmatrix} 15 \\ 30 \end{bmatrix}$

d. $\begin{bmatrix} 4 & -2 \\ 2 & -1 \end{bmatrix} \begin{bmatrix} x_1 \\ x_2 \end{bmatrix} = \begin{bmatrix} 12 \\ 4 \end{bmatrix}$
e. $\begin{bmatrix} 5 & -15 \\ 3 & -6 \end{bmatrix} \begin{bmatrix} x_1 \\ x_2 \end{bmatrix} = \begin{bmatrix} 5 \\ 12 \end{bmatrix}$
f. $\begin{bmatrix} 3 & -2 \\ 6 & -4 \end{bmatrix} \begin{bmatrix} x_1 \\ x_2 \end{bmatrix} = \begin{bmatrix} 3 \\ 24 \end{bmatrix}$

21. Solve the following systems of equations using the matrix inverse. If the system cannot be solved, explain why.

a. $\begin{aligned} 2x - y &= 11 \\ -3x + 7y &= -33 \end{aligned}$
b. $\begin{bmatrix} 4 & 0 \\ -2 & 1 \end{bmatrix} \begin{bmatrix} x_1 \\ x_2 \end{bmatrix} = \begin{bmatrix} -8 \\ 11 \end{bmatrix}$
c. $\begin{aligned} 7x - 3y &= 1 \\ -2x - 4y &= 6 \end{aligned}$

d. $\begin{aligned} x + 4y - 2z &= -8 \\ -3x + 2y + z &= 13 \\ -2y + z &= 3 \end{aligned}$
e. $\begin{aligned} x + 4y - 2z &= 2 \\ 3x + 2y - z &= -1 \\ -2y + z &= 3 \end{aligned}$
f. $\begin{bmatrix} 4 & -2 & 1 \\ -1 & 0 & 2 \\ 0 & 4 & -3 \end{bmatrix} \begin{bmatrix} x \\ y \\ z \end{bmatrix} = \begin{bmatrix} -23 \\ 7 \\ 5 \end{bmatrix}$

22. Given the circuit and equations shown in Figure 3.16 with $R_1 = 1\Omega$, $R_2 = 2\Omega$, $R_3 = 3\Omega$, $R_4 = 4\Omega$, $R_5 = 5\Omega$, $R_6 = 6\Omega$, $V_1 = 8V$, and $V_2 = 9V$, find the currents I_1, I_2, and I_3.

23. Given the circuit and equations shown in Figure 3.16 with $R_1 = 5\Omega$, $R_2 = 1\Omega$, $R_3 = 3\Omega$, $R_4 = 2\Omega$, $R_5 = 10\Omega$, $R_6 = 2\Omega$, $V_1 = 8.5V$, and $V_2 = 8V$, find the currents I_1, I_2, and I_3.

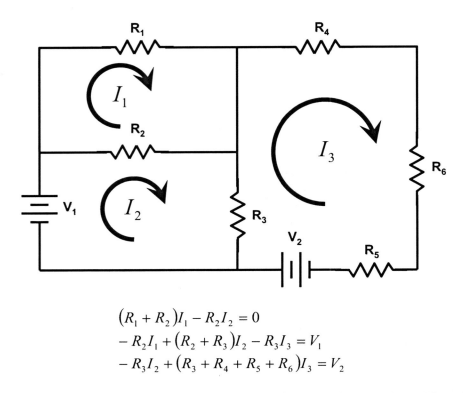

$$\left(R_1 + R_2\right)I_1 - R_2 I_2 = 0$$
$$- R_2 I_1 + \left(R_2 + R_3\right)I_2 - R_3 I_3 = V_1$$
$$- R_3 I_2 + \left(R_3 + R_4 + R_5 + R_6\right)I_3 = V_2$$

Figure 3.16 The circuit schematic and equations for use with problems 22 and 23.

UNIT 2

Electricity
and
Electronics

Chapter 4. Electrical Engineering Concepts

Chapter Outline

4.1   The International System of Units

In the previous chapters, we have spent some time looking at a few of the basic mathematical concepts that will be encountered in our study of electrical engineering. As we begin looking at engineering concepts in more detail in the following chapters, it will be vitally important that we understand how to describe the physical quantities that we encounter. So, when you want to describe an object, how do you go about doing it? Well, the most common way of describing an object to someone who has not seen it before it to compare it to some other object that they are familiar with. This is exactly how physical **units** work. Each unit of measure is derived from a set of basic measures that are widely known and accepted as standards. This is important because using a system of measurements that is not based upon standards would lead to large uncertainty. For example, if you describe an object to a friend by saying that its length is equivalent to one arm length, your friend cannot obtain an accurate measure from your description. While you are describing the length in terms that you can easily measure and your friend can understand and perhaps even draw some general conclusions from, it is impossible to make any definite conclusions without knowing whose arm length was used to measure the object. However, if you both possess an object, let's say a metal bar, that is some specified length, you can tell your friend that the new object is 3 times the length of the metal bar. Your friend, using his metal bar, can easily reproduce this distance and, thus, your friend can obtain an exact picture of the length you are describing. Thus, by using a series of agreed upon and easily reproducible standard values to compare unknown quantities to, we can describe these values to others and they will be able to understand exactly what we are taking about.

Currently, there are two main sets of measurement standards in use. These are the U.S Customary System and the International System (Système International or SI for short). If you live in the United States, you are probably very familiar with units such as the foot (length) and the pound (force/weight). These units belong to the U.S Customary System of measures. Outside the United States, however, this system sees very little use. Instead the primary system of measure for the rest of the world is called the international system of units. In this system, units such as the meter (length), kilogram (mass), and Newton (force) are used (Since the SI system uses the meter as its base unit of length, it is often referred to as the metric system). So, which of these two systems should we use?

The scientific and engineering community has long been using the international system of measures. There are several important reasons for this choice. First, the units of the international system are base ten units. In other words, to express a measurement given in terms of a large unit, for example the meter, in terms of a much smaller one, for example the millimeter, requires only a multiplication by a power of ten (1 meter = 1000 millimeters, so 1.362 meters = 1362 millimeters). Likewise, moving from a small unit to a larger one requires only a division by a

Table 4.1. Some commonly used prefixes for the international system of units.

Prefix	Symbol	Power of 10	Prefix	Symbol	Power of 10
exa	E	10^{18}	deci	d	10^{-1}
peta	P	10^{15}	centi	c	10^{-2}
tera	T	10^{12}	milli	m	10^{-3}
giga	G	10^{9}	micro	μ	10^{-6}
mega	M	10^{6}	nano	n	10^{-9}
kilo	k	10^{3}	pico	p	10^{-12}
hecto	h	10^{2}	femto	f	10^{-15}
deka	da	10^{1}	atto	a	10^{-18}

power of ten (800 millimeters = 0.8 meters). This makes the conversion between units much simpler than the U.S. Customary system. In the U.S. system, units are not powers of ten of each other. For example, 1 foot is equal to 12 inches and one yard is equal to three feet. Thus, the use of the international system greatly simplifies unit conversions. Another good reason for using the international system is the way in which units are expressed with prefixes. Instead of naming each size

Table 4.2. The seven basic SI units.

Measured Quantity	Unit	Symbol
Length	Meter	m
Mass	Kilogram	kg
Time	Second	s
Electric Current	Ampere	A
Thermodynamic Temperature	Kelvin	K
Amount of a Substance	Mole	mol
Luminous Intensity	Candella	cd

unit for a given measurement with different names, for example, inches, feet, and yards, the metric system uses prefixes added to the base unit to express the larger and smaller units for the same measure, for example millimeters, meters, and kilometers. Table 4.1 gives some of the commonly used prefixes in the international system. The final reason that the engineering and scientific community chose to use the international system is its international recognition. Science and engineering research is a global effort, and thus, the use of the more universally accepted international system is a logical choice for sharing results and publishing findings. As a result, except in the case of some rare situations, most scientific and engineering work will be expressed in units taken from the international system. Table 4.2 gives the basic units of the international system. These seven units are not based on any other unit, but rather a physical definition. Table 4.3 displays some of the important derived units that you should know. These units are usually not based on a physical definition, but rather can be expressed in terms of the base units shown in table 4.2.

Table 4.3. Some important derived SI units and their symbols.

Measured Quantity	Unit	Symbol	Measured Quantity	Unit	Symbol
Angle	Radian	rad	Electric Potential	Volt	V
Frequency	Hertz	Hz	Resistance	Ohm	Ω
Force	Newton	N	Conductance	Siemens	S
Pressure	Pascal	Pa	Inductance	Henry	H
Energy/Work/Heat	Joule	J	Capacitance	Farad	F
Power	Watt	W	Magnetic Flux	Weber	Wb
Electric Charge	Coulomb	C	Magnetic Flux Density	Tesla	T

4.2 The Basics of Electricity

Now that we have established the needed mathematical background and a basic understanding of the units that we will be using to measure quantities, we can begin looking at some actual engineering concepts. Since we are interested in the area of electrical engineering, we will begin by examining a concept that is at the heart of our field: electricity. As electrical engineers, we must have a complete understanding of what electricity is and how it behaves. Whether we are dealing with static electricity, such as that generated by walking across carpet on a dry day, or the electricity generated in power plants that runs our lights and household appliances, the concepts are essentially the same. The truth is that, today electricity is so commonplace that we often take it for granted. While everyone makes use of it, few people actually understand how it works.

4.2.1   Electrical Charge

To get a clear picture of how electricity works, we must turn to the simplest particle of any element, the atom. As you know, a typical atom consists of 3 basic sub-particles, the proton, which possesses a positive charge, the neutron, which holds no charge, and the electron, which possesses a negative charge.

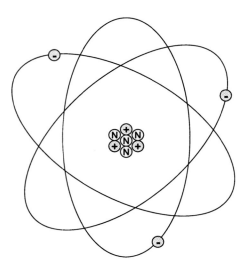

As is the case with magnetic forces, electrical charges of the same type repel one another while opposite electrical charges attract. The strength of this attraction or repulsion is a function of the square of the distance between the charges. For example, if the distance between the charged particles is doubled, the force between them is reduced by a factor of four. This is known as the inverse square law for electrical charges. In an atom, the protons and neutrons cluster together in the center, or nucleus, while the electrons, being attracted to the protons, swarm, or orbit, around the nucleus. It is important to note that the protons in the nucleus would typically repel one another since they have similar electrical charge. However, these particles are held together by

Figure 4.1. A conceptual drawing of an atom. Notice the positions of the Protons (+), the Neutrons (N), and the Electrons (-).

another force known as the strong nuclear force. Figure 4.1 shows a concept drawing of an atom. You will notice the protons (+) and Neutrons (N) form the nucleus, while the electrons (-) orbit around it.

A stable atom will have an overall neutral charge. This means that it will have the same number of positive particles (protons) and negative particles (electrons). By possessing an equal number of each particle, the overall charges equalize and the atom can achieve a net neutral charge. But what happens when the atom does not have the same number of both positive and negative particles? Well, as you would guess, when an atom does not have the same number of protons and electrons (this usually happens because the atom has lost or gained electrons) the atom will have an overall positive or negative charge. Such an atom is referred to as an ion. If the atom gains an extra electron, it will have a net negative charge, and thus, will be called a negative ion, or **anion** (Figure 4.2). If the atom loses an electron, however, it will possess a net positive charge and will be referred to as a positive ion, or **cation** (Figure 4.3).

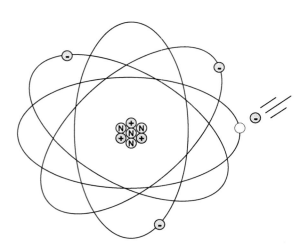

Figure 4.2. A negative ion is created when an atom gains an extra electron, leaving it with more electrons than protons.

When an electron is removed from an atom, whether this is done by mechanical friction, light, heat, or some form of chemical reaction, the removed electrons become free electrons. This means that they are not currently bound to any particular atom. Free electrons may take one of three basic actions after they are removed from their atoms. They may rest on a surface, giving the surface a net negative charge, they may travel through space or matter (this occurs at speeds near that of light), or they may be captured by a nearby positive ion, thus creating a neutral atom.

Free electrons at rest create a negative electrical charge that is stationary. This is often referred to as a negative "static" electrical charge. Likewise, the atoms from which the free electrons escaped become stationary positive ions. This creates a positive "static" electrical charge. It is important that we do not focus exclusively on the fact that these charges are stationary. Instead, we should view stationary electrons much like scientists view water. Water, even when stationary is capable of exerting force (Water Pressure). Likewise, the study of electrostatics is

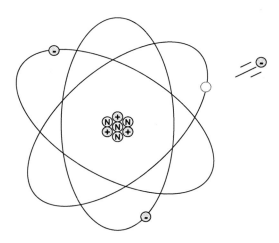

Figure 4.3. A positive ion is created when an atom loses one of its electrons, leaving it with more protons than electrons.

not the study of stationary electrons, but rather the study of electrical charges and the forces they exert.

We have all experienced some form of "static" electricity before. One easy way to generate "static" electricity is to run a comb through your hair. As the comb rubs against the individual strands of hair, some (millions) electrons are mechanically removed from the hair and rest on the comb. Thus the comb takes on a negative electrical charge while each strand of hair takes on a positive charge. Since similar electrical charges repel each other, the individual strands of hair will stand up in an attempt to separate from neighboring strands with the same electrical charge. Thus, we can say that the comb is negatively charged and the strands of hair are positively charged. When dealing with electrical charge, it is important to always remember that overall electrical charge is conserved. By this we mean that, charge is neither created nor destroyed, but rather moved from place to place. As we saw in the comb example, the act of combing your hair can result in electrons being physically removed from the strands of hair and deposited on the comb. This gave the comb a net negative charge and the hair a net positive charge. Thus, the comb's negative charge gain is offset by the hairs' equal positive charge gain, and the total charge remains constant. This is known as the **law of conservation of charge** and is one of the fundamental laws of nature.

With all this talk about charge, you may be asking yourself: What do we mean exactly by "charged" and can we quantitatively measure charge? By charged, we simply mean that a given object has a surplus of a specific (either positive or negative) electrical charge carrier. The total number of these charge carriers that are present on the object gives us a quantitative value for the

---

Definition: Electrical Charge

Electrical Charge, denoted by Q, can be thought of as the total number of excess charge carrying particles (protons or electrons) present in an object. Charge can be either positive or negative and is usually measured in coulombs. The **fundamental unit of charge**, e, is the charge associated with a single electron or proton and is equal to 1.60×10^{-19} coulombs.

Definition: Coulomb

The Coulomb is the basic unit of electrical charge. Denoted by C, one Coulomb of charge consists of approximately 6.24×10^{18} electrons (or protons).

---

charge (Charge is typically denoted by the capital letter Q). In other words, if an object has 2 million excess electrons it will have a larger negative charge than an object with 1 million excess electrons. As you might imagine, the charge carried by a single electron or proton is very small. So, in order to make the numbers a bit more manageable, we use a much larger unit called a coulomb (C) when referring to electrical charge. A coulomb consists of approximately 6.24×10^{18} electrons. In other words, there are over 6 billion billion (six quintillion) electrons in a coulomb. Using this information, we find that a single electron must have a charge of 1.60×10^{-19} coulombs. This value is known as the **fundamental unit of charge** and is denoted by the symbol e. Since protons and electrons have the same quantity of charge, we simply use the sign to determine which particle we are discussing. A proton's charge is represented as e and an electrons charge is $-e$. We can also relate the number of excess electrons in an object to the total net charge by using the following equation:

$$Q = \frac{\text{number of electrons}}{6.24 x 10^{18} \text{ electrons/C}}$$

Example 4.1

Problem: How many coulombs of negative charge do 756.23×10^{17} electrons represent.

Solution: Here we simply divide the number of charge carriers (electrons) present by the number of charge carriers per coulomb. We can use the equation that we defined above to do this.

$$Q = \frac{\text{number of electrons}}{6.24 x 10^{18} \text{ electrons/C}}$$

$$Q = \frac{756.23 x 10^{17} \text{ electrons}}{6.24 x 10^{18} \text{ electrons/C}}$$

$$Q = 12.12 \, C$$

So, 756.23×10^{17} electrons is approximately equal to 12.12 coulombs.

Example 4.2

Problem: Approximately how many electrons would there be in 7 coulombs of
negative charge.

Solution: Here we are given the amount of charge present and asked to
determine how many electrons we are dealing with. If we rearrange the formula
used in example 4.1 we can write:

$$\text{number of electrons} = 6.24 \times 10^{18} \text{ electrons/C} * Q$$

Plugging in to this equation, we get:

$$\text{number of electrons} = 6.24 \times 10^{18} * 7$$

$$\text{number of electrons} = 4.368 \times 10^{19} \text{ electrons}$$

So, we find that 7 coulombs consists of approximately 4.36×10^{19} electrons.

Historical Note: Charles Augustin Coulomb

The coulomb is named after the French military engineer and
scientist Charles Augustin Coulomb (1736 – 1806). Coulomb is best
known for developing the inverse square law governing the force
exerted between two separated electrical charges.

4.2.2   Electrical Forces and Fields

As was mentioned before, electrical charges exert forces on each other. These forces tend to
repel objects with a similar charge and attract objects with different charge. The strength of these
forces is dependent on several factors. These include the amount of charge, the distance of
separation, the shape of the objects, and the type of material separating them. If we assume that
we are dealing with objects that have a size that is much smaller than the distance of separation,
we can ignore the complication introduced by object shape and assume we are working with point
charges. By making this assumption, we can represent the force between two charged particles
with the following equation, known as Coulomb's Law:

$$\mathbf{F} = k \frac{q_1 q_2}{r^2} \hat{\mathbf{r}}_{12}$$

In this equation, F is the force between the charges, q_1 and q_2 are the charges at the two points, r
is the distance of separation, k is a proportionality constant, known as **Coulomb's constant**, and
$\hat{\mathbf{r}}_{12}$ is a unit vector pointing from q_1 to q_2. For charges in a vacuum, we can assume $k=8.99 \times 10^9$

Definition: Electrostatic Force

Electrostatic Force is the force exerted by one body of charge on another. This force can be calculated for simple point charges by using the following equation.

$$\mathbf{F} = k\frac{q_1 q_2}{r^2}\hat{\mathbf{r}}_{12}\ \text{N}$$

In this equation, F is the force between the charges, q_1 and q_2 are the charges at the two points, r is the distance of separation, k is a proportionality constant, known as **Coulomb's constant**, and $\hat{\mathbf{r}}_{12}$ is a unit vector pointing from q_1 to q_2. Like other forces, electrostatic force is reported in Newtons.

Definition: Coulomb's Constant

Coulomb's constant, k, is a constant of proportionality that is used in the calculation of electrostatic force. Its value is dependent on the type of material that is separating the charges. In the case of two points in a vacuum, we use $k=8.99 \times 10^9\ \text{Nm}^2/\text{C}^2$.

Nm^2/C^2. If we are concerned only with the magnitude of the force between the two points, we can ignore the unit vector and write our equation as follows:

$$F = k\frac{q_1 q_2}{r^2}$$

With this basic information, we should be able to calculate the force between two point charges in space.

Example 4.3

Problem: Given two point charges in a vacuum, $q_1=1.5$ coulombs and $q_2=2$ coulombs, what magnitude of force will be exerted between the charges if they are separated by a distance of 5 centimeters? What happens if the distance is doubled to 10 centimeters? What happens if the first charge is negative?

Solution: We have been given all the information that we need to solve the problem. We simply plug our values into the force equation and compute the answer.

$$F = 8.99x10^9\ \frac{Nm^2}{C^2}\frac{(1.5C)(2C)}{(0.05m)^2}$$

$$F = 8.99x10^9\ \frac{Nm^2}{C^2}\frac{3C^2}{0.0025m^2}$$

$$F = 1.08x10^{13} N$$

So, we find that two point charges exert a very strong repulsive (positive) force on each other. Now we will increase the distance to 10 centimeters.

$$F = 8.99x10^9 \frac{Nm^2}{C^2} \frac{3C^2}{0.01m^2}$$

$$F = 2.70x10^{12} N$$

Here we find that by doubling the distance between the charges, we have reduced the repulsive force by a factor of four, just as the inverse square law predicts. Now, we will change the sign of the first charge to make it negative.

$$F = 8.99x10^9 \frac{Nm^2}{C^2} \frac{(-1.5C)(2C)}{0.0025m^2}$$

$$F = 8.99x10^9 \frac{Nm^2}{C^2} \frac{-3C}{0.0025m^2}$$

$$F = -1.08x10^{13} N$$

As expected, giving the two points opposite charges resulted in an attractive (negative) force between the two points.

So far, we have discussed the forces that can be generated between two charges or bodies of charge that are fixed in place. Now, what would happen if we made the second point charge moveable? We could then use this second point to measure the force exerted by the fixed point on our movable point at a large number of locations around the fixed point. This information would give us a good picture of what magnitude and direction of force a particle would experience in regions around the fixed charge. The problem is that, this would give us information for only a particle identical in charge to our movable test point. To generalize the information, we need to divide the force at each location by the charge of our test point. By doing this we get a value of proportionality between the force exerted on the particle and its charge. If we call our movable test point q_0, we can express this proportionality value as:

$$\mathbf{E} = \frac{\mathbf{F}}{q_0}$$

Where $\mathbf{E}$ is the value of proportionality, $\mathbf{F}$ is the force vector acting on the test charge, and q_0 is the charge of the test point. Now, if we want to introduce a particle with a different charge, q_1, we can find the force between this new particle and the fixed point simply by multiplying the proportionality value by the charge of the new particle.

$$\mathbf{F}_1 = q_1 \mathbf{E}$$

Definition: Electric Field

All electrical charges or bodies of charge produce an electric field. It is this field that exerts forces on other charged particles. The value of the electric field at a given point is a proportionality value that describes the force that will be exerted on a charged object per unit of charge (N/C). For a given positive test charge q_0, the electric field strength is given as:

$$E = \frac{F}{q_0}$$

So, if we measure the force generated between a fixed charge point and a test point at all possible locations around our fixed point charge (This would be a infinite number of points) and determine the proportionality value by dividing the force by the charge, we will have a continuous field of values describing the forces that a charged particle will experience around our point charge. This field is known as the **electrical field**, or E-Field of the fixed point. Using this field information, we can determine the forces exerted on any charged particle at any point within the field.

Example 4.4

Problem: While trying to determine the electric field in a particular area of space, you place a test charge of 3-nC at that point and measure the force exerted on your test charge to be 7.5×10^{-3} N in the *y* direction. What is the electric field at your test point? What force would be exerted on a single electron placed in the same location?

Solution: From our definition of the electrical field, we know that we can find our solution by dividing the force exerted on our test charge by the charge of the test particle. Doing this yields:

$$E = \frac{(7.5 \times 10^{-3}\,\text{N})\mathbf{j}}{3 \times 10^{-9}\,\text{C}} = (2.5 \times 10^6\,\text{N/C})\mathbf{j}$$

Here we see that the electric field points in the *y* direction just as our force did. Now, let us determine the force on a single electron placed in the same position as our test charge. Since we already know the value of the electric field at this point, finding the force on the electron is as simple as finding the product of *–e*, the charge of an electron, and **E**, the electric field value at the test point.

$$\mathbf{F} = (-e)(\mathbf{E}) = (-1.6 \times 10^{-19})(2.5 \times 10^6\,\text{N/C})\mathbf{j}$$

$$\mathbf{F} = (-4 \times 10^{-13}\,\text{N})\mathbf{j}$$

Thus, once the electric field has been determined, it is a simple task to find the force acting on our electron, or any other charge for that matter. As was expected, the force acting on the electron is an attractive force. This is opposite of the repulsive force experienced by the initial positive test charge.

In many cases, you will find the electric field of a charge or system of charges depicted graphically as a series of lines. Figure 4.4 illustrates the electric field around a single positive point charge. Note that the field is represented by directed lines pointing away from the point charge. If the point charge had been negative, the lines would point in the opposite direction. The closer the lines are to each other, the stronger the field is at that point. When looking at illustrations of electrical field lines, it is helpful to keep the following guidelines in mind.

- Field lines for an electric field will always begin on a positive charge and end on a negative charge. If there is no negative charge to terminate on, the lines will continue out to infinity.

- The number of field lines exiting a positive charge or entering a negative charge will be proportional to the size of the charge.

- The density of the field lines at a given point is proportional to the magnitude of the field at that particular point.

- Field lines cannot cross.

Knowledge of the electric field allows us to determine other important information as well. For example, let us say we have a fixed negative point charge. Now, we introduce a free positive charge near the fixed negative charge. Obviously, we will expect the negative fixed charge to draw the free positive charge toward it, but how will the free charge move and how much work is required to move it? In a situation like this, the electric field of the point charge behaves much like the gravitational field of the earth. If you raise an object to a given height and then drop it, the earth's gravitational field does work on the object to draw it to the ground. Obviously, as the object is drawn closer to the earth, the gravitational field will exert a greater force. However, when dealing with gravity near the surface of the planet the increase is so small that it is often ignored. Thus, if you drop a book from your desk, the gravitational force acting on it for the short distance between the desktop and the floor is assumed to be a constant value equal to the mass of the book multiplied by the value of acceleration due to gravity (typically 9.81 m/s^2).

Here we can immediately draw some parallels between the way objects behave in the earth's gravitational field and the way charged particles behave in an electric field. If we

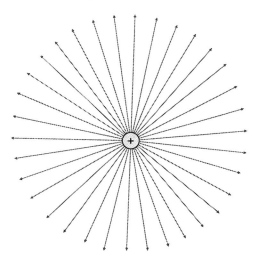

Figure 4.4. An illustration of the electrical field lines around a single positive point charge.

assume that the center of the earth is equivalent to a fixed negative point charge and that all the objects on earth surrounding that core are positively charged, the systems will work in much the same way. In this system, the acceleration due to gravity at a given point would be equivalent to the electric field value and the mass of the object would be equivalent to its amount of positive charge. So, the downward force exerted on an object due to gravity is much like the attractive force exerted by a fixed negative charge on a free positive charge (It is important to note that while the behavior of these forces is similar, electrical forces are far stronger than gravitational forces).

If we continue our comparison, we find that many of the relationships describing motion of objects in a gravitational field also work in a similar way in an electric field. If we think back to some of these basic relationships we will recall that an object in a gravitational field possesses two basic types of energy, that is kinetic energy, or energy of motion and potential energy. For example, if we consider the floor to be our reference point and we have a book suspended 2 meters above the floor, the book will have no kinetic energy since it is at rest. However, it has a potential energy $U = mgh$, where U is the potential energy, m is the book's mass, g is the acceleration due to gravity (gravitational field value), and h is the book's height above the reference point (the floor). Now, if the book is released from its resting position, it will begin to fall (the potential energy becomes kinetic energy). The reason that the book falls is because the gravitational field of the earth is acting on it and performing work equivalent to $W = -mgd$, where W is the work performed by the gravitational field, m is the mass of the object, g is the gravitational field value, and d is the distance that the book falls. From this, we can see that an object's potential energy is simply equal to the negative of the work done by the gravitational field to move the book from its resting point to the floor.

Now let's consider a similar scenario in an electric field. If we have a fixed negative point charge and a free positive charge, how can we determine the free charge's electrostatic potential energy? First, we recall that in a gravitational system, potential energy was calculated as $U = mgh$. From our discussion, we also discovered that mass in a gravitational system was similar to charge in an electrical system and that the gravitational field value was similar to the electrical field value. Now, if we consider the distance between the fixed point charge and our free charge to be equivalent to the height of an object in a gravitational system, we can write:

$$U_{ES} = q_0 \mathbf{E} r$$

where U_{ES} is the **electrostatic potential energy**, q_0 is the charge of the free point, $\mathbf{E}$ is the electric field value at the free point's position, and r is the distance between the fixed and free charges. So, as the electric field does work on the free charge to draw it closer, the electrostatic potential energy will decrease. Electrostatic potential energy is a useful quantity, but it gives us information for just the charges involved at the time. A more useful measure called the **electric potential** is

---

Definition: Electrostatic Potential Energy

Electrostatic Potential Energy is the potential energy possessed by a given charged particle or object at a specific point in space. It depends on the charge of the object, q_0, the electric field strength at the object's position, $\mathbf{E}$, and the object's distance, r, from the reference point (location of zero electrostatic potential energy).

$$U_{ES} = q_0 \mathbf{E} r$$

Definition: Electric Potential

Electric Potential is the measure of electrostatic potential energy per unit charge for a given point in space.

---

obtained by dividing the electrostatic potential energy by the charge q_0, thus giving us a measure of electrostatic potential energy per unit charge. Just like the potential energy of an object in a gravitational field decreases as it nears the reference point, electric potential will decrease as the free positive charge gets closer to the fixed negative charge. So, if we measure the potential of the free charge at two different locations, we will get two different values for its electric potential. The difference between the initial electric potential and the final potential measurements is known as **potential difference** and is measured in joules per coulomb, or **volts**. Thus, the potential difference is commonly referred to as **voltage**. If you think about a common 9-volt battery, the positive terminal will have an electric potential that is 9 volts higher than the negative terminal. Since charges want to move from an area of high potential toward one of lower potential, voltage can also be referred to as the **electromotive force**, or EMF. In other words, it indicates how strongly motivated the charge is to move from its current position to the final position.

Another way to picture voltage is to think of it as the pressure that is being placed on electrons in an attempt to move them from one point to another. If we think of electrons as water in a pipe, voltage would be similar to the pressure that is placed on that water by a pump.

4.2.3   Electric Current

At this point, we should be comfortable with several basic aspects of electricity. First, we are aware of the concept of charge. We know that a quantity of charge, measured in coulombs, consists of a large number of small charge carrying particles called electrons and protons. We

---

Definition: Volt

The volt is the unit of measure for both electric potential and potential difference. Since both of these measures involve energy per unit charge (joules per coulomb), one volt is simply equal to one joule per coulomb. The unit of the volt is named after Alessandro Volta, an Italian scientist

$$1\,Volt = \frac{1\,Joule}{1\,Coulomb}$$

Definition: Voltage

Voltage, also known as potential difference or electromotive force, is the difference in potential between the starting and ending positions of a quantity of charge. Voltage measures the motivation or "push" behind charge carriers that move within an electric field (charge carries want to move from areas of high potential to areas of lower potential). Voltage can also be defined as the negative of the work done by the electric field to move a quantity of charge between two points.

Historical Note: Alessandro Volta

The volt is named after the Italian scientist Alessandro Volta (1745 – 1827). His many accomplishments include the construction of a device to generate static electricity, and the discovery of methane gas. His biggest contribution to the field of electronics was the development of the first battery in 1800.

also are aware that areas of charge are capable of exerting forces on one another through their electric fields. Finally, we know that a charge in an electric field can possess a varying amount of electric potential based on its position in the field. This difference in electric potential or voltage is the motivation or "push" behind the movement of charges in an electric field. So, now that we know that forces can push quantities of charge around, our next concern should involve quantifying this charge flow. How do you count or measure something that is moving? The simple answer is: By counting what passes a fixed point. This is how the flow rates of both electrical charge and water are measured. In the case of water, think about your kitchen sink. If you want to know what kind of flow rate your faucet is capable of, how could you measure it? One simple method would be to place a container under the faucet. The faucet could then be turned on for a specified period of time. Once you turn off the faucet, you can measure the water that has collected in the container and determine how much water has flowed through your faucet in the given amount of time. Let us say that you tried this experiment. The faucet is turned on for 5 minutes and, during that time, 25 liters of water is collected in the container. Now, if we divide the amount of water collected by the amount of time the faucet was open, we get 5 liters per minute. This is the flow rate of the faucet in this experiment.

So, what does this have to do with electricity? Well, as we have seen previously, the way in which we measure quantities of water and electricity can be quite similar. Let us say now that instead of a kitchen sink we have a piece of wire that is stretched between the two terminals of a

Definition: Electric Current

Electric Current, symbolized by I, is the quantitative measure of the flow rate of electric charge carriers. It is measured by determining the number of coulombs of charge that pass a specific point in the period of one second. The unit of electric current is the Ampere.

Definition: Ampere

The ampere, or amp for short, is the unit of measure for electric current. One ampere of current is equivalent to one coulomb of charge passing a given point in one second. The unit of the ampere is named after the French scientist André Marie Ampère.

$$1 \, Ampere = \frac{1 \, Coulomb}{1 \, Second}$$

Historical Note: André Marie Ampère

The ampere is named after the French scientist André Marie Ampère (1775 – 1836). He developed an important early theory of electricity and magnetism in 1820, and was the first person to build a device to measure the flow of electrical charge (electric current).

battery. We know that, much like a pump pressurizes a pipe and causes water to flow, the potential difference (voltage) between the terminals of the battery causes charge to flow through the wire, but how do we quantify this charge? If we think back to the kitchen sink example, we found the flow rate of our sink by measuring how much water moved past the end of the faucet in a given amount of time. Likewise, we measure the flow of charge in a wire by determining how many charge carriers pass a given point in a set period of time. More specifically, we measure how many coulombs of charge pass a given point in one second. This flow rate is known as **electric current**, I, and it is measured in units called amperes, A.

$$i = \frac{dq}{dt}$$

Where i is the electric current, q is charge that moves past a given point, and t is the time required for that charge to move. One ampere of electric current is equivalent to one coulomb of charge passing a point in the period of one second.

Example 4.5

Problem: A piece of wire is connected across the terminals of a battery for a period of 3 seconds. During this time, it is determined that 18 coulombs of charge move through the wire. What is the current in the wire while it is connected to the battery terminals?

Solution: Here we are given a period of time and an amount of charge that has been moved. To find the current in the wire, all that is required is to plug into our definition of current and determine how much charge is moving per second. Thus, we can write:

$$I = \frac{18C}{3Sec} = 6A$$

So, in this case, we find that the current is 6 amps.

4.2.4   Electrical Conductance

Up to this point, we have been talking about electric charge either running through a metal wire or isolated in a vacuum. Obviously, electric charge is capable of traveling through many different materials. The ease at which charge carriers can move through these different materials is determined by the material's **conductance**. Some materials allow charge to flow easily, while

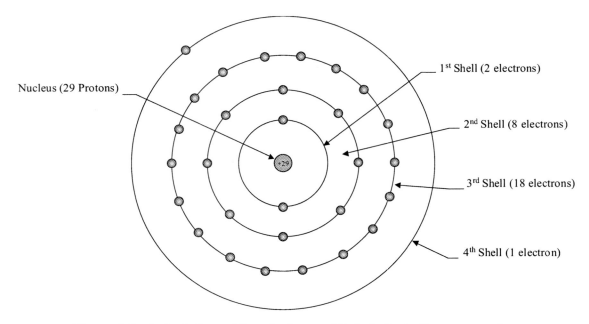

Nucleus (29 Protons)

1st Shell (2 electrons)

2nd Shell (8 electrons)

3rd Shell (18 electrons)

4th Shell (1 electron)

Figure 4.5. A depiction of the electron orbital shells in the copper atom.

others require much more effort. We call materials that allow for the easy flow of charge carriers **conductors,** and materials that impede this flow are called **insulators**. Since, in most cases, the charge carrier that is traveling through the material is an electron, a good indicator of a material's ability to conduct electricity is how easily it can gain and lose electrons.

To understand this concept, we must once again turn our attention to the atom. Figure 4.5 displays the atomic structure for one of the most common conductors, copper. As was mentioned earlier, the atom consists of a nucleus containing particles with positive and neutral charge (protons and neutrons respectively) and an outer cloud of particles with a negative charge (electrons). These electrons are organized into a series of orbital shells based on their distance from the nucleus. The further an electron is from the nucleus, the higher its energy. The outermost shell of electrons is called the **valence shell**. These electrons are the ones that interact with other atoms to form bonds during chemical reactions. They are also the electrons that are capable of most easily escaping from the atom and becoming free electrons. As we can see in figure 4.5, the copper atom has a single electron in its outer shell. Since it is easier to ionize atoms with a small number of electrons in the valence shell, atoms with one to three valence electrons will usually be conductors. The fact that the copper atom has only one valence electron indicates that it should be a very good electrical conductor. In fact, it takes only a sufficient amount of thermal energy (heat) to free the outer electron from copper and produce a free electron. Copper at room temperature contains a very large number of these free electrons, making the establishment of a current in copper an easy task. Most metals have a similar electron

---

Definition: Conductance

Conductance is a material's ability to sustain an electric current. The higher the conductance of a material, the less potential difference is required to maintain the current.

configuration (a small number of valence electrons) and, thus, make good conductors. Silver is the best conducting metal and copper works almost as well. We use copper in most of our wires simply because it is cheaper and more abundant than silver.

On the other hand, the electron bonds in materials like wood, rubber and glass, are much stronger. Thus, a very large potential difference is required to ionize, or pull the electrons away from the atoms in these materials. Materials such as these are called insulators. Insulators are often used to prevent the flow of current into areas where it is not wanted. For example, the copper wires in your home are insulated with a plastic or rubber coat to prevent contact with other conductors and, of course, to prevent electrical shocks. If you were to look at the valence shell of the atom of an insulator, you would typically find five or more valence electrons.

While most materials can be classified into one of the previous two groups, there are some materials that do not belong in either category. To accommodate these additional materials, there exist two additional categories. The first of these is the **semiconductor**. As the name suggests, this group of materials has a conductance level that lies between conductors and insulators. Semi-conducting materials, such as silicon and germanium are usually poor electrical conductors, but under certain special circumstances, these materials can be useful conductors. In fact, most modern-day electronics, such as the computer, radio, and television, are based on semi-conductor technology (Diodes, Transistors and Microchips).

The final category is for materials that are perfect conductors of electricity. In other words, it is possible to maintain a current in these materials without applying a constant potential difference. Once a current is created in a closed loop of this type of material, it will continue to flow forever. Such materials are very rare and are typically only possible in extremely cold environments. Materials capable of these properties are called **superconductors**. While superconductors have great promise for application in electrical systems, their difficult construction and low operating temperatures currently prohibit their widespread implementation. Table 4.4 summarizes the four basic conductance categories.

Table 4.4. An overview of the four conductance categories.

Category	Description	Examples
Insulator	A material that restricts the flow of current. Large potential differences are required to push electrical current through these materials.	Wood, Rubber, Plastic, and Glass
Semiconductor	A material that is usually viewed as a poor conductor, but under special circumstances it can be a useful conductor	Silicon and Germanium
Conductor	A material that allows for the easy establishment of a current with a minimal applied voltage	Silver, Copper, Aluminum, and most other metals
Superconductor	A material that is capable of conducting electrical current with no resistance when placed below a certain temperature. This means that electrical currents can exist without the presence of an electric field.	Mercury below 4.15K and barium-yttrium-copper-oxide below 92K K = Degrees Kelvin

4.2.5   Resistance

Now that we understand conductance and how the properties of a material can determine how easily a current can flow, we can look at the opposite of conductance, **resistance**, R. While conductance was the degree to which a material allowed current to flow through it, resistance is the degree to which a material opposes current flow. Thus, materials with high conductance have

---

Definition: Resistance

The opposite of conductance, resistance is a measure of a material's opposition to current flow. The higher the resistance of a material, the more potential difference is required to maintain a current. Resistance is measured in units called ohms.

---

low resistance and those with low conductance have high resistance. This resistance is measured in units called ohms, Ω.

As was the case with conductance, the resistance of a material is related to the material's atomic structure. If there is not an abundance of free electrons or electrons that can be easily removed from their orbits, a large voltage will be required to move electric charge carriers through the material. All materials with the exception of superconductors, possess at least a small amount of resistance. In the case of short segments of copper wire, the resistance is so small that it is usually ignored. However, in the case of less conductive materials and extremely long lengths of wire, resistance must be taken into account.

To get a little better picture of how resistance influences the flow of electrons, we will return to our water analogy. Take a look at the pipe shown in Figure 4.6. The beginning and end of this pipe have a large diameter and, thus, these portions of the pipe do not restrict the flow of water to a great extent. You will see that the center of the pipe, however, has a much narrower diameter. Thus, when compared to the other sections of the pipe, it will have a much higher resistance to water flow. Now, if we connect a pump to one end of this pipe and push water through it, what do you think will happen? As you probably expect, the water will flow easily through the first section of the pipe. At the smaller center section, however, it will become harder to move water through the pipe. As a result, pressure will build up behind this section and the water will be forced through the smaller section. When the water reaches the end of the small section, there is suddenly plenty of room again. Thus, the pressure on that end of the pipe will be lower. Now, if the pump simply provides a constant pressure of water to the pipe, the smaller center section of the pipe will limit the overall flow rate of the water. If, on the other hand, the pump can be adjusted to ensure a constant flow rate, the pressure difference between the beginning and end of the pipe will increase in proportion to the desired rate of flow.

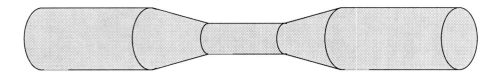

Figure 4.6. This water pipe illustrates the concept of resistance. The smaller center section of the pipe has a larger resistance to water flow.

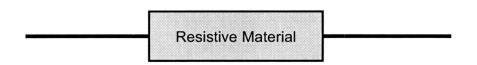

Figure 4.7 The electrical equivalent of the water-based example given in figure 4.6. A resistive material connected to two conducting copper wires.

Definition: Ohm's Law

Ohm's Law gives a relationship between a material's resistance, R, the voltage across it, V, and the current flowing through it, I. This relationship can be found written in any of the following three arrangements:

$$R = \frac{V}{I} \qquad I = \frac{V}{R} \qquad V = IR$$

Historical Note: Georg Simon Ohm

The unit of resistance gets its name from Georg Simon Ohm (1787 – 1854). Born in Bavaria in 1787, Ohm is credited with the work that he did in the formulation of a relationship between resistance, current, and voltage. This relationship, also named after him, is known as Ohm's law.

Electrical resistance can be viewed in much the same way. If we have two pieces of copper wire connected to a material with a high resistance, as shown in figure 4.7, we will find that if we connect a fixed potential difference to the copper wires, the resistive material will limit the current that can flow through the wire. In fact, we will find that as the resistance increases, the current will decrease. Likewise, if we have a material of a given fixed resistance, we will find that the current will increase as we increase the voltage. This relationship between voltage, current, and resistance is known as **Ohm's Law**. It can be written in any of the following forms:

$$R = \frac{V}{I} \qquad I = \frac{V}{R} \qquad V = IR$$

where R is the resistance of the material, V is the potential difference or voltage between the two ends of the material, and I is the current running through the material. By using Ohm's law, we can easily determine either the voltage, current, or resistance as long as we have been given, or are able to measure, the other two quantities. Let's look at an example:

Example 4.6

Problem: You have just applied a voltage of 7 volts to a material with a total resistance of 25 ohms. How much current should be running through the material? What voltage would be required to achieve a current of half an ampere?

Solution: For this problem, we simply apply the second form of Ohm's law given above and solve for our current. Doing so yields:

$$I = \frac{V}{R} = \frac{7V}{25\Omega} = 0.28A = 280mA$$

So, we find that in this case, our current is 280 mA. Now, if we keep our resistance constant, we want to figure out what voltage we need to achieve one ampere of current. To solve this problem, we can use the third form of Ohm's law given above. Doing this yields:

$$V = IR = (0.5A)(25\Omega) = 12.5V$$

Thus, 12.5V would be required to achieve a current of half an ampere through a resistance of 25 ohms.

4.2.6 Capacitance

Earlier in this chapter, we discussed the concept of an electric field. We found that a charge can possess different levels of electric potential based upon its location in the electric field. Thus we learned that this difference in potential, or voltage, is what causes charge to move, producing a current. Another useful application of the electric field is the temporary storage of energy. This is achieved through the use of a special device called a capacitor. So, what is this device, and how does it use an electric field to store energy? The capacitor is actually a surprisingly simple device. Until now, we have looked at what happens when a material, such as a metal wire connects two areas of differing potential. We know that, based upon the potential difference and the resistance of the material, a certain amount of current will flow. However, what would happen if, instead of connecting the points of differing potential directly to each other, we instead connect each point to a metal plate. These plates do not connect, but instead are separated by an insulating material, called a **dielectric** (for example, air), as shown in Figure 4.8. Since the two plates are insulated from each other, a current cannot flow. Instead, one metal plate will gain an excess of free electrons and become negatively charged, while the other plate will gain a net positive charge. This difference in charge continues to build until the potential difference between the two plates is equal to the potential difference between the points to which the plates are connected. For example, if this configuration is attached to the terminals of a 9 volt battery, the potential difference between the plates will increase until it reached 9 volts. Since the plates are not connected, and thus cannot pass charge between each other, an electric field forms between them, as shown in Figure 4.9.

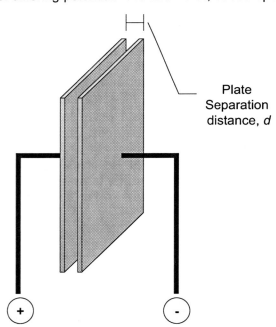

Plate Separation distance, *d*

Figure 4.8. A conceptual example of a basic capacitor. A capacitor consists of two metal plates separated by an insulating dielectric material.

So, by storing electric charge carriers on the plates of the capacitor, an electric field is formed. These charge carriers will remain on the plates of the capacitor, even after they are disconnected from the source of potential difference (the battery for example). Thus, the electric field between the two plates stores energy until such a time as the plates are connected. When this occurs, the electric field collapses and the negative charge carriers (free electrons) flow to the opposite, positively charged plate and neutralize the charge difference.

Now that we know how a capacitor stores energy, is there any way we can measure some of the aspects of a capacitor? The answer to that question is: yes. In fact, there are two measurements that are of interest when dealing with capacitors. These are the amount of charge stored on the plates of a capacitor and the amount of energy that is stored in the electric field between the plates. We will begin by looking at the capacitors ability to store charge.

As we have just seen, when a voltage is applied to the two plates of a capacitor,

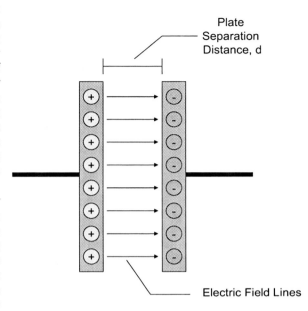

Figure 4.9. An illustration of the electric field lines existing between the plates of a capacitor. The energy that is stored in a capacitor is stored in this electric field.

some quantity of charge is placed on the plates. The ability of a capacitor to store these charge carriers is described in a measure called **capacitance** (C). Capacitance, measured in units called farads (F), basically describes the number of coulombs of charge that can be placed on a capacitor's plates per unit of voltage applied to the capacitor. Thus, if the application of one volt of potential difference to the plates of a capacitor results in one coulomb of charge being stored, we can say that the capacitor has a capacitance of one farad. In reality, a capacitor with a capacitance this high is rarely seen. Most electrical devices contain capacitors with much smaller capacitance ratings, usually on the order of microfarads (μF) and picofarads (pF). The formula for capacitance of a capacitor (C) in terms of the voltage applied (V) and the charge stored (Q) can be expressed as:

$$C = \frac{Q}{V}$$

Definition: Capacitance

Capacitance, simply stated is the amount of charge that a capacitor is capable of holding per unit of voltage applied. From this relationship, we can write an important equation relating voltage, charge, and capacitance:

$$C = \frac{Q}{V}$$

Where Q is the charge, V is the voltage, and C is the capacitance.

Historical Note: Michael Faraday

The unit of capacitance, the Farad, gets its name from the English physicist and chemist Michael Faraday (1791 – 1867). Faraday made huge contributions to the field, particularly in the area of electromagnetics. He discovered that electricity could be produced my moving a magnet in a coil of wire. In addition, he built the first electric generator, the first electric motor, and the first transformer.

The physical construction of the capacitor has a large impact on its overall capacitance. There are several factors that play a role in determining the capacitance value. These include the area of the metal plates, the distance the plates are separated, and the type of material that is separating the plates. We will look at how each of these contributes to the capacitance value, starting with the plate area.

The area of the plates of the capacitor determines how much space exists to cram charge carriers onto. Thus, as you would expect, it is easier to fit a large number of charge carriers on a large capacitor plate than it is to fit them on a small one. As the size of the plate shrinks, more voltage is required to cram them into the tiny area available. Thus, the capacitance of a capacitor is directly proportional to the area of its plates.

Another important factor is the distance of separation between the plates. As the separation between the plates decreases, the amount of voltage needed to achieve a specific charge decreases, and, thus, the capacitance of the capacitor increases. As the distance grows larger, the required voltage grows, and, the capacitance value will decrease. So, we find that the capacitance of a capacitor is inversely proportional to the distance between the plates.

Finally, the material that separates the plates is of importance. The material between the plates, called a dielectric, serves two purposes. First, by placing a better insulator than air between the plates, we can bring them closer together without worrying about the voltage breaking down the material and shorting out the two plates. All dielectric materials have a given voltage that, when exceeded, results in the material breaking down and conducting electricity. If we can use a material that requires a larger voltage per unit distance to break down, we can move the plates closer together or increase the maximum operating voltage of the capacitor. The breakdown voltages of dielectric materials are usually given in volts per mil (a mil is equal to 0.001 inches or around 0.0254 mm). The breakdown voltages of some common materials can be found in Table 4.5. Secondly, a dielectric material placed between the plates of a capacitor has the ability to weaken the electric field between the two plates because the dielectric material will form an electric field in opposition to the one created between the capacitor plates. By weakening the electric field, the voltage required to place a given amount of charge on the plates is reduced, and,

Table 4.5. Dielectric constants and breakdown voltages for some common materials.

Material	Dielectric Constant, κ	Breakdown Voltage/mil
Air	1.00059	80
Glass (Pyrex)	5.6	360
Mica	5.4	3800 - 5600
Mylar®	3.2	7500
Neoprene	6.9	300
Paper	3.7	410
Plexiglass	2.8	990
Quartz	3.8	1000
Styrofoam	2.6	500 - 700
Teflon®	2.1	1000 - 2000

thus, the capacitance value of the capacitor increases. This increase in capacitance is determined for each material in terms of a value called a **dielectric constant**. The dielectric constant of a vacuum, for example, is exactly equal to one, and the constant for air is very close to one (it is usually assumed to be one). So, the dielectric capabilities of other materials are often expressed relative to the dielectric constant of a vacuum. Thus, a material, such as mica, with a dielectric constant of 5.4, provides a capacitance that is 5.4 times better than what could be achieved using air. Some common dielectric constants can be found in table 4.5. It should be noted that these dielectric constants have no units. This is because dielectric constants are reported as ratios. These ratios allow us to describe the benefit of the material without having to write the actual, more complicated value. For example, the true dielectric value for a vacuum is 8.85×10^{-12} F/m. So, to make things easier, we report the dielectric values of other materials as a ratio of their true dielectric value over the dielectric value of a vacuum (that is why the dielectric constant ratio of a vacuum is one).

As we have just seen, the capacitance of a capacitor relies on several of its physical characteristics. As a result, we can use the relationships between each of the physical characteristics and their effects on the overall capacitance of the capacitor to arrive at an expression for capacitance in terms of the physical aspects of the device. Such an equation would be written as follows:

$$C = \frac{A\kappa(8.85 \times 10^{-12}\,\text{F/m})}{d}$$

where A is the area of the plates, κ is the dielectric constant of the dielectric material between the plates, and d is the distance that the plates are separated. Now we have a relationship that can be used to determine the capacitance of any parallel plate capacitor given its physical characteristics.

Definition: Capacitance Based on Physical Traits

The capacitance of a capacitor can be determined based on the physical traits of the capacitor. The size of the plates, the distance of separation, and the type of material between the plates all play a role in this determination. The following equation describes the relationship of these attributes in the calculation of the capacitance:

$$C = \frac{A\kappa(8.85 \times 10^{-12}\,\text{F/m})}{d}$$

Where C is the capacitance, A is the plate area, κ is the dielectric constant, and d is the distance between the plates.

Example 4.7

Problem: You have just constructed a parallel plate capacitor. You used two square metal plates with each side measuring 20 centimeters. You decide to separate the plates by 1 millimeter. What is the capacitance of your capacitor? What would the capacitance be if you fill the gap between the plates with Glass?

Solution: To find the capacitance, all you need to do is plug the given information into the equation for capacitance that was given above. Since we are dealing with square metal plates measuring 20 centimeters, the plate area will be $(0.2 \text{ m})^2 = 0.04 \text{ m}^2$. Our first calculation assumes that the plates are separated by air, so $\kappa \approx 1$. Thus, we find:

$$C = \frac{(0.04 \text{m}^2)(1)(8.85x10^{-12}\,\text{F/m})}{0.001\text{m}} = \frac{3.54x10^{-13}\,\text{Fm}}{0.001\text{m}} = 3.54x10^{-10}\,\text{F} = 354\,\text{pF}$$

So, the capacitance of our capacitor is 354 pF in this case. Now, let us see what happens when we insert a dielectric of glass between the plates. In this case, our equation becomes:

$$C = \frac{0.04 \text{m}^2(5.6)(8.85x10^{-12}\,\text{F/m})}{0.001\text{m}} = \frac{1.9824x10^{-12}\,\text{Fm}}{0.001\text{m}} = 1.9824x10^{-9}\,\text{F} = 1.98\text{nF}$$

In this case, the capacitance was increased to 1.98 nF. This value is 5.6 times the value achieved by using air. So, we see that the addition of a dielectric will indeed increase the capacitance.

As we mentioned before, the second important measure relating to a capacitor is the amount of energy that can be stored in the electric field between its plates. We already know that the electric field between the two plates is formed as the result of charge being moved. The negative charge carriers (free electrons) are moved from one plate to the other, resulting in one plate being negatively charged and the other being positively charged. The process of moving these charge carriers from one plate to the other required that some amount of work be done. Some of the energy that was used to perform this work is stored in the electric field between the two capacitor plates. The question then becomes, how much energy is stored in that field? To answer this question, we must look at the work that is done during the charging process. First, let us say that a given small amount of charge, q, is moved through a potential difference, V. The result of this is that, the potential energy of the charge is increased by the amount qV. Thus, as more charge is moved onto the capacitor, the voltage between the plates approaches the voltage applied to the device, and the electrostatic potential energy between the two plates grows. If we then think of q as the amount of charge transferred at some point during the charging of the capacitor, we can then use our relationship between charge, voltage, and capacitance to find that the potential difference between the plates at that point will be:

$$V = \frac{q}{C}$$

Now, if we assume that our capacitor has a fixed capacitance, C, we can plot our voltage, V, versus the amount of charge transferred at a given point, q. This plot is shown in Figure 4.10. As is shown in the figure, our capacitor begins with no net charge on its plates. However, as more charge, q, is transferred to the negative plate, the potential difference, V, rises. When our final amount of charge, Q, has been transferred, we have reached our maximum potential difference between the plates. The potential energy that we are trying to determine is simply the area of the

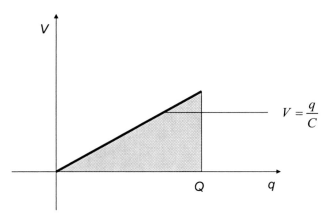

Figure 4.10. A plot of the charge vs. voltage relationship between the plates of a capacitor. The energy stored in the electric field of the capacitor is equal to the area under the triangle.

triangle shown in the figure. This triangle has a base of the final charge, Q, and a height of the final voltage, V. Since the area of a triangle is simply half the product of the base and the height, we find that the energy stored in the electric field of the capacitor, U, is equal to:

$$U = \frac{1}{2}QV$$

If we now note from our capacitance, voltage, and charge relationship that, Q = CV, we can rewrite this expression in the more common form of:

$$U = \frac{1}{2}CV^2$$

By using this expression, we can determine the energy stored in the electric field between the plates of a capacitor by using the easily measurable values of capacitance and voltage. The energy stored in the capacitor is reported in units of joules.

Example 4.8

Problem: You have been given a capacitor with a capacitance rating of 500 pF. How much energy is stored in this capacitor if it is charged by connecting it to a 9 volt battery?

Solution: We are given both the capacitance and the voltage of the capacitor in question. All that we need to do is plug these values into the equation for the energy stored in the capacitor. Doing this yields:

$$U = \frac{1}{2}CV^2 = \frac{1}{2}(500x10^{-12}\,\text{F})(9\text{V})^2 = 2.025x10^{-8}\,\text{J} = 20.25\,\text{nJ}$$

So, we have discovered that 20.25 nJ of energy are stored in the electric field of the capacitor.

Definition: Energy Stored in a Capacitor

The energy stored in a capacitor is stored in the electric field that develops between the plates. The energy is the result of the work done to move the charge carriers through a potential difference from one plate to the other when the capacitor is charged. The amount of energy stored, measured in joules, can be determined using one of the following two equations:

$$U = \frac{1}{2}QV \quad \text{or} \quad U = \frac{1}{2}CV^2$$

Where U is the energy stored, Q is the charge, V is the voltage, and C is the capacitance.

4.2.7   Inductance

As we just saw in the previous section, the electric field plays a critical roll in the storage of energy within a capacitor. Of course, throughout this chapter, we have seen that electric fields are fundamental to many aspects of electronics. However, by examining just the electric field, we do not see the whole picture. There is another field involved in electronics that always accompanies any changing electric field and is of equal importance. This field, known as the magnetic field, is caused by charges in motion (for example, a current running through a wire). Before we discuss the roll of the magnetic field in electronics, it is important to have a clear understanding of what this field is like and how it is produced.

If you have ever worked with a permanent magnet, as most of you probably have, you have dealt with a magnetic field. To better understand what is going on, we need to look at a magnet in more detail. A typical bar magnet consists of two poles, labeled north and

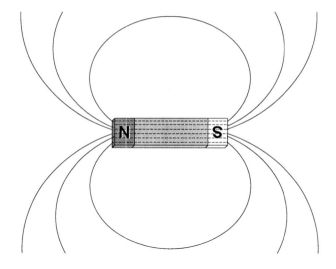

Figure 4.11. An illustration of the magnetic field of a typical bar magnet. Note that the magnetic field flows from the north pole to the south pole of the magnet and then back to the north pole through the material.

south. These poles exist at opposite ends of a bar of material (thus the term bar magnet). As is the case with electric charges, opposite values will attract and similar values will repel. Thus, we find that if we place the north poles of two magnets near each other, they will tend to push apart while the north and south poles will pull together. These forces are produced by the magnet's magnetic field. This field consists of lines of magnetic force, sometimes called flux lines that radiate in a circular pattern. These circular lines begin at the north pole of the magnet, move through space to the south pole, and then return to the north pole by passing through the magnetic material itself. This field is illustrated in Figure 4.11. In this figure, the number of lines drawn was reduced so that the flow of the lines of magnetic flux could be more clearly seen. Figure 4.12 more clearly illustrates the magnetic field around a bar magnet as evidenced by the positions of iron filings.

In addition to the naturally occurring magnetic fields found in permanent magnets, magnetic fields can be produced by quantities of moving electrical charge. For example, as a current runs through a piece of wire, a magnetic field is generated around the wire, as shown in Figure 4.13. The north to south direction followed by the flux lines in such a magnetic field can be determined through a simple "right-hand rule." This rule dictates that if you were to grab a current-carrying conductor (hopefully an insulated one) with your right hand such that your thumb pointed in the direction of current flow, then your fingers would curl around the wire in the direction of the magnetic field.

Magnetic fields created by current moving through a wire have many practical applications including electromagnets and electric motors. However, the field produced around a straight piece of wire is often too weak to be of any practical use. To remedy this problem, let's look at what happens when we wrap the wire into a coil, often called a solenoid, and then pass current through it. As figure 4.14 illustrates, the magnetic fields formed around each turn of the coil complement each other and form a more powerful magnetic field around the coil. The magnetic field of each turn of wire in the coil has an additive effect to the overall magnetic field of the solenoid. Provided that the length of the solenoid is much greater than its diameter, the resulting magnetic field within the solenoid can be easily determined by examining the physical characteristics of the coil along with the current running through it. These physical characteristics include the length of the coil, ℓ,

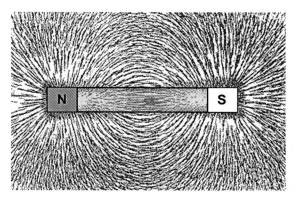

Figure 4.12. A more detailed illustration of the magnetic field around a bar magnet.

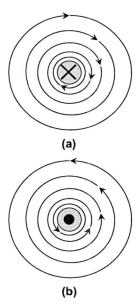

(a)

(b)

Figure 4.13. An illustration of the magnetic field produced around a conductor that is carrying a current. (a) displays the field direction for a current running into the page and (b) displays the field direction for a current running out of the page.

the number of turns of wire, N, and the permeability of the core material, μ. Thus, the magnetic field, B, of a long solenoid can be easily calculated as follows:

$$B = \frac{\mu N I}{\ell}.$$

Table 4.6. The maximum values for the magnetic permeability of some common core materials.

Material	Maximum Relative Permeability
Cobalt, 99% pure	250
Nickel, 99% pure	600
Cold rolled steel	2,000
Iron, Annealed	5000
78 Permalloy	100,000
Superpermalloy	1,000,000

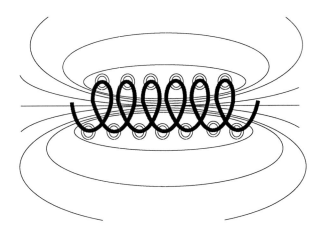

Figure 4.14. An illustration of the magnetic field produced around a solenoid.

Assuming that the core material is simply air, we can use the permeability of free space, $\mu = \mu_0 = 4\pi \times 10^{-7} \ \mathrm{H/m}$. If we place some other material in the center of our solenoid, we must take that material's permeability into account. Most tables of magnetic permeability give permeability values relative to the permeability of free space. As a result, we need to use $\mu = \mu_r \mu_0$ when determining the magnetic field. Table 4.6 provides the maximum relative permeability measurements for several different core materials. To better illustrate the calculation of the magnetic field in a solenoid, let's look at an example.

Example 4.9

Problem: Assume you have a long solenoid with a length of 30 cm. This solenoid consists of 300 turns of wire through which a current of 1 amp will run. Find the magnetic field within the solenoid if the core of the coil remains empty. What will the magnetic field be if the core is filled with annealed iron that has a relative permeability of 1200?

Definition: Magnetic Field (Flux Density) within a Long Solenoid

The magnetic field found within a long solenoid is dependent on the physical characteristics of the solenoid as well as the current passing through it. These physical characteristics include the length of the coil, ℓ, the number of turns of wire, N, and the permeability of the core material, μ. Using this information the magnetic field, B, is calculated as:

$$B = \frac{\mu N I}{\ell}$$

Solution: The first part of this problem requires us to use the equation for finding a magnetic field in a solenoid. If we plug the values provided by the problem into this equation and use the permeability of free space as our core permeability, we can determine the result as follows:

$$B = \frac{\left(4\pi \times 10^{-7} \text{ H/m}\right)\left(300 \text{ Turns}\right)\left(1\text{A}\right)}{0.3\text{m}} = 1.3 \times 10^{-3} \text{ T}$$

Thus, our magnetic filed within the solenoid is equal to 0.013 Teslas. Note that a Tesla is the SI unit for magnetic field. Now, lets' see what happens to the magnetic field if we place an annealed iron core in our solenoid. In this case, we must determine the new core permeability by multiplying the relative permeability by the permeability of free space to yield:

$$\mu = \mu_r \mu_0 = \left(1200\right)\left(4\pi \times 10^{-7} \text{ H/m}\right) = 1.5 \times 10^{-3} \text{ H/m} .$$

By plugging this value into our equation for magnetic field we will arrive at our solution as follows:

$$B = \frac{\left(1.5 \times 10^{-3} \text{ H/m}\right)\left(300 \text{ Turns}\right)\left(1 \text{ A}\right)}{0.3 \text{ m}} = 1.5 \text{ T}$$

So, we have discovered that by simply placing an annealed iron core in our solenoid, we have greatly amplified the magnetic field.

Now that we know how to determine the magnetic field within a solenoid, we will turn our attention to another important measure. This measure is known as magnetic flux, ϕ_m. The value for the magnetic field that we just calculated in the previous example is also known as the magnetic flux density. In other words, it is a measure of the number of magnetic field lines per unit area within the solenoid. Thus, the measure of magnetic flux is simply a determination of the total number of magnetic field lines that run through a given cross-sectional area. So, for a single turn of wire the magnetic flux is simply determined as the product of the magnetic field (flux density), B, and the cross-sectional area of the turn, A_c. This can then be logically extended to a solenoid by multiplying by the number of turns to yield:

$$\phi_m = NBA_c .$$

This equation can be combined with the magnetic field equation and rewritten in terms of the physical qualities of the solenoid as follows:

$$\phi_m = \frac{\mu N^2 A_c I}{\ell}$$

It should be noted that magnetic flux is measured in units called Webers, Wb. A Weber can be defined as 1×10^8 maxwells or magnetic field lines. One Tesla is equivalent to one Weber per square meter. Let's look at an example of how the magnetic flux is determined for a given solenoid.

Definition: Magnetic Flux through a Long Solenoid

The magnetic flux through a long solenoid, ϕ_m, is a measure of the number of magnetic field lines passing through the solenoid's cross-sectional area. It can be determined by multiplying the magnetic field, or flux density, B, of the solenoid by its cross-sectional area A_c and the total number of turns in the solenoid, N. Thus, either of the following formulas can be used to find the magnetic flux through the solenoid.

$$\phi_m = NBA_c \qquad\qquad \phi_m = \frac{\mu N^2 A_c I}{\ell}$$

Example 4.10

Problem: You have been given a long solenoid with a length of 35 cm, a radius of 2 cm, 800 turns of wire, and an iron core with a permeability of 1100. Determine the magnetic flux passing through the solenoid if a current of 0.2 amps is run through the wire.

Solution: Here we can determine the solution by either; first calculating the magnetic field and then multiplying by the cross-sectional area and number of turns, or by plugging directly into our second equation for determining magnetic flux. In this case, it is easier to plug directly into the magnetic flux equation as follows:

$$\phi_m = \frac{(1100)(4\pi \times 10^{-7} \text{ H/m})(800 \text{ Turns})^2 \pi (0.02 \text{ m})^2 (0.2 \text{ A})}{0.35 \text{ m}} = 0.635 \text{ Wb}$$

Thus, we see that the magnetic flux through our solenoid is 0.635 Webers.

If you carefully examine the equation used in the previous example, you may notice that, with the exception of the current passing through the solenoid, all of the values used to determine the magnetic flux are directly related to the physical construction of the device. As a result, these values will not change after the solenoid is built. As a result, we see that the physical characteristics of the device determine a proportionality constant for the device which, when multiplied by the applied current will determine the magnetic flux. This proportionality constant is very important and is often referred to as the self inductance, L, of the solenoid. Thus, we can write:

$$L = \frac{\phi_m}{I} = \frac{\mu N^2 A_c}{\ell}.$$

So, much like capacitance, inductance is determined entirely from the physical characteristics of the device. However, unlike capacitance, which is dependant on the voltage applied to the

---

Definition: Self Inductance of a Solenoid (Coil)

The self inductance of a solenoid is the proportionality constant that, when multiplied by the current passing through the coil, yields the magnetic flux through the coil. Thus, the self inductance of a solenoid can be determined either by dividing the magnetic flux by the current or purely from the physical characteristics of the device. So, either of the following equations can be used to determine self inductance. The decision as to which one should be used depends on the information that is available.

$$L = \frac{\phi_m}{I} \qquad\qquad L = \frac{\mu N^2 A_c}{\ell}$$

In these equations L represents self inductance, ϕ_m represents magnetic flux through the solenoid, I represents current through the wire, μ represents the permeability of the core material, N is the number of turns of wire, A_c is the cross-sectional area of the solenoid, and ℓ is the length of the coil.

---

device (electric field), inductance is dependant on the current running through the device (magnetic field). Inductance is measured in units called Henrys (H), where one Henry is equal to one Weber of flux per ampere of applied current. Let's calculate the self inductance for an example solenoid.

Example 4.11

Problem: Find the self inductance of a long solenoid with a length of 10 cm, a radius of 2mm, and 500 turns of wire. Assume an iron core with a permeability of 1000 is being used.

Solution: In this problem, we simply plug the physical characteristics of the solenoid into our equation for self inductance. Doing this yields:

$$L = \frac{(1000)(4\pi \times 10^{-7}\ \text{H/m})(500\ \text{Turns})^2\ \pi(0.002\ \text{m})^2}{0.1\ \text{m}} = 0.0395\ \text{H} = 39.5\ \text{mH}$$

So, here we have found that a solenoid with the given physical characteristics can be expected to have a self inductance of 39.5 mH.

Now that we have seen how a current moving through a wire can be used to generate a magnetic field, let's look at the reverse case. Since moving charge produced a magnetic field, one might suspect that there is probably some way in which a magnetic field is capable of causing a current flow. This is in fact the case, but, it is not simply the presence of a magnetic field that can lead to the flow of electric current. In fact, it is changes in magnetic flux that are capable of inducing voltages across loops or coils of wire. The easiest way to show this is to take a coil of wire and hook it up to a voltmeter. Next, pass a permanent bar magnet through the center of the coil. If you watch the voltmeter while you do this, you will observe a brief surge in the voltage across the coil. If you stop moving the magnet, however, the voltage will return to zero. This is because it is not the presence of the magnetic field that is creating the voltage difference, but rather the change in the flux through the coil that is a result of moving the magnet. This method of inducing a voltage across a coil of wire is similar to the way in which power is generated in power plants. We will look at this process in more detail in the next chapter, but for now, be aware of the fact that a change in the magnetic flux that passes through a coil of wire will induce a voltage in the coil.

Now, let's assume that we have a coil of wire and that we are running a given current through that coil. As we now know, this current produces a magnetic field around the coil. But, what would happen if we suddenly changed the amount of current that we were sending through the coil. This would cause the magnetic field to change. As the current increased, the electric field would expand, and as the current decreased the electric field would collapse. So, what goes on inside the coil while this is taking place? Let's start out by assuming that we are not passing any current through our coil. Now, as we begin increasing the current flow, a magnetic field will begin expanding around the coil. This growing magnetic field means that the magnetic flux through the coil is increasing. This increase in flux will in turn induce a voltage in the coil. As it turns out, this induced voltage is of the opposite polarity of the voltage that is associated with our current flow. Thus, the induced voltage resists the increase in current, and likewise, the change in the magnetic field. Once the current reaches its final constant value, the magnetic field stabilizes, and there is no longer any change in the magnetic flux through the coil. As a result, there is no longer any induced voltage. The only resistance provided by the inductor at this point is the resistance of the wire itself. So, what happens if we begin to reduce the amount of current that we pass through the coil? In this case, the reduction in current causes the magnetic field to begin collapsing. The shrinking magnetic field causes a change in the magnetic flux through the coil and induces a voltage. This time the polarity of the voltage is the same as the voltage associated with the current flow. Thus, as the current flow is reduced, the collapsing magnetic filed induces a voltage that attempts to sustain the current flow and slow the collapse of the magnetic field. It is interesting to note here that, if we were to change the current very quickly, for example, if we turned off the circuit, the magnetic field would collapse extremely quickly and a very large voltage would be induced across the coil. In many cases voltages in the hundreds of volts can be produced. As a result, most equipment that employs devices with coils, such as inductors and relays, is specially designed to prevent these large voltage spikes from damaging other circuit components.

From what we have just seen here, we can think of inductors much like mechanical flywheels which take advantage of the law of inertia. In other words, an object at rest tends to want to stay at rest and requires effort to get moving. The work employed in getting the flywheel moving is stored in its angular momentum. Once it is moving, it wants to stay moving and resists changes in its angular velocity. An inductor is an electronic version of this principle. The work that is required to get the current up to its final value is stored in the magnetic field around the coil. At this point, the inductor will use this stored energy to resist any efforts to change the current much like a flywheel will use its momentum to resist efforts to change its angular velocity. The amount of energy that is stored in the magnetic field of an inductor is proportional to the inductance of the coil and the square of the current passing through it. Thus, the energy stored in the magnetic field of an inductor is written as follows:

$$U = \frac{1}{2}LI^2$$

This equation has many similarities to the equation representing energy stored in the electric field of a capacitor. We saw that the energy stored in a capacitor was dependent on the capacitance of the device and the square of the voltage across the plates. In the case of an

Definition: Energy Stored in the Magnetic Field of an Inductor

The energy stored in the magnetic field of an inductor is proportional to the self inductance of the coil, L, and the current running through the coil, I. The energy within the magnetic field is written as:

$$U = \frac{1}{2}LI^2$$

inductor, we must look at the inductance of the device and the square of the current running through it. Let's look at a quick example of the calculation of the energy stored in the magnetic field around a given inductor.

Example 4.12

Problem: Given an inductor with a stated inductance of 75 mH and a current of 0.5 A, determine the energy stored in the magnetic field of the inductor.

Solution: Here we must simply plug our given inductance and current values into our energy equation for the magnetic field of an inductor. This gives us:

$$U = \frac{1}{2}(0.075)(0.5)^2 = 18.7 \, \text{mJ}$$

Thus, the given inductor stores 18.7 micro Joules of energy in its magnetic field when the current flowing through it is 0.5 amperes.

So, based upon our study of capacitance and inductance, we note several important differences between inductors and capacitors. These are that: 1) capacitors store energy in an electric field while inductors use a magnetic field, 2) capacitors are voltage dependant devices and inductors are current dependent devices, 3) capacitors resist changes in voltage while inductors resist changes in current, and 4) the collapsing electric field in a capacitor produces a current while the collapsing magnetic field of an inductor produces a voltage.

4.2.8   Power

One of the most important aspects of electricity is that, like any form of energy, it has the ability to perform work. Our goal is to harness this work and direct it so it can be used for our benefit. In the case of electricity, we learned the voltage between two points in an electric field represents the work required to move a unit of charge from one point to the other. We also know the current running through a circuit represents the number of units of charge that pass a fixed point in one second. Using this information we can easily determine the power, or the amount of work that is performed per unit of time as follows:

$$P = IV \, ,$$

Definition: Instantaneous Power

The instantaneous power, P of an electrical device is defined as the work that is done per unit of time. In terms of the input voltage, V, and current, I, power can be defined as:

$$P = IV$$

---

where P represents the power, I represents the current, and V represents the voltage. The unit of power, which is described as one joule per second, is called the watt. Thus, when you pick up a 100 watt light bulb, you know that the 100 watt rating means the electricity passing through the bulb performs 100 joules of work every second. In a light bulb, the work performed is converted into the radiation of light and the production of heat, but the work can be done in multiple ways depending on the design of the electronic device.

Example 4.13

Problem: An electronic device requires an input voltage of 12 volts and draws a current of 3 amperes. What is the power consumption of the device?

Solution: Here we are given both the voltage and the current. As a result, we simply need to multiply these values together to obtain the power. Doing so yields:

$$P = IV = (3 \, \text{Amps})(12 \, \text{Volts}) = 36 \, \text{Watts}$$

So, here we see that the given device requires 36 watts of power to operate. In other words, the electricity supplied to the device must perform 36 joules of work for each second that the device operates.

4.2.9   Alternating Current

Up until this point, we have discussed electrical currents resulting from relatively constant potential differences, such as the potential differences that exist between the terminals of a

---

Definition: Alternating and Direct Current

Direct Current is an unvarying current that is associated with a stable potential difference. It flows in only one direction at a constant rate.

Alternating Current is a fluctuating current that is associated with a changing potential difference. Alternating current flows at a fluctuating rate and switches direction periodically. The most common alternating current pattern is associated with a sinusoidal change in voltage.

---

battery. Since the potential difference between two points remains constant, the flow of current between the points is in one direction. Such a current is often referred to a direct current, or DC. But, what would happen if the electric field were to change over time so that the potential difference between two points ranged from a peak positive voltage to a peak negative voltage? Such an oscillation would result in current that briefly flows in one direction when the potential difference is positive and then flows in the opposite direction when the voltage becomes negative. This type of current is known as alternating current, or AC. Alternating current is widely used in electrical applications, particularly in the electric power industry for the transmission of electrical power to homes and businesses. The generation and transmission of AC will be covered in more detail in the next chapter.

The fluctuations in voltage associated with an alternating current can follow many patterns including the sinusoidal, triangular and square wave patterns shown in Figure 4.15. Since the sine wave approach results in the most efficient transfer of energy, it is by far, the most commonly used. In fact, the electricity delivered to most homes and businesses uses a sinusoidal voltage variation. When dealing with AC systems in this configuration, we can describe the voltage mathematically as a function that varies in amplitude over time. This can be written as:

$$V(t) = A\sin(2\pi f t),$$

where A is the amplitude of the waveform and f is the frequency at which the sinusoidal wave oscillates. When dealing with such a voltage, there are three ways that we can describe it. The first and most simplistic is to simply state the maximum value, or amplitude of the voltage waveform. This is often referred to as the peak voltage, V_p. The second approach is to describe the difference between the maximum and minimum voltages. This voltage description, known as peak-to-peak voltage can be obtained as follows:

$$V_{p-p} = (+A) - (-A) = 2A.$$

The final and most commonly used approach is to describe the RMS voltage. So, what is meant by RMS? RMS is an abbreviation for root mean square. As the name suggests, the RMS voltage is the square root of the mean (average) value of the squared voltage over a given period of time (usually one period of the waveform). In the case of a voltage that varies in a sinusoidal pattern, the RMS voltage can be easily determined as follows:

$$V_{RMS} = \frac{A}{\sqrt{2}}.$$

The main benefit of describing the voltage using the RMS measurement is seen when the voltage is connected to a given device (load). In this case,

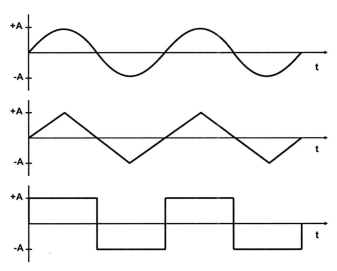

Figure 4.15 Examples of some of the most common AC waveforms. From top to bottom these are sinusoidal, triangle, and square waves.

the power that is delivered to the device by both an AC and DC source will be the same if the RMS voltage of the AC source is equal to the DC voltage. Thus, the RMS value of an AC source describes the DC voltage that would be required to provide the same amount of power. Because of this, the RMS voltage is the most common method of describing AC voltages. Let's sum things up with an example.

Example 4.14

Problem: The AC voltage measured at a typical power outlet is described by the equation $156\sin(2\pi60t)$. Answer the following questions:

    a. What is the frequency of the AC source?
    b. What is the peak voltage?
    c. What is the peak-to-peak voltage?
    d. What is the RMS voltage?

Solution: In this example, we must use what we have learned about alternating current to answer the questions. The first question asks what the frequency of the AC source is. From the equation of a typical sinusoidal signal, we can easily determine that the frequency of the measured voltage is 60 Hz. The second question requires the peak voltage. We know that the peak voltage is simply the amplitude of the sinusoidal signal. Thus, the peak voltage is 156 volts. The third question asks us to determine the peak to peak voltage. Since this is simply twice the amplitude, we know that $V_{p-p} = 2(156) = 312$ volts. The final question asks for the RMS voltage. Since the signal is sinusoidal, we can easily find the RMS voltage by dividing the peak voltage by the square root of two. Doing so gives us:

$$V_{RMS} = \frac{A}{\sqrt{2}} = \frac{156}{\sqrt{2}} = 110.3 \text{ Volts}.$$

Thus, the RMS voltage in this case is approximately 110 volts. These values are typical of those found at the electrical outlets in most homes in the United States.

4.2.10 Electrical Circuits

Another important concept in electronics is that of the circuit. The word circuit is often used to represent any connection of electronic components, such as resistors, capacitors, and inductors, but in its simplest form, a circuit is just a closed path through which a current of electrons can flow. Thus, the simplest possible circuit would be a direct connection between two areas of differing potential, an area of high electron concentration, known as a source, and an area with a shortage of electrons, known as a sink. This connection is typically made using some form of conducting wire. In more practical terms, we would also want the flowing electrons to perform some useful work while they travel. So a simple circuit is usually constructed of three basic items. First, there must be an electrical source, a battery for example. Next, we need some type "load" or component in which the electrons will perform useful work, for example, a light bulb. Finally, we need a means of connecting the source to the load so that a current of electrons can flow from the source, through the load, and into the sink. Copper wire works well for this. Thus, the process of connecting a light bulb to the positive and negative terminals of a battery results in the creation of a simple electrical circuit. An illustration of this simple example circuit is shown in Figure 4.16.

Most practical circuits are much more complicated than the previous example. In fact, they can consist of hundreds of components. Because of this, we need a simple and standard way of representing the interconnection of the components in a circuit so that a given circuit is easy to understand, analyze, and reproduce. To do this, we turn to schematic diagrams. In a schematic diagram, each electrical component is represented by a standard symbol and the interconnections of the components are illustrated by lines drawn between the symbols. The schematic diagram of our example simple circuit is shown in Figure 4.16.

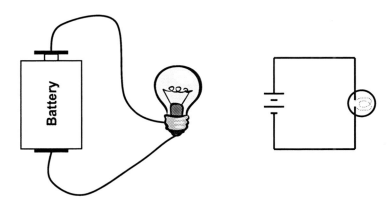

Figure 4.16 An illustration of a simple circuit and its associated schematic diagram.

4.3   Electrical Components

In the previous sections, we discussed the basic theory behind many of the components that are used in electronics applications today. In the following sections, we will look at the more practical side of modern circuit components. Each section will discuss a specific circuit component by briefly looking at the way in which the component is constructed, how it may be physically recognized on a circuit board, and how it would appear on a schematic diagram for a piece of equipment.

4.3.1   Sources

One of the most important components in an electric circuit is the source. It provides the potential difference that causes electrons to flow through the rest of the circuit elements. There are many ways to construct an electrical source. Each particular type of source transforms some form of energy, such as chemical, mechanical, or heat energy into electricity.

A battery for example, relies on a chemical reaction that produces electrons. Such a reaction is known as an electrochemical reaction. This reaction produces excess electrons at the negative terminal of the battery. When a path is provided (a complete circuit) the electrons will flow from the negative terminal to the positive terminal. As the electrons flow, the chemical reaction continues. After a period of time, the chemical reaction reaches a point at which it can no longer provide sufficient amounts of electrons. At this point, the battery is no longer useful and must be replaced. If, however, the electrons cannot flow between the battery terminals, the chemical reaction stops. This is why batteries can last a long time in storage if they are not used. The rate at which the chemical reaction can produce electrons determines the maximum current that the

Figure 4.17. Examples of some common electrical sources. From left to right, you will find batteries, a generator, and a solar panel.

battery can produce. Thus, based upon its physical configuration, a battery produces a fixed voltage and has a maximum amount of current that it can supply. If larger currents or voltages are needed, multiple batteries can be connected to provide the desired effect. Connecting the batteries in parallel provides the same voltage, but allows for additional current to be produced. On the other hand, connecting the batteries in series results in the same maximum current, but the voltage provided will be the sum of the voltages of the batteries that are connected. Both a series and parallel configuration are shown in figure 4.18.

Another common electrical source is the generator. A generator, like the one in Figure 4.17, converts mechanical energy into electricity. This is typically done by harnessing some mechanical force to turn a coil of wire within a magnetic field. While the generator in Figure 4.17 relies on hand cranking, the mechanical force used by most generators is typically produced by moving steam or water. The steam or water moves through a turbine which in turn rotates the coil in the generator. Wind and tidal forces, although less common, can also be employed to turn the coil within a generator.

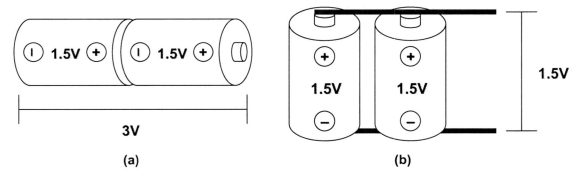

Figure 4.18. An example of batteries connected in both series (a) and parallel (b) configurations.

Solar energy is becoming an increasingly attractive source of energy for the production of electricity. Solar cells can also be used directly as electrical sources. In this case, the solar energy is directly converted into electricity by the photovoltaic material in the cell. Such a cell is shown in Figure 4.17. In other cases, the solar energy is focused and used to produce heat. This heat is used to boil water and produce steam which, in turn, moves a turbine to generate electricity.

So, as we have seen, there are many different ways to provide electrical energy for use in a circuit. There are also several symbols that can be used in schematic diagrams to illustrate the electrical source. Some of the most common of these symbols are shown in Figure 4.19.

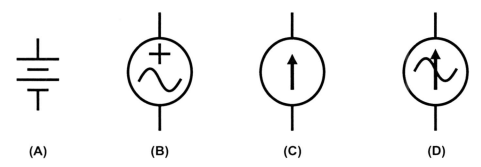

Figure 4.19. Some of the more common schematic symbols for electrical sources. From left to right, they are (A) a DC voltage source, (B) an AC voltage source, (C) a DC current source, and (D) an AC current source.

4.3.2   Switches

Another important element of a circuit is a means of easily interrupting the current flow or routing the current through different circuit branches. The component that is typically used to perform this task is the switch. Several common switches are displayed in Figure 4.20. A basic switch contains one or more sets of metal contacts. These contacts are physically brought together to complete the circuit or separated to break the circuit. When describing switches there is some basic terminology that must be understood. First, there are two basic configurations for switches. These are typically referred to as normally open, NO, or normally closed, NC. A normally open switch is one that maintains an open or broken circuit until the switch is activated. A normally closed switch, on the other hand, maintains a complete or closed circuit until the switch is activated.

Switches are also described using the terms "pole" and "throw." A pole is simply a set of switch contacts used for a single circuit. In other words, a single pole (SP) switch can connect or break a single circuit, and a double pole (DP) switch can open or close two circuits. Thus, the

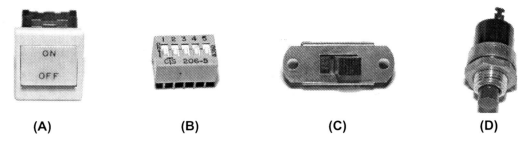

Figure 4.20. Some commonly used switches. From left to right they are (A) a rocker switch, (B) a DIP (Dual Inline Package) switch, (C) a slide switch, and (D) a pushbutton switch.

"pole" count of a switch is simply the number of circuits that the switch is designed to work with. A "throw" represents one of the positions that the switch can assume. In other words, the number of "throws" represents the number of circuits that can be switched between. A single throw (ST) switch, for example is capable of routing current through a singe circuit. The circuit is either connected or it is not. A double throw (DT) switch, on the other hand, can route current through one of two different circuits. Thus, the source is connected to either circuit A or circuit B. The schematic symbols for some common switch configurations can be found in Figure 4.21.

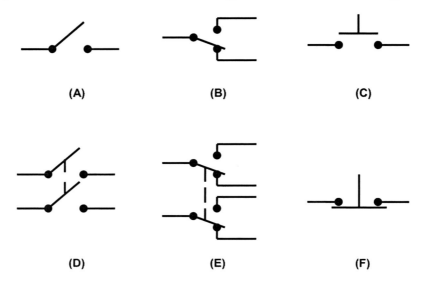

| (A) | (B) | (C) |

| (D) | (E) | (F) |

Figure 4.21 Schematic symbols for some common switches. These symbols represent (A) a SPST toggle switch, (B) a SPDT toggle switch, (C) a normally open pushbutton switch, (D) a DPST toggle switch, (E) a DPDT toggle switch, and (F) a normally closed pushbutton switch.

The switches that have been discussed up to this point are all manually activated. In other words, they require a physical action to open or close their contacts. However, there are some switching devices that do not require an individual to manually change the switch position. These devices are called relays. A relay consists of a set of switch contacts and a coil. One of the contact sets is fixed and the other set is movable. When a current flows through the coil, the resulting magnetic field pulls the movable contacts toward the coil. This action either connects or separates the contact pairs resulting in either a closed or open circuit. Relays are useful tools when it is necessary to remotely control dangerous high voltage or current connections. A small current applied to the coil can be used to control much larger voltage and current sources. An example of a typical relay and the associated schematic symbol can be found in Figure 4.22.

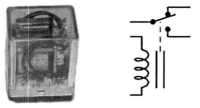

Figure 4.22. An example of a common relay and its associated schematic symbol.

4.3.3   Resistors

When designing a circuit it is often necessary to insert fixed amounts of resistance in various branches of the design. For example, if you need to limit the current that flows through a branch of the circuit, this can be done by placing the appropriate amount of resistance in that branch. To do this, components called resistors are typically used. When building a circuit, there are many types of resistors that can be used. These include fixed value resistors, manually variable resistors, and environmentally variable resistors.

Figure 4.23. A wirewound resistor.

Fixed Resistors, which are the most commonly used, can be built in a number of different ways. One simple way is to wind a length of resistive wire, for example nickel chromium (nichrome) around a ceramic core. The required resistance is obtained by wrapping the correct length of wire around the core. This type of fixed resistor is known as a wirewound resistor. An example of such a resistor can be found in Figure 4.23.

Another common method of construction is to deposit thin films of resistive material onto an insulating core. The length of the track of resistive material determines the resistance value of the component. The most common type of thin film resistor is the carbon film resistor, but metal oxide and metal film resistors are also used when highly accurate resistance values are needed.

The final common type of fixed value resistor construction is carbon composition. In this approach, a resin is mixed with a carbon powder. Depending on the amount of carbon powder used, a component with a specific resistance can be manufactured.

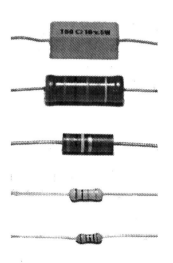

Figure 4.24. Several common types of fixed value resistors.

Each of these resistor design methods bring with them specific strengths and weaknesses. Some types, for example, provide highly accurate resistance values while others can vary from component to component by large percentages. The cost of constructing the different varieties of resistor can also vary widely. So, while there are a wide variety of resistor construction techniques, the type of resistor that is used for a particular application is totally up to the circuit designer and is often based upon circuit design and cost considerations.

Another issue must be considered when working with resistors and that is power dissipation. As might be expected, current running through a resistive material generates heat. As a result, resistors are manufactured to be able to dissipate only a given amount of power before they burn out. The power handling capability is usually reflected by the size of the device. In most cases a resistor that is physically larger can withstand more power before burning out. An arrangement of resistors of various sizes can be found in Figure 4.24.

Since most resistors are physically small devices, it is difficult to print their resistance values directly on them. Instead, most resistors use a standard color code to indicate their resistance value. Most standard fixed resistors use four bands to denote their value. In this case, the first two bands indicate numeric values, the third band represents a multiplier (power of ten), and the fourth band indicates the tolerance (How close to the indicated value the resistor will be). High precision resistors sometimes use five bands. On these resistors, the first three bands are numerical values, the fourth band is the multiplier, and the fifth is the tolerance. Table 4.7 lists the

Table 4.7. A list of the meanings of the color bands on 4 and 5 band resistors. The numeric 3 column is only used in 5 band resistors.

Color		Numeric 1	Numeric 2	Numeric 3	Multiplier	Tolerance
Black		0	0	0	1	
Brown		1	1	1	10	±1%
Red		2	2	2	100	±2%
Orange		3	3	3	1K	
Yellow		4	4	4	10K	
Green		5	5	5	100K	±0.5%
Blue		6	6	6	1M	±0.25%
Violet		7	7	7	10M	±0.10
Gray		8	8	8		±0.05%
White		9	9	9		
Gold					0.1	5%
Silver					0.01	10%

values associated with the various color bands on resistors. Let's use this table to see how one would determine the value of a resistor from the color bands.

Example 4.15

Problem: You take a random resistor from a bin on the desk. The resistor has four bands with the following colors: { Brown, Black Red, Gold }. What is the value of this resistor and what is its tolerance range?

Solution: In this case, you have been given the four color bands and must determine the value and tolerance of the resistor. From Table 4.7 we know that the gold band indicates a tolerance of ±5%. We also see that the two numeric values are: Brown -> 1 and Black -> 0. Thus, the base number is 10. Next, we see that the multiplier is Red -> 100. So, the value of this resistor is $10 \times 100 = 1000\Omega = 1K\Omega$. Therefore, this resistor's actual value will fall within five percent of one thousand ohms.

Sometimes it is necessary for the resistance in a circuit to be easily changed. For example, if we want to adjust the volume on our radio, we do not want to have to replace a resistor. The method of changing the resistance in the circuit, and thus the volume, must be as simple as possible. For this type of application, circuit designers turn to manually variable resistors. These resistors are sometimes referred to as rheostats or potentiometers depending on their configuration and the way in which they are connected in the circuit. The basic method used to construct these devices is to deposit a thin film of resistive material on an insulating

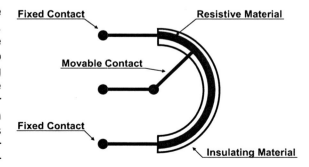

Figure 4.25. An illustration of the construction of a variable resistor.

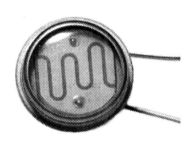

Figure 4.26. A photograph of several common variable resistors.

Figure 4.27. A photograph of a photo resistor.

surface. A stationary contact is connected to each end of the resistive film and a movable contact is designed to slide along the surface of the resistive material. As the movable contact slides along the film's surface, the length of resistive material between the movable contact and the fixed end contact changes. This changes the resistance between these contacts. Figure 4.25 illustrates this concept and Figure 4.26 displays some typical variable resistors.

Occasionally, we need to change resistance values based upon the environment in which the circuit is located. For example, we may need to change the resistance in response to changing temperatures or light levels. In cases such as these, we employ special materials that can change in resistance when these environmental values fluctuate. The two most common resistors in this category are the thermistor which changes resistance based on temperature and the photoresistor which changes resistance based upon light levels. A picture of a typical photoresistor can be found in Figure 4.27.

Resistors are easy to recognize on schematic diagrams. The symbol can look somewhat different based upon whether the resistor is fixed or variable, but the overall pattern is always easily recognized. Some of the most commonly used schematic symbols for the resistors that we have looked at are shown in Figure 4.28.

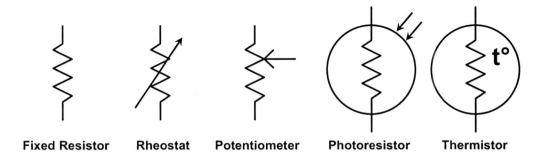

Figure 4.28. Schematic symbols for some common types of resistors.

4.3.4　Capacitors

　　　Capacitors are commonly used components in electronic circuit applications. Their uses range from the temporary storage of charge to their ability to block DC and pass AC signals. As we learned previously, capacitors are constructed from parallel conducting plates. To construct capacitors that are of a practical size, the plates are often stacked or rolled. One of the most common ways of building a capacitor is by separating the two plates, typically made of a metal foil with some form of insulating material. These insulated plates are then rolled into a cylinder and coated with an insulating material. This construction technique is illustrated in Figure 4.29. The main way of differentiating between the different types of capacitor is by describing the type of insulating dielectric material that is used between the plates. Commonly used materials include paper, plastic, ceramic, and polyester. Another common variety of capacitor is known as the electrolytic capacitor. In this variety of capacitor, two thin foil plates are used. The first is coated with an insulating oxide layer. In addition to the metal foil plates, a paper spacer soaked in an electrolyte solution is employed. The metal oxide layer becomes the dielectric material between the plates. Since this oxide layer is extremely thin but highly resistive to conduction, high capacitance values are possible. The main difference between electrolytic capacitors and most others is that they have a polarity. If they are connected incorrectly, they can sustain damage or even explode. The correct polarity is always marked on electrolytic capacitors, usually with large arrows indicating the negative leg of the device. Several of the commonly used capacitor designs are displayed in Figure 4.30.

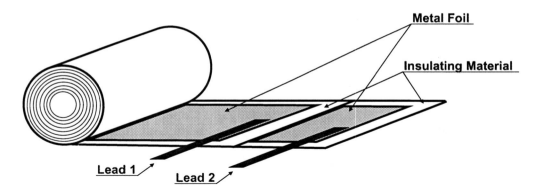

Figure 4.29. An illustration of a common capacitor construction technique.

Figure 4.30. Some common capacitor designs. From left to right these are ceramic disk, axial lead electrolytic, standard lead electrolytic and polyester.

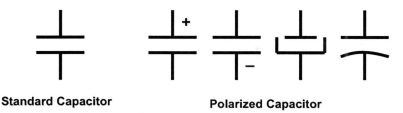

Standard Capacitor   **Polarized Capacitor**

Figure 4.31. Typical schematic symbols used for capacitors

The schematic symbols used to represent capacitors are easy to spot. It is easy to see that the symbol is designed to represent the two parallel plates used in the construction of a capacitor. Several common schematic symbols are shown in Figure 4.31.

4.3.5   Inductors

Inductors are also commonly found in electronics applications. Their ability to pass DC signals while impeding AC signals is useful in a variety of applications. As we discussed earlier, the construction of an inductor is simple. A wire must simply be wound around a core material. The number of turns of wire and the type of core material influence the value of the inductor. So, commercially available inductors are simply constructed by creating a coil with a specific number of turns. These inductors can be constructed either with or without a central core material. A picture of some common inductors is shown in Figure 4.32. The schematic symbol for the inductor is modeled after the wire coils from which it is constructed. The schematic symbols that are typically used to represent inductors are shown in Figure 4.33.

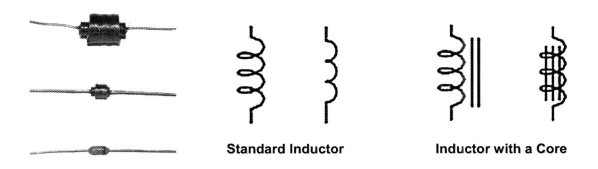

Standard Inductor   **Inductor with a Core**

Figure 4.32 An example of some common inductors.

Figure 4.33 The most common schematic symbols used to represent inductors.

4.3.6   Transformers

When we studied inductors, we found that a magnetic field is produced when current flows through a coil of wire. We also learned that a changing magnetic filed can induce a voltage in a coil of wire. The transformer operates by combining these two concepts. A typical transformer is constructed from two coils of wire wrapped around a common core material, usually soft iron. If a changing current is applied to one coil, for example an AC source, the resulting magnetic field will fluctuate in response to the changing current. The changing magnetic field will induce a voltage in the second coil. The relative size of the induced voltage depends on the ratio of the number of turns in the two coils. If the second coil has more turns than the input coil, the induced voltage will be higher, and if the turn count is lower, the induced voltage will also be lower. While the voltage can be changed by a transformer, the overall power that is transferred must not change. As we know, power is defined as the product of current and voltage. Thus, if the voltage increases, the current must decrease proportionally so that the overall power is conserved. The voltage on the output of a transformer can be easily determined by multiplying the voltage of the input signal (V_{in}) by the ratio of the number of turn in the output coil (N_{out}) to the number of turns in the input (N_{in}). This can be written mathematically as

Figure 4.34 An example of a typical transformer.

$$V_{out} = \left(\frac{N_{out}}{N_{in}} \right) V_{in} .$$

Let's take a look at an example.

Example 4.16

Problem: You have been given two different transformers. They both have 400 turns on their input coils, but the first had 100 turns on its output coil and the second has 1200 turns on its output coil. Determine the voltages that will be measured across the output coils of each transformer if you apply an AC voltage of 120 V to the input coils.

Solution: Here we simply apply the equation for finding the output voltage of a transformer to each case. Doing this will give us

$$V_{out} = \left(\frac{100}{400} \right) 120V = 30V$$

for the first transformer. This means that the first transformer is a step down transformer. Turning our attention to the second transformer, we find

$$V_{out} = \left(\frac{1200}{400} \right) 120V = 360V .$$

This means that the second transformer is a step up transformer.

So, we can see that as long as we are dealing with an input that varies over time, transformers can provide us with the very useful ability to increase and decrease the voltage of our signal. A picture of a typical transformer can be found in Figure 4.34.

The schematic symbol used for a transformer is modeled after the two coils employed in its construction. The symbols used to represent transformers can be seen in Figure 4.35.

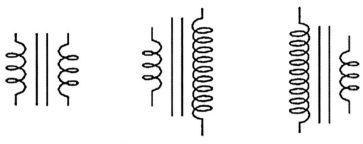

Coupling Transformer   Step Up Transformer   Step Down Transformer

Figure 4.35 Common schematic symbols used to represent transformers.

4.3.7   Diodes

The components that we have examined up to this point can be easily constructed from basic materials such as coils of wire, strips of resistive material, and metal plates. The remaining devices are known as semiconductor devices. This is because they are constructed by using pieces of semiconducting material, typically silicon that has been "doped" with specific impurities. When dealing with devices of this type, one must be familiar with the two basic types of silicon that are used. These are P-type and N-type silicon. P-type silicon is silicon that contains impurities such that, it contains a large number of atoms that are missing electrons or "holes" and N-type silicon is silicon with impurities that result in an excess number of electrons that can be easily broken free from their atoms to produce a current. The semiconductor devices that we will describe in the following sections all make use of combinations if these two types of material.

A simple and extremely useful component called a diode can be constructed by forming a simple junction between a P-type and N-type material. The diode basically functions as a one-way valve for electrons. In other words, the diode allows electrons to easily flow in one direction, but it blocks flow in the other direction. Such a component is extremely useful, particularly in the construction of power supplies that convert alternating current to direct current. So, how does the diode operate? Let's take a look at how the PN junction works when it is exposed to a voltage. The diagram in Figure 4.36 shows a PN junction with no voltage applied. As you can see, the holes in the P-type material and the charge carriers in the N-type material are relatively equally distributed.

Now, what happens if we attach the negative terminal of a source to the N-type material and the positive terminal of a source to the P-type material? As Figure 4.37 demonstrates, the negative terminal of the source will repel the electrons in the N-type material and move them toward the P-type material. Likewise the positive terminal of the

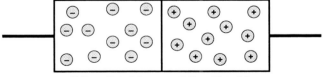

N-Type Material   P-Type Material

Figure 4.36 An illustration of the charge carriers in a diode that is not connected to a voltage source.

source pulls electrons from the P-type material. This results in even more holes that move toward the junction with the N-type material. At the junction, the electrons and holes combine and neutralize each other. So, as more electrons flow into the N-type material and are pulled out of the P-type material, a sustainable current is achieved. When this occurs, the diode is in what is known as a forward biased state.

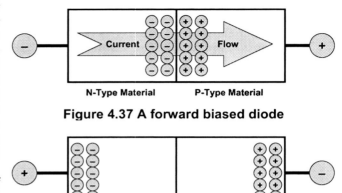

Figure 4.37 A forward biased diode

Now, we will look at what happens if we attach the negative source terminal to the P-type material and the positive source terminal to the N-type material. In this case, the negative charge of the negative source terminal pulls the holes

Figure 4.38 A reverse biased diode

toward the edge of the device and away from the junction. Likewise the positive charge from the positive source terminal pulls the electrons in the N-type material toward the edge and away from the junction. This process is illustrated in Figure 4.38. Because the charge carriers are pulled away from the junction they cannot combine and no current can flow through the device. This is known as the reverse biased state of the diode.

As is the case with any component, if your push hard enough, you can make it do something it is trying to prevent. This is true with the diode. If you apply enough voltage when the diode is in its reverse biased state, you can force a current to flow. Doing this will usually damage the diode. The current flowing in the reverse direction leads to a large power dissipation within the diode, and most diodes cannot handle the resulting heat that is generated. The voltage that must be applied to overcome the blocking ability of the diode is known as the "breakdown voltage." In most cases, putting the diode in a breakdown situation is to be avoided. However, some diodes are designed to operate in a breakdown condition. These diodes, known as zener diodes, block current flow for any voltage below their breakdown voltage, but when the voltage goes above the breakdown voltage, the voltage that is measured across the leads of the diode will remain constant at the breakdown voltage. This desirable trait is often used in power supplies to regulate the output voltage at a set reference voltage. Figure 4.39 displays some typical diodes. The top two diodes are normal diodes and the lower two are zener diodes.

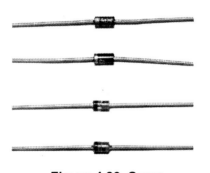

Figure 4.39 Some common diodes

Figure 4.40 An example of some commonly used light emitting diodes.

The junctions in many diodes reach such high energy levels when they are forward biased that they emit photons of light. In most cases, these photons are reabsorbed or are of a frequency that the human eye cannot detect. If the right materials are used in the diode's construction, however, these photons are visible. These light emitting diodes, or LEDs, can be built to emit light of a wide variety of colors from red to blue. Due to the long lifespan of LEDs, they have become a common device in electrical applications, and in many cases have replaced the use of light bulbs. Some common light emitting diodes are shown in Figure 4.40.

The schematic symbol used for diodes is representative of their one-way nature. The symbol consists of an arrow and a blocking line. This illustrates that the current will pass through in one direction and will be blocked in the other. Some typical diode schematic symbols are shown in Figure 4.41.

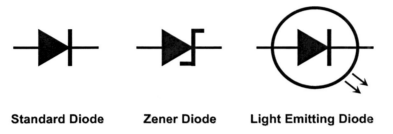

Standard Diode   Zener Diode   Light Emitting Diode

Figure 4.41 The schematic symbols used for some common types of diode.

4.3.8   Transistors

The transistor is an extremely important electronic component and can be found in most modern electrical circuitry. As was the case with the diode, the transistor is constructed from silicon that has been "doped" with various impurities to form P-type and N-type materials. By building on the principles discovered in the construction of the diode, the transistor is capable of allowing a small current or voltage to control a much larger one. Thus, one important use of the transistor is signal amplification. In this type of application, a small signal, for example, the signal received by a portable radio, can be used by a transistor to produce a much larger signal that can be sent to a speaker.

Transistors come in a variety of external packages, as shown in Figure 4.42, but internally, there are only a few ways in which they are constructed. Overall, there are two common transistor design approaches. The first of these is very similar to the design approach for the diode. This design is known as a bipolar junction transistor, or BJT. Such a transistor is constructed by separating two common types of doped silicon by a thin, lightly doped section of the opposite type

Figure 4.42 Some common transistor packages that might be found in basic electronics.

of material. Based upon this, there are two basic BJT arrangements. These are NPN, which consists of two sections of N-type material separated by a thin, lightly doped section of P-type material, and PNP, which is built by separating two sections of P-type material with a thin, lightly doped section of N-type material. So, as we can see, this device consists of two PN junctions. Under normal operating conditions one junction is typically forward biased while the other is reverse biased. Thus, we get the term bipolar junction transistor. Regardless of which of the two material arrangements is used, a lead is connected to each of the sections of material. The lead connected to the thin section of the opposite material type is called the "base" and the leads connected to the two other sections of material are called the "collector" and the "emitter." The basic design concept for the BJT is displayed in Figure 4.43. In this design, the current that flows through the base lead controls the current that is allowed to flow from the collector to the emitter.

The second transistor design is known as the field effect transistor, or FET. This transistor is built by using one of the two types of doped silicon to construct a channel. The "source" and "drain" leads are connected to each one of the ends of this channel material. Then, a section of the opposite type of material is placed around the channel, as shown in Figure 4.44. This material is connected to the "gate" lead. So, as you can see, we can construct either an N-channel or a P-channel FET. In this design, the voltage placed upon the "base" lead controls the current that runs through the channel. The FET design is the approach most often used in the construction of power amplifiers and on microchips.

It should be noted that transistors can be used in electronic circuits in two different ways. First, they can be used to allow small voltages and currents to control smaller ones. This is the approach used in analog equipment like amplifiers. The second approach is to use them like a switch. By driving the transistor in such a way that it either blocks current flow (off) or lets it freely flow (on) the transistor can be used for digital applications. This is the way in which transistors are used in digital logic and computer designs.

The schematic diagram used for the transistor depends on both the design type that is used and the P and N material arrangement. Some typical schematic symbols are shown in Figure 4.45. As you can see, they reflect the design of the device itself.

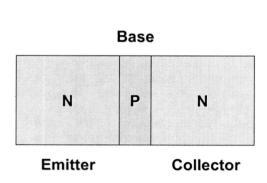

Figure 4.43 The basic design of a BJT transistor. The design displayed here is for the NPN arrangement. The PNP arrangement simply reverses the material types.

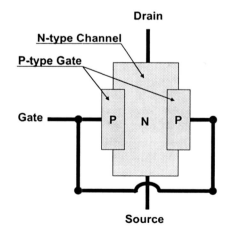

Figure 4.44 The basic design for an N-channel junction FET. A P-channel FET would have opposite material types.

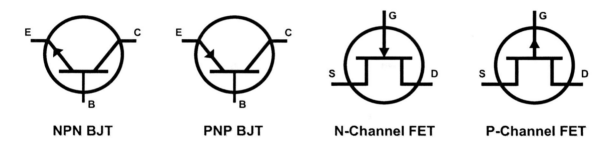

NPN BJT   **PNP BJT**   **N-Channel FET**   **P-Channel FET**

Figure 4.45 Schematic symbols for some common transistors.

4.3.9   Integrated Circuits

The final component that we will discuss is the integrated circuit or IC. Integrated circuits, like diodes and transistors are silicon-based devices. In fact, an integrated circuit often contains very large numbers of diodes and transistors. Basically an integrated circuit, as the name suggests, is an entire circuit within a single package. This circuit can be very simple, such as a logic gate, or extremely complex, such as a microprocessor. Integrated circuits are seeing increasing use in electronics applications since their very small size and ease of use make the construction of complicated devices much easier. When displayed in a schematic diagram, most integrated circuits are shown as a rectangular box with numbered or labeled connections called "pins" The pins on a typical integrated circuit package are numbered from a reference point. For example, pin number one on the devices shown in Figure 4.46 is marked by a small indentation or a dot. Example schematic symbols used for basic integrated circuits are also shown in Figure 4.46.

Figure 4.46 Some basic integrated circuits and their schematic symbols.

4.4   Measurement Devices

When dealing with electrical circuits, knowledge of the various components that you encounter is very important. As we discovered in the previous sections, there are a wide variety of components that we may encounter. It is equally important, however, to understand and be able to use the equipment that is available to perform measurement. In many cases, you may want to know what the exact resistance of a device or circuit branch is, or you may need to find the voltage drop across or current flowing through a given component. All of these quantities can be obtained by using the appropriate measurement device. These devices will be described in the following sections. Before we look at the measurement devices, however, let's take a look at the concepts of measurement error, accuracy, precision, and resolution.

4.4.1   Measurement Errors

Naturally, when we take a measurement, we want it to be as **accurate** as possible. When we say accurate, we mean that we want our measurement to approach the actual value as closely as possible. Of course, depending upon the application, an acceptable margin of error can be set. It might be acceptable to have an error of a few grams while weighing vegetables, but definitely not while weighing gold!

We often talk of **precision** instruments. What is precision? Precision is the ability of the instrument to return the same numerical measured value on successive attempts to measure the same quantity. In other words, it is the ability to be consistent, or reliable. Precision is the measure of the degree to which successive measurements of a quantity differ from one another. If the measurement device possesses high precision, the successive measurements will be extremely similar.

An easy way to understand accuracy and precision is to compare measuring a quantity with a measurement device to shooting arrows at a target. The bullseye on the target represents the exact value of the quantity being measured. Thus, if an arrow strikes the target near the bullseye, it is considered to be more accurate than one that strikes further from the bullseye. After several shots have been fired, we can determine precision. If all of the shots fall within a very small area, we can say that the shooter is very precise. The larger the area over which the shots fall, the less precise the shooter. It should be noted that the grouping only needs to be small to provide high precision. It does not need to be near the bullseye. Thus, a measurement device can be precise without being accurate.

Another concept that should be understood at this point would be that of **significant figures**. When a value is expressed using significant figures, all of the measured digits of the value that we are certain of are provided. Then one digit that we are slightly uncertain of is added. For example, if we express a weight as 125.3 grams, we are indicating that we are absolutely certain that the item weighs 125 grams, but we are slightly uncertain about the final measurement of 3 tenths of a gram. Since we are uncertain about the tenths of a gram measurement, it would be useless to add any further figures to the measurement as they would depend on the accuracy of the uncertain digit. As might be expected, an indication of the precision of a measurement can be obtained from the number of significant figures in which the result is expressed. Significant figures convey actual information regarding the magnitude and the measurement precision of a quantity. The more significant figures in the value, the greater the precision of measurement. For example, if a resistor is specified as having a resistance of 50 ohms, its resistance should be closer to 50 ohms than to 49 ohms or 51 ohms. However, if the value of the resistor is described as 50.0 ohms, it means that its resistance is closer to 50.0 ohms than it is to 49.9 ohms or 50.1 ohms. When we write 50 ohms, we only supply two significant figures, but in 50.0 ohms we specify three. Thus, when we provide more significant figures, we are more certain of our answer. This implies a higher level of precision in our measurement device.

Resolution is the smallest division that an instrument can measure. Perhaps a scale, used to weigh vegetables, is calibrated in grams. So it has a resolution of 1 gram. However, the scale used by a scientist would likely be calibrated in milligrams (or even smaller units). Thus, if there is a change of even 1 milligram on the scientific scale, it will register a different reading, but the change of 1 milligram will be lost on the grocer's scale. As you might expect, the resolution of the measurement device has an impact on the precision of the final measurement. If the scale has a resolution in grams, a measured weight is uncertain within a range of ± 0.5 grams. If the scale is reporting values in milligrams, however, our uncertainty range is reduced to ± 0.0005 grams.

Now that we have an understanding of accuracy, precision, and resolution, we can discuss the important topic of error. **Error**, simply put, is the deviation of the measured value from the true value. In order for our measurements to result in any meaningful information, we need to attempt to minimize errors. The study of error allows us to determine the accuracy of a given instrument.

The errors that are regularly encountered can be broken down into three basic types. These are: 1) Gross Errors, 2) Systematic Errors, and 3) Random Errors.

The first of these categories, **gross errors**, result from human mistakes. Misreading instruments, incorrect adjustment and computational mistakes all fall under this category of error. Thus, to combat gross errors, an individual must simply be sure that they are performing their job correctly.

Systematic errors stem from shortcomings of the measuring instrument itself, for example worn out parts and the effects of the environment on the instrument and the user. Systematic errors can be further subdivided into two other kinds of errors: 1) Instrumental error and 2) Environmental error.

Instrumental errors are errors that come into play because of an inherent problem with the instrument itself. Many measuring devices are electromechanical devices and rely on springs, magnets and gears. These parts will exhibit wear over time. As a result, they cannot function in the same way that they did when they were new. So, unless the instrument is carefully calibrated to account for these changes, error creeps in. Another common cause of instrument error is overloading the instrument. Any measuring instrument is designed to measure its respective quantity within a certain limit. For example, a scientific balance can weigh substances in the order of milligrams while an airport check-in counter can weigh bags in the order of hundreds of kilograms. So, if you were to place a large suitcase on a delicate scientific balance, you can be sure that it would never measure milligrams again! (or anything else, for that matter!) This is, of course, a worst-case scenario, but it is still important to use equipment that is designed to handle the range of values that you expect to be dealing with. Instrument calibration, or marking the scale on the instrument, is another important area of consideration. The instrument may measure perfectly, without any error, but if the pointer points on an erroneous value, the measurement conveyed to the instrument's operators is also erroneous. Thus, the instrument must be calibrated by comparing its results against those of another perfectly calibrated, standard instrument. Most of the time, it is not possible to discard an instrument, which provides some minor errors, as they are quite expensive. In such a case, the exact amount of error that occurs while using that instrument is measured and a correction factor is calculated. This correction factor is incorporated into each quantity measured by the instrument.

Environmental errors are caused by the surroundings in which the instrument is placed. Changes in temperature, humidity, pressure, and magnetic or electrostatic fields, can all cause erroneous readings in instruments. Some instruments come with certain prescribed conditions in which they must be operated. For example, a sensitive scientific balance requires that it is placed on a hard level surface in order to provide correct readings. In addition, any vibration transmitted to the instrument will also distort the reported weight. As a result, these instruments are often placed on carefully prepared surfaces (such as heavy slabs of marble) to minimize these environmental sources of error. If instead, the instrument is placed on an uneven folding table, the user should not expect accurate results. The instrument may be of extremely high quality, but if the operator does not correctly control the environment in which it is placed, the full potential of the instrument cannot be realized.

Random errors, as the name suggests, are caused by random variations in the quantity under measurement or in the system of measurement. These errors have unknown origin and occur even when all systematic and gross errors have been accounted for. The design of the measurement device often plays an important role in determining the existence of random errors. A well-designed instrument or experiment reduces the chance of random errors. In most normal cases, random errors are not significant and can be ignored, but in high-precision work, they play a major role and must be addressed in some way.

One of the best ways to address random error is through repetition. When making any measurement, one should repeat the same procedure a few times and note the value each time. Then the arithmetic mean (average) of all the independent readings can be calculated. This approach will serve to make the reading more accurate and reduce the effect of errors that may creep into the measuring process. The best way to counter random errors is to increase the number of measurements and use statistical methods to obtain the best approximation of the true value of the quantity under measurement. Thus, when dealing with important measurements, a careful investigator will employ **statistical analysis**.

Statistical analysis allows an analytical determination of the uncertainty of the final test result. Statistical methods depend upon large quantities of data that have been carefully measured and collected over long periods of time. This information can help us to make an accurate approximation that reduces the error of a given measurement. Some common statistical methods are described in the following paragraphs.

One of the simplest statistical approaches is the **arithmetic mean**. When a series of measurements is taken, the most probable value for the quantity being measured is the arithmetic mean (average) of those readings. The most accurate value would be obtained if an infinite number of readings were available. However, since that is not possible, we generally take as many readings as is practical. Since arithmetic mean of these values is simply their average, we can write the formula for the arithmetic mean as:

$$\bar{x} = \frac{x_1 + x_2 + x_3 + ... + x_n}{n},$$

where $x_1, x_2, \cdots, x_n$ represent the values of the n readings that were taken.

In many cases, it is useful to know how much a given measurement of a quantity differs from the average reading that was taken. Such a measure is known as the **deviation from the mean**. This quantity gives us an indication of the difference between that particular reading and the value to which the entire measurement set is tending. If the deviation of an arbitrary reading, x_i, is called d_i, then d_i can be defined as:

$$d_i = x_i - \bar{x}.$$

It should be noted that that the deviation may be either positive or negative, depending on the value of the reading in question. If the reading is less than the mean, the resulting deviation will be negative, while a reading larger than the mean will result in a positive deviation.

Taking the idea of deviation from the mean a step further, we arrive at the **average deviation**. The average deviation gives us an indication of the precision of the instrument used in making the measurement. Highly precise instruments yield a low average deviation between their readings. The average deviation is calculated as the sum of the absolute values of the deviation between the mean value and each measurement divided by the number of readings. Thus, it can be written as:

$$\bar{d} = \frac{|d_1| + |d_2| + \cdots + |d_n|}{n},$$

where $|d_1|, |d_2|, \cdots, |d_n|$ represent the absolute values of the n deviations.

Another measurement of deviation, known as the **standard deviation** can also be calculated. The standard deviation, σ is a specific kind of deviation calculated on the root-mean-square of the values. Thus, it is also sometimes called the root-mean-square deviation. It is defined as the square root of the sum of all the individual deviations squared, divided by the number of readings. This can be expressed mathematically as:

$$\sigma = \sqrt{\frac{d_1^2 + d_2^2 + d_3^2 + \cdots + d_n^2}{n}}$$

where $d_1, d_2, \cdots, d_n$ represent the deviation of each of the n readings from the mean value.

The final statistical measurement that is commonly used when dealing with random errors is **variance**, or mean square deviation. This measure is almost the same as the root-mean-square deviation, except that the root is not calculated. Thus, variance, σ^2, is simply the square of the mean square deviation. This can be written as:

$$\sigma^2 = \frac{d_1^2 + d_2^2 + d_3^2 + \cdots + d_n^2}{n}.$$

The use of these and other statistical analysis methods can help us to deal with the inevitable random errors that are encountered when measuring quantities. Let's sum up our look at measurement errors with an example of the use of these statistical analysis approaches.

Example 4.17

Problem: A given voltage was measured 5 different times. The following values were obtained from these measurements: $\{V_1 = 5.56v, V_2 = 5.49v, V_3 = 5.55v, V_4 = 5.52v, V_5 = 5.61v\}$.

    a.   What is the arithmetic mean of the measurements?
    b.   What is the deviation from the mean for each value?
    c.   What is the average deviation?
    d.   What is the standard deviation?
    e.   What is the variance?

Solution: In this example, we have been given five voltage measurements. We are asked to determine some statistical information about this set of data. For part a, we simply need to calculate the average value. Doing this yields:

$$\overline{V} = \frac{5.56v + 5.49v + 5.55v + 5.52v + 5.61v}{5} \approx 5.55v$$

Next, we need to determine the deviation of each measurement from this mean value. Thus, we write:

$$d_1 = 5.56v - 5.55v = 0.01v$$
$$d_2 = 5.49v - 5.55v = -0.06v$$
$$d_3 = 5.55v - 5.55v = 0.00v$$
$$d_4 = 5.52v - 5.55v = -0.03v$$
$$d_5 = 5.61v - 5.55v = 0.06v$$

Using these values, we can find the average deviation.

$$\overline{d} = \frac{0.01v + 0.06v + 0.00v + 0.03v + 0.06v}{5} \approx 0.03v$$

Next, we can determine the standard deviation as:

$$\sigma = \sqrt{\frac{(0.01v)^2 (-0.06v)^2 (0.00v)^2 (-0.03v)^2 (0.06v)^2}{5}} \approx 0.04v$$

Finally, we can find the variance by squaring the standard deviating.

$$\sigma^2 = (0.04v)^2 = 0.0016$$

So, as we have seen, error is something that must be considered when measuring physical quantities. Now that we are aware of the potential problems, let's take a look at some of the devices that are used to measure quantities related to electronics.

4.4.2   Ammeters

The first measurement device that we will discuss is the ammeter. The ammeter as the name suggests is a device that measures the current in amps that is flowing within a circuit. Older ammeters relied on the use of the magnetic field generated by the current to deflect a needle. The degree to which this needle was deflected determined the amount of current flowing through the circuit. This meant that the correct polarity and expected range of current flow must be anticipated before taking a measurement. Using the wrong polarity could result in damage to the meter by forcing the needle backwards and possibly bending it. Likewise, using the wrong measurement range for the amount of current could lead to the needle being displaced too far and

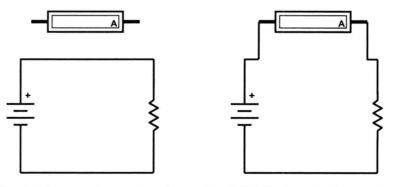

Circuit Before Inserting the Ammeter   **Circuit With the Ammeter Properly Inserted**

Figure 4.47 An illustration of the proper way to use an ammeter to measure current.

slamming into the stop peg at the end of the meter's range. This could also bend the needle. Modern ammeters are a bit easier to use. Since they offer digital readouts, they can easily handle negative values and many of them select the correct range automatically.

When working with an ammeter, it is important to remember that since it measures current running through the circuit, the circuit must be broken and the ammeter must be put into the circuit as shown in Figure 4.47. In this figure, the current running through the resistor is being measured.

4.4.3   Voltmeters

Another important electrical quantity that we need to measure is the voltage, or potential difference between two points within a given circuit. For this measurement, we turn to the voltmeter. As was the case with ammeters, older voltmeters use a coil in a magnetic field to displace a needle in order to indicate the voltage that is detected. This coil is similar to the coil used to construct analog ammeters, except in this case, a large resistance is placed in series with the coil. This large resistance limits the current that can flow through the coil to a very small portion of the overall current in the circuit. This is done to minimize the disturbance that attaching the device to the circuit might introduce. Since these analog voltmeters employ a coil and needle, they suffer from many of the same problems as analog ammeters. The correct polarity and voltage range must be selected to avoid damaging the device. Newer digital voltmeters do not suffer from these problems and are often capable of selecting the appropriate range for voltage display automatically.

When using a voltmeter, one must remember to connect the voltmeter in parallel with the section of the circuit across which the voltage is to be measured. The idea is to measure the change in potential between two points in the circuit, so each lead of the voltmeter is attached to one of these points of interest. This connection arrangement is illustrated in Figure 4.48. In this figure, the voltage drop that occurs across the resistor is being measured.

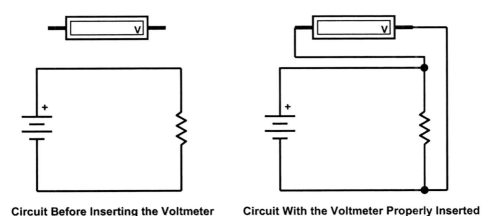

Circuit Before Inserting the Voltmeter   Circuit With the Voltmeter Properly Inserted

Figure 4.48 An illustration of the proper way to use an voltmeter to measure voltage.

4.4.4   Ohmmeters

As you will recall, ohm's law made use of three different electrical quantities. These were voltage, current, and resistance. Using voltmeters and ammeters, we are able to measure the first two of these three quantities. In order to be able to fully use ohm's law to analyze an electrical circuit, we must also have a means of measuring resistance. For this measurement, we turn to the ohmmeter. The ohmmeter actually employs ohm's law to determine the resistance of a given component or circuit segment. This is done in one of two ways. One early approach involved the application of a known voltage. Once the voltage is applied, the resulting current is measured. Since the voltage is the product of the current and resistance, a fixed voltage produces a current that is proportional to the resistance. The other approach works by passing a known current through the component or circuit segment, and then measuring the resulting voltage. By dividing this voltage by the constant current, the resistance can be obtained.

When measuring the resistance of a component, such as a resistor, it is important to remember that the component must be isolated from the rest of the circuit. If it remains connected, other components in the circuit can influence the resulting measurement. The proper method of measuring the resistance of a resistor is shown in Figure 4.49.

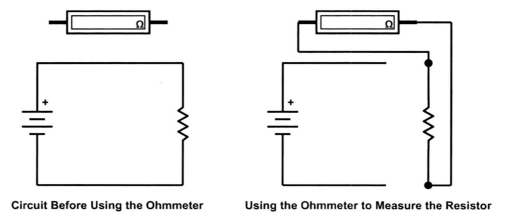

Circuit Before Using the Ohmmeter   Using the Ohmmeter to Measure the Resistor

Figure 4.49 An illustration of the proper way to use an ohmmeter to measure resistance.

4.4.5   Multimeters

While independent devices do exist for measuring current, voltage, and resistance, it is far more convenient to have the ability to measure all three of these physical quantities in one compact device. This is the idea behind the multimeter. Multimeters, as the name implies, provide the ability to measure multiple physical quantities. The quantity being measured and the expected range are often set by switches or knobs on the device. The unit will then operate as if it were a stand-alone device for measuring the selected physical quantity. Some of the more advanced multimeters provide additional features. These may include continuity tests to indicate if there are breaks in a wire or to check if a diode is operating properly. Since multimeters provide the ability to measure all of the physical characteristics necessary to use ohm's

Figure 4.50 An example of a typical handheld multimeter.

law to analyze a circuit, the individual measurement tools are seldom used. A typical handheld multimeter is shown in Figure 4.50.

4.4.6   Oscilloscopes

The oscilloscope is one of the fundamental instruments in electronics. As the name implies, an oscilloscope is used to view oscillations in a signal. In other words, it allows us to observe how the voltage in a given part of a circuit changes over time. As a result, we can use an oscilloscope to visually see the kind of waveforms being generated in any part of a given circuit.   A typical oscilloscope, much like the one shown in Figure 4.51, has a set of probes, which allow it to be connected to the part of the circuit that we are interested in. Once connected to the circuit, the oscilloscope will display the voltage level on the screen. Most oscilloscopes possess a number of features to allow the user to fine-tune the output display. Modern oscilloscopes can even keep the voltage waveforms in memory to be analyzed later, and some of them can be interfaced to a digital computer to take advantage of its resources to better process the waveforms.

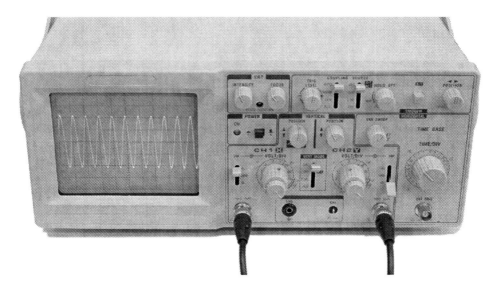

Figure 4.51 A picture of a typical oscilloscope.

By utilizing the measurement instruments described in the previous sections along with knowledge of ohm's law and circuit theory, it is possible to analyze most circuits that you will encounter. We will wrap up this chapter on electronics and electrical theory by looking at an example of a practical circuit.

4.5   Practical Example – AC to DC Conversion

When we discussed the differences between direct and alternating current earlier in the chapter, we noted that alternating current is the distribution method used to supply electricity to most homes and businesses. While many electrical devices, such as your kitchen appliances, have no problem using alternating current as a power source, other devices like your television, radio, and computer contain circuitry that requires direct current to operate. So then how can you plug these devices into the wall and get them to operate. The reason that you can plug these devices into an alternating current source is that they contain circuitry to convert the signal provided by the wall outlet into a direct current source that can be used by the device's internal circuitry. This circuitry is usually grouped together and called a DC power supply. So how do these supplies work? In this section we will look at how one could go about building a simple version of such a device. For the sake of this example, let us define the desired outcome. In this particular example, we will be designing a simple device to convert the 110V AC wall voltage to 5V DC voltage. This is a common voltage used to power digital logic circuits.

Our first step is to convert the input wall voltage into something that is a bit closer to the voltage that we are interested in for our DC application. The easiest way to do this is with a step down transformer. There are a wide variety of transformers on the market designed to step down the wall voltage to one of several standard lower voltages. In this case, we will use a transformer that provides a 9V output. The result of implementing this transformer is shown in Figure 4.52.

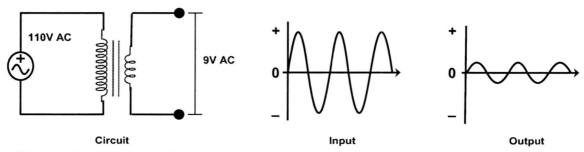

Figure 4.52 The result of using a transformer to step the wall AC voltage down to 9V AC.

Now that we have a manageable AC voltage, our next task is to devise a way to convert it to direct current. The first step in this process is known as rectification. Our goal here is to modify the AC waveform so that it no longer goes negative. This is done by using diodes. As you recall, diodes allow current to flow in only one direction. So, if we connect a diode to the positive output of the transformer, it would allow the current to flow in only one direction and block it completely in the other. The end result in this case is that the negative half of the AC waveform is reduced to zero. This process, known as half-wave rectification is illustrated in Figure 4.53.

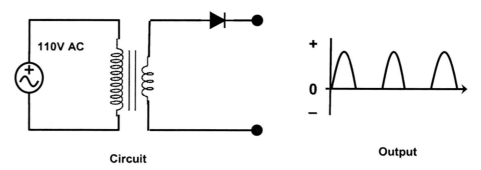

Figure 4.53 The result of half-wave rectification.

As you can see, half-wave rectification is an effective method of limiting the AC voltage to positive values. Unfortunately, it also removes half of the waveform that could be utilized. To remedy this problem, we will use full-wave rectification. If you think about it, an AC signal is a voltage source whose polarity switches back and forth during a given cycle. So, if we could use a diode arrangement to redirect the positive polarity to the positive output terminal during both halves of the cycle we could make use of the full waveform. To do this we will make use of a diode arrangement known as a bridge rectifier. The bridge rectifier is a four diode structure that is arranged so the diodes complete a circuit for the positive portion of the input cycle and then flips the polarity on the negative half of the cycle. This concept is illustrated in Figure 4.54. As you can see in the figure, the same DC output polarity is maintained through both cycles of the input AC waveform. By adding the bridge rectifier to our power supply circuit, we will get an output similar to that shown in Figure 4.55.

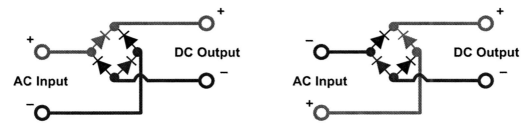

Figure 4.54. An illustration of the operation of a bridge rectifier during both polarities of the AC cycle.

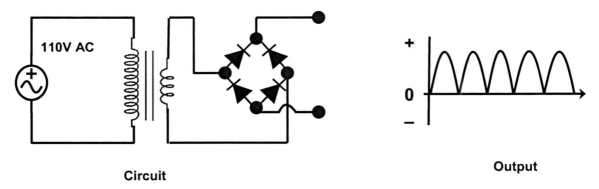

Circuit　　　　　　　　　　　　　　　　　　　Output

Figure 4.55 The result of full-wave rectification using a bridge rectifier.

As you can see, we now have an output with a constant current direction, but a varying voltage. As a result, our next task is to smooth out the voltage. If you think back to our discussion of capacitance, you will recall that capacitors resist changes in voltage by releasing stored charge to try to maintain the existing voltage. So, if we place a capacitor across our output, it will charge up during the rising portion of the waveform and discharge as the voltage drops off. This will result in a smoothed version of the waveform. The result of this can be seen in Figure 4.56.

The figure demonstrates that the addition of the capacitor does indeed smooth the output, but it does not provide a perfectly smooth result. Some amount of ripple voltage remains in the output. Increasing the size of the capacitor will decrease the ripple, but usually a ripple of less than 10% is considered acceptable for most electronics. However, we can take one final step to insure a smooth output voltage. We need to make sure that the output is regulated to the five volts that we initially specified. One of the easiest ways to do this is by using a zener diode. As you recall, a zener diode is designed to maintain a specific voltage drop across it. So, if we place a zener diode that is designed to break down at five volts across the output and a current limiting

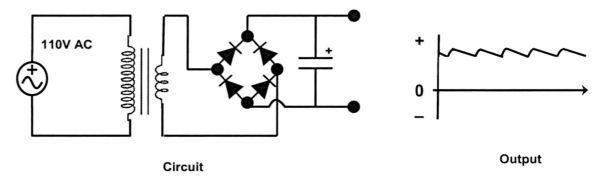

Circuit   **Output**

Figure 4.56 An illustration of the smoothing provided by adding a capacitor to the rectified output.

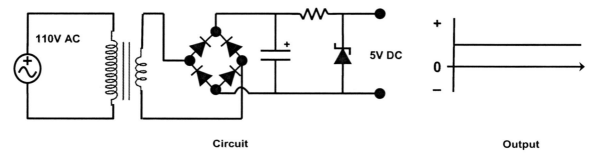

Circuit   **Output**

Figure 4.57 An illustration of a regulated power supply that generates a 5 volt DC output from the AC signal taken from a typical wall outlet.

resistor in series with it, we can maintain a five volt output. This arrangement is implemented in Figure 4.57.

So, we have just seen how to construct a simple power supply that is capable of taking the AC source from a wall outlet and converting it into a 5V DC source for use with electronics. This example illustrates that by using the understanding that we gained within this chapter, we can begin to understand how circuits operate and get a general idea of how to approach a circuit design problem.

4.6 Chapter Summary

In this chapter we discussed many of the basic topics relating electricity and electronics.

- We began by discussing the international system of measurements. We learned that in order to properly describe the measurements of physical quantities, we need to have a common system of standard units so that the measurements that we report can be reproduced and precisely understood by others.

- Next, we began our discussion of basic electrical concepts with a discussion of electrical charge. We learned that electrical charge results from an excess of charge carrying atomic particles, namely electrons or protons. We discovered that the SI unit of the coulomb is composed of 6.24 x 10^{18} charge carrying particles. In relation to the coulomb, we also discovered that the fundamental unit of charge, denoted as e is defined as 1.60 x 10^{-19} coulombs of charge.

- Since we found that similar electrical charges repel and differing electrical charges attract, we next looked at the concept of electrical forces. We found that the force exerted between two charged particles can be computed as

$$\mathbf{F} = k\frac{q_1 q_2}{r^2}\hat{\mathbf{r}}_{12}$$

where k is coulomb's constant (8.99 x 10^9 Nm^2/C^2), q_1 and q_2 are the charges of the two points, r is the distance of separation, and $\hat{\mathbf{r}}_{12}$ is a unit vector pointing from q_1 to q_2.

- Once we understood electrical forces, we turned our attention to electrical fields. We found that all electrical charges or bodies of charge produce an electric field. It is this field that exerts forces on other charged particles. The value of the electric field at a given point is a proportionality value that describes the force that will be exerted on a charged object per unit of charge (N/C). For a given positive test charge q_0, the electric field strength is given as:

$$\mathbf{E} = \frac{\mathbf{F}}{q_0}.$$

- We then found that when a charged particle exists in an electric field, it possesses an electric potential energy. This potential can be calculated as the product of the value of the electric field strength at the particles position in the field and the distance from the particle's location to the reference point (point of zero potential energy).

- Next, we discovered that the difference in potential energy between two points in an electric field is defined as the potential difference or voltage. The voltage measure described that motivation or push behind charge carrying particles within the electric field. One volt of potential difference is equivalent to one joule of energy per coulomb of charge.

- When charges move within an electric field, we need to understand the concept of an electric current. Electric Current, symbolized by I, is the quantitative measure of the flow rate of electric charge carriers. It is measured by determining the number of coulombs of charge that pass a specific point in the period of one second. The unit of electric current is the ampere. One ampere is defined as one coulomb of charge per second.

- Certain materials allow charge carriers to move more freely than others. The measure of a material's ability to sustain an electric current is known as conductance. The higher the conductance of a material, the less potential difference is required to sustain a current.

- The opposite of conductance is resistance. Resistance is a measure of a material's opposition to current flow. The higher the resistance the more voltage is required to maintain a current. Resistance is measured in ohms.

- With an understanding of the concepts of voltage, current, and resistance, we examined a very important relationship between them. This relationship, known as ohm's law, can be expressed in one of the following thee ways:

$$R = \frac{V}{I} \qquad I = \frac{V}{R} \qquad V = IR .$$

- Next, we directed our attention to the concept of capacitance. Capacitance is a measure of the amount of charge that can be stored within a capacitor. A basic capacitor can be constructed by connecting leads to two metal plates that are separated by a small distance. When a voltage is applied to the plates, charge carriers are stored on the plates and an electric field forms between them. Capacitance measures the amount of charge that can be stored in a given capacitor per unit of voltage applied.

- Capacitance may also be determined from the physical characteristics of the capacitor. By using the plate area, A, the dielectric constant, κ, and the distance between the plates, d, the capacitance value can be calculated using

$$C = \frac{A\kappa(8.85x10^{-12} \text{ F/m})}{d} .$$

- The energy stored in a capacitor is stored in the electric field that develops between the plates. The energy is the result of the work done to move the charge carriers through a potential difference from one plate to the other when the capacitor is charged. The amount of energy stored, measured in joules, can be determined using one of the following two equations:

$$U = \frac{1}{2}QV \quad \text{or} \quad U = \frac{1}{2}CV^2$$

Where U is the energy stored, Q is the charge, V is the voltage, and C is the capacitance.

- A capacitor operates by resisting changes in voltage. The device charges or stores charge when voltage is applied and will discharge to try to maintain the applied voltage.

- While capacitors store energy in an electric field, inductors use a magnetic field. We learned that a current flowing through a wire generates a magnetic field and that if we wind the wire into a coil, or solenoid, the magnetic field is concentrated. The magnetic field found within a long solenoid is dependent on the physical characteristics of the solenoid as well as the current passing through it. These physical characteristics include the length of the coil, ℓ, the number of turns of wire, N, and the permeability of the core material, μ. Using this information the magnetic field, B, is calculated as:

$$B = \frac{\mu N I}{\ell}$$

- The magnetic flux through a long solenoid, ϕ_m, is a measure of the number of magnetic field lines that pass through the solenoid's cross-sectional area. It can be determined by multiplying the magnetic field, or flux density, B, of the solenoid by its cross-sectional area A_c and the total number of turns in the solenoid, N. Thus, either of the following formulas can be used to find the magnetic flux through a solenoid.

$$\phi_m = NBA_c \qquad \phi_m = \frac{\mu N^2 A_c I}{\ell}$$

- The self inductance of a solenoid is the proportionality constant that, when multiplied by the current passing through the coil, yields the magnetic flux through the coil. Thus, the self inductance of a solenoid can be determined either by dividing the magnetic flux by the current or purely from the physical characteristics of the device. So, either of the following equations can be used to determine self inductance. The decision as to which one should be used depends on the information that is available.

$$L = \frac{\phi_m}{I} \qquad L = \frac{\mu N^2 A_c}{\ell}$$

In these equations L represents self inductance, ϕ_m represents magnetic flux through the solenoid, I represents current through the wire, μ represents the permeability of the core material, N is the number of turns of wire, A_c is the cross-sectional area of the solenoid, and ℓ is the length of the coil.

- The energy stored in the magnetic field of an inductor is proportional to the self inductance of the coil, L, and the current running through the coil, I. The energy within the magnetic field is written as:

$$U = \frac{1}{2} L I^2$$

- An inductor operates by resisting changes in the current flowing through it. As the current through the inductor's coil changes, the resulting change in the magnetic field around the coil induces a counter voltage to try to maintain the existing current.

- Next, we examined the concept of electrical power. The instantaneous power, P of an electrical device is defined as the work that is done per unit of time. In terms of the input voltage, V, and current, I, power can be defined as:

$$P = IV$$

- It is important to understand the difference between direct and alternating current. Direct Current is an unvarying current that is associated with a stable potential difference. It flows in only one direction at a constant rate. Alternating Current is a fluctuating current that is associated with a changing potential difference. Alternating current flows at a fluctuating rate and switches direction periodically. The most common alternating current pattern is associated with a sinusoidal change in voltage.

- An electrical circuit is defined as a path for current to flow between two points. Most electrical circuits consist of three basic parts, a voltage source, a load, or device in which work is done, and a path for the charge carriers to follow.

- Once we understood the concept of the circuit, we looked at a wide variety of electrical components that might be encountered in typical electronics applications. We examined the basic construction of the component, looked at examples of each component, and learned what the schematic symbols looked like. The components we examined included: sources, switches, resistors, capacitors, inductors, transformers, diodes, transistors, and integrated circuits.

- Another important concept when dealing with electrical circuits is understanding measurement. We looked at the concepts of precision, accuracy, and resolution. Then we examined some of the errors that can occur when measuring electrical quantities. These include gross, systematic, environmental, and random errors. We then saw that we could use various statistical analysis tools to help reduce the impact of these errors.

- Finally, we looked at a practical example of some of what we had learned in this chapter. We examined how we could construct a power supply that was capable of converting an AC electrical source into a DC electrical source using some of the components that we learned about.

4.7 Review Exercises

Section 4.1 – The International System of Units
_____

1. Explain why a standard system of measurement units is important.

2. Describe the international system of units. What benefits does it have over the US customary system?

3. Give the SI unit that is used for the following measured quantity.

    a. Length   b. Electric Current   c. Electric Potential

    d. Frequency   e. Power   f. Energy

    g. Resistance   h. Capacitance   i. Inductance

4. Provide the measured quantity associated with each of the following SI units.

    a. Meter   b. Hertz   c. Ohm   d. Tesla   e. Joule

    f. Watt   g. Henry   h. Farad   i. Newton   j. Volt

5. Provide the SI prefix for the following powers of ten.

    a. 10^{-6} b. 10^{3} c. 10^{12} d. 10^{-12} e. 10^{6}

    f. 10^{-2} g. 10^{-9} h. 10^{9} i. 10^{-3} j. 10^{18}

6. State the power of ten associated with the following SI prefixes.

    a. centi   b. giga   c. pico   d. tera   e. milli

    f. mega   g. nano   h. kilo   i. micro   j. peta

Section 4.2 – The Basics of Electricity
_____

7. What is the term used to describe an atom that gains an extra electron?

8. What is the term used to describe an atom that is missing an electron?

9. Describe how "static" electrical charge is developed when a comb is run through someone's hair. What charge does the comb take on? What charge does the hair take on?

10. Define electrical charge. What symbol is used to represent electrical charge?

11. Describe the fundamental unit of charge. What symbol is used to represent this quantity?

12. Define the coulomb.

13. Determine how many coulombs of negative electrical charge the following numbers of electrons represent.

    a. 6.24×10^{19} b. 186.24×10^{17} c. 35×10^{22} d. 18.72×10^{17}

    e. 1.56×10^{18} f. 9.05×10^{19} g. 3.75×10^{15} h. 13.75×10^{25}

14. Approximately how many electrons are represented by the following values?

    a. 1.75 C   b. 7.75 C   c. 0.2 C   d. 15 C   e. 1.18 C

    f. 9.22 C   g. 107 C   h. 0.01 C   i. 1.03 C   j. 3.45 C

15. Define Coulomb's constant.

16. Given the following charge pairs and distances of separation, determine the force that is exerted between the charges.

    a.   Charge 1 = 1.75 C   Charge 2 = -1.5 C   Distance = 2 cm
    b.   Charge 1 = 0.75 C   Charge 2 = 2.5 C   Distance = 10 cm
    c.   Charge 1 = 3 C   Charge 2 = -2 C   Distance = 22 cm
    d.   Charge 1 = 5 C   Charge 2 = 3 C   Distance = 2 mm
    e.   Charge 1 = -0.25 C   Charge 2 = -0.5 C   Distance = 5 mm
    f.   Charge 1 = -6.75 C   Charge 2 = 1.55 C   Distance = 3 cm

17. Assume that you are using a test charge to define an electric field. Given the following test charges and the forces detected at a given point in the field, determine the electric field value at that point.

    a. Test Charge = 2 nC   Force = $1.35 \times 10^{-3} \mathbf{j}$ N

    b. Test Charge = 75 nC   Force = $3.54 \times 10^{-2} \mathbf{i}$ N

    c. Test Charge = 25 mC   Force = $3.25\mathbf{i} + 7.25\mathbf{j} - 1.5\mathbf{K}$ N

    d. Test Charge = 1.35 mC   Force = $-2 \times 10^{-2}\mathbf{i} + 3 \times 10^{-2}\mathbf{j} - 5.3 \times 10^{-2}\mathbf{k}$ N

18. List some of the rules to keep in mind when looking at illustrations of electric field lines.

19. What is electrostatic potential energy? When measuring the electrostatic potential of a charge in an electric field, what quantities does it depend on?

20. What is the relationship between electrostatic potential energy and electric potential?

21. What is meant by voltage? What are some other terms used to describe the voltage?

22. What is an electric current? What unit is used to measure electric current? How is this unit defined?

23. Given the following times and quantities of charge, determine the current that is present.

    a. 15 C of charge pass a given point in 5 seconds.
    b. 115C of charge pass a given point in 30 seconds.
    c. 1.5 C of charge pass a given point in 0.25 seconds.
    d. 4.75 C of charge pass a given point in 1.82 seconds.

24. Define conductance.

25. List the four categories of conductance, describe them, and provide an example of a material in each category.

26. Define resistance.

27. What is Ohm's Law? Why is it so important?

28. Given the following electrical quantities, use Ohm's Law to solve for the missing value.

    a. Voltage = 24 V   Current = 2 A   Resistance = ? Ω
    b. Voltage = ? V   Current = 1.5 A   Resistance = 100 Ω
    c. Voltage = 100 V   Current = ? A   Resistance = 25 Ω
    d. Voltage = 9 V   Current = 5 A   Resistance = ? Ω
    e. Voltage = ? V   Current = 7.5 A   Resistance = 200 Ω
    f. Voltage = 110 V   Current = 10.5 A   Resistance = ? Ω
    g. Voltage = 12 V   Current = ? A   Resistance = 5 KΩ
    h. Voltage = ? V   Current = 0.5 A   Resistance = 300 Ω

29. How is a basic capacitor constructed?

30. What does capacitance measure?

31. What is a dielectric material? What purpose does it serve in a capacitor? Give an example of a dielectric material.

32. A given capacitor has a capacitance of 50 mF. How much charge can the capacitor hold when a voltage of 10 V is applied?

33. Calculate the capacitance of a capacitor that stores a charge of 0.75 C of charge when a voltage of 25 V is applied.

34. A capacitor with a capacitance of 500μF is currently storing 50 C of charge. What is the voltage that is being applied to the capacitor?

35. You are designing a capacitor with a plate separation of 1mm and using air as a dielectric. What plate area is required to achieve a capacitance of 55 nF? What is the required plate size if Mylar® is used instead of air for the dielectric material?

36. Calculate the capacitance of a capacitor with the following characteristics.

   a. Plate Area = $2m^2$ Plate Separation = 1 mm   Dielectric Material = Mica
   a. Plate Area = $0.75m^2$ Plate Separation = 0.5 mm   Dielectric Material = Air
   a. Plate Area = $0.2m^2$ Plate Separation = 2 mm   Dielectric Material = Paper
   a. Plate Area = $0.05m^2$ Plate Separation = 1 mm   Dielectric Material = Glass

37. How do capacitors store energy?

38. A given capacitor has a capacitance of 50 mF and has a voltage of 30 volts applied to it. How much energy is stored in the electric field of the capacitor?

39. How is the magnetic field related to electricity? How can electricity form an electric field?

40. What happens to the electric field when a wire is arranged in a long coil or solenoid?

41. Calculate the magnetic field of a solenoid of length 20 cm with 100 turns of wire and a core of air if you pass a current of 1.5 A through it. What would the field strength be if you insert a core of annealed iron with a permeability of 1200?

42. If you are constructing a coil that has a length of 5 cm and no core, how many turns of wire will be required to achieve a magnetic field of 3 mT when a current of 2 A is passing through it?   How many turns would be needed in an annealed iron core with a permeability of 1100 is used?

43. You have been given a long solenoid with a length of 25 cm, a radius of 1 cm, 500 turns of wire, and an iron core with a permeability of 1100. Determine the magnetic flux passing through the solenoid if a current of 0.2 amps is run through the wire.

44. Find the self inductance of a long solenoid with a length of 20 cm, a radius of 1mm, and 700 turns of wire. Assume a core of air is used. What is the self inductance if an iron core with a permeability of 1000 is being used?

45. Describe how an inductor stores energy. How does an inductor react to changes in current?

46. Given an inductor with a stated inductance of 100 mH and a current of 0.75 A, determine the energy stored in the magnetic field of the inductor.

47. If an inductor with a stated inductance of 75 mH has 125 mJ of energy stored in its magnetic field, how much current is passing through the coil?

48. Define instantaneous power.

49. Using the given equation for power, show that power can also be computed as the product of the resistance and the square of the current.

50. An electronic device requires an input voltage of 9 volts and draws a current of 1.5 amperes. What is the power consumption of the device?

51. A resistor with a resistance of 200 ohms has a current of 0.75 amperes passing through it. How much power is being dissipated by the resistor?

52. Describe the difference between alternating and direct current.

53. A given AC voltage is described by the equation $340\sin(2\pi 50t)$. Answer the following questions:

   a. What is the frequency of the AC source?
   b. What is the peak voltage?
   c. What is the peak-to-peak voltage?
   d. What is the RMS voltage?

54. Describe what is meant by an electrical circuit.

Section 4.3 – Electrical Components
---

55. List some common electrical sources.

56. What happens to the voltage and current when sources are placed in series? How about when they are placed in parallel?

57. What is the purpose of a switch.

58. How does a relay work?

59. Describe some of the techniques used to construct resistors.

60. Find the value and tolerance of the following resistors given their color band pattern.

   a. { Green, Brown, Yellow, Gold }
   c. { Brown, Brown, Black, Gold }
   e. { Red, Yellow, Brown, Orange, Red }
   g. { Violet, Blue, Red, Red, Green }

   b. { Red, Orange, Green, Silver }
   d. { Green, Black, Violet, Gold }
   f. { White, Orange, Brown, Gold }
   h. { Orange, Orange, Brown, Blue }

61. Describe how a variable resistor works.

62. Describe the construction techniques used to build most capacitors.

63. Describe the construction of inductors.

64. Explain how transformers work.

65. You have been given a transformer. It has 800 turns on its input coil, and 2000 turns on its output coil. Determine the voltages that will be measured across the output coil of the transformer if you apply a voltage of 120 V to the input coils.

66. What function does a diode perform? Explain what happens when the diode is reverse biased. What happens when it is forward biased?

67. Describe the two types of BJT.

68. How is a FET constructed?

69. What can a transistor be used for in electronic applications?

70. What is an integrated circuit? What purpose do they serve?

Section 4.4 – Measurement Devices
_____

71. Describe the difference between accuracy and precision.

72. What are significant figures?

73. What is a gross error?

74. What is a systematic error? List two types of systematic error and describe them.

75. What is a random error?

76. What purpose does statistical analysis serve?

77. You take seven different measurements of a given voltage and record the following values:

$$\{V_1 = 5.50v, V_2 = 5.62v, V_3 = 5.49v, V_4 = 5.30v, V_5 = 5.62v, V_6 = 5.82v, V_7 = 5.22v\}$$

    a.   What is the arithmetic mean of the measurements?
    b.   What is the deviation from the mean for each value?
    c.   What is the average deviation?
    d.   What is the standard deviation?
    e.   What is the variance?

78. What does an ammeter measure? How should it be used?

79. What does a voltmeter measure? How should it be used?

80. What does an ohmmeter measure? How should it be used?

81. What benefit does a multimeter provide?

82. What is the purpose of an oscilloscope?

Section 4.5 – Practical Example – AC to DC Conversion
_____

83. Why would we need to convert AC to DC?

84. Describe the process of converting AC to DC.

Chapter 5. Electric Power

Chapter Outline

5.1   Introduction

When the average person thinks about electricity, the first thing that typically comes to mind is the electricity that comes from a wall outlet. This electricity, which is commonly provided to homes and businesses, is essential to most people's daily lives. Without it, most people would be, at the least, very inconvenienced. For evidence of this, one only needs to experience a power outage. Without lights and computers, most businesses would grind to a halt. Even with our heavy reliance on electrical power, few people stop to think about all of the infrastructure and engineering required to generate and supply this power to their homes. While the average person may take the electrical power system for granted, it is an important area of study for electrical engineers. In fact, the initial focus of electrical engineering was the design and construction of power generation and distribution systems and electrical power is still an important branch of electrical engineering today. In this chapter, we will take a quick look at the beginning of the electrical power industry in the United States and examine how a typical electrical power distribution system works today. With this understanding, you will be able to better appreciate what must be done to provide power to your home or business.

5.2   A Brief History of Electrical Power

As you might imagine, a great deal of effort went into the development of today's electric power grid. So, we will begin our study of electric power by looking at how the current industry developed.

Table 5.1 Descriptions and examples of some common types of energy.

Energy	Description	Example
Kinetic	Energy of motion.	Any object in motion possesses kinetic energy
Potential	Energy of position.	An object suspended above the ground possesses potential energy.
Chemical	Energy stored in chemical form.	The energy in materials such as wood and other fuels is stored in chemical form.
Thermal	Heat energy.	Heat produced by burning fuels.
Radiant	Energy transmitted in waves.	Sunlight is considered radiant energy. Any other electromagnetic waveform, like radio waves, is also a form of radiant energy.
Nuclear	Energy stored in the forces binding atoms together.	Nuclear energy is stored in all atoms. When atoms are split apart (fission) or fused together (fusion) some of this energy is released.
Electrical	Energy involving charged particles, typically electrons.	The most common form of naturally-ocurring electrical energy is lightning. Other forms of electrical energy can be created by people by converting other forms of energy to electrical energy.

From mankind's earliest history, an attempt has been made to harness the environment around us to make our lives easier. In its simplest form, this equates to the development of tools and methods that can be employed to convert or direct naturally occurring sources of energy into something that does work for us. For example, we discovered early in our history that, we could use fire to convert the energy stored in wood into both radiant energy in the form of light and thermal energy. Some of the types of energy that we can harness are described in Table 5.1. Of these types of energy, electrical energy has some major advantages. First, electrical energy is relatively easy to generate from other energy sources. In addition, it is easy to control energy in its electrical form. It can be switched and routed from place to place with ease and travels at the speed of light. Finally, electrical energy can be easily transformed back into other forms of energy, particularly radiant, thermal, and kinetic energy.

Methods for converting energy between electrical and the other available forms were not always available, however. In fact, electrical energy was seldom used outside of the laboratory until the 1800's. People typically relied on sunlight, fire, and candles to provide them with heat and light, and they relied typically on horse-drawn carriages for transportation. In the early 1800's, gas lighting began to become popular in many large cities. These lights replaced many of the candles and kerosene lamps that had been used before and remained popular through much of the 1800's. From about 1840 through the 1870's scientists were hard at work exploring the concepts of electricity and magnetism. During this time Michael Faraday developed his famous law relating electricity and magnetism. This scientific study led to the development of some of the earliest electrical generators. These primitive devices used magnetic induction to produce electric current. With the advent of the electric generator, there was now a means of producing electrical energy as it was needed. However, in the first half of the 1800's there was little application for electrical power, and without a commercial application, there would be little motivation to spend large amounts of time and money to refine the technologies related to electric power generation. Luckily, the 1870's brought the commercialization of the arc light. The arc light produces an extremely bright source of light by separating two carbon electrodes. When a sufficient voltage is applied, the air between the electrodes is ionized and an intense arc of electricity is formed. This bright light source found a valuable use in street lighting applications and as a result, arc lights produced one of the first motivations to generate and provide large amounts of electrical power.

While arc lighting provided an application for the generation of electric power and resulted in the construction of some early generators and power plants, the intense brightness of these devices made them impractical for use inside the home. For indoor applications a different type of electrical lighting would be required before widespread acceptance of electrical illumination could occur. Early studies had shown that by passing electricity through a thin piece of wire or other conducting material the material could be heated up to the point that it would give off light. This light was far less intense than that of the existing arc lights, and if a way could be found to get a material to glow for a long period of time without burning up, it could be easily used for indoor electrical lighting. The difficulty was finding a substance that would glow for a reasonable amount of time before it burned up. Numerous individuals, motivated by the amazing potential of such a device worked relentlessly to discover a usable material. It was not until October of 1879, however, that a breakthrough was made. A young inventor named Thomas Edison discovered that if he placed a carbonized cotton thread in a clear glass bulb, removed the air, and applied an electric current to the thread, it would glow for a little over 40 hours before the thread burned out. This proved that the creation of a practical indoor light source was possible. Edison went on to improve on his light bulb design and developed a new highly efficient generator to power it.

It was in 1882, only three years after this groundbreaking discovery, that the first power system designed to generate and sell power to consumers in the United States for the purpose of indoor lighting was built. Housed at Pearl Street Station in New York City, this system provided a total power of around 30kW of direct current to customers and was the beginning of today's commercial electrical power system. As the use of Edison's electric light became more widespread, more and more "Illuminating" companies were formed to sell electric power. However, these companies soon encountered one of the major problems faced by all power companies. This was the problem of fluctuating demand. They found that the demand for power would be greatest in the early evening hours and drop off quickly over the late night hours. This meant that they were not taking full advantage of all of the power that they could generate during these off-peak times. As a result, they needed to find applications other than lighting for which their power could be used. For this, they turned to industrial applications and the electric motor. The new availability of electric power led to the development of numerous other devices, such as the electric iron, all designed to use electricity to make our lives easier.

As demand for electric power increased, another problem was encountered. Since the existing electric power stations used DC, the power received by customers would drop off as the distance between the power plant and the customer grew. In fact, the early DC plants, which supplied power at a voltage of approximately 200 volts, were only effective for a radius of about a half mile. Since it was too expensive and impractical to construct a new power station every mile or so, the most obvious solution to this problem was to begin transmitting the power from existing stations at higher voltages. However, it was found to be unsafe to supply these high voltages to the home. Thus, the best solution would be to transmit high voltage power over large spans and drop this voltage down for use in local areas. Unfortunately, there is no easy way to do this with direct current. The voltages of alternating current, on the other hand, can be easily changed using a transformer. To make the AC solution look even more attractive, another scientist by the name of Nikola Tesla devised methods for both generating AC power and for using it to drive electric motors. By addressing these two problems, Tesla cleared the way for the widespread adoption of AC power in the United States. Using Tesla's designs, the entrepreneur George Westinghouse built AC electric generators at Niagara Falls in 1893. These generators produced an AC voltage at a frequency of 25 Hz and helped to supply the city of Buffalo with power. While both the power output and frequency have changed, the power station at Niagara Falls is still a critical part of the United States power system today.

Now that power companies were generating AC power, the extended range over which they could supply power resulted in many areas of overlap between competing power companies. In fact, in some areas, power for lighting, transportation, and industrial applications were all supplied by different companies using different specifications. Duplicating resources to supply power in this way was an inefficient waste of power production resources. To reduce redundancies and allow power companies to interconnect their power systems to share resources, a standard needed to be developed. As we just learned, the power station at Niagara Falls generated power at 25 Hz. While this was an efficient frequency at which to generate power, it produced a noticeable flicker in electric lights. As a result, a higher frequency was needed so that the human eye could not detect the flickering. After some time and negotiations, a compromise was reached and a standard of 60 Hz was agreed upon in the United States (50 Hz became the standard in Europe). Thanks to this standardization, it became possible for power companies to interconnect their systems so that they could help provide power to other grids during peak usage hours. This early interconnection was the beginning of today's advanced power grid. Since that time, there have been numerous refinements in the efficiency and methods of power generation, but the basic approach to power distribution has remained relatively unchanged. To this end, the power grid in the United States is unlike that of most other countries. While the country's power grid is interconnected, it is not federally owned. Instead it is owned by hundreds of locally and government owned companies. This cooperation between multiple corporations has led to an advanced and adaptable power system that provides millions of homes and businesses with power with minimal interruptions. So, while the future will provide a wide array of challenges, the power system will continue to adapt and provide reliable service.

5.3   Layout of a Typical Electrical Power System

Now that we have an understanding of how today's power system developed, we can turn our attention to the way in which a modern power system is laid out. In other words, let's look at what it takes to get power from a generator to the wall outlet in your home. This process can typically be broken down into six different steps. These steps are illustrated in Figure 5.1. The first step involves the generation of the electrical power in a power plant.

Next, in the transmission phase, the AC voltage produced by the generator is stepped up to an extremely high voltage, typically in the hundreds of thousands of volts, by a transformer. This high voltage signal is then sent long distances over high power lines. These lines connect to large sub-stations, which once again use transformers to step the

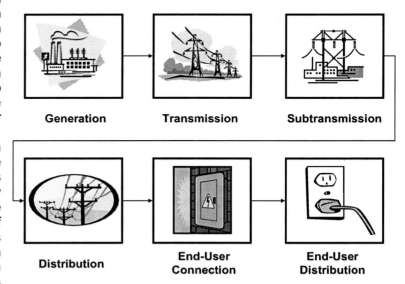

Generation   **Transmission**   **Subtransmission**

Distribution   **End-User Connection**   **End-User Distribution**

Figure 5.1 An illustration of the layout of a typical modern power distribution system.

voltage down slightly and retransmit the power to multiple distribution points. These distribution points are smaller sub-stations that take the incoming high-voltage power and distribute it at lower voltages. These distribution sub-stations typically service a relatively small region, such as a neighborhood. From the distribution point, the power reaches a transformer outside the home and the voltage is stepped down for the final time by a transformer, typically located on the pole outside the home or underground. From there it connects to the home's distribution point, the fuse or breaker box. Here, multiple circuits are connected to feed the outlets, lights, and appliances in the home. So, as you can see, the process of getting power from the power station to the individual outlets and lights in your home requires the careful integration of multiple elements. In the followings sections, we will look at each of the six parts of the power system in more detail.

5.3.1 Generation

The first step in providing power is generation. Most naturally occurring forms of electrical energy, like the lightning shown in Figure 5.2, are too unpredictable and extreme for us to use to provide reliable electrical power. As a result, we must obtain our electrical energy by converting other forms of energy into electrical energy. The most commonly harnessed energy forms are thermal, kinetic, and chemical. For example, the chemical energy in coal can be converted into thermal energy by burning it. This thermal energy can then be used to boil water to produce steam which, in turn, can be harnessed to turn a turbine. The kinetic energy produced by the turbine is then finally converted into electrical energy by a generator. Power generation using oil and nuclear methods is similar to coal-based generation in that water is heated and the steam is used to turn a turbine. Hydroelectric power is somewhat different because the water is used to directly turn the turbine

Figure 5.2 Naturally occurring electrical energy.

instead of being converted into steam. Since a large amount of water is needed to keep the turbines turning, hydroelectric power generation facilities are usually housed in dams so that the entire reservoir of water behind that dam can be used for power generation if necessary. One such dam is shown in Figure 5.3.

Regardless of the method used to turn the turbine, the end result is rotational kinetic energy. So, how is this kinetic energy converted into electrical energy? The answer to this lies in the design of the generator. If you recall the discussion of induction in the previous chapter, you will remember that if a coil of wire is moved through a magnetic field, a current is

Figure 5.3 A typical dam housing hydroelectric power generation facilities.

induced in the coil. This basic idea, illustrated in Figure 5.4, is employed in simple AC generators. As you can see when the coil of wire is rotated in a magnetic field, the amount of magnetic flux passing through the loop changes. Figure 5.5 illustrates the points in the rotational cycle during which there will be a maximum and minimum amount of magnetic flux passing through the coil. The way in which the magnetic flux passing through a rotating coil changes over time results in the production of an induced alternating current. So, by using the rotational kinetic energy that is provided by the turbine to turn the coil of wire within the generator, it is possible to generate an AC signal.

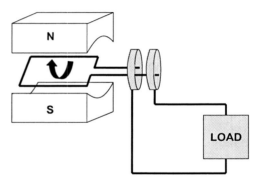

Figure 5.4 An illustration of simple generator operation.

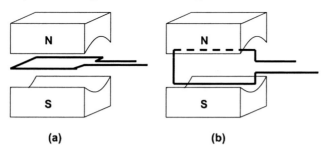

(a)   (b)

Figure 5.5 An illustration of the generator coil positions when the magnetic flux passing through the coil is at (a) its maximum and (b) its minimum.

A single AC signal like the one shown in Figure 5.5a is acceptable for simple applications such as lighting and running basic household appliances. However, it is less than ideal for industrial applications such as running large motors. This makes sense if you simply look at the AC waveform in Figure 5.5a. As you can see, the power provided by a "single-phase" AC waveform such as this falls to zero at the beginning, end, and center of each cycle. This results in a lack of efficiency since the power drops to zero at multiple points in

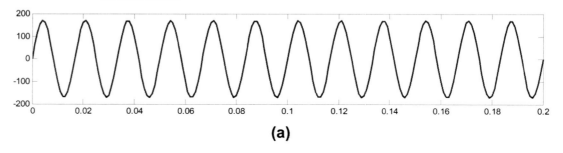

(a)

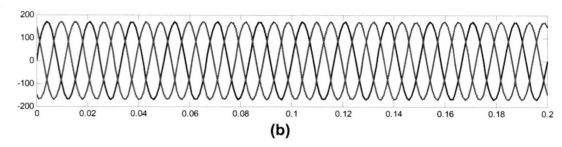

(b)

Figure 5.6. Example waveforms produced by (a) a single-phase generator and (b) a three-phase generator.

each cycle. As a result, modern power plants generate what is known as "three-phase" power. In such an arrangement, the generator possesses three coils, each separated rotationally by 120 degrees as shown in Figure 5.7. As the generator is turned, three separate AC waveforms are produced. Each waveform is 120 degrees out of phase with the previous one. This results in an output that looks like that shown in Figure 5.6b. The benefit of such a design is that there is never a point in time that the power drops to

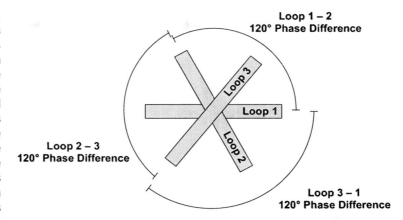

Figure 5.7. An illustration of the wire coil arrangement within a three-phase generator.

zero. There is always at least one phase of the output that is not at zero at a given time. Thus, a large gain in overall efficiency is achieved. So, why not add even more coils to the generator and generate power with an even higher number of phases? The answer to this is that it is not practical. While there is a large efficiency gain when the system moves from single-phase to two-phase and from two-phase to three-phase, there is not an appreciable gain in efficiency when using more phases. In addition, the complexity of the system that would be needed to support this type of generation would not justify the minor gain in efficiency that does occur. As a result, most power plants supply power to the power grid by using four conductors, one common or ground wire and one wire for each of the three phases. Once the electrical power is produced by the generator, it passes on to the next phase of the power system, the transmission phase.

5.3.2   Transmission

Simply generating power is of little value if we cannot make use of it. In most cases, however, the power generation plants are located many miles from the locations that use the power that they generate. As a result, it is necessary to transmit the generated power from the power plant to these remote locations. This is achieved through the use of long wires run either through underground tunnels or above ground on the familiar high voltage towers, much like the ones shown in Figure 5.7.

One of the most important problems that must be addressed by the transmission phase of the power system is the dissipation of power along the transmission lines. As you recall, early power transmission systems could only operate over small areas because the voltage would drop off as the customer location moved further from the power station. While metal wire is a very good conductor of electricity, it does have a small resistance. As the length of the wire increases, so does the resistance encountered by the current. The key to overcoming this difficulty is the transformer.

A transformer, as we learned in the previous chapter, is capable of changing the voltage and current of an AC electric signal while maintaining nearly the same overall power as the input (some power is lost in the transformer). Remember that electrical power can be expressed as $P = IV$, where I is the current and V is the voltage. Thus, if we use a transformer to increase the voltage, the current will be reduced and vice versa. Now, let's think about the

Figure 5.8 A picture of typical high-voltage transmission lines.

problem of power being dissipated on power lines. Recall that power dissipated across a resistance is defined as $P = I^2R$, where I is the current and R is the resistance. This tells us that the amount of power lost on the transmission lines will be highly dependent on the current passing through them. So, if we want to minimize the power lost during transmission, we need to minimize the current of our transmitted power. The only way to do this while maintaining the same level of power transmission is to use a transformer to increase the voltage, and thus, decrease the current of the generated power. As a result, the voltages used on transmission lines are extremely high, often in the hundreds of thousands of volts. Some overhead high voltage transmission lines carry power at voltages as high as 765,000 volts.

As you might imagine, transmitting power at extremely high voltages, while more efficient, can be dangerous if not performed with care. Voltages of this magnitude can ionize air and arc over large distances. This is why the high voltage towers that are used to support the transmission cables have such long insulators. Switching at these voltages is also challenging. Large gaps must be used to break the circuit. Safely creating these circuit breaks is much more difficult than with lower voltages. This is why very high voltages are only used for long distance transmission of large amounts of power, typically to connect power stations to each other (for load balancing) or to connect power generation facilities to large consumption zones, such as cities.

5.3.3   Sub-transmission

In many cases, the point at which a high voltage transmission line enters a city or other consumption area is still a fairly long distance from many power consumers. For example, a transmission line that is designed to feed several small towns or suburbs of a large city may end at a station somewhere between the towns it is to serve. The power arriving at this station must be stepped down and retransmitted to stations that are more local to each area of service. Since the remote stations do not require the full power provided by the transmission line, the

Figure 5.9. A picture of typical subtransmission power lines. Note that due to the lower voltages, the insulators are much shorter than those on the high voltage towers shown in Figure 5.8.

Sub-transmission station also allows for the ability to switch and reroute power between many remote distribution sites as needed to satisfy demand. Due to their shorter length and smaller power requirements, sub-transmission lines, like those in Figure 5.9, typically carry power at a much lower voltage than transmission lines (typically in the tens of thousands of volts). In many cases, the sub-transmission lines that are in use today were at one time full transmission lines, but as demand increased, higher capacity transmission lines had to be constructed. Instead of wasting the outdated transmission runs, they were converted to serve in a sub-transmission capacity. Each sub-transmission line that is split off from a main transmission line will typically end at a distribution station where the power it carries will then be sent to the consumers.

5.3.4 Distribution

Figure 5.10 A picture of a power distribution substation.

Each of the sub-transmission lines described in the previous section will end at a distribution station, much like the one shown in Figure 5.10. These substations perform several important functions. First, they step the voltage received from the sub-transmission line down to more manageable voltages, usually less than 10,000 volts. Next, this voltage is conditioned, and finally used to energize a series of local circuits that supply houses and businesses with electrical power. These stations usually receive all three phases of the generated power. Based upon the needs of the remote customers, however, not all of these phases will be provided. For example, a basic home does not require high efficiency power to run lights and appliances, but many industrial consumers often need to run heavy machinery that requires efficient three-phase power. As a result, circuits supplying purely residential neighborhoods are usually supplied with only one phase. This way, each incoming phase can be used to power different circuits. In other cases, wires for all three phases are strung, but only one of these phases will be tapped to provide power to a home while all three can be used for an industrial location. This way, the possibility exists to provide both single and three-phase connections on the same circuit.

When connecting homes and businesses to a substation, there are two basic methods that are used. The first of these, illustrated in Figure 5.11a is called the radial approach. In this configuration, a single connection radiates from the substation and power flows to the connected loads in one direction. The second approach, illustrated in Figure 5.11b is known as the loop approach. In this configuration, the power connection path forms a loop. The loop configuration makes it possible for power to flow to the loads from multiple directions. This makes it less likely that a break in the line will cause a large number of customers to lose power. However, the configuration is not always practical. For example, when connecting homes along long rural

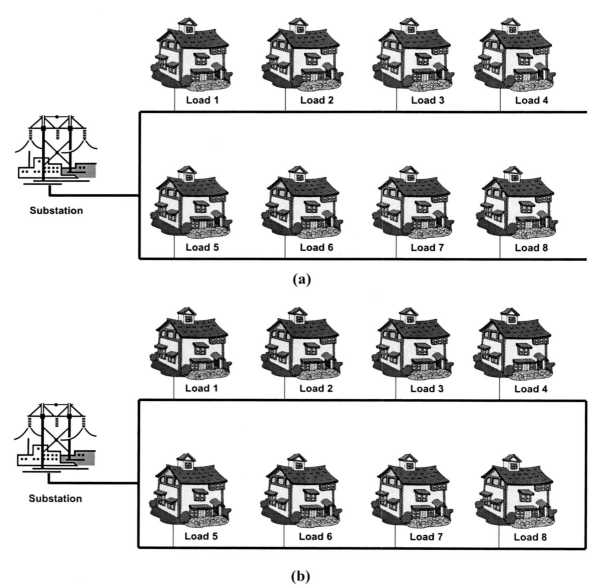

Figure 5.11 An illustration of the two methods used to connect customers to a substation. Connections can be made in a (a) radial or a (b) loop configuration.

roads, it can be difficult to form looping structures. As a result, loop configurations are more often implemented in cities and subdivision neighborhoods.

5.3.5   Connection to the End User

The distribution phase of the power system carries the power as far as an individual home or business, but before that power can be used, it must be conditioned and connected to the customer's building. This is done by using a transformer, much like the one shown in Figure 5.12, to step the voltage down one final time. The voltage output of this transformer is dependent upon the type of customer, but is typically set at 120, 240, or 480 volts for most applications in the United States. If you closely examine the residential transformer shown in the figure, you will notice that it has three output wires. This is because the transformers used for most residential connections have a center-tap arrangement. This means that the windings on the output of the transformer are tapped midway through. This allows for more versatile use of the voltage output. For example, if the center tap is not used, a single AC waveform is available. In most residential applications, this AC output has a voltage of 240 volts. It is used to run household appliances that require large amounts of power like the stove, electric dryer, and heat and air unit. On the other hand, by using the center tap as a common point, two AC waveforms are generated 180 degrees out of phase. These outputs will have a voltage that is half of the voltage that is obtained when the center tap is not used. So for most applications, this will be 120 volts. This is the voltage that is supplied to most electrical outlets and lights. Thus, by using a center-tapped transformer, one phase of electrical power can be used to power both the large and small electrical devices in a typical household. The method used to supply power to industrial

Figure 5.12 A picture of a typical residential step-down transformer.

customers is similar, but either three phase or multiple single phase transformers are often employed.

Figure 5.13 An example of an electric power meter.

Once the voltages are stepped down to the appropriate levels, the power can be connected to the consumer's building. Electric power companies are in business to make money, however. As a result, these companies need to have a way to fairly bill customers for their power consumption. This is done through the use of a power meter. An example of such a meter is shown in Figure 5.13. These meters keep track of the amount of energy that a given customer has used. As you know, energy is defined as the amount of power used over a period of time. Because of this, power meters measure usage in kilowatt hours. One kilowatt hour is the amount of energy required to provide one thousand watts of power for an hour. Thus, a single 100 watt light bulb that is lit for 10 hours will consume 1 kilowatt hour of energy. Keeping this in mind, let's look at an example.

Example 5.1

Problem: Assume that a given power consumer heats a room of their home using a portable heater with a power rating of 1,500 watts. If they run this heater 12 hours a day for a month (30 days), how many kilowatt hours of energy will they consume? If power in their area costs 15 cents per kilowatt hour, how much will running the heater add to their monthly power bill?

Solution: The first step in solving this problem is to determine how many kilowatt hours are used. To do this, we must simply multiply the power rating of the heater by the number of hours that the heater is used in the 30 day period. Thus, we discover that the heater will consume

$$1,500W \times 12 \times 30 = 1,500W \times 360 \text{ Hours}$$
$$= 540,000 \text{ Watt Hours}$$
$$= 540 \text{ Kilowatt Hours}.$$

So, running the heater requires 540 kilowatt hours of energy. Next, we see that using this much energy at a cost of 15 cents per kilowatt hour will result in a cost of

$$540 \text{ Kilowatt Hours} \times \$0.15 = \$81.00$$

So, we have now seen how the electric power company generates power and transmits it to an individual user's home. The final step involved in the process of getting the electric power from the generator to your wall outlet is the responsibility of the home or business owner.

5.3.6   End User Internal Distribution

Once the power is connected to a customer's meter, it passes on to a central distribution point for the building. From this point, different circuits are broken off to feed the various electrical loads within the structure. For example, circuits are run to supply power to heating and air conditioning units. Other circuits supply power to wall outlets and lights. Thus, one important function of the central distribution point is to provide a central point of control. In other words, it allows the various circuits in the structure to be turned on and off from a single location.

Another important function of the central distribution point is to regulate the current flow through the various circuits. While the voltages used in residential structures are considerably lower than transmission level voltages, they are still very dangerous and can be deadly if not properly regulated. Circuits that are run from the central distribution point are insulated to prevent the

Figure 5.14 A picture of the breakers located at a typical distribution point for a small structure.

wires from coming into contact with a ground source which would cause large amounts of current to flow. If large currents were permitted to flow, the distribution wires would become hot and could cause a fire. This is why each circuit is usually protected by a fuse or breaker. A fuse is an element that is inserted into a circuit that is designed to burn out quickly if a specific current flow is exceeded. Once a fuse burns out, the circuit is broken until the fuse is replaced with a new one. Breakers, like those shown in Figure 5.14 are current sensitive switches that are designed to turn off when a given current is exceeded. The benefit of breakers is that they can be reset and do not need to be replaced when they break a circuit. Thus, by limiting the current flow, the central distribution point ensures that any accidental overloading of a given circuit will not result in dangerous amounts of current overheating the wiring. In addition, the breakers can be used to reduce the duration of accidental human exposure to electricity. These special breakers, known as ground fault breakers are often used in kitchens and bathrooms. They work by monitoring the current flow on both the live and ground side of the circuit. If there is an imbalance between the current being sent and the current returning, it means that the current has found a different path to ground. Since this path could involve a person, the breaker will open the circuit to reduce the chance that severe injury or death will result from electric shock.

So, as we have just seen, the process of providing electric power is a complicated process involving electrical engineering at many levels. The process of providing electric power was one of the first motivating factors in the early development of electrical engineering and it continues to be an important aspect of engineering today.

5.4   Chapter Summary

In this chapter we discussed the basics of the electrical power system and the steps required to generate and route power to the wall outlet in our homes.

- We began our look at the power system by investigating its conception and the evolutionary process it underwent as it developed into its current form within the United States. We looked at some of the benefits of electrical energy over other forms of energy and found that these benefits were one of the main motivations in the electrical power industry.

- Next, we explored the makeup of the current electrical power system. We learned that the process of generating electrical power and getting that power to the wall outlet in our homes requires six distinct steps.

- Electrical power is produced by using a generator to convert rotational kinetic energy into electrical energy. This is the generation step. We found that typical generators use three evenly spaced generation coils to produce three-phase power as opposed to the simpler, but less efficient single phase power.

- Once power has been generated, the voltage is increased using a transformer allowing the power to be transmitted over long distances. The increased voltage and subsequently reduced current results in a minimization of the power dissipated by the transmission lines. The process of transmitting power over long distances using high voltages is known as the transmission step.

- In many cases, the power received by the substation at the end of a transmission line must be divided up and retransmitted to feed a large area. This retransmission is carried out using lower voltages than those of the initial transmission phase. The process involved in dividing up and retransmitting power to distribution centers is known as the sub-transmission step.

- Each distribution substation connected to a sub-transmission line has the task of providing power to the customers in the area around it. In this phase, known as the distribution phase, nearby homes and businesses are supplied with power using either a radial or loop connection configuration. Once again, distribution occurs at a lower voltage than both transmission and sub-transmission.

- When power arrives at a home or business, the voltage is reduced one final time and a meter is inserted into the circuit to determine the amount of energy used by each customer. Energy usage is measured in kilowatt hours.

- The final phase involved in getting power to an outlet in a customer's home is the user's internal power distribution system. In this phase, a number of circuits are connected to a central distribution point. This point provides both a central control point and means of limiting the current flow on individual circuits to protect against overloading and electrical shock related injuries.

5.5   Review Exercises

Section 5.2
___

1. List and describe three types of energy.

2. Describe some of the benefits of electrical energy.

3. What was the first major use of electrical power?

4. Who was instrumental in the development of practical indoor lighting?

5. What was the biggest limitation of early DC electrical power systems?

6. Why was AC a better choice for electrical power distribution?

7. Who developed much of the early AC power generation and distribution system technology?

8. What AC frequency became the standard for power generation in the United States?

Section 5.3
___

9. List and describe the purpose of the six stages of the electrical power distribution system.

10. Describe how a generator works.

11. Describe the difference between single and three-phase power. Why is three-phase power more efficient?

12. Describe the benefit of using transformers to transmit power over long distances at very high voltages.

13. What is the purpose of sub-transmission lines?

14. Describe the two methods of connecting loads to a distribution substation. Which is more fault-tolerant?

15. How do power companies determine how much to bill their customers? What unit is energy consumption measured in?

16. Assume that a given power customer uses five 60 watt light bulbs and three 100 watt light bulbs to light their house. If these lights are on for an average of 6 hours per day for a 30 day period, determine the amount of energy that the customer used. If electricity costs $0.12 per KWh, determine what the cost of this energy will be.

17. An electric power customer, who normally uses ten 60 watt light bulbs to light their house, replaces them with high efficiency fluorescent lights that consume 28 watts each. If the lights are lit for an average of eight hours a day for a 30 day period, what reduction in energy use will result from the use of the high efficiency light bulbs? If power costs $0.15 per KWh, how much money will the customer save by using the new bulbs?

18. Describe how the power company can provide both 120V and 240V power to a home using only one transformer.

19. Describe the benefits of using a central distribution point for the wiring within a home or business.

20. Describe how fuses and breakers operate. Why are they used?

Chapter 6. Basic Circuit Analysis

Chapter Outline

6.1   Introduction

Now that we have reviewed the basics of electrical theory and have examined the large-scale application of electricity in the electric power system, we will turn our attention to smaller-scale applications, namely the analysis of DC electrical circuits. While the AC power system performs the important service of providing power to your home, smaller electronics also play an important roll. Your television, radio, telephone, and computer all involve DC electrical circuits. While the full analysis and design of the circuits used in these applications is beyond the scope of this text, we will provide an introduction to the basic theory and present some of the approaches that can be used in the analysis of electric circuits. This introduction will focus primarily on the analysis of circuits containing only resistors, but the approaches that will be presented can be easily extended to handle more complicated components like capacitors and inductors.

We will begin our study by presenting some of the basic concepts necessary to understand and analyze circuits. We will then turn our attention to some common analysis methods and look at some example circuit analysis problems. By following this approach you will be able to understand and analyze basic electrical circuits.

6.2   Basic Concepts

Before we can approach the analysis of electric circuits, we must first gain an understanding of some basic theory. By applying this theory we will then be able to obtain a detailed view of the way in which the circuits that we investigate operate. For example, we will need to answer questions involving current, voltage, and resistance. How can we find the numerical value of one of these quantities if we are given the other two? If we have a group of interconnected resistors, how can we determine the effective resistance of the group? Given a circuit with a voltage source, how can we analyze that circuit for the voltage loss across the individual resistors? To be able to answer these questions, and many others, we need to develop a series of tools and rules that we can apply in specific situations. Let's begin our exploration of basic concepts by reviewing a relationship that we first encountered in chapter four, Ohm's Law.

6.2.1   Ohm's Law

One of the most fundamental tools that we can use to analyze circuits is Ohm's Law. As you may recall, Ohm's Law describes a relationship between current (I), voltage (V), and resistance (R). Essentially, it states that voltage is directly proportional to both current and resistance, current is directly proportional to voltage and inversely proportional to resistance, and resistance is directly proportional to voltage and inversely proportional to current. So, if we write out this relationship mathematically, we arrive at one of the following three expressions.

1. $V = IR$

2. $I = \dfrac{V}{R}$

3. $R = \dfrac{V}{I}$

Thus, we see that by using Ohm's law, we can find an unknown voltage, current, or resistance as long as we know the value of the other two quantities. To gain a better understanding of the way in which we can apply the power of this tool to circuit analysis, let's look at a few simple examples.

Example 6.1

Problem: Find the current flowing through the following circuit.

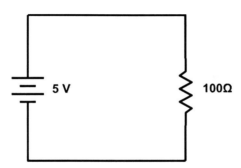

Solution: As we can see from the schematic diagram, we have a simple circuit with a 5 volt source and a 100 ohm load. Since we know the voltage across the load is 5 volts and we know the resistance of the load, we can use Ohm's law to easily determine the current flowing through the circuit. Using our second form of Ohm's law, we can write

$$I = \frac{V}{R} = \frac{5\,\text{V}}{100\,\Omega} = 0.05\,\text{A} = 50\,\text{mA}.$$

Thus, we see that the current flowing through this simple circuit is 50 mA.

Example 6.2

Problem: Find the resistance of the load in the following circuit.

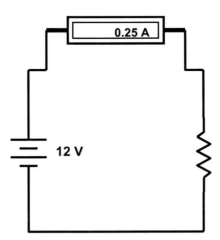

Solution: In this case, we have a circuit with a known voltage source and current, but we do not know the value of the resistance. Thus, if we use our third form of Ohm's law, we can find this resistance by writing

$$R = \frac{V}{I} = \frac{12 \text{ V}}{0.25 \, A} = 48 \, \Omega$$

So, in this case, we used Ohm's law to find that the resistance in this circuit is 48 ohms.

Example 6.3

Problem: Find the voltage of the source in the following circuit.

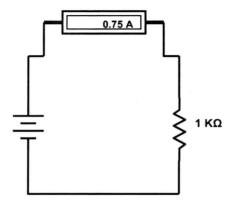

Solution: This problem can also be easily solved using Ohm's law. In this case, we can use our first form of the equation since we know the current and resistance. Thus, we can write

$$V = IR = (0.75 \text{ A})(1000 \, \Omega) = 750 \text{ V}$$

So, we find that to push a current of 0.75 A through a 1 KΩ resistance we need a voltage of 750 volts.

As the previous examples illustrate, Ohm's law can be a very powerful tool for finding an unknown voltage, current, or resistance.

6.2.2   Watt's Law

While the relationship illustrated in the previous section is very powerful, Ohm's law can be further extended to provide a relationship between voltage, resistance, current and the amount of power dissipated by a load. For example, in Chapter 4, we found that power can be expressed as the product of voltage and current. This expression, often referred to as Watt's law, can be written as

$$P = IV.$$

Thus, if we know the voltage across a load and the current flowing through it, we can determine the amount of power that the load dissipates. If we take advantage of Ohm's law, we can use the fact that $V = IR$ to write

$$P = I(IR) = I^2 R.$$

In addition, if we take advantage of the fact that $I = \dfrac{V}{R}$, we can write

$$P = \left(\frac{V}{R}\right)V = \frac{V^2}{R}.$$

So, for Watt's law, we have three forms that can be used.

1. $P = IV$

2. $P = I^2 R$

3. $P = \dfrac{V^2}{R}$

As you can see in the following examples, the form of Watt's law that should be used depends upon what is known about the circuit in question.

Example 6.4
_____

Problem: Determine the power that is dissipated by the resistor in the following circuit.

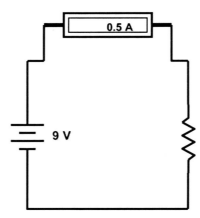

Solution: In this problem we have been given the voltage placed across the resistance as well as the current flowing through it. In order to find the power dissipated in the resistor, we simply need to plug these values into the first form of Watt's law. Doing this yields

$$P = IV = (0.5 \text{ A})(9 \text{ V}) = 4.5 \text{ W}$$

So, in this case, the resistor is dissipating 4.5 watts of power as heat.

Example 6.5

Problem: Assume that a current of 2.5 amps is flowing through a 20 foot length of wire. If this wire has a resistance of 0.5 ohms per foot, how much power is dissipated in the wire?

Solution: In this case, we are given a current and the means to calculate a total resistance. From this information, we can use the second form of Watt's law to determine the power dissipation. First, let's calculate the resistance of 20 feet of wire.

$$R = (0.5 \text{ }\Omega/\text{Foot})(20 \text{ Feet}) = 10 \text{ }\Omega$$

Now, we can use Watt's law to calculate the power dissipation as follows

$$P = I^2 R = (0.5 \text{ Amps})^2 (10\Omega) = 2.5 \text{ W}$$

Thus, this particular length of wire will dissipate 2.5 watts of power when a current of 2.5 amps flows through it.

Example 6.6

Problem: Determine the power dissipation in the following circuit.

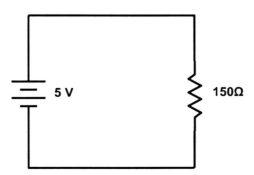

Solution: This problem provides us with the supply voltage and load resistance and asks for the dissipated power. This power can be easily determined by using the third form of Watt's law. Plugging into the equation we find

$$P = \frac{V^2}{R} = \frac{(5V)^2}{150\Omega} = \frac{25V^2}{150\Omega} \approx 0.1667 \text{ W}$$

So, we see that the power dissipation in this circuit is approximately 0.1667 watts.

While the preceding examples illustrate some of the potential of Watt's law, the relationships of both Watt's and Ohm's laws can be combined and used to form multiple expressions that can be used to find current, voltage, resistance, and power given a wide variety of different known conditions. The circular diagram in Figure 6.1 demonstrates all of the different relationships that can be formed.

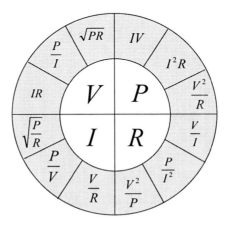

Figure 6.1 An illustration of the relationships that can be formed between power (P), voltage (V), current (I), and resistance (R) using Ohm's and Watt's laws.

6.2.3   Resistances in Series

The analysis of circuits with a single voltage source and a single resistance is not very complicated. We discovered this through our analysis of the circuits in the previous sections. Most circuits however, contain numerous resistances and can often contain multiple voltage and current sources. As a result, it is important to know how to simplify resistor configurations when possible. One of the most common and most easily simplified configurations in which resistors can be found is the series configuration. In a series configuration, which is illustrated in Figure 6.2, resistors are connected end-to-end.

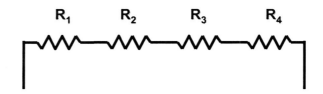

Figure 6.2 An illustration of resistors arranged in series.

As you can see, the figure displays four resistors in a series arrangement, each with different resistances R_1 through R_4. The total equivalent resistance of resistors in a series configuration can be computed by simply adding up the values of each resistance. So, why can we compute the equivalent resistance by just adding up the values of the resistors in a series arrangement? The answer to this question is simple. If you think about it, in a series arrangement, there is only one path for the current to follow from the beginning of the circuit to the end. Therefore, as additional resistors are added to this path, the degree to which the flow of the current is impeded is increased by an amount proportional to the resistance of each new resistor. Thus, for an arrangement of n resistors placed in series, the equivalent resistance, R_{eq} can be computed as

$$R_{eq} = R_1 + R_2 + \cdots + R_n.$$

Let's take a look at an example of how we can apply this equivalent resistance calculation method to help us analyze a circuit.

Example 6.7

Problem: Determine the current flowing through the following circuit.

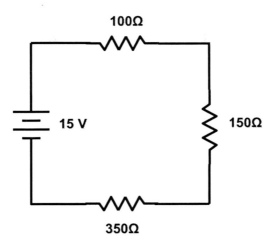

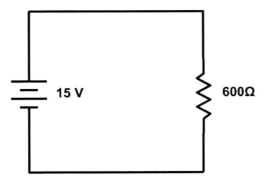

Solution: In this problem, we have a single voltage source, but the voltage is supplied across three resistors that are wired in a series arrangement. In order to find the correct amount of current that is flowing through the circuit, we must first determine the total resistance that the voltage is supplied across. Since the resistors in this circuit are in a series arrangement, we can find the total equivalent resistance that is experienced by the current by adding the resistances up. Doing this gives us

$$R_{eq} = R_1 + R_2 + R_3 = 100\Omega + 150\Omega + 350\Omega = 600\Omega$$

If we redraw our circuit to illustrate the newly determined equivalent resistance, we get the following circuit.

Now that we know the total resistance, we can use Ohm's law to easily determine the current.

$$I = \frac{V}{R} = \frac{15V}{600\Omega} = 0.025A = 25mA$$

So, we have used both the series equivalent resistance relationship and Ohm's law to find out that the current in this circuit is 25 milliamps.

This example illustrates the benefit of being able to simplify a circuit containing multiple resistors in series. However, we can use what we know to discover even more information about a circuit. Let's look at another example.

Example 6.8

Problem: Find the voltage across the output terminals of the following circuit.

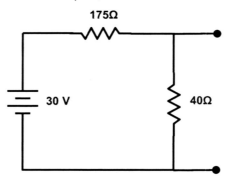

Solution: In this problem we have been given the voltage of the source and the resistances of the two resistors in the circuit. The two resistors are in a series arrangement, but the value that we are searching for is the voltage measured across the 40 ohm resistor. So, how do we find this voltage? From Ohm's law, we know that the voltage dropped across a resistance is equal to the product of the current and the resistance. Therefore, if we can find the current that is flowing through the circuit, we can use that information to find the voltage dropped across the 40 ohm resistor. Since the two resistors in the circuit are in series, they both contribute to the resistance encountered by the current. Thus, we must first find the equivalent resistance. Doing this, we find that

$$R_{eq} = R_1 + R_2 = 175\Omega + 40\Omega = 215\Omega$$

Now, using this equivalent resistance in conjunction with Ohm's law we can find the current.

$$I = \frac{V}{R} = \frac{30}{215} \approx 0.1395A$$

Since we now know that the current is approximately 0.1395 amps, we can plug this information into Ohm's law to find the voltage dropped across the 40 ohm resistor.

$$V = IR = (0.1395A)(40\Omega) \approx 5.58V$$

So, the voltage that is dropped across the 40 ohm resistor is approximately 5.58 volts, but what happened to the rest of the 30 volts that were supplied to the circuit? To answer that question, let's use the current that we calculated to see how much voltage will drop across the 175 ohm resistor. Doing this yields

$$V = IR = (0.1395A)(175\Omega) \approx 24.42V.$$

This tells us that the remaining 24.42 volts was dropped across the 175 ohm resistor. Thus, the entire source voltage is accounted for.

A circuit like the one shown in Example 6.8 is known as a voltage divider since the total voltage provided by the source is divided between the two resistances.

6.2.4   Voltage Division

Voltage division is a useful method of providing a lower voltage to a load within the circuit when the original circuit is supplied with a higher voltage. For example, let's say that we have a circuit that is connected to a 12 volt battery, but a sub-section of that circuit requires a 3 volt source. We can use a voltage divider relationship to provide the appropriate voltage. So, how exactly do we develop a single expression that will provide us with the voltage dropped across one of the resistors in a voltage divider configuration given the source voltage and the resistances? To develop this relationship, let's look at the typical voltage divider configuration shown in Figure 6.3. We know from Ohm's law that

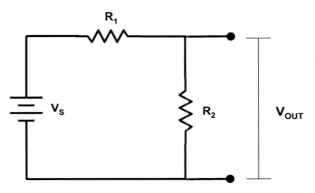

Figure 6.3 A typical voltage divider configuration.

$$V_{OUT} = IR_2,$$

but we don't know what I is for this particular circuit. Luckily, Ohm's law tells us that the current flowing through this circuit is equal to the source voltage divided by the total resistance encountered. Since the resistors are in series, we can write

$$I = \frac{V_S}{R_1 + R_2}.$$

Now that we know the current, we can multiply that value by the output resistance, R_2 to obtain

$$V_{OUT} = \frac{V_S R_2}{R_1 + R_2} = \left(\frac{R_2}{R_1 + R_2}\right) V_S.$$

So, as we have just discovered, the voltage dropped across the second resistance is the product of the source voltage and the ratio of the second resistance over the total resistance. To better understand how we can make use of this relationship, let's look at an example.

Example 6.9

Problem: Find the value of R_1 that will result in a voltage drop of 3V across R_2 in the following voltage divider circuit.

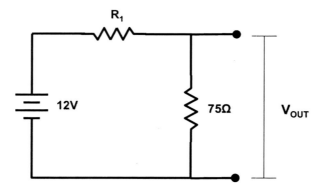

Solution: Here we can see that the source voltage is 12V, and that the second resistance is 75 ohms. If we use this information in our voltage divider equation, we can find the resistance required to give us the desired 3 volts across the second resistance. So, we will begin by writing

$$3V = \left(\frac{75\Omega}{R_1 + 75\Omega} \right) 12V.$$

Now, if we multiply both sides by $R_1 + 75\Omega$ we will get

$$(R_1 + 75\Omega)3V = (75\Omega)12V.$$

Then we can divide both sides by 3V to yield

$$R_1 + 75\Omega = (75\Omega)(4).$$

Finally, we can subtract 75Ω from both sides to solve the equation for R_1. Doing this gives us

$$R_1 = (75\Omega)(4) - 75\Omega.$$

$$R_1 = 225\Omega$$

So, we have discovered that we can get our desired 3V voltage drop across the second resistance in our voltage divider if we use a resistance of 225 ohms for the first resistance. Thus, we have used the voltage divider relationship to solve a basic circuit analysis problem.

The voltage divider illustrates another important concept of a series circuit. That is, that the entire voltage supplied to the series loop is dropped across the resistances. This knowledge can also be very useful when analyzing circuits.

6.2.5   Kirchhoff's Voltage Law

Gustav Kirchhoff, a German physicist noted what we have just seen. In other words, he realized that the voltage drops around a series circuit loop were equal to the voltage supplied to the circuit. From this observation, he formulated what is known as Kirchhoff's Voltage Law. This law states that for any closed loop in a circuit, the algebraic sum of the voltages must be equal to zero. This means that if we add the voltage gains introduced by voltage sources in the circuit and subtract the voltage drops caused by the resistances in the circuit, we should end up with a total sum of zero. This law will prove extremely useful when analyzing more complex circuits later in the chapter, but it can also prove useful when working with a simple single-loop circuit. To get a better understanding of the power of this simple law, let's look at an example.

Example 6.10

Problem: Find the current flowing through the following circuit.

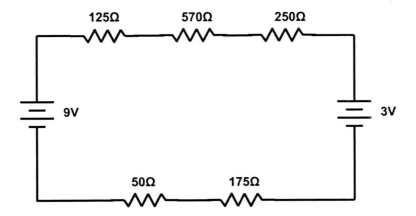

Solution: In the circuit provided in this problem, we have several resistors in series and two voltage sources. When analyzing a circuit using Kirchhoff's Voltage Law (KVL), we need to take a series of steps.

1.   First, we need to choose a direction for the current in the circuit loop. The direction that is chosen is not critical. It simply provides us with a direction to move around the

circuit. If the direction that we choose turns out to be incorrect, we will find a negative current. This simply indicates that the current flows around the loop in the opposite direction. In our sample circuit, we will choose a clockwise current flow.

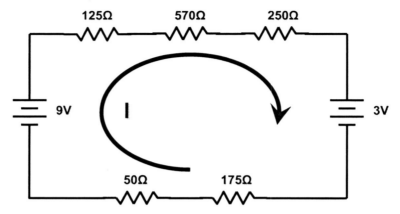

2. Next, we need to label the polarity of each component in the circuit loop. In other words, for each component we need to mark the end with a higher voltage relative to ground with a plus sign (+) and the side with lower voltage relative to ground with a negative sign (-). This helps us to determine whether a voltage should be added or subtracted when performing the algebraic summation for the loop. If we do this for the circuit that we were given we will get the following.

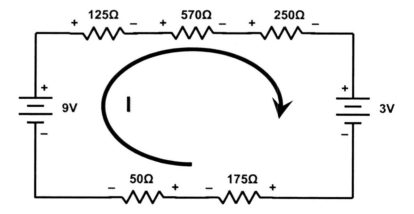

As you can see, the labeling for the voltage sources is easy since they already have a positive and negative polarity. The labeling of the resistors is a bit trickier. If you think about it a little bit, however, it is not too difficult. Since we have assumed clockwise current flow, we simply label the side of each resistor that the current flows into with a positive polarity (+) and the side of the resistor that the current flows out of with a negative polarity (–).

3. Now that we have our polarities labeled, we need to use Kirchhoff's Voltage Law to help us find the current, but how do we do that when we don't know the voltages dropped by the individual resistors? The answer is that we don't have to know the

numerical values for the voltages. We know from Ohm's law that the voltage dropped across a resistor is equal to the product of the current and the resistance. Thus, the voltage drop for each resistance can be described as the product of the unknown current, I, and the resistance of the resistor. So, now that we can represent each voltage drop as a product, how do we use Kirchhoff's Voltage Law to find the current? Well, we simply need to sum the voltage drops and gains around the loop and set that equal to zero. Let's start on the positive side of the 9V source and move in the direction of our selected current flow. As we move around the circuit, we need to subtract the voltage for each component that results in a + to − transition (voltage drop) and add the voltage for each component that results in a − to + transition (voltage gain). If we do this for our sample circuit and set the result equal to zero, we will arrive at the following expression.

$$-(125\Omega)I - (570\Omega)I - (250\Omega)I - 3V - (175\Omega)I - (50\Omega)I + 9V = 0$$

Now, we can rewrite this expression to isolate the unknown current value.

$$(125\Omega + 570\Omega + 250\Omega + 175\Omega + 50\Omega)I = 6V$$

Summing up the resistances gives us

$$(1170\Omega)I = 6V.$$

Finally, we can calculate our current by dividing by the resistance value. This gives us

$$I = \frac{6V}{1170\Omega} = 0.005128A = 5.128mA$$

So, we have discovered that the current in the given circuit is equal to 5.128 mA.

This example illustrates some of the potential power of Kirchhoff's Voltage Law. We will explore its use in more detail when we analyze more complicated circuits containing multiple loops.

6.2.6   Resistances in Parallel

While understanding the behavior of series circuits greatly improves our ability to successfully analyze circuits, only the most simplistic circuits are laid out in a single series loop. Thus, to improve our ability to work with circuits even further, we must look at the other common basic circuit arrangement. In this arrangement, the resistances in the circuit are not connected end-to-end, but instead, a common point is used to connect each side of the resistive loads as shown in Figure 6.4. As this figure illustrates, one end of each resistor is connected to one common point while the other end of each resistor is connected to a second common point. This means that all of the resistors experience the same voltage on each side, and thus, they must account for the same amount of voltage drop.

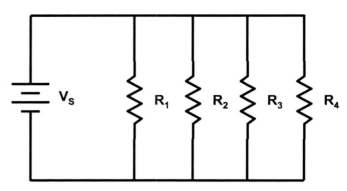

Figure 6.4 Four resistors placed in a parallel arrangement.

Unlike the resistors placed in series, the resistances cannot be simply added up to determine the total equivalent resistance for a parallel arrangement. As the current flows around a series circuit from the source, it must pass through each resistor in the circuit before it reaches the opposite terminal of the battery. In a parallel circuit, however, the current can take multiple paths. As a result, the current will split up amongst the different resistances with more of it flowing through the lower resistance branches than the higher resistance ones. To understand this, let's assume for a moment that we have a simple two resistor parallel configuration. If we return to the analogy of electricity behaving much like water, we can think of the two resistors as two smaller diameter pipes that branch off from a larger pipe as illustrated in Figure 6.5.

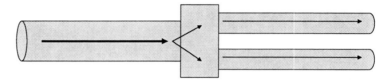

Figure 6.5 An illustration of the way in which current behaves in a circuit with a parallel branch.

In the figure, you will note that the water (current) flowing through the larger pipe is divided between the two smaller pipes. Since the pressure of the water (voltage) in the larger pipe remains constant, the diameters (resistances) of the two smaller pipes will determine how much of the water (current) that each pipe can carry. If the two pipes have the same diameter, the water will be divided evenly between them. From this illustration, we can also see that dividing the water between the two output pipes allows more water to flow through the system than if we were to provide only one pipe of that diameter. This is also the way that current behaves when it encounters parallel resistors. Since there are multiple paths over which to divide the current, the actual resistance encountered by the current as a whole will be lower than the smallest single resistance in the parallel arrangement. To illustrate this, look at the parallel resistor arrangement in Figure 6.6. As you can see, the current flowing through this circuit will be divided between the two resistances. If the two resistances are the same, the current will be divided evenly and half will flow through each resistor. As a result, the total resistance experienced by the current will be half of what it

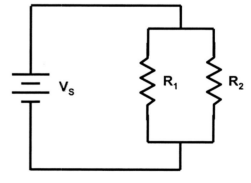

Figure 6.6 A simple two resistor parallel arrangement.

would have experienced with just one of these resistors. So, what happens if the resistances of the two resistors are different? In the case where the two resistances are different, the current will be split unevenly and more of it will flow through the smaller resistance. The resistance experienced by the current in this case will be less than the smaller of the two resistances.

The exact resistance that the current experiences can be calculated with a little thought. First, from Ohm's law, we know that the total current flowing through the circuit will be equal to the source voltage divided by the total resistance experienced by the current. In addition, we know that the total current in the circuit is divided up between the different branches, and that each resistor drops the entire source voltage across itself. Thus, using Ohm's law again, we know that the current in each branch will be equal to the source voltage divided by the resistance of that branch. So, if we set the total current equal to the sum of the currents in each of the parallel branches, we can write a mathematical expression. Doing this for our two resistor case, we will find that

$$I_{Total} = I_1 + I_2 .$$

Using Ohm's law, we can then write

$$\frac{V_s}{R_{eq}} = \frac{V_s}{R_1} + \frac{V_s}{R_2} .$$

Factoring V_s out of both sides of the expression yields

$$V_s \left(\frac{1}{R_{eq}} \right) = V_s \left(\frac{1}{R_1} + \frac{1}{R_2} \right) .$$

Now, we can divide both sides by the source voltage to obtain

$$\frac{1}{R_{eq}} = \frac{1}{R_1} + \frac{1}{R_2} .$$

Finally, we can solve for the equivalent resistance R_{eq}, to yield the expression

$$R_{eq} = \frac{1}{\frac{1}{R_1} + \frac{1}{R_2}}$$

for our two resistor case. A similar technique can be used to derive an expression for a larger number of resistors. So, in the case of a parallel arrangement of n resistors, the equivalent resistance can be expressed as

$$R_{eq} = \cfrac{1}{\cfrac{1}{R_1} + \cfrac{1}{R_2} + \cdots + \cfrac{1}{R_n}} \quad .$$

Thus, if we have a parallel arrangement containing two 100 ohm resistors, we will find that the total resistance experienced by the current is

$$\cfrac{1}{\cfrac{1}{100\Omega} + \cfrac{1}{100\Omega}} = \cfrac{1}{\cfrac{2}{100\Omega}} = \frac{100\Omega}{2} = 50\Omega \; .$$

So, as we would expect, the resistance provided by two 100 ohm resistors in parallel is exactly half of the resistance provided by one 100 ohm resistor. In the special case of only two resistors in parallel, we can derive an easier expression to find the equivalent resistance. If we start from our original equation and find a common denominator for the two resistances we will get

$$R_{eq} = \cfrac{1}{\cfrac{1}{R_1} + \cfrac{1}{R_2}} = \cfrac{1}{\cfrac{R_2}{R_1 R_2} + \cfrac{R_1}{R_1 R_2}} = \cfrac{1}{\cfrac{R_1 + R_2}{R_1 R_2}} = \frac{R_1 R_2}{R_1 + R_2} \; .$$

This is often called the product over sum method and is usually a quicker way to calculate equivalent resistance for two resistors in parallel. To verify that it will give the same result, we can apply this expression to our two 100 ohm resistances. By doing this, we will once again find that

$$R_{eq} = \frac{(100\Omega)(100\Omega)}{100\Omega + 100\Omega} = \frac{10000\Omega^2}{200\Omega} = 50\Omega \; .$$

Thus, we can use this expression as a quicker way to simplify parallel arrangements containing only two resistors. To get a better feel for how we can use these new parallel resistance relationships to work with circuits, let's take a look at some examples.

Example 6.11

Problem: Find the total current flowing through the following circuit.

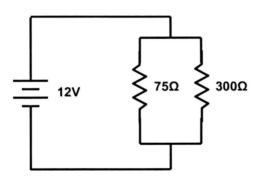

Solution: In this problem, we have been given a circuit with two resistors in parallel and asked to find the total current. To do this, we must first find the total resistance seen by the current. Since there are only two resistors in this parallel circuit, we can use the product over sum approach to calculate the equivalent resistance. Doing this yields

$$R_{eq} = \frac{R_1 R_2}{R_1 + R_2} = \frac{(75\Omega)(300\Omega)}{75\Omega + 300\Omega} = \frac{22500\Omega^2}{375\Omega} = 60\Omega .$$

Thus, the total equivalent resistance experienced by the current is only 60 ohms. As we would expect, this is less than the smallest resistor in the arrangement. Using the equivalent resistance, we can now redraw our circuit.

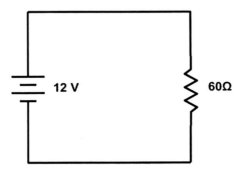

Now, we can easily use Ohm's law to determine the current.

$$I = \frac{V}{R} = \frac{12}{60} = 0.2A .$$

So, we have found that the total current in this circuit is 0.2 amps.

___

Example 6.12

Problem: Find the value for R_2 that will result in a total current of 0.5 A in the following parallel circuit arrangement.

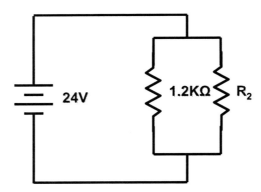

Solution: In this problem we are given a parallel circuit with one unknown resistance and asked to find what that resistance should be in order to yield a total current flow of 0.5 A. Our first goal in this case is to use Ohm's law to discover what total resistance is necessary to produce a current of 0.5 A with a voltage source of 24 V. Thus, we can write

$$R_{eq} = \frac{V_s}{I_{Total}} = \frac{24V}{0.5A} = 48\Omega \,.$$

Since we now know that we need an equivalent resistance of 48 ohms, we can use our parallel equivalent resistance relationships to determine the necessary resistance value. In this case, we can use the product over sum version of the relationship because we have only two resistances. Plugging in our known values, we can write

$$48\Omega = \frac{(1200\Omega)(R_2)}{1200\Omega + R_2} \,.$$

Solving for R_2 we find

$$(1200\Omega)(48\Omega) + (48\Omega)(R_2) = (1200\Omega)(R_2)$$

$$(1200\Omega)(R_2) - (48\Omega)(R_2) = 57600\Omega^2$$

$$(1152\Omega)(R_2) = 57600\Omega^2$$

$$R_2 = 50\Omega \,.$$

So, if we place a resistor with a value of 50 ohms in place of R_2, the equivalent resistance will be 48 ohms and the total current in the circuit will be equal to 0.5 amps.

Example 6.13

Problem: Find the total current flowing in the following circuit. What will the currents be in each parallel branch?

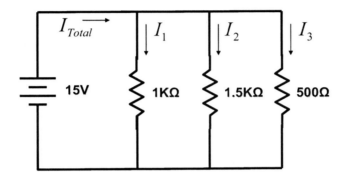

Solution: This problem, once again, asks us to find the total current in the circuit. In addition, we need to find the amount of current that is running through each of the branches. Let's start by finding the total current. To do this we need to find the equivalent resistance of the three resistors. Using the equivalent resistance relationship for more than two parallel resistors we find

$$R_{eq} = \cfrac{1}{\cfrac{1}{R_1}+\cfrac{1}{R_2}+\cfrac{1}{R_3}} = \cfrac{1}{\cfrac{1}{1000\Omega}+\cfrac{1}{1500\Omega}+\cfrac{1}{500\Omega}} = \cfrac{1}{\cfrac{11}{3000\Omega}} = \cfrac{3000\Omega}{11} \approx 272.73\Omega .$$

Now that we know the equivalent resistance of our three parallel resistors, we can use Ohm's law to find the total current. Thus, we write

$$I_{Total} = \frac{V_S}{R_{eq}} = \frac{15V}{272.73\Omega} \approx 0.055A = 55mA .$$

So, the total current is approximately 55 milliamps. Now, we need to find the currents in each of the branches. Since we know that the entire source voltage is dropped across each branch of the parallel circuit, we can use Ohm's law to find the current through each resistor. For the first resistor, we can write

$$I_1 = \frac{V_s}{R_1} = \frac{15V}{1000\Omega} = 0.015A = 15mA.$$

Now, for the second resistor we compute

$$I_2 = \frac{V_s}{R_2} = \frac{15V}{1500\Omega} = 0.01A = 10mA.$$

Finally, we an compute the final current by writing

$$I_3 = \frac{V_s}{R_3} = \frac{15V}{500\Omega} = 0.03A = 30mA.$$

So, from our results, we see that the total current is 55mA, and if we add up the current running through each of the branches of the parallel circuit, we can account for all 55mA of the current.

6.2.7   Current Division

As we just saw in the previous example, the current that flows into a parallel arrangement of resistances is divided up between those branches. We showed this by first finding the total current, and then using Ohm's law to calculate the currents in each branch. By doing this, we found that the currents in the branches added up to the total input current. So, just as resistors in series divide up the total voltage, resistors in parallel divide up the total current. We can make use of this relationship to help us when analyzing circuits. For example, it was a simple task to find the currents in each of the circuit branches of the last example because we knew what the source voltage was, but how would we find these currents if we knew only the value of the total current? To do this we need to take advantage of several relationships that we have already developed. Let's start by looking at a two parallel resistor arrangement like the one in Figure 6.7. The schematic in the figure does not provide any information about the voltage, but instead supplies a total current. We know from Ohm's law that the source

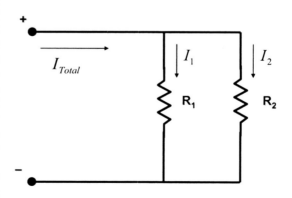

Figure 6.7 An example of a two path current divider circuit.

voltage is equal to the total current multiplied by the total equivalent resistance of the circuit. So, the source voltage can be represented as

$$V_s = \left(I_{Total}\right)\left(R_{eq}\right).$$

Likewise, we know from our parallel resistance studies that the equivalent resistance of a two resistor parallel arrangement can be written using the product over sum method as

$$R_{eq} = \frac{R_1 R_2}{R_1 + R_2}.$$

Combining these two expressions, we can write

$$V_s = \left(\frac{R_1 R_2}{R_1 + R_2}\right)I_{Total}.$$

Now, we also know from Ohm's law that the voltage in the two branches of the circuit can be expressed as

$$V_1 = I_1 R_1$$

and

$$V_2 = I_2 R_2.$$

Since the total source voltage is dropped across each branch of the parallel arrangement, we can set each of our expressions for the voltage in the branches equal to the expression for the source voltage to obtain

$$I_1 R_1 = \frac{R_1 R_2}{R_1 + R_2}I_{Total}$$

and

$$I_2 R_2 = \frac{R_1 R_2}{R_1 + R_2}I_{Total}.$$

If we solve for the currents, I_1 and I_2, we end up with expressions for the two branch currents in terms of only the two resistances and the total current. Doing this yields

$$I_1 = \frac{R_2}{R_1 + R_2} I_{Total}$$

and

$$I_2 = \frac{R_1}{R_1 + R_2} I_{Total}.$$

By using a similar derivation, it is possible to find an expression for the current in an arbitrary branch, n, of a parallel arrangement containing more than two branches. Thus, the current in such a branch can be found to be

$$I_n = \left(\frac{R_{eq}}{R_n} \right) I_{Total}.$$

So, the current in an arbitrary branch can be found by dividing the total equivalent resistance of the circuit by the resistance of the branch in question and then multiplying by the total current in the circuit. With this in mind, let's look at a couple of example problems.

Example 6.14

Problem: Find I_1 and I_2 in the following circuit.

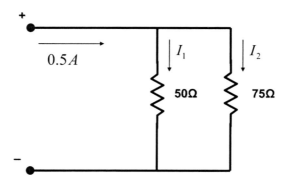

Solution: In this problem we are given a two resistor current divider circuit and asked to find the currents in the two branches. To do this we simply need to employ the relationships that we found for determining the branch currents in a two resistance current divider. Plugging into our equation for the first current we find

$$I_1 = \frac{R_2}{R_1 + R_2} I_{Total} = \left(\frac{75\Omega}{50\Omega + 75\Omega}\right) 0.5A = (0.6)0.5A = 0.3A .$$

At this point, we could simply subtract the first current from the total current to discover that the final branch current is 0.2A, but for the sake of completeness, we will use the formula in this example. So, plugging into our expression for the second current, we find that

$$I_2 = \frac{R_1}{R_1 + R_2} I_{Total} = \left(\frac{50\Omega}{50\Omega + 75\Omega}\right) 0.5A = (0.4)0.5A = 0.2A .$$

Thus, we have discovered that the two currents in this circuit are 0.3 amps and 0.2 amps respectively.

Example 6.15

Problem: Find I_1, I_2, and I_3 in the following circuit.

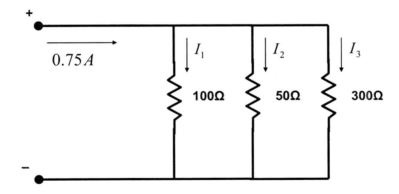

Solution: Here we have a current divider with three branches. As a result, we must use the expression for branch current in a current divider with more than two branches. This requires that we first determine the equivalent resistance of the parallel resistors. Doing this we find

$$\frac{1}{R_{eq}} = \frac{1}{R_1} + \frac{1}{R_2} + \frac{1}{R_3} = \frac{1}{100\Omega} + \frac{1}{50\Omega} + \frac{1}{300\Omega} = \frac{10}{300\Omega}$$

$$R_{eq} = 30\Omega .$$

Now, we can plug this value in to find our three branch currents. Thus, we find that the first current is

$$I_1 = \left(\frac{30\Omega}{100\Omega}\right)0.75A = (0.3)(0.75) = 0.225A = 225mA .$$

Next, we find the second branch current by calculating

$$I_2 = \left(\frac{30\Omega}{50\Omega}\right)0.75A = (0.6)(0.75) = 0.45A = 450mA .$$

Finally, we find the third branch current by writing

$$I_3 = \left(\frac{30\Omega}{300\Omega}\right)0.75A = (0.1)(0.75) = 0.075A = 75mA .$$

Thus, we have determined all three of the branch currents and if we add them up we will find that they sum to the total current of 0.75 amps.

6.2.8   Kirchhoff's Current Law

In addition to his investigation of the voltage drops in series circuits, Gustav Kirchhoff also studied the behavior of current in electrical circuits. He noticed, just as we did, that when a current encounters a parallel structure in the circuit, it splits up among the different branches. From this and other observations, he formulated what is now known as Kirchhoff's current law (KCL). This law states that for any point in a circuit, the current flowing into that point must equal the current flowing out of that point. Thus, if we think of currents entering as having positive values and currents leaving as having negative values, we can say that the algebraic sum of the currents associated with the point must be equal to zero. This statement holds true for circuit arrangements other than just simple parallel configurations and can prove

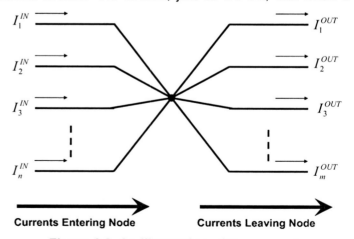

Figure 6.8 An illustration of the currents described by Kirchhoff's current law.

very useful when analyzing a circuit.   The illustration in Figure 6.8 demonstrates the general concept of Kirchhoff's current law.   In the figure, we see a node with n input currents and m output currents.   Thus, based upon Kirchhoff's current rule, we can write

$$I_1^{IN} + I_2^{IN} + I_3^{IN} + \cdots + I_n^{IN} = I_1^{OUT} + I_2^{OUT} + I_3^{OUT} + \cdots + I_m^{OUT}$$

or

$$I_1^{IN} + I_2^{IN} + I_3^{IN} + \cdots + I_n^{IN} - I_1^{OUT} - I_2^{OUT} - I_3^{OUT} - \cdots - I_m^{OUT} = 0.$$

Knowledge of this relationship can help us to quickly determine information about a circuit that might otherwise require more complicated calculations to discover.   Let's look at an example of how this tool can help us analyze circuits.

Example 6.16

Problem: Find I_1, I_2, and I_3 in the following circuit.

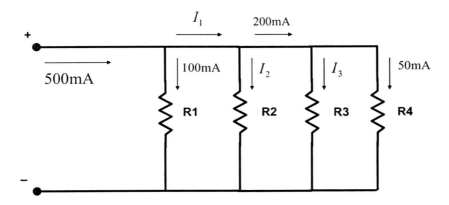

Solution: In this problem, we are given a circuit with only some of the current measurements. We know nothing about any of the voltages or resistances within the circuit. As a result, we must make use of Kirchhoff's current law to find the missing currents. Starting with I_1, we know that the total input current to the circuit is 500 mA. We also know that 100 mA of current flows through the first resistor. If we create a Kirchhoff current node at the point where the first resistor connects to the top part of the circuit, we can write

$$500mA = 100mA + I_1.$$

If we solve this expression for I_1, we will find

$$I_1 = 500mA - 100mA = 400mA.$$

Next, we can find I_2 by creating a node where the second resistor connects to the upper part of the circuit and write

$$I_1 = 200mA + I_2.$$

Since we know that I_1 is equal to 400 milliamps, we can write

$$400mA = 200mA + I_2,$$

and solving for I_2 will yield

$$I_2 = 400mA - 200mA = 200mA.$$

Now, we can use the remainder of our information to solve for I_3. Creating a node at the point where the third resistor connects to the top of the circuit, we can write

$$200mA = I_3 + 50mA.$$

Solving this equation for I_3 gives us

$$I_3 = 200mA - 50mA = 150mA.$$

So, Kirchhoff's current law has allowed us to discover that the three unknown currents are 400 milliamps, 200 milliamps, and 150 milliamps respectively.

As was the case with Kirchhoff's voltage law, Kirchhoff's current law is helpful when analyzing simple circuits, but it will be even more beneficial when analyzing even more complex arrangements.

6.3   Circuit Analysis Methods

Now that we have developed some useful tools for analyzing circuits, we can look at the application of these tools in more complicated circuits. The simple circuits that we examined while developing the initial tools and relationships in the previous section were specifically designed to illustrate the desired approach. The circuits that you will normally encounter will be much more complex and will often require you to employ several of the tools and relationships that you have learned. In the following sections, we will look at some systematic ways of applying these tools to help analyze some of these more complicated circuits. Let's begin by looking at a concept that will help to make complex circuit analysis a little bit more manageable, circuit simplification.

6.3.1   Circuit Simplification

Circuit simplification is the process of replacing parallel and series resistance combinations in a circuit with a single equivalent resistance. For example, if we want to find the total current running through a complex arrangement of resistors, we do not need any information about the voltage dropped across or the current flowing through any of the specific resistors, we are only concerned with the total current. So, if we simplify the resistor arrangement down into just one equivalent resistance, we can easily use Ohm's law to find the information that we need. Complex resistor arrangements can often be simplified by simply applying the series and parallel resistor relationships that we developed in the previous section. The following three steps are often helpful when approaching such a simplification:

1. Replace any parallel resistor arrangements in the circuit with a single equivalent resistance.

2. Combine any series resistances into a single equivalent resistance.

3. Return to step 1 and continue the simplification process until a single equivalent resistance has been determined.

In order to illustrate this approach, let's take a look at an example.

Example 6.17
_____

Problem: Find the total current, I in the following circuit. How much power will this circuit dissipate?

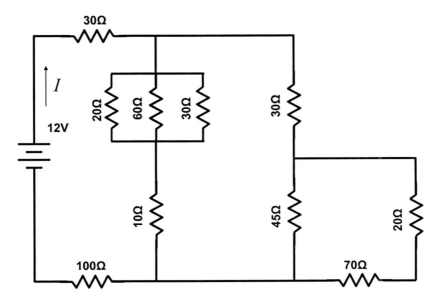

Solution: In this example, we have a complex resistor arrangement. We need to find both the total current and power consumption of the circuit. The easiest way to do this is to simplify the resistor arrangement down to just one equivalent resistance. Then we can use Ohm's and Watt's laws to find the information that we need. So, let's start simplifying this circuit by following our three step process. We first want to simplify any parallel groupings. In this case, we have three resistors in parallel, a 20 ohm, 60 ohm, and 30 ohm resistor. Applying our parallel relationships, we will find that the equivalent resistance for these three resistors is

$$\frac{1}{R_{eq}} = \frac{1}{20\Omega} + \frac{1}{60\Omega} + \frac{1}{30\Omega} = \frac{3}{60\Omega} + \frac{1}{60\Omega} + \frac{2}{60\Omega} = \frac{6}{60\Omega}$$

$$R_{eq} = 10\Omega.$$

Next, we look for series groupings that we can reduce. In this case, we have two groupings. The first is the 10 ohm equivalent resistance that we just calculated in series with another 10 ohm resistance. Using our series relationship we find

$$R_{eq} = 10\Omega + 10\Omega = 20\Omega.$$

The second series group is the 70 ohm and 20 ohm resistors in the lower right portion of the circuit. Using our series relationship on these resistors we find

$$R_{eq} = 70\Omega + 20\Omega = 90\Omega.$$

After our first simplification pass, we can redraw our circuit as follows.

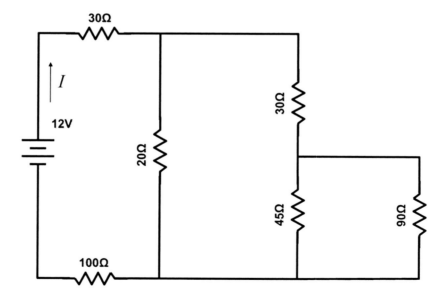

Now, we can begin the second pass through the circuit. Searching for parallel arrangements, we find that we now have a 45 ohm and 90 ohm resistance in parallel in the lower right portion of our circuit. Using our product over sum parallel relationship we can write

$$R_{eq} = \frac{(45\Omega)(90\Omega)}{45\Omega + 90\Omega} = \frac{4050\Omega^2}{135\Omega} = 30\Omega.$$

Since there are no other parallel arrangements, we look for series arrangements. The only series group we can simplify at this point is the 30 ohm equivalent we just calculated and an additional 30 ohm resistance. Using the series relationship we find

$$R_{eq} = 30\Omega + 30\Omega = 60\Omega.$$

So, after our second pass, we can redraw our simplified circuit as follows.

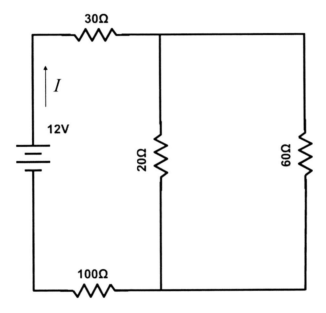

Taking another pass through our circuit, we search for parallel groupings and find that we have a 60 ohm and 20 ohm resistance in parallel on the right side of the circuit. Using the product over sum parallel relationship, we can write

$$R_{eq} = \frac{(60\Omega)(20\Omega)}{60\Omega + 20\Omega} = \frac{1200\Omega^2}{80\Omega} = 15\Omega.$$

Since there are no more parallel groupings in the circuit, we look for series groups. In this case, the 15 ohm equivalent resistance that we just found is in series with the two remaining resistances in the circuit. Using the series relationship, we can write

$$R_{eq} = 15\Omega + 30\Omega + 100\Omega = 145\Omega.$$

So, we have now reduced our entire complex resistor arrangement down to just one equivalent resistance. We can redraw the circuit one last time as follows.

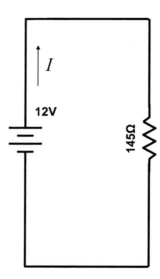

Now that we have a simple single loop circuit, we can answer the question that the problem requires. Using Ohm's law we can find the total current

$$I = \frac{V}{R} = \frac{12V}{145\Omega} \approx 0.0828A = 82.8mA.$$

The power consumption of the circuit can then be found using Watt's law to write

$$P = \frac{V^2}{R} = \frac{144V^2}{145\Omega} \approx 0.9931W = 993.1mW.$$

So, we have just seen how we can use our equivalent resistance relationships to simplify and solve a more complicated circuit problem.

6.3.2   Mesh Analysis

We have just seen that we can use resistor equivalent simplification to find information about a circuit when we are not worried about the voltage and current information in the branches of the circuit. However, if we are interested in this information, circuit simplification by itself is not always the most efficient approach. For example, if we are interested in finding the currents in each of the circuit branches, we could use Ohm's law to solve for our information by working backwards from the simplified circuit, but it is often better to go beyond basic circuit simplification and apply some more advanced techniques. These techniques employ the Kirchhoff circuit laws that we developed in the previous sections. One simple technique that makes use of Kirchhoff's voltage law to find the branch currents is known as loop or mesh analysis. This approach is particularly useful when the circuit in question contains multiple voltage sources and we are interested in finding the currents within the circuit. So, how do we employ this method? The easiest approach is to follow a series of steps.

1. Identify each mesh within the circuit. A mesh is a loop or closed path in the circuit that does not contain any other loops.

2. For each mesh, define a current direction. This direction is typically assumed to be clockwise within each mesh. This may not be the actual current direction, but at this point that is not important. If the resulting current is negative valued, we will know that we chose the wrong direction. Regardless of the direction chosen, the magnitude of the current will be correctly determined.

3. Use Kirchhoff's voltage law to write an equation for each mesh. Remember that the voltage drops across the resistors in the circuit will depend on the currents flowing through each mesh that contains the resistance. Thus, the voltage drops in a mesh may involve several unknown currents. There should be the same number of equations as there are unknown currents within the circuit.

4. Solve the system of equations that you just created for the unknown currents. You can use the matrix-based methods presented in chapter 3 to do this.

The easiest way to illustrate how this all works is to look at an example.

Example 6.18

Problem: Find I_1, I_2, and I_3 in the following circuit.

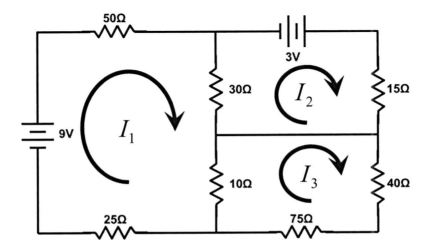

Solution: As you can see in the schematic diagram, the given circuit contains three meshes. These are associated with the three currents that we need to find. Since the meshes and the current directions have already been labeled in the circuit, we can begin our approach at step three. Thus, we need to use Kirchhoff's voltage law to write an equation for the voltage drops around each mesh. Let's start with the mesh associated with the first current. As you can see, this mesh borders the second mesh along the 30 ohm resistor and the third mesh along the 10 ohm resistor. As a result, the voltage drops in these resistors will be impacted by the respective currents. The 50 and 25 ohm resistors rely only on the current in mesh one. If we use Ohm's law, we can write the voltage drops on these two resistors as

$$V_{50\Omega} = (I_1)(50\Omega)$$

$$V_{25\Omega} = (I_1)(25\Omega).$$

Now, looking at the 30 ohm resistor we see its voltage drop will rely on both I_1 and I_2. Since the currents run through the resistor in different directions, the voltage drop associated with the second current will be subtracted from the voltage drop associated with the first current. Thus, we can write

$$V_{30\Omega} = (I_1)(30\Omega) - (I_2)(30\Omega) .$$

Using similar reasoning, we can compute the voltage drop across the 10 ohm resistor as

$$V_{10\Omega} = (I_1)(10\Omega) - (I_3)(10\Omega).$$

Now, we can use Kirchhoff's voltage law to write an equation for this mesh.

$$9V - V_{50\Omega} - V_{30\Omega} - V_{10\Omega} - V_{25\Omega} = 0$$

$$9V - (I_1)(50\Omega) - [(I_1)(30\Omega) - (I_2)(30\Omega)] - [(I_1)(10\Omega) - (I_3)(10\Omega)] - (I_1)(25\Omega) = 0$$

$$9V - I_1(50\Omega + 30\Omega + 10\Omega + 25\Omega) + I_2(30\Omega) + I_3(10\Omega)$$

$$9V - I_1(115\Omega) + I_2(30\Omega) + I_3(10\Omega).$$

Rearranging, we can write our first equation as

$$(115\Omega)I_1 - (30\Omega)I_2 - (10\Omega)I_3 = 9V .$$

Now, let's look at the second mesh. In this case, the mesh shares the 30 ohm resistor with mesh one and does not share the 15 ohm resistor. Thus, we can write the following expressions for the voltage drops in these two resistors.

$$V_{15\Omega} = (I_2)(15\Omega)$$

$$V_{30\Omega} = (I_2)(30\Omega) - (I_1)(30\Omega).$$

Using Kirchhoff's voltage law, we can now write an equation for the second mesh.

$$3V - (I_2)(15\Omega) - [(I_2)(30\Omega) - (I_1)(30\Omega)] = 0$$

$$3V + I_1(30\Omega) - I_2(15\Omega + 30\Omega) = 0.$$

Rearranging, we can write our second equation as

$$-(30\Omega)I_1 + (45\Omega)I_2 = 3V.$$

Finally, let's take a look at the third mesh. Here current I_3 flows through the 40 and 75 ohm resistors while currents I_1 and I_3 both flow through the 10 ohm resistor. Thus, we can write

$$V_{40\Omega} = (I_3)(40\Omega),$$

$$V_{75\Omega} = (I_3)(75\Omega),$$

and

$$V_{10\Omega} = (I_3)(10\Omega) - (I_1)(10\Omega).$$

Using Kirchhoff's voltage law, we can now write an equation for the third mesh as

$$-(I_3)(40\Omega) - (I_3)(75\Omega) - [(I_3)(10\Omega) - (I_1)(10\Omega)] = 0.$$

Simplifying this expression yields our third equation

$$(10\Omega)I_1 - (125\Omega)I_3 = 0.$$

Now that we have our three equations, we can solve the system of equations that we have developed. In this case, we will use the matrix methods that we studied in chapter three to solve the system for our three currents. Rewriting the system of equations in matrix form yields

$$\begin{bmatrix} 115 & -30 & -10 \\ -30 & 45 & 0 \\ 10 & 0 & -125 \end{bmatrix} \begin{bmatrix} I_1 \\ I_2 \\ I_3 \end{bmatrix} = \begin{bmatrix} 9 \\ 3 \\ 0 \end{bmatrix}.$$

Finding the inverse of the coefficient matrix yields

$$\mathbf{A}^{-1} = \begin{bmatrix} 5/471 & 10/1413 & -2/2355 \\ 10/1413 & 571/21195 & -4/7065 \\ 2/2355 & 4/7065 & -19/2355 \end{bmatrix}.$$

Solving this system of equations thus gives us

$$\begin{bmatrix} I_1 \\ I_2 \\ I_3 \end{bmatrix} = \begin{bmatrix} 5/471 & 10/1413 & -2/2355 \\ 10/1413 & 571/21195 & -4/7065 \\ 2/2355 & 4/7065 & -19/2355 \end{bmatrix} \begin{bmatrix} 9 \\ 3 \\ 0 \end{bmatrix} = \begin{bmatrix} 55/471 \\ 386/2671 \\ 22/2355 \end{bmatrix} \approx \begin{bmatrix} 0.1168A \\ 0.1445A \\ 0.0093A \end{bmatrix} = \begin{bmatrix} 116.8mA \\ 144.5mA \\ 9.3mA \end{bmatrix}.$$

Thus, we find that the three currents are 116.8 milliamps, 144.5 milliamps, and 9.3 milliamps respectively.

So, as we have just seen, Kirchhoff's voltage law can be employed in mesh analysis to help us solve complex circuit analysis problems.

6.3.3   Nodal Analysis

As we just discovered, we can use Kirchhoff's voltage law in conjunction with a series of meshes to determine the currents in a given circuit, but what if we want to know the voltages? We could use the currents that we found from our mesh analysis to calculate the voltages, but the voltage drops across resistors that are contained in multiple loops depends on more than one current. As a result, finding these voltages can get complicated. So, in cases where we have multiple sources and we want to find voltages, we can turn to a different type of circuit analysis known as nodal analysis. In this approach, we make use of Kirchhoff's current law to describe the behavior of the currents leaving and entering the nodes in the circuit. When performing this type of analysis, we typically follow the following steps.

1. Pick a node to use as a reference node. It usually works best to select the node in the circuit that has the most connections.

2. Assign current directions for each branch connecting to the remaining nodes. The direction of these currents is not important, but once they are assigned they must not be changed for the rest of the analysis.

3. Write an expression for the current in each of the branches. These expressions are written in terms of the voltages of neighboring nodes and resistances that the currents flow through.

4. Use Kirchhoff's current law to write an equation for each node in the circuit.

5. Solve the system of equations to discover the voltages at each node.

As was the case with mesh analysis, the easiest way to learn how to perform nodal analysis is to work through an example problem.

Example 6.19

Problem: Find the voltages at points **a** and **b** in the following circuit.

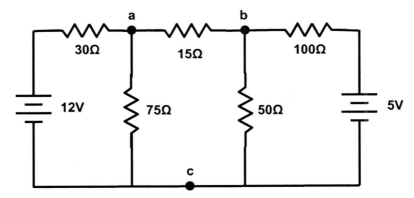

Solution: In this circuit, we are interested in learning the voltages at two particular points. As we can see, there are three branching nodes in this circuit. These are labeled **a**, **b**, and **c**. So, our first task in analyzing this circuit using nodal analysis is to pick a reference node. Since we are interested in the voltages at nodes **a** and **b**, and since node **c** connects to the most circuit branches, we will choose to use it as the reference node. Now that we have a reference node, we need to pick directions for the current in the branches connecting to both of the other two nodes. The direction is not critical. We chose the directions labeled in the circuit below simply for the sake of convenience.

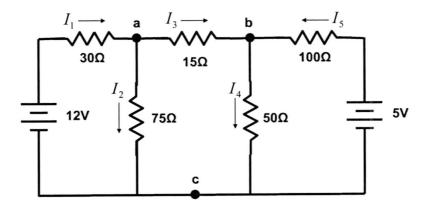

With our current directions labeled, we can begin to write expressions for each of the currents. Since all we know at this point is the resistances and the source voltages, we need to write these expressions in terms of resistances and voltages. Let's begin by looking at the first current. It flows into node **a** by passing through a 30 ohm resistor. The voltage dropped across the resistor will be equal to the voltage on the side of the resistor from which the current is flowing minus the voltage on the side of the resistor to which the current is flowing. In this case, that will be the source voltage of 12 volts minus the voltage at point **a**. Thus, using Ohm's law, we can write

$$I_1 = \frac{V_s - V_a}{30\Omega} = \frac{12V - V_a}{30\Omega}.$$

Next, we can write an expression for the second current by noting the resistance of 75 ohms and the voltage drop. In this case, the voltage drop will equal the voltage at point **a** minus the voltage at point **c**. Since the voltage at point **c** is at zero, we can write

$$I_2 = \frac{V_a}{75\Omega}.$$

The third current is calculated in much the same way as the first. In this case, the voltage drop will be the difference between the voltage at point **a** and point **b**. Using Ohm's law, we write

$$I_3 = \frac{V_a - V_b}{15\Omega}.$$

The fourth current can be written as

$$I_4 = \frac{V_b}{50\Omega},$$

and the fifth current can be expressed as

$$I_5 = \frac{5V - V_b}{100\Omega}.$$

Now that all of the currents have been expressed in terms of voltages and resistances, we can use Kirchhoff's current law to write node equations for the two nodes, **a** and **b**. If we assume that currents flowing into a node are positive and currents flowing out are negative we can write an equation for node **a** as

$$I_1 - I_2 - I_3 = 0$$

$$\frac{12V - V_a}{30\Omega} - \frac{V_a}{75\Omega} - \frac{V_a - V_b}{15\Omega} = 0$$

$$\frac{12V}{30\Omega} - \frac{V_a}{30\Omega} - \frac{V_a}{75\Omega} - \frac{V_a}{15\Omega} + \frac{V_b}{15\Omega} = 0.$$

Grouping similar terms, we can write

$$V_a\left(\frac{1}{30\Omega} + \frac{1}{75\Omega} + \frac{1}{15\Omega}\right) - V_b\left(\frac{1}{15\Omega}\right) = \frac{12V}{30\Omega},$$

and simplifying we get

$$V_a\left(\frac{17}{150\Omega}\right) - V_b\left(\frac{1}{15\Omega}\right) = 0.4A.$$

Now, for node **b**, we can write

$$I_3 - I_4 + I_5 = 0$$

$$\frac{V_a - V_b}{15\Omega} - \frac{V_b}{50\Omega} + \frac{5V - V_b}{100\Omega} = 0$$

$$\frac{V_a}{15\Omega} - \frac{V_b}{15\Omega} - \frac{V_b}{50\Omega} + \frac{5V}{100\Omega} - \frac{V_b}{100\Omega} = 0.$$

Grouping similar terms we can write

$$-V_a\left(\frac{1}{15\Omega}\right) + V_b\left(\frac{1}{15\Omega} + \frac{1}{50\Omega} + \frac{1}{100\Omega}\right) = \frac{5V}{100\Omega},$$

and simplifying the expression, we get

$$-V_a\left(\frac{1}{15\Omega}\right) + V_b\left(\frac{29}{300\Omega}\right) = 0.05A.$$

At this point, we have two node equations and two unknown voltages. If we solve the system of equations, we can find these two voltages. Rewriting the system in matrix form, we get

$$\begin{bmatrix} 17/150 & -1/15 \\ -1/15 & 29/300 \end{bmatrix}\begin{bmatrix} V_a \\ V_b \end{bmatrix} = \begin{bmatrix} 0.4 \\ 0.05 \end{bmatrix}.$$

Taking the inverse of the coefficient matrix yields

$$\begin{bmatrix} 4350/293 & 3000/293 \\ 3000/293 & 5100/293 \end{bmatrix},$$

and using the inverse, we find

$$\begin{bmatrix} 4350/293 & 3000/293 \\ 3000/293 & 5100/293 \end{bmatrix}\begin{bmatrix} 0.4 \\ 0.05 \end{bmatrix} = \begin{bmatrix} 1890/293 \\ 1455/293 \end{bmatrix} \approx \begin{bmatrix} 6.45V \\ 4.97V \end{bmatrix}.$$

So, the voltage at point **a** is 6.45 volts and the voltage at point **b** is 4.97 volts.

Thus, we have seen that we can use Kirchhoff's current rules to help us solve complicated circuit analysis problems. So, by using the techniques that you have learned in this chapter, you will have a good starting point when approaching circuit analysis problems in your future work.

6.4   Chapter Summary

In this chapter we developed important tools that can be used in future circuit analysis work. We looked at several important relationships and approaches to circuit analysis.

- We began our study by reviewing Ohm's law. We saw that it provides a relationship between voltage, resistance, and current that can be helpful when analyzing circuits. It can be expressed in any of the three following forms based upon what values are known.

$$V = IR \qquad\qquad I = \frac{V}{R} \qquad\qquad R = \frac{V}{I}$$

- Next, we expanded upon Ohm's law by looking at Watt's law. This law provides a relationship between resistance, current, voltage, and the power dissipated in a device. This law can be used to express power in the following three ways.

$$P = IV \qquad\qquad P = I^2 R \qquad\qquad P = \frac{V^2}{R}$$

- We also saw that Ohm's law and Watt's law can be combined to produce a variety of expressions for power, resistance, voltage, and current.

- Resistors in series arrangements were explored next. It was discovered that the equivalent resistance of resistors in this arrangement can be calculated by simply summing the individual resistances. Thus, for *n* resistors in series, we can find the equivalent resistance by calculating

$$R_{eq} = R_1 + R_2 + \cdots + R_n.$$

- When we looked into resistors in series, we discovered that each resistor drops a small portion of the total voltage supplied by the source. Thus, resistors in series can be used to divide up the voltage. For two resistors in series, we found that we could find the voltage across the second resistor as

$$V_{OUT} = \left(\frac{R_2}{R_1 + R_2} \right) V_S.$$

- Gustav Kirchhoff developed a relationship that explains the way in which the voltage drops across the resistors in a series circuit work. This relationship states that the voltage drops and gains around the loop of a series circuit must sum to zero. Or, in other words, the voltage drops caused by the resistors must equal the voltage gains provided by the voltage sources.

- Next, we explored resistors arranged in parallel. We found that the equivalent resistance provided by n resistors arranged in parallel can be expressed as

$$R_{eq} = \frac{1}{\dfrac{1}{R_1} + \dfrac{1}{R_2} + \cdots + \dfrac{1}{R_n}}.$$

- For parallel arrangements of only two resistors, the equivalent resistance can be expressed in a simpler form, known as the product over sum approach. It can be written as

$$R_{eq} = \frac{R_1 R_2}{R_1 + R_2}.$$

- Just as resistors in series divide up the voltage, we saw that resistors in parallel divide up the current. Thus, the current in the resistors of a two resistor parallel arrangement can be expressed as a ratio of resistances multiplied by the total current.

$$I_1 = \frac{R_2}{R_1 + R_2} I_{Total} \qquad\qquad I_2 = \frac{R_1}{R_1 + R_2} I_{Total}.$$

- Kirchhoff also developed a relationship to describe the way in which current behaves in a circuit. This relationship, known as Kirchhoff's current law states that the total current entering a point in a circuit must be equal to the current leaving that point.

- Next, we looked at techniques to approaching more complex circuit analysis problems. The first of these was circuit simplification. We saw that by employing both parallel and series relationships, we could simplify a complicated resistor network down into a single equivalent resistance.

- The second technique that we looked at was mesh analysis. This approach involved the use of Kirchhoff's voltage law. An equation is written for each mesh in the circuit and then these equations are simultaneously solved to find the currents in the meshes.

- The final approach we looked at uses Kirchhoff's current law. Equations are written for each node in the circuit. These equations describe the current flow for that node in terms of node voltages and branch resistances. These equations are solved simultaneously to find the voltages at the nodes in the circuit.

6.5   Review Exercises

Section 6.2 – Basic Concepts

1. Using the circuit in Figure 6.9 and the values provided, find the voltage dropped across the resistor.

   a. $R = 250\Omega$ and $I = 250$ mA
   b. $R = 75\Omega$ and $I = 150$ mA
   c. $R = 835\Omega$ and $I = 75$ mA
   d. $R = 65\Omega$ and $I = 1.5$ A

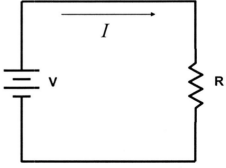

2. Using the circuit in Figure 6.9 and the values provided, find the current flowing through the resistor.

   a. $R=75\Omega$ and $V = 25$ V
   b. $R=120\Omega$ and $V = 5$ V
   c. $R=65\Omega$ and $V = 9$ V
   d. $R=1.7$ KΩ and $V = 12$ V

Figure 6.9 Schematic diagram for use with problems 1 – 5.

3. Using the circuit in Figure 6.9 and the values provided, find the value of the resistor.

   a. $I = 75$ mA and $V = 12$ volts
   b. $I = 4.5$ mA and $V = 9$ volts
   c. $I = 20$ mA and $V = 5$ volts
   d. $I = 1.5$ A and $V = 15$ volts

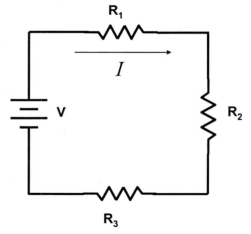

4. Using the circuit in Figure 6.9 and the values provided, find the power dissipated by the resistor.

   a. $I = 25$ mA and $R = 250$ Ω
   b. $V= 120$ V and $R = 10$ Ω
   c. $I = 1.5$ A and $V = 25$ V
   d. $I = 75$ mA and $V = 5$ V

Figure 6.10 Schematic diagram for use with problems 6 – 10

5. If the circuit shown in Figure 6.9 has a supply voltage of 20 volts and the power dissipated in the resistor is 5 watts, what is the resistance of the resistor? What is the current flowing through the circuit?

6. Given the circuit in Figure 6.10 with $R_1 = 25\Omega$, $R_2 = 450\Omega$, $R_3 = 35\Omega$, and V = 15 V, find the following values:

   a. Find the equivalent total resistance in the circuit.
   b. Find the current flowing through the circuit.
   c. Find the voltage dropped across each of the three resistors.
   d. Find the power dissipated by the circuit.

7. Given the circuit in Figure 6.10 with $R_1 = 125\Omega$, $R_2 = 50\Omega$, $R_3 = 735\Omega$, and V = 9 V find the following values:

    a. Find the equivalent total resistance in the circuit.
    b. Find the current flowing through the circuit.
    c. Find the voltage dropped across each of the three resistors.
    d. Find the power dissipated by each of the resistors.

8. Given the circuit in Figure 6.10 with $R_1 = 35\Omega$, $R_3 = 15\Omega$, and V = 5 V find the value of R_2 necessary to create a current flow of 50 mA.

9. Given the circuit in Figure 6.10 with $R_2 = 75\Omega$, $R_3 = 100\Omega$, and V = 50 V find the value of R_1 necessary to provide a power dissipation of 5 watts.

10. If the circuit in Figure 6.10 has resistances of $R_1 = 175\Omega$, $R_2 = 15\Omega$, $R_3 = 50\Omega$, and a current of 1.5 amps, what voltage is being supplied?

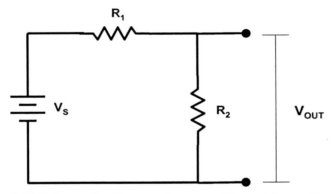

Figure 6.11 Schematic diagram for use with problems 11 – 14

11. Given the voltage divider circuit in Figure 6.11, determine the output voltage if $R_1 = 175\Omega$, $R_2 = 55\Omega$, and $V_s = 35V$. Find the current flowing through the circuit.

12. Given the voltage divider circuit in Figure 6.11, determine the output voltage if $R_1 = 430\Omega$, $R_2 = 335\Omega$, and $V_s = 5V$. What is the current in the circuit?

13. Using the circuit in Figure 6.11, assume that $R_1 = 300\Omega$ and $V_s = 12V$. Determine what resistance needs to be used for R_2 to obtain the following voltages for V_{OUT}.

    a. 9 volts
    b. 6 volts
    c. 3 volts
    d. 4 volts
    e. 5 volts

14. Given the circuit in Figure 6.11, determine what the values of the two resistors need to be in order to provide an output voltage of 15 volts and a current of 50 mA if the source voltage is 25 volts.

15. Explain Kirchhoff's voltage law.

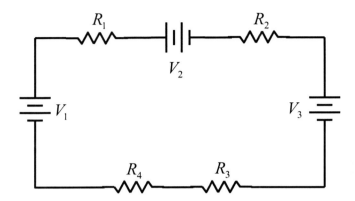

Figure 6.12 Schematic diagram for use with problems 16 - 17

16. Using Kirchhoff's Voltage Law and the circuit in Figure 6.12 find the current given the following values

a. $R_1 = 25\Omega$, $R_2 = 55\Omega$, $R_3 = 90\Omega$, $R_4 = 10\Omega$, $V_1 = 5V$, $V_2 = 12V$, and $V_3 = 3V$

b. $R_1 = 70\Omega$, $R_2 = 155\Omega$, $R_3 = 10\Omega$, $R_4 = 50\Omega$, $V_1 = 3V$, $V_2 = 6V$, and $V_3 = 4V$

c. $R_1 = 5\Omega$, $R_2 = 200\Omega$, $R_3 = 85\Omega$, $R_4 = 30\Omega$, $V_1 = 11V$, $V_2 = 4V$, and $V_3 = 9V$

17. Use Kirchhoff's Voltage Law to help find the voltage dropped across each resistor in Figure 6.12 given that

a. $R_1 = 15\Omega$, $R_2 = 50\Omega$, $R_3 = 40\Omega$, $R_4 = 85\Omega$, $V_1 = 3V$, $V_2 = 10V$, and $V_3 = 2V$

b. $R_1 = 50\Omega$, $R_2 = 95\Omega$, $R_3 = 100\Omega$, $R_4 = 30\Omega$, $V_1 = 8V$, $V_2 = 5V$, and $V_3 = 9V$

c. $R_1 = 40\Omega$, $R_2 = 65\Omega$, $R_3 = 65\Omega$, $R_4 = 90\Omega$, $V_1 = 6V$, $V_2 = 3V$, and $V_3 = 12V$

18. Given the following resistances, determine the equivalent resistance that would be produced if they were configured in a parallel arrangement.

a. $R_1 = 35\Omega$, $R_2 = 50\Omega$.

b. $R_1 = 175\Omega$, $R_2 = 500\Omega$.

c. $R_1 = 150\Omega$, $R_2 = 500\Omega$, and $R_3 = 440\Omega$.

d. $R_1 = 85\Omega$, $R_2 = 70\Omega$, $R_3 = 20\Omega$, $R_4 = 175\Omega$.

19. Given the following resistances, determine the equivalent resistance that would be produced if they were configured in a parallel arrangement.

    a. $R_1 = 80\Omega, \ R_2 = 120\Omega$.
    b. $R_1 = 1.5\,\text{K}\Omega, \ R_2 = 700\Omega$.
    c. $R_1 = 850\Omega \ R_2 = 400\Omega,$ and $R_3 = 240\Omega$.
    d. $R_1 = 885\Omega, \ R_2 = 7\,\text{K}\Omega, \ R_3 = 200\Omega, \ R_4 = 475\Omega$.

20. Show that the equivalent resistance for three resistors in parallel can be expressed as

$$R_{eq} = \frac{R_1 R_2 R_3}{R_1 R_2 + R_1 R_3 + R_2 R_3}.$$

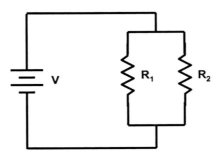

Figure 6.13 Schematic diagram for use with problems 21 – 22

21. Given the circuit in Figure 6.13 find the current given the following voltages and resistances.

    a. $R_1 = 125\Omega$ and $R_2 = 500\Omega$ with a voltage of 25 volts.
    b. $R_1 = 15\,\text{K}\Omega$ and $R_2 = 5\,\text{K}\Omega$ with a voltage of 9 volts.
    c. $R_1 = 125\Omega$ and $R_2 = 500\Omega$ with a voltage of 25 volts.

22. If the circuit in Figure 6.14 has a voltage of 12 volts and $R_1 = 2\,\text{K}\Omega$, find the value of the second resistor so that a current of 12 mA will flow through the circuit.

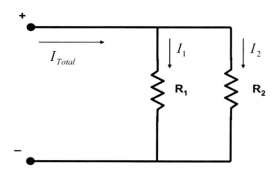

Figure 6.14 Schematic diagram for use with problem 23

23. Given the circuit in Figure 6.14 find currents I_1 and I_2 with the following conditions

    a. $I_{Total} = 1.7 A$, $R_1 = 75\Omega$, and $R_2 = 125\Omega$
    b. $I_{Total} = 75 \, mA$, $R_1 = 575\Omega$, and $R_2 = 825\Omega$
    c. $I_{Total} = 750 \, mA$, $R_1 = 1.5 \, K\Omega$, and $R_2 = 625\Omega$
    d. $I_{Total} = 2 A$, $R_1 = 7.5 \, K\Omega$, and $R_2 = 525\Omega$

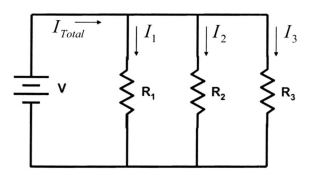

Figure 6.15 The schematic diagram for use with problem 24 .

24. Given the circuit shown in Figure 6.15, find currents I_1, I_2 and I_3 with the following conditions

    a. $R_1 = 250\Omega$ $R_2 = 100\Omega$, $R_3 = 720\Omega$, and $I_{Total} = 500 \, mA$.
    b. $R_1 = 50\Omega$ $R_2 = 500\Omega$, $R_3 = 220\Omega$, and $I_{Total} = 350 \, mA$.
    c. $R_1 = 350\Omega$ $R_2 = 120\Omega$, $R_3 = 60\Omega$, and $I_{Total} = 75 \, mA$.

25. Explain Kirchhoff's current law.

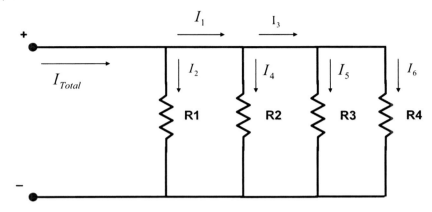

Figure 6.16 The schematic diagram for use with problem 26.

26. Given the circuit in Figure 6.16, use Kirchhoff's current law to fill in the missing currents given that

   a. $I_{Total} = 1.5\text{A}, \quad I_2 = 500\,\text{mA}, \quad I_3 = 250\,\text{mA}, \text{and } I_5 = 100\,\text{mA}.$
   b. $I_{Total} = 755\,\text{mA}, \quad I_1 = 600\,\text{mA}, \quad I_5 = 250\,\text{mA}, \text{and } I_6 = 100\,\text{mA}.$
   c. $I_{Total} = 1.5\text{A}, \quad I_2 = 500\,\text{mA}, \quad I_3 = 250\,\text{mA}, \text{and } I_6 = 100\,\text{mA}.$
   d. $I_{Total} = 3\text{A}, \quad I_2 = 450\,\text{mA}, \quad I_5 = 650\,\text{mA}, \text{and } I_6 = 560\,\text{mA}.$

Section 6.3 – Circuit Analysis Methods

27. Find the equivalent resistance of the following resistor network.

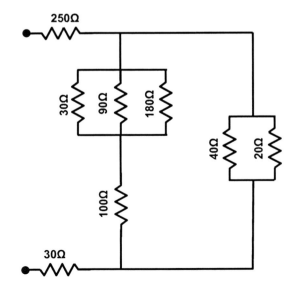

28. Find the equivalent resistance for the following resistor network as seen by the voltage source. Determine the total current flowing through the circuit.

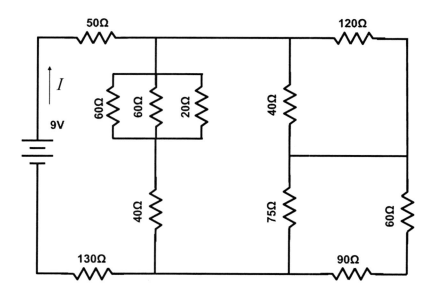

29. When is it useful to use mesh analysis on a circuit?

30. Describe how mesh analysis is performed.

31. Use mesh analysis to find the two mesh currents in the following circuit.

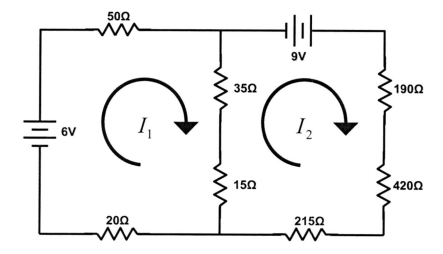

32. Use mesh analysis to find the three mesh currents in the following circuit.

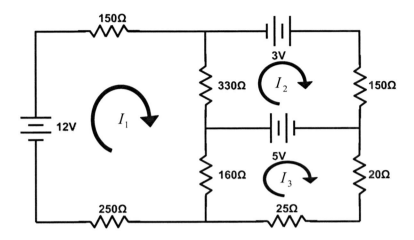

33. When is it beneficial to use nodal analysis?

34. How is nodal analysis performed?

35. Using node c as a reference point, employ nodal analysis to find the voltages at nodes **a** and **b** in the following circuit.

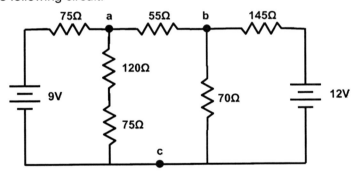

36. Using node d as a reference point, employ nodal analysis to find the voltages at nodes **a**, **b**, and **c** in the following circuit. Find the voltages dropped across each resistor in the circuit.

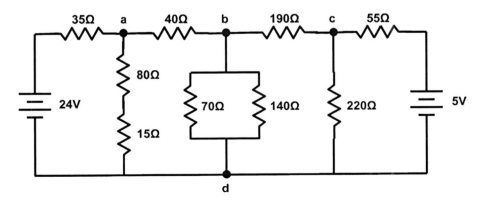

UNIT 3

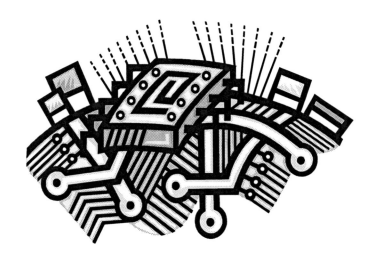

Digital Logic
and
Computers

Chapter 7. Digital Logic

Chapter Outline

7.1   Introduction to Digital Logic

In the previous chapter, we investigated circuits that worked in the realm of analog. In other words, circuits that could take an infinite range of input voltages, currents, etc. and supply an infinite number of output values. For example, if we connected a 9-volt battery to a circuit, we could calculate the values that we would expect to see at various points throughout the circuit. These values would all be represented in terms of real numbers, for example 1.375V or 0.359A.

While these analog circuits play a vital role in most electronics, there is another family of electronics that has become increasingly popular. These circuits, unlike their analog counterparts, work only with digital values. What do we mean by digital values? Essentially what we mean when we refer to a digital value is a value that is represented completely as a series of binary digits. In other words, instead of taking a whole range of voltages like analog circuits, digital circuits will accept only combinations of two states, on or off, as inputs and will provide output in the same manner. Thus, digital circuits operate like networks of switches. An input can be viewed as a switch that is either turned on or off, and the combination of a series of input switches can determine whether a given output is turned on or off. So, now that we know what a digital circuit is, why do we care about them? The reason for concerning ourselves with digital circuits is the fact that they can be extremely useful. The practice of processing information in digital form has become increasingly widespread, particularly in the arena of signal processing and communications. In fact, digital circuits are critical to the entire branch of computer engineering. Over the years, engineers have discovered that by representing and transmitting information digitally, we can achieve better data compression as well as signals error detection and correction. For example, these days it is difficult to find music in an analog form (tapes or records), instead most music is found on CD's and computer music files, both digital representations. In addition, since computers operate in the digital realm, it is essential to have a good grasp on basic digital circuit theory to fully understand their operation.

Before we begin our examination of digital circuits, we will need to be aware of some notations. In order to describe the operation of digital circuits, we need to make use of some new concepts. These include truth tables and Boolean expressions. These will be explored in the next two sections. Once we have these concepts covered, we will begin to look at the basic components of digital circuits. First, however, we need to look at how we label the inputs and outputs of a digital circuit. As we established before, the inputs and outputs of digital circuits can have only two states and are much like switches. As a result, there are several ways we can label the inputs. The most obvious would be to use the labels on and off, or we could use the logical values of true and false. While both of these notations are acceptable, and in some situations preferred, the use of the binary digits 1 and 0 are the most common notations. Finally, we must define a standard way to refer to a specific input and output. In this text, since we will be dealing primarily with no more than five inputs and one output, we will refer to the inputs with the capital letters ranging from A to E and the output with the capital letter F. Figure 7.1 illustrates this configuration.

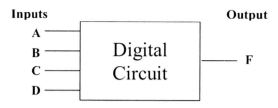

Figure 7.1. An illustration of the labeling method used for digital circuits in this text.

7.2   Truth Tables

Now that we can describe which inputs we are referring to and what logic state they possess, we need a way to describe the overall performance of our logic circuit. To truly understand how a logic circuit will perform, it is necessary to be able to see all of the possible inputs for the circuit coupled with the output that they will generate. This is usually done in a tabular form. The possible inputs for the circuit are listed in ascending binary order. Beside each of the possible inputs, a column is provided for the circuit's output. Thus by glancing at one simple table, we get a global picture of what the circuit does. This kind of tabular description of the logic circuit is known as a "Truth Table." Figure 7.2 demonstrates the layout for the truth table representing a typical three input circuit.

Input Combinations			Output
A	B	C	F
0	0	0	1
0	0	1	0
0	1	0	0
0	1	1	1
1	0	0	1
1	0	1	0
1	1	0	1
1	1	1	0

Figure 7.2. An example truth table.

7.3   Boolean Expressions

Another common way of describing the operation of a digital circuit is through the use of Boolean expressions. In a Boolean expression, the inputs and outputs of the circuit are represented by their alphabetic letter designation. Combinations of these input and output values in either their normal (A, B, C, etc.) or complimented (A', B', C', etc.) form are called Boolean expressions. We will take a closer look at how these expressions work as we progress through the chapter, but it is important to be able to recognize one now. Figure 7.3 illustrates a typical Boolean expression.

$$F = A'B' + AB' + AB$$

Figure 7.3. An example of a typical Boolean expression.

7.4   Basic Logic Gates

Now that we have a basic understanding of the notation used in digital logic circuits, we can begin to investigate some of the components that we will find in a typical circuit and how each of these components behaves. Most digital logic circuits consist of a series of special integrated circuits called "gates." These gates form the building blocks of the circuit and can be combined to construct a circuit to satisfy practically any situation. Let's take a look at some of these building blocks now.

7.4.1   NOT Gates

The simplest of all logic gates is the NOT gate. This gate is often referred to as an inverter, and as that name suggests, it simply "inverts" the logic value that it is given. In other words, if the NOT gate receives a True (1) input, its output will be False (0). Likewise, if it receives a false (0) input, its output will be true (1). When this gate is drawn in circuits, it is illustrated as a triangle with a small circle at one end. Figure 7.4 displays the schematic symbol for the

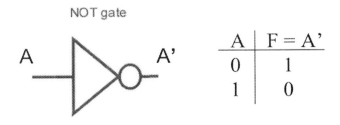

A	F = A'
0	1
1	0

Figure 7.4. The schematic symbol and truth table for a NOT gate.

NOT gate and summarizes its behavior in a truth table. It is also important to note how the not gate changes the Boolean value of the input. From the figure, we see that the input is represented as the normal Boolean value, A, but the output is represented as the compliment, A'. Thus, we see that in a Boolean expression, a compliment is like the inverse of the normal value. So, a complimented Boolean value typically means that the original input was passed through a NOT gate.

7.4.2   AND Gates

Another important logic gate is the AND gate. Unlike the NOT gate, the AND gate has two inputs, and as the name suggests, the AND gate requires that both of the inputs have a logic true (1) value for the output to be true (1). All other combinations of inputs will result in a logic false (0) output. One way to think of an AND gate is as two switches wired in series. Figure 7.5 shows such a configuration. It is easy to see from the figure that if we want to light the bulb, we must complete the circuit by closing both of the switches. If either switch (or both switches) is left open, the circuit will be broken and the bulb will not light.

The schematic symbol for an AND gate is shown in Figure 7.6 along with the truth table representing its operation. From this figure, we can also see how an AND gate is represented in a

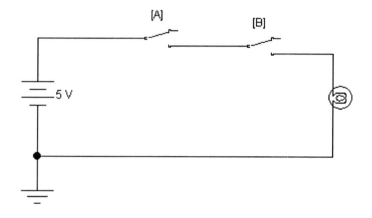

Figure 7.5. A circuit diagram illustrating the characteristics of an AND gate. In order to light the bulb, both switches must be closed. Likewise, to get a logic true (1) output from an AND gate, both inputs must be true (1).

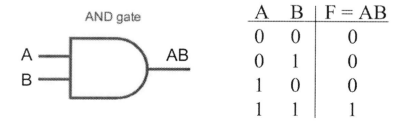

A	B	F = AB
0	0	0
0	1	0
1	0	0
1	1	1

Figure 7.6. The schematic symbol and truth table for a two input AND gate.

Boolean expression. The output of an AND gate is represented as the product of its inputs. This makes sense if you think about it a little. If we multiply the logic values for the input signals, the only way we can get a 1 as the result is if both of the inputs are 1. If either of the inputs is a zero, the product is equal to zero. It should be noted that AND gates can have more than two inputs. When this occurs, the AND gate simply requires that all of the inputs must be true (1) before a true (1) output can be obtained.

7.4.3   OR Gates

Like AND gates, OR gates have two or more inputs. However, OR gates only require that at least one input be true (1) in order to give a true (1) output. Therefore, no matter how many inputs there are with a false (0) value, one true (1) value input will result in a true (1) result. So, the only way to get a false (0) input from an OR gate is to insure that all of the inputs are false (0). If we think of the OR gate in terms of a standard circuit, it would consist of multiple switches wired in parallel. Thus, the light bulb could be lit by closing any one of the switches, but the only way to turn it off would be to open all of the switches. Figure 7.7 demonstrates this circuit layout. Note that no matter how many open switches there are, one closed switch will light the bulb.

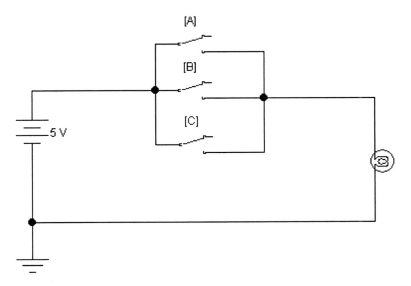

Figure 7.7. A circuit diagram illustrating the operation of an OR gate. The bulb can be lit by closing any of the switches, but can be extinguished only by opening all of the switches.

The schematic symbol for the OR gate is shown in Figure 7.8 along with the truth table for a two input version of the gate. It should also be noted that in a Boolean expression, an OR gate is represented with addition. The output of an OR gate is described as the sum of its inputs. However, this sum cannot be larger than one, so if more than one input is true (1), the output simply sums to true (1). Thus, it is perfectly acceptable for a Boolean expression such as 1+1=1 to be written.

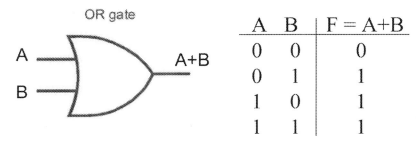

Figure 7.8. The schematic symbol and truth table for a two input OR gate.

With the addition of the OR gate, we have a sufficient number of gates to design a circuit to fit any desired truth table. The NOT, AND, and OR gates are considered to be the basic gates in digital logic from which all other gates can be derived. While these gates are sufficient for any design, they are not the only commonly encountered gates. Before we investigate ways of putting the gates together, let's look at a few more important gates that are derived from these three gates.

7.4.4   NAND Gates

Another useful logic gate is the NAND gate. A NAND gate is basically a normal AND gate followed by an inverter, or NOT gate. Thus, the output of a NAND gate is exactly opposite that of an AND gate. Figure 7.9 illustrates the combination of an AND and NOT gate that would be required to achieve the same output as a NAND gate.

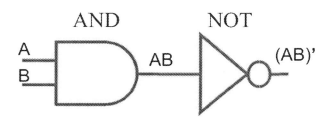

Figure 7.9. An illustration of the combination of an AND and a NOT gate. This combination has the same output as a NAND gate.

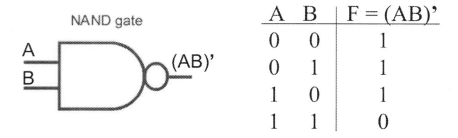

Figure 7.10. An illustration of the schematic symbol for the NAND gate and its truth table.

As the figure illustrates, the incoming logic values first pass through an AND gate. The output of this AND gate, (AB), is then passed through a NOT gate, resulting in a truth table that is the exact opposite of that of the AND gate, (AB)'. The actual schematic symbol used for the NAND gate is a combination of the AND gate and the circle from the NOT gate. Figure 7.10 provides the scematic symbol and truth table for a two input NAND gate.

7.4.5   NOR Gates

Much like the NAND gate, the NOR gate is composed of a previously defined gate followed by an inverter. In this case, it is an OR gate. Thus, as we would expect, the output of a NOR gate is the exact opposite of that of a typical OR gate. Figure 7.11 illustrates this logic combination.

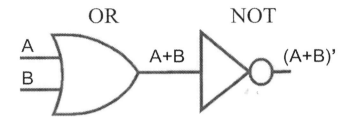

Figure 7.11. An illustration of the combination of an OR and NOT gate that yields the same output as a NOR gate.

From the figure, we can see that the inputs pass through an OR gate. The output of this OR gate, (A+B), is then passed through a NOT gate to achieve the resulting truth table, (A+B)'. As was the case with the NAND gate, the schematic symbol for the NOR gate consists of the symbol for an OR gate followed by the circle from an inverter. The schematic symbol and truth table for a two input NOR gate can be found in Figure 7.12.

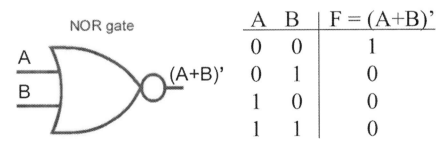

A	B	$F = (A+B)'$
0	0	1
0	1	0
1	0	0
1	1	0

Figure 7.12. The schematic symbol and truth table of a two-input NOR gate.

It is worth noting that by combining NAND and NOR gates, any logic circuit can be constructed. Thus, NAND and NOR gates form a complete logic set just as the NOT, AND, and OR gates did. In fact, it is possible to construct these three gates using combinations of NAND and NOR gates. Figure 7.13 illustrates these combinations.

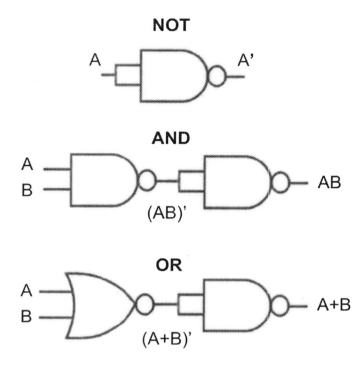

Figure 7.13. An illustration of some possible configurations of NAND and NOR gates that yield the same outputs as the standard NOT, AND, and OR gates.

7.4.6   XOR Gates

One final gate that is worth looking at is the exclusive OR gate, usually written as XOR. This gate is a slight variation of the OR gate. As we know, the OR gate returns a logic true (1) output as long as one or more of the input values are true (1). The exclusive OR gate is a bit more picky. It requires that only one of the inputs is true (1). If more than one input value is true (1), the gate will return a false (0) output. Figure 7.14 provides the schematic symbol and truth table for a two input version of the XOR gate.

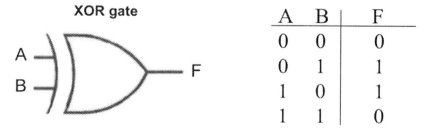

A	B	F
0	0	0
0	1	1
1	0	1
1	1	0

Figure 7.14. The schematic symbol and truth table for a two input XOR gate.

The XOR gate can also be defined in terms of a combination of NOT, AND, and OR gates. In the case of the XOR gate, however, it is a little more complex than the simple combinations of the NAND and NOR gates. The circuit diagram for the XOR gate in terms of NOT, AND, and OR gates has been provided in Figure 7.15. We will learn how to combine logic gates together to form circuits such as this later in the chapter.

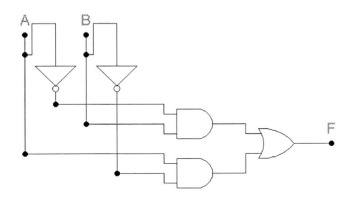

Figure 7.15. A circuit diagram for the XOR gate in terms of NOT, AND, and OR gates.

7.5   Combinations of Logic Gates

Now that we have examined some of the more common logic gates, we can investigate more complex arrangements of these gates. The first aspect of these complex arrangements that should be investigated is how we can determine the behavior of a given circuit. In other words, if we come upon an existing logic circuit, how do we determine what its output will be for a given situation? The best way to make this determination is to develop a truth table for the circuit. By doing so, we will have a picture of its output for all possible input combinations. There are several ways to arrive at a truth table. We will look at two of these methods, beginning with the most basic (and time consuming) approach, circuit tracing.

7.5.1   Finding a Truth Table by Circuit Tracing

When we come across a logic circuit that performs some unknown operation, our first task should be to examine the circuit to determine what its function is. To understand the circuit's operation, we must construct a truth table, and one simple (but time consuming) way to do this is by performing circuit tracing. The way this works is very simple.   First, we write out our truth table, leaving the results for each of the inputs blank for now. Next, we place the first input combination onto the inputs of the circuit and trace the logic through the entire circuit to determine what the output will be. Once we have an output value, we place that result in the truth table and move to the next input combination. If we repeat the steps for all of the input combinations we can complete our entire truth table and will know exactly how the circuit will behave in all cases. Let's take a look at how this works.

Example 7.1

Problem: Given the following logic circuit, determine its truth table by circuit tracing.

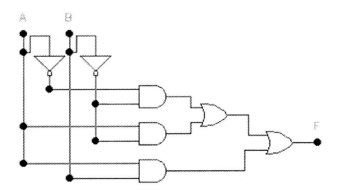

Solution: As we can see, the circuit we were given has two inputs. Thus, our first step is to write out a blank truth table for a two input logic circuit. Since we have two inputs, there will be four binary possibilities to examine. So, our truth table should look like this:

A	B	F
0	0	
0	1	
1	0	
1	1	

Now, we must trace through the circuit for each of the input combinations. Since there are four combinations, we must trace through the circuit four times. A=0, B=0 is the first input combination. So, let's begin by tracing those inputs through the circuit. While moving through the circuit, it is usually helpful to note the resulting logic level after each gate. This makes determining the input combinations for the next gate easier. When we arrive at the output from the last gate, we will have the circuit output for the given input combination. This value is then recorded in the truth table. By doing this we get:

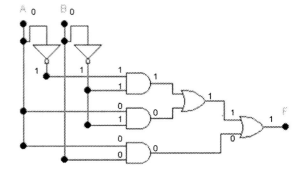

A	B	F
0	0	1
0	1	
1	0	
1	1	

Now that we have a result for the first input combination, let's repeat the process to get the results of applying the other three input combinations. This is done as follows:

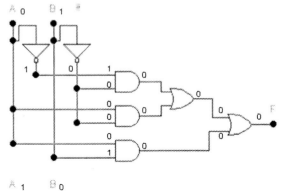

A	B	F
0	0	1
0	1	0
1	0	
1	1	

A	B	F
0	0	1
0	1	0
1	0	1
1	1	

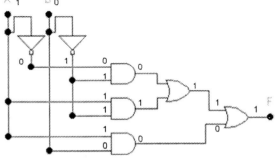

A	B	F
0	0	1
0	1	0
1	0	1
1	1	1

So, the truth table describing the behavior of our circuit is:

A	B	F
0	0	1
0	1	0
1	0	1
1	1	1

Thus, we have determined the behavior of our circuit.

As example 7.1 illustrates, the truth table of a circuit can be obtained by tracing all of the possible input combinations through the circuit and then recording the results. While this method will yield a correct result, it is a long and time-consuming approach. This is especially true when the circuit has more than just two input lines. This method could take a considerable amount of time if, for example, used to analyze a circuit with four input lines (16 possible combinations). In addition, the use of this method leaves a lot of room for clerical or careless error to produce an incorrect result. Because of this, the circuit tracing method is not typically used unless the circuit is very small and simplistic. Luckily, there is a better way to analyze the more complex circuits. This method involves the use of the Boolean expressions that we noted earlier. By using the Boolean expression associated with each gate type, we can look at the circuit and write a Boolean expression that describes the conditions under which the circuit will evaluate to a logic true (1) and then use this expression to build a truth table for the circuit.

7.5.2 Finding a Boolean Expression for Gate Combinations

Before we can find the truth table for a logic circuit from a Boolean expression, we must derive a Boolean expression for the logic circuit that we are analyzing. The Boolean expression that describes a logic circuit is built up by using the Boolean expressions for each of the individual gates and chaining them together to form one larger expression. To do this, we must remember the expressions that are associated with each of our gates. Table 7.1 summarizes these expressions.

Table 7.1 A Summary of the main logic gates and their Boolean expressions.

Gate	Input(s)	Boolean Expression
NOT	A	A'
AND	A,B	AB
OR	A,B	A+B
NAND	A,B	(AB)'
NOR	A,B	(A+B)'

You should remember that we also discussed one other gate that is not shown in Table 7.1, the XOR gate. This particular gate was derived from a slightly more complex circuit than the NAND and NOR gates, and no Boolean expression was given for it. Let's turn our attention back toward the XOR gate and use what we know about the other gates to derive a Boolean expression for it now.

Example 7.2

Problem: Derive a Boolean expression for the circuit representing the XOR gate shown in Figure 7.15.

Solution: First, let's look back at the circuit diagram in Figure 7.15. In this diagram we see that there are two NOT gates, two AND gates, and an OR gate. Now, from Table 7.1 we find that a NOT gate is described by the Boolean expression A', an AND gate is represented as AB, and an OR gate is represented as A+B, where A and B are the inputs to the gates. If we start at the circuit inputs (A and B) and note the output of the inverters, we will get A' for the first inverter and B' for the second. Using these outputs as the inputs of the next stage, we find that the top AND gate receives A' and B as its two inputs. The second AND gate receives A and B'. Using this information and the knowledge that the output of an AND gate is the product of its inputs we know that the top AND gate's output will be A'B and the bottom gate's output will be AB'. So, our final OR gate will see A'B and AB' as its two inputs. Since the output of an OR gate is the sum of its inputs, we can write the final Boolean expression for our XOR gate as F=A'B+AB'. This process is illustrated below.

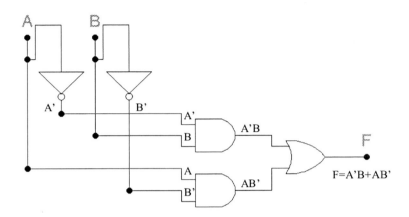

Now we have a Boolean expression for the XOR gate, F=A'B+AB', and we only had to trace through the circuit once. Adding this result to our reference table, we get Table 7.2.

Table 7.2 A Summary of the main logic gates and their Boolean expressions.

Gate	Input(s)	Boolean Expression
NOT	A	A'
AND	A,B	AB
OR	A,B	A+B
NAND	A,B	(AB)'
NOR	A,B	(A+B)'
XOR	A,B	A'B+AB'

From this example, we find that the creation of a Boolean expression requires us to trace through the circuit only once. The main difference between this approach and circuit tracing is that instead of determining the output for a specific input combination, we are building an equation that describes the output for any arbitrary input combination. Before we continue, however, let's look at one more example.

Example 7.3

Problem: Given the following digital logic circuit schematic, determine a Boolean expression describing its operation.

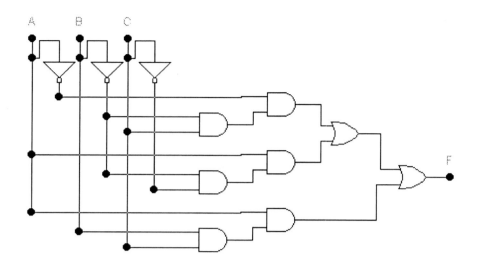

Solution: In this case, we have been given a three input circuit. If we were to perform our analysis using circuit tracing, we would have to look at eight different input combinations. Instead, we will build a Boolean expression to describe the general operation of the circuit. This operation will require only one pass through the circuit and will prove to be a far more useful result. We will generate our Boolean expression in much the same way we did in the previous example. We simply start with the inputs and trace the Boolean values through to the end. Doing this yields the following:

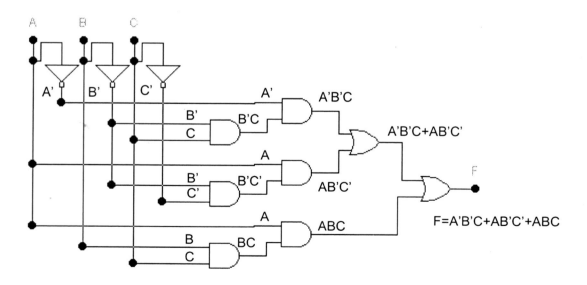

Thus, the Boolean expression representing the output of this circuit is F=A'B'C+AB'C'+ABC.

Now that we understand how we can obtain a Boolean expression for a circuit schematic we can turn our attention to obtaining the truth table. Once the expression is determined, it is a simple task to fill in the truth table. So, let's look at what steps we need to take to do that.

7.5.3   Converting a Boolean Expression into a Truth Table

Building a truth table for a circuit is a simple task once the Boolean expression has been found. All that we need to do is figure out under what conditions the expression will evaluate to true and make these input combinations down in the truth table with a true (1) result. Once all the combinations resulting in a true (1) output are found, the results for the rest of the input combinations can be set to false (0). As an illustration of this concept, let's think about the three basic logic gates, NOT, AND, and OR. Starting with the NOT gate, we know that its Boolean expression is A'. So, what value of A would make its compliment, A', evaluate to true? Obviously the answer to that question is that a value of logical false (0) would be true when complimented. If we examine the truth table for the NOT gate, we will find that the output for A=0 is true (1) and the rest of the entries are false. Next look at the AND gate. Its Boolean expression is AB. What values of A and B are needed to make this Boolean statement evaluate to true (1). It is easy to see that only A=1 and B=1 will work in this situation. Thus, if we look at the truth table for an AND gate we will find that the output for A=1 and B=1 is true (1) and all the other results are zero. Finally, look at the OR gate. Its Boolean expression is A+B. What values of A and B are needed here? In this case, three possible combinations will work. As long as either A or B is true (1) we will get a true (1) output. In addition, if both A and B are true (1) we will get a true (1) output. Thus, if we examine the OR truth table we will find a true (1) result in those three locations and a false (0) in the remaining position. So, what we have discovered is that if we examine the Boolean expression and find the cases that make it true (1), we can make those in the truth table and leave the rest as false (0). This process can be broken down into five simple steps:

1. Determine the number of inputs involved in the Boolean expression.
2. Create an empty truth table for a circuit of the determined number of inputs.
3. Determine what conditions cause the Boolean expression to evaluate true (1).
4. Set the outputs for the determined input conditions to true (1) in the truth table.
5. Set the remaining outputs in the truth table to false (0).

Now, let's apply these five steps by looking at how the truth table for the XOR gate can be determined from its Boolean expression.

Example 7.3

Problem: Determine the truth table for the following Boolean expression.

$$F=A'B+AB'$$

Solution: Here we have been given a simple Boolean expression. A quick examination of the structure of the expression shows us that we are dealing with a two input system (only two different letters are found in the expression). Since we know the number of inputs, we can now construct our blank truth table as follows:

A	B	F = A'B+AB'
0	0	
0	1	
1	0	
1	1	

Now that we have our blank truth table, we can also note that within the expression we have a series of products that are added together. Since products represent AND gates and sums represent OR gates, we see that as long as one or more of the products (AND gates) evaluates to true, the result will be true. Thus we can focus on finding the cases that make each AND gate true. The two cases that cause the AND gates to evaluate true will be the two entries in the truth table that will be true (1). Looking at the first product we see that we need to find that input combination will make A'B true (1). We can easily see that the only input combination that will do that is A=0, B=1. So we place a one in the truth table for the output in the row with A=0 and B=1. Now, looking at the second product we must find the combination to make AB' evaluate to true (1). It can easily be seen that this requires inputs of A=1 and B=0. So, now we can place a one in that row as well. Since there are only two products (AND gates) that are summed, we know there are no other combinations that will allow for a true (1) output. Thus, we fill in the rest of the positions with false (0). Doing all of this yields the following truth table:

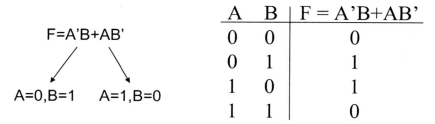

A	B	F = A'B+AB'
0	0	0
0	1	1
1	0	1
1	1	0

As can be verified by looking back at Figure 7.14, we have developed the truth table for an XOR gate from its Boolean expression.

As we found in Example 7.3, conversion from a Boolean expression to a truth table is a relatively simple task. Let's look at one more example before we move on.

Example 7.4

Problem: Determine the truth table for the following Boolean expression.

$$F= AB'C+A'B'C +A'B$$

Solution: In this Boolean expression we note that there are three inputs involved. Therefore, our blank truth table will contain eight rows. Building our blank truth table gives us:

A	B	C	F
0	0	0	
0	0	1	
0	1	0	
0	1	1	
1	0	0	
1	0	1	
1	1	0	
1	1	1	

Now, we must examine the expression to determine what conditions will cause it to evaluate to a logic true (1). Once again we are dealing with a series of products that are summed together. Thus, we will, once again, focus on finding the situations that make each of the products (AND gates) evaluate true (1). Since the first two terms contain all three of the inputs we know that each of them will correspond to exactly one row in the truth table. We see that the only combination of the inputs that will make the first product true (1) is A=1, B=0, C=1. Likewise, the second term requires A=0, B=0, C=1. Therefore, we can make the output of those rows as true (1). The final term is a little bit trickier. Since it contains only two of the input values, A and B, its output is not dependent on the value of C. As a result, there will be two lines in the truth table that will work with this term. Since the value of C does not matter, it could be a 0 or a 1 as long as the value for A and B are correct. Thus, the values A=0, B=1, C=0 and A=0, B=1, C=1 will both satisfy the expression. As a result, we will mark both of the associated rows in the truth table as true (1). With all the possible true (1) cases identified, we can fill in the rest of the truth table results with false (0). Doing this yields the following truth table:

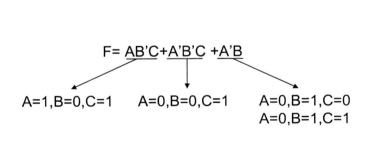

A	B	C	F
0	0	0	0
0	0	1	1
0	1	0	1
0	1	1	1
1	0	0	0
1	0	1	1
1	1	0	0
1	1	1	0

So, we have found the truth table associated with our given Boolean expression.

In the last two examples, we have seen how to convert a Boolean expression into a truth table. Now that we can do that, we can easily find a truth table for a circuit by extracting the circuit's Boolean expression and then converting that expression to a truth table. This method is far easier and less error prone than circuit tracing, especially when dealing with circuits with more than two inputs. By using the techniques outlined here it is possible to develop a truth table for practically any logic circuit you may encounter.

7.5.4   Putting the Gates Together – Creating a Useful Logic Circuit

In the previous section we discussed the process of analyzing an existing digital logic circuit to determine its behavior in the form of a truth table. We found that this could be done either by tracing through the system for each input combination, or by building a Boolean expression for the circuit and using that expression to populate the truth table. While the ability to analyze an existing circuit is very important, if all we could do is analyze existing circuits, nothing new would ever be built. As a result, we will now look at what is needed to convert a truth table describing the desired characteristics of a logic circuit into an actual circuit diagram that will satisfy a given requirement. There are actually several ways that a truth table can be converted into a logic circuit. In this text, however, we will focus on one of these methods, the sum of products or minterm method.

7.5.4.1   Sum of Products / Minterms

Since we are going to use the minterm method to convert a truth table to an actual circuit, we need to know what a minterm is. Basically, a minterm of an arbitrary number of variables n is the product of those n variables in which each variable appears only once. The variable can appear in either true (for example A) or complimented (for example (A') form. So, for a given logic circuit, there are as many possible minterms as there are rows in the circuit's truth table. In the case of a two input circuit, we will have four possible minterms, one corresponding to each row of the truth table. Table 7.3 displays these four possible minterms.

Table 7.3 A list of the four possible minterms in a two input system.

A	B	Minterm	Read
0	0	A'B'	NOT A, NOT B
0	1	A'B	NOT A, B
1	0	AB'	A, NOT B
1	1	AB	A, B

As the table shows, the minterm that is associated with a given row of the truth table is simply the Boolean expression representing an AND gate that would evaluate true (1) when given the associated input values. So, now that we know what a minterm is, how can we use it to create a logic circuit? The use of minterms to build a circuit is a two step process.

1.   Use minterms to create a Boolean expression for the desired circuit
2.   Build a circuit from the Boolean expression

Let's look at the first step. We already know that there is a minterm associated with each row of a given truth table. We also know that the minterm basically represents the Boolean expression for an AND gate that would evaluate true (1) when given the input values from its associated row in the truth table and false (0) with any other inputs. So, if we find all the rows in our truth table that we want to evaluate to a true (1) output and create a Boolean expression that consists of the sum of all of these minterms, we will have a Boolean expression that corresponds to the truth table. In other words, this Boolean expression would evaluate to true (1) only for the rows that are true in the truth table and false for all other input combinations. To clarify the procedure, let's take a look at an example.

Example 7.5
_____

Problem: Construct a minterm expression for the following truth table.

A	B	C	F
0	0	0	1
0	0	1	0
0	1	0	0
0	1	1	1
1	0	0	0
1	0	1	1
1	1	0	1
1	1	1	0

Solution: In this example, we have been given a truth table and asked to find a corresponding minterm expression. To complete this task, we must perform three basic steps.

1. Find the rows in the truth table with true (1) results.
2. Determine the minterms for these rows
3. Create a Boolean expression consisting of the sum of these minterms.

By glancing at our truth table, we see that we have four rows that evaluate to a true (1) output. The minterm values for these rows are A'B'C', A'BC, AB'C, and ABC'. This is illustrated on the truth table below.

A	B	C	F	
0	0	0	1	◄─── A'B'C'
0	0	1	0	
0	1	0	0	
0	1	1	1	◄─── A'BC
1	0	0	0	
1	0	1	1	◄─── AB'C
1	1	0	1	◄─── ABC'
1	1	1	0	

Now that we have the minterms for all the rows that have a true (1) output, we simply need to create a Boolean expression for the circuit by summing them together. Doing so gives us the following minterm expression:

$$A'B'C'+A'BC+AB'C+ABC'$$

Thus, we have determined the minterm expression for the truth table.

Now that we can easily create a Boolean expression to represent a truth table, we need a way to convert this minterm expression into an actual circuit schematic. Luckily this is an easy thing to do.

7.5.4.2   Assembling Gates to Satisfy a Minterm Expression

If we take a look at the structure of the minterm expressions that we have seen so far, we quickly see that they consist of a sum of products. Thinking back to the way that our logic gates operate, we know that a product typically represents the operation of an AND gate and a sum usually indicates the operation of an OR gate. By using AND gates to represent the products, OR gates to represent the sums, and NOT gates to represent the complimented terms, we can easily construct our circuit. The best way to understand the way in which we need to assemble these gates is by looking at an example. So, let's do just that.

Example 7.6

Problem: Design a digital logic circuit to implement the following minterm expression.

$$F=A'B'+A'B+AB'$$

Solution: We will begin constructing our logic circuit by focusing on the product terms. We will use one AND gate to represent each of the product terms. Thus, we will use an AND gate with the inputs A' and B', one with the inputs A' and B and one with the inputs A and B'. Lining these gates up in a column, we get the following diagram:

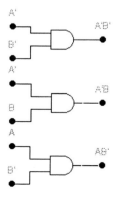

Now, we need to represent the summation of these terms by running the outputs of the AND gates into an OR gate (in this case a three-input OR gate). This will give us:

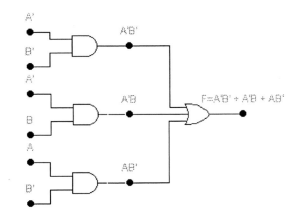

Finally, we need to take care of the complimented terms that are present on the inputs to our AND gates. This is done by using a NOT gate for each input that is complimented. The addition of these final NOT gates will yield the following complete schematic.

$$F = A'B' + A'B + AB'$$ $\longrightarrow$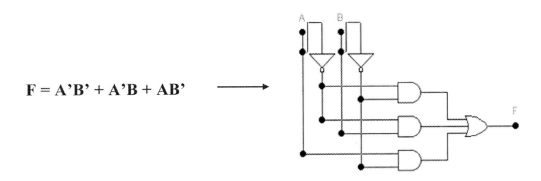

So, we have created a digital logic circuit schematic that satisfies our provided minterm expression.

As we have seen, we now have the ability to convert a truth table into a logic circuit by using minterms. The circuit that is created through this process is only one possible circuit that will satisfy the truth table, however. There could exist a circuit arrangement that would be simpler or more efficient. Thus, the minterm method gives us a working circuit that can serve as a starting point in the process of designing a system based on digital logic.

7.6   Simplification

The next step in creating a useful digital logic circuit is to make sure the circuit is the best possible design for the application in which it will be used. Typically this means that we want to use the simplest possible circuit design. For example, consider the two circuit diagrams in Figure 7.16.

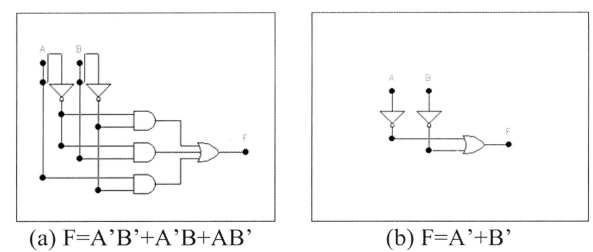

(a) F=A'B'+A'B+AB'   (b) F=A'+B'

Figure 7.16. An example of a circuit developed using minterms (a) and a simplified version of that circuit (b). Both circuits represent the same truth table.

If you were to build a truth table for both of these circuits you would find that these truth tables were identical. However, the second circuit only uses three logic gates while the first uses six. Therefore, we would say that the second circuit is a simplified version of the first circuit. But, if they both produce the same output, why do we want to take the time to simplify our circuits? There are several reasons we might want to do this. Three of the most important are:

1. **Cost :** Every gate that we use in our circuit design costs money. Thus, if we reduce the number of gates that we must use, we can reduce the cost involved in building our device.
2. **Power :** Each gate in the circuit also requires power to operate. As a result, a reduction in the number of gates used can decrease the power requirements for the circuit. This is especially important in battery-powered devices where we desire to maximize battery life.
3. **Speed :** Every logic gate requires some time to produce an output value. While this time is usually very short, on the order of nanoseconds, it can have a large impact on system performance. The fewer gates that a signal needs to pass through, the faster the circuit design can produce an output. The amount of time required to get a valid output from a circuit determines how quickly the input values can change without causing the circuit to provide bad output. If the device being designed is for use in high speed environments, such as computers of digital communication systems, this becomes a critical consideration.

So, now that we understand why we would want to simplify our circuits, how do we go about doing it? Once again, there are many ways that a circuit can be simplified. In this text we will examine two common approaches: Boolean algebra and Karnaugh maps.

7.6.1   Boolean Algebra

One of the most common approaches to digital logic circuit simplification is through the use of Boolean algebra. We are already somewhat familiar with Boolean algebra from our earlier use of Boolean expressions. For example, we know that Boolean algebra uses variables such as A, B, and C to represent the inputs and outputs of a digital system. We also know that these variables can take on only one of two numeric values, namely one or zero. We are already familiar with many of the basic operations of Boolean algebra from our examination of the different logic gates earlier in the chapter. These basic operations are summarized to the right.

In addition to these basic concepts, Boolean algebra follows a series of theorems and laws, just like normal algebra. Some of the most important basic theorems in Boolean algebra are listed below.

Basic Operations of Boolean Algebra

Compliment Notation

0' = 1 (read as NOT 0 is equal to 1)
1' = 0 (read as NOT 1 is equal to 0)

If   A=1,   then A'=0.
If   A'=1,   then A=0.
Therefore, (A')' = A

Multiplicative Behavior:
(AND Gates)

0*0 = 0 (0 AND 0 = 0)
0*1 = 0 (0 AND 1 = 0)
1*1 = 1 (1 AND 1 = 1)

Additive Behavior
(OR Gates)

0 + 0 = 0 (0 OR 0 = 0)
0 + 1 = 1 (0 OR 1 = 1)
1 + 1 = 1 (1 OR 1 = 1)

Some Basic Boolean Algebra Theorems

Theorem 1.	A+0=A	If zero is ORed (added) with a variable, the result will be the value stored in the variable. (See Additive Behavior).
Theorem 2.	A+1=1	If one is ORed (added) with a variable, the result will always be one. (See Additive Behavior).
Theorem 3.	A+A=A	Any variable ORed (added) with itself will result in the original variable. If A=1, the result would be 1+1 = 1, if A=0, the result would be 0+0 = 0. (See Additive Behavior).
Theorem 4.	A*1=A	A variable ANDed (multiplied) by one will be unchanged. (See Multiplicative Behavior).
Theorem 5.	A*0=0	A variable ANDed (multiplied) by zero results in zero. (See Multiplicative Behavior).
Theorem 6.	A*A=A	Any variable ANDed (multiplied) with itself will result in the original variable. If A=1, the result would be 1*1 = 1, if A=0, the result would be 0*0 = 0. (See Multiplicative Behavior).
Theorem 7.	(A')'=A	Complimenting a variable twice yields the original variable. (See Compliment Notation)
Theorem 8.	A+A'=1	A variable ORed (added) with its compliment is always 1. If A=0, 0+1=1, if A=1, 1+0=1. (See Additive Behavior)
Theorem 9.	A*A'=0	A variable ANDed (multiplied) by its compliment is always 0. If A=0, 0*1=0, if A=1, 1*0=0. (See Multiplicative Behavior)

Along with the theorems outlined above, there are several critical laws in Boolean algebra that will help with our simplification work. Many of these laws you will be comfortable with since you have, no doubt, used them for some time in your work with regular algebra. The laws governing the commutative, associative, and distributive properties that are critical to working in Boolean algebra are outlined to the right. In addition, you will find DeMorgan's theorems, which are instrumental in the task of converting between digital logic circuits that use AND and OR gates and those that use NAND and NOR gates. Most of the laws probably make sense, with the possible exception of the second distributive law. Since it is a little bit unusual, let's take a look at how it works. First, we will start by looking at the right hand side of the expression. We will apply some of our theorems as well as a couple of the other laws to transform this expression into the one on the left.

Important Laws and Theorems of Boolean Algebra

Commutative Laws

$$AB = BA$$
$$A+B = B+A$$

Associative Laws

$$(AB)C = A(BC)$$
$$(A+B)+C = A+(B+C)$$

Distributive Laws

$$A(B+C) = AB + AC$$
$$A+BC = (A+B)(A+C)$$

DeMorgan's Theorems

$$(A + B)' = A'B'$$
$$(AB)' = (A' + B')$$

Starting with $(A+B)(A+C)$, we can use the first distributive law to write:

$$(A+B)(A+C) = AA+AB+AC+BC$$

Next, using theorem 6, we can write:

$$(A+B)(A+C) = A+AB+AC+BC$$

Now, we can use theorem 4 to write:

$$(A+B)(A+C) = A(1)+AB+AC+BC$$

Next, we will use the first distributive law in reverse to factor an A out of the first three terms. Doing this will yield:

$$(A+B)(A+C) = A(1+B+C)+BC$$

From theorem 2, we know that adding 1 to a variable always results in 1, so we can write:

$$(A+B)(A+C) = A(1)+BC$$

Finally, from theorem 4, we know that anything multiplied by one is unchanged, so we write:

$$(A+B)(A+C) = A+BC$$

Thus, we have arrived at the expression for the second distributive law. This relationship can also be shown by moving from the left hand side of the expression to the right, but that is left as an exercise for the reader.

Now that we have seen how Boolean algebra works, we can look at how we can use it to simplify our digital logic circuits. Our basic approach to simplification is to begin with the minterm expression and use our knowledge of Boolean algebra to simplify the expression. When doing this, there are several things to keep in mind.

1. Try to find variables that can be factored out in such a way as to leave terms that will simplify to zero or one. For example AB' + AB = A(B'+B) = A(1)=A.
2. When looking for terms to factor values out of, it is best to find groups that are powers of two, for example, factoring A out of 2 or 4 terms as opposed to 3. Factoring a value out of groups that are not powers of two will sometimes help, but that is not usually the case.
3. If it helps you recognize patterns, rearrange the order of the terms in the expression to group terms that have common variables together.
4. Keep the second distributive law in mind. It is used a great deal and situations where it would help are often missed by people new to Boolean expression simplification.

Keeping these things in mind will help, but the easiest way to see how this works is to do it. So, let's walk through the simplification of our expression from Example 7.6.

Example 7.7

Problem: Given the following minterm expression, use Boolean algebra to derive a simplified expression.

$$F=A'B'+A'B+AB'$$

Solution: We have been given a minterm expression and asked to simplify it using Boolean algebra. Let's begin by looking at the expression and seeing if any terms have variables in common. A quick examination shows us that the first and second term have A' in common and the first and last terms have B' in common. Factoring out the common term in either case will result in a sum that will simplify to one, so we could use either one. Let's use the first distributive law to factor A' out of the first two terms. This will give us:

$$F=A'(B'+B)+AB'$$

Now, from theorem 8, we know that B'+B = 1. Thus, we can write:

$$F=A'+AB'$$

At first glance, it may look like there is no more simplification that can be done, but that is not the case. We can use the second distributive law to write:

$$F=(A'+A)(A'+B')$$

Now we have a product of two sums. The first sum can be simplified by theorem 8, yielding:

$$F=(1)(A'+B')$$

Finally, from theorem 4, we can write:

F=A'+B'

Thus, we have simplified our minterm expression in such a way that it now only requires two NOT gates and an OR gate to implement.

As we just saw, Boolean algebra can be a powerful tool for logic circuit simplification. However, it can, at times, be difficult to see the groupings from which to factor values. This can make the simplification of Boolean expressions a time consuming activity. For this reason, a more graphical approach is available. This approach, called Karnaugh mapping, organizes the terms of the expression into a graphical table or map that makes it easy to see what terms have the potential for simplification

7.6.2   Karnaugh Maps

The main purpose of using a Karnaugh map in the simplification process is to help organize the Boolean expression so that relationships between terms are easy to see. Thus, a typical Karnaugh map consists of a grid that contains as many squares as there are entries in the truth table. So, for a three input system, the Karnaugh map will have eight squares, and a four input system will have sixteen. Karnaugh maps can be used to simplify systems with more than four inputs, but since we will not be dealing with systems that large in this text, we will only cover Karnaugh maps for up to four inputs. Figure 7.17 shows the Karnaugh map layouts for two, three, and four input logic circuits.

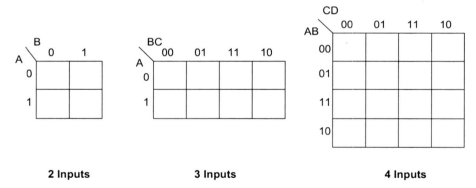

2 Inputs   3 Inputs   4 Inputs

Figure 7.17. Examples of Karnaugh map layouts for 2, 3, and 4 input logic circuits.

As the figure shows, we can label the horizontal and vertical axes after up to two of the input variables. We then label the blocks along that axis with the possible value combinations that those variables can take. One important limitation in this labeling is that only one variable can change when moving between adjacent squares. That is why the squares for two inputs are labeled 00, 01, 11, 10 instead of in binary order.

Once our map grid is drawn out and labeled, we have a grid representing all of the possible input combinations for the system. Using this grid we need to place a one in each box that represents a term in our Boolean expression. Once all of the terms in the expression have been represented, the other blocks can be populated with zeros. Figure 7.18 shows some examples of populated Karnaugh map grids.

Now, we need to search our completed map for groupings that can be simplified. These groups are identified as groups of non-diagonally adjacent ones in the grid. In order to be simplified, the groups must consist of a number of ones that is a power of two, for example 2, 4, or 8 ones. To provide the best simplification, you should avoid overlapping between groups, but if

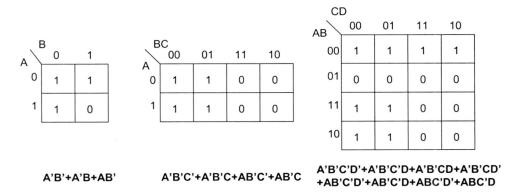

A'B'+A'B+AB'   A'B'C'+A'B'C+AB'C'+AB'C   A'B'C'D'+A'B'C'D+A'B'CD+A'B'CD'
                                                     +AB'C'D'+AB'C'D+ABC'D'+ABC'D

Figure 7.18. Examples of the Karnaugh maps associated with some Boolean expressions.

there is no other way to create an appropriately sized grouping and one group does not contain the other, overlapping is permitted. The groupings associated with the example maps in Figure 7.18 are shown in Figure 7.19 along with their simplified expressions.

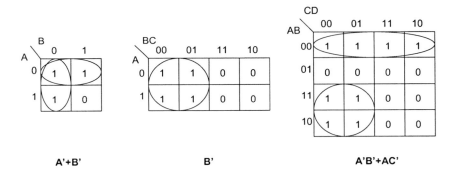

A'+B'   B'   A'B'+AC'

Figure 7.19. An example of the groupings found in the Karnaugh map examples from figure 6.18 along with their simplified expressions.

So, how did we determine the simplified expressions in Figure 7.19? The answer is quite simple. For each circled group, look at the values it contains. If one variable is constant in the grouping write it down. If it changes, leave it out of the new expression. The resulting simplified Boolean expression is the sum of the terms associated with each grouping. This can be a bit difficult to understand at first, so let's look at an example.

Example 7.8

Problem: Use Karnaugh maps to simplify the following minterm expression.

A'B'C'D'+A'B'C'D+A'BC'D'+A'BC'D+A'BCD+AB'CD+AB'CD'+ABC'D+ABCD

Solution: We have been given a four variable minterm expression in this example. So, first, we need to draw out a Karnaugh map in a 4 x 4 grid, label it and fill in ones for the terms in our expressions. Doing this will give us the following Karnaugh map.

CD

AB \	00	01	11	10
00	1	1	0	0
01	1	1	1	0
11	0	1	1	0
10	0	0	1	1

A'B'C'D'+A'B'C'D+A'BC'D'+A'BC'D+A'BCD
+AB'CD+AB'CD'+ABC'D+ABCD

So, now that we have our map drawn out, we need to circle our groupings of ones. By seeking out groups with minimum overlapping, we can find two groups of four and one group of two. Circling these three groups we get the following:

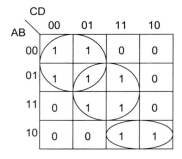

Examining each of the groupings and looking for the variables within each group that do not change gives us the following Boolean products:

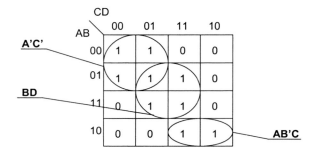

Now, by adding our three Boolean products together, we arrive at our final simplified expression:

A'C'+BD+AB'C

So, here the use of the Karnaugh map has greatly aided us in simplifying our minterm expression.

___

This example illustrates the power of Karnaugh maps. Instead of spending a lot of time working through the simplification in regular Boolean algebra, the map helped us to organize the problem in a way that made it very easy to solve.

One final consideration to keep in mind when working with Karnaugh maps is to remember that the maps technically wrap around. This can lead to unusual groupings like those shown in Figure 7.20.

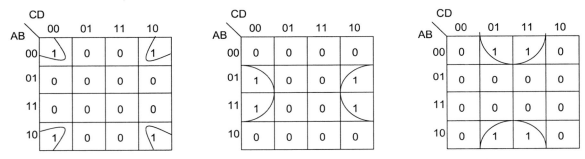

Figure 7.20. An example of some unusual Karnaugh map groupings.

7.7   Logic Word Problems

So far in this chapter, we have examined methods that are useful for analyzing a given circuit and for developing circuits when provided with a truth table. However, outside the educational lab environment, you will seldom encounter problems in this form. Instead, you will encounter situations where you find that you need to design a logic circuit to solve a specific problem. In cases such as these, you must examine the situation and derive a truth table based on your goals and available information. Then, from this truth table, you can design your circuit. In this section, we will examine one such situation.

Example 7.9

___

Problem: You have been assigned the task of developing an electronic lock to protect the front entrance to a store. The store owner provides you with the following information:

1. The store's normal operating hours are from 8:00 AM to 7:00 PM Monday through Friday. During this time, the door should be automatically unlocked.
2. The store is closed on weekends and holidays. The door needs to remain locked to the public on these days.
3. The shopkeeper needs to be able to open the door at any time to access his store. He wants to use a keypad entry code to open the lock.

While researching the problem, you discover three off-the-shelf systems that will help you in your design.

1. A digital clock system that can be set to provide a logic high (1) between a set start and stop time and a logic low (0) otherwise.
2. A digital programmable calendar system that can be set up to provide a logic high (1) on weekdays that the store should be open and a logic low (0) otherwise.
3. A keypad entry system that provides a logic high (1) for several seconds after a valid code is entered and a logic low (0) otherwise.

Your job is to integrate the three systems by creating a digital logic circuit that can determine, based upon the outputs from the three component systems, whether the door should be locked or unlocked. You should design a simplified circuit for this application.

Solution: In this example, we have been given a problem that is much closer to a situation that an engineer would encounter in the real world. Much of the hard work has been done in the creation of the three component systems, but you must still properly integrate these systems together to achieve your goal. To approach a problem such as this one, it is important to summarize what you know about the problem. In this case, we know that we have three input systems that provide us with a given input for a specific situation. These situations and their resulting outputs are shown below.

Input A - Clock	Input B - Calandar	Input C - Keypad
0 = Not Business Hours	0 = Weekend or Holiday	0 = No Valid Code Entered
1 = Business Hours	1 = Weekday	1 = Valid Code Provided

Output - Door Open/Locked
0 = Door Locked
1 = Door Open

Now that we can see what input each system is providing, we can develop a truth table to describe what we want our system to do given each of the possible input combinations from these systems. Based upon the information that we were given we can look at each row of our truth table. In the first row, we have the situation where it is not business hours (0), it is a holiday or weekend (0), and there has been no valid code provided (0). In this case, the door should be locked (0). In the second row, we find that it is not business hours (0) and it is a weekend or holiday (0). However, in this case, a valid access code has been provided (1). So, we want to open the door (1). The third row provides us with a situation where it is not business hours (0) on a weekday (1) and no code was provided (0). Thus the door is locked (0). The fourth row is just like the third (0 1), except in this case we have a valid code (1). Thus the door opens (1). In the fifth row, we find that it is during business hours (1) on a weekend (0), and no code is provided (0). In this case we want the door locked (0). The sixth case is just like the fifth (1 0), except there is a valid code (1). The door must be opened in this case **(1)**. **The** seventh case occurs during business hours (1) on a weekday (1) with no access code **(0)**. **In** this case, the door will be open (1). Finally, we find that if we encounter a case where **it is** during business hours (1) on a weekday (1), and a valid code is entered (1). While it **would be** unusual to enter an access code when the door would already be open from the first two **conditions**, it must be considered. In this case, the door must be open (1). Examining all these **cases gives** us the following truth table:

A	B	C	F
0	0	0	0
0	0	1	1
0	1	0	0
0	1	1	1
1	0	0	0
1	0	1	1
1	1	0	1
1	1	1	1

From the truth table, we see that there are five possible situations when the door should be opened. To build our circuit to implement this table, we must start by constructing a minterm expression from our truth table. In this case, we can expect our minterm expression to be the sum of five products. Creating the minterm expressions yields:

A	B	C	F	
0	0	0	0	
0	0	1	1	← A'B'C
0	1	0	0	
0	1	1	1	← A'BC
1	0	0	0	
1	0	1	1	← AB'C
1	1	0	1	← ABC'
1	1	1	1	← ABC

A'B'C+A'BC+AB'C+ABC'+ABC

At this point, we have enough information to build a working logic circuit. Given the five term minterm expression, we can expect our circuit to have a column of five AND gates followed by an OR gate. Constructing the circuit from our minterm expression will yield:

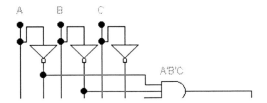

So, our circuit consists of nine gates, three inverters, five three-input AND gates, and a five-input OR gate. At this point, we could build this circuit, hook it up and be done. However, the problem requires us to design a simplified circuit for this project. So, we need to simplify our Boolean expression. Let's do this with a Karnaugh map. If we build a three-input Karnaugh map and populate it with the values from our minterm expression, we will have the following:

BC	00	01	11	10
A				
0	0	1	1	0
1	0	1	1	1

A'B'C+A'BC+AB'C+ABC'+ABC

Now, on this map, we can easily see that there are two groups of ones. There is a group of four and a group of two. If we group these terms together we will get the following simplified solution:

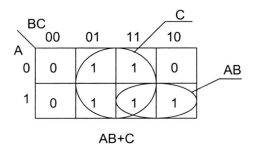

AB+C

So, we have discovered that we can implement this design using only an AND gate and an OR gate. The design is much simpler and easier to construct than the minterm design. If we sketch this circuit design out we will get the following circuit diagram:

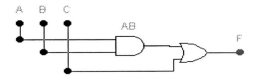

Now, all that is left to do is to physically build this circuit, connect the three input systems to its input lines and connect the door locking mechanism to the output. So, we have used what we have learned in the previous sections to design a system with an actual application.

While the above example is definitely a simplified case, it illustrates the approach that must be taken when dealing with a problem of this type. If the problem is approached systematically and care is taken to insure that all of the initial requirements are satisfied, there is virtually no limit to what can be developed.

7.8   Practical Considerations

As we near the end of our discussion of digital logic circuits, there are a few practical considerations that you should be aware of. The most important of these is that the circuits that we have been designing in terms of individual gates are not really as simple as we make them appear. When building actual logic circuits, there are no "gate" components. Instead, each of the gates in a logic circuit represents a semiconductor circuit. These gate circuits are found on integrated circuits, microchips. Thus, when you actually construct the circuits that you design, you must take the physical components into account. In addition, the logic levels of zero and one are actually represented by a range of voltage levels. When the integrated circuit checks the logic level of an input line, it actually compares the voltage to a required level. If it is within one voltage range it is considered logic true, if it is in a different range it is considered false. Since different applications could require different voltage ranges and because there are multiple ways to construct the gates, several families of logic integrated circuits exist. Of these families, there are two that are widely used. These are TTL (Transistor-Transistor Logic) and CMOS (Complementary Metal-Oxide Semiconductor). Each of these families has different voltage ranges and uses different construction approaches. However, as long as you stick to devices in the same family, you may easily connect them to form logic circuits. A typical TTL device will consider an input voltage of 0 to 0.8 volts to be a logic low and an input in the range of 2 to 5 volts to be logic high. In contrast, devices in the CMOS family consider inputs in the range of 0 to 3 volts to be a logic low and inputs in the range of 7 to 10 volts to be logic high. Any inputs outside of these voltage ranges cause unpredictable behavior and, thus, are forbidden input values. Figure 7.21 shows a diagram of the voltage ranges for the TTL and CMOS logic families.

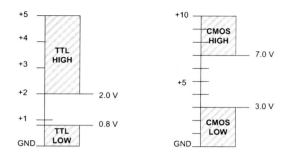

Figure 7.21 Diagrams of the voltage ranges for the TTL and CMOS logic families.

Now that we understand the voltages of the two common families, let's take a look at what the actual integrated circuits look like. Each family has its own IC package, but in both cases, multiple copies of a gate are included on one physical IC. This is done mostly because it would be a waste of space to put only one copy on the chip, and in most cases, a project will require more than one copy of the gate anyway. In the case of TTL two-input gates, four are included per chip, and there are six TTL inverters per chip. Figure 7.22 shows the layouts of the TTL AND, OR, and NOT gates.

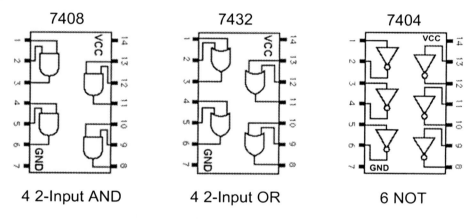

Figure 7.22 Examples of logic gate ICs for the TTL family.

Now that we can see how these chips are laid out, let's look at one final example.

Example 7.10

Problem: Given the simplified circuit developed in example 7.10, design the physical circuit layout for the TTL 74* series integrated circuits.

Solution: In this example, we have been asked to use our new knowledge of the basic TTL IC gates to determine a physical circuit layout for our simplified circuit from example 7.9. Looking back at this design we see that it was a very simple design consisting of just one AND gate and one OR gate. The circuit diagram is shown below:

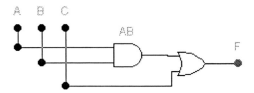

Now that we see our basic circuit design, we need to develop the physical IC design. From the circuit we see that we will need two ICs to implement this design, one 7408 AND gate and one 7432 OR gate. These gates not only need to be wired together in the same way as the diagram above, but the ICs must be powered as well. By looking at the chip design for the 7408 we can see that pins 1 and 2 are the inputs for the first AND gate on the chip. These will be connected to the circuit inputs A and B. We will connect input A to pin 1 and input B to pin 2 of the 7408. Next we take pin 3 of the 7408, the output of the first AND gate, and connect it to pin 1 of the 7432. This pin is the first input to the first OR gate on the chip. We can connect pin 2, the second input, to the circuit, input C. Now, our circuit output, F is connected to pin 3 of the 7432. To finalize our circuit design, we must take care of powering the ICs. To do this we connect our positive voltage (5 volts for TTL) to pin 14 on both the 7408 and 7432, and the ground is connected to pin 7 on both ICs. Doing all of this will yield a design similar to the following:

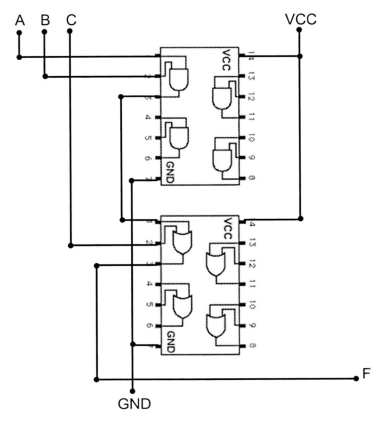

So, here we have a physical layout for our digital logic design.

With this example, we have shown the final step in taking a logic circuit from the design phase to a physical implementation. By carefully mapping the logic schematic onto the available ICs within a specific design family, we can build any circuit that we need.

7.9   Chapter Summary

In this chapter, we covered a wide range of topics related to digital logic circuit analysis and design.

- We began by introducing two of our most important logic analysis tools, the truth table and the Boolean expression.

- Next, we introduced the six basic logic gates (NOT, AND, OR, NAND, NOR, and XOR) and their truth tables.

- Using this information we looked at how we could analyze a given truth table by either tracing all possible inputs through the circuit or by building a Boolean expression and using it to populate a truth table.

- Once we had learned how to analyze existing circuits, we turned our attention to building new circuits. For this, we learned about the minterm or sum or products method. A minterm is a product in which each input to the logic circuit appears only once in true or complimented form. In this method, each input combination that resulted in a true output became a term in the minterm expression. These terms are then summed to form a Boolean expression for the logic circuit.

- We then construct our circuit schematic from this expression by using a column of AND gates to take care of the minterm products in the sum and an OR gate to take care of the summation. Finally, inverters were used to take care of the complimented terms.

- Now that we had a technique for the development of logic circuits from a truth table, we began looking at ways to improve our design. We found that in many cases the circuit designed by the minterm method could be simplified.

- We discussed our motivation for simplification. The number of gates in the circuit was related to the cost, power consumption and speed of the resulting device.

- Two methods of simplification were explored. The first approach involved the direct use of Boolean algebra to simplify the Boolean expression generated by the minterm method. We then learned that we could use the graphical approach of Karnaugh maps to make the simplification easier by graphically illustrating relationships between the terms in the Boolean expression.

- We noted that in most cases logic problems would not present themselves as a circuit diagram or truth table. Instead, a problem or situation will arise that will require the use of digital logic techniques to build a suitable product. We looked at one such example.

- Finally, we discussed some practical considerations. We learned that the gates we had been dealing with in our schematics were really circuits on integrated circuit chips and that our logical true and false values were actually voltage ranges. We looked at these voltage ranges for two common gate families, TTL and CMOS. In addition, we looked at the IC layout of three common chips in the TTL family. Finally, we looked at an example of a circuit that had been implemented using the TTL family of chips.

7.10 Exercises

Section 7.2 – Truth Tables

1. Explain what a truth table is. Why are truth tables helpful?

Section 7.3 – Boolean Expressions

2. Explain what a Boolean expression is.

Section 7.4 – Basic Logic Gates

3. Describe a NOT gate. What does its schematic symbol look like? By what other name is it known? What is its truth table?

4. Describe an AND gate. What symbol is used to represent an AND gate in a circuit diagram? What does the truth table for a two input AND gate look like? What about a 3 input AND gate?

5. Describe the OR gate. Sketch its schematic symbol. Write out the truth tables for both two and three input OR gates.

6. Describe the NAND gate. What is its schematic symbol? What two other gates can be combined to produce a NAND gate? Write out the truth tables for both two and three input NAND gates.

7. Describe the NOR gate. What symbol is used to represent the NOR gate in a schematic diagram? What two other gates can be combined to form a NOR gate? Write out the truth tables for both the two and three input NOR gates.

8. Describe the XOR gate. What is its schematic symbol? White out a truth table for a two input XOR gate. How does this truth table differ from that of a two input OR gate?

9. There are two sets of logic gates that can be used to satisfy any truth table. What gates are in each of these groups?

Section 7.5 – Combinations of Logic Gates

10. Determine the truth tables for the following logic circuits.
    a.

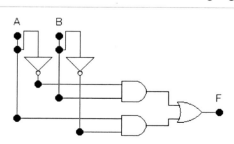

b.

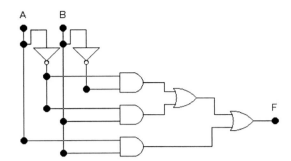

c.

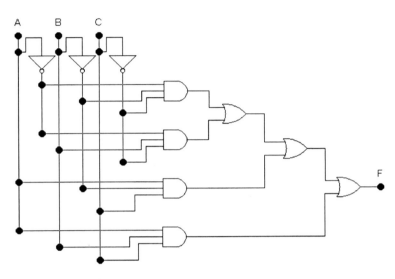

d.

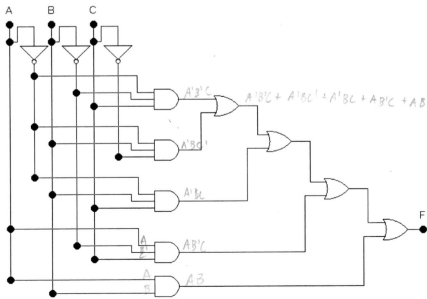

11. Given the following Boolean expressions, construct a truth table.

   a. A'B'+A'B + AB   b. A'B + AB'

   c. A'B'C'+A'BC + AC   d. A'BC'+A'BC + AB'C + ABC'

   e. A'+ABC   f. A'C + AB

   g. A'B'C'+BC + AC   g. A'B'C'+A'BC'+A'BC + AB'C'+ABC'

12. Given the following truth tables, construct their minterm expressions.

a.

A	B	F
0	0	1
0	1	0
1	0	1
1	1	1

b.

A	B	F
0	0	1
0	1	1
1	0	0
1	1	1

c.

A	B	C	F
0	0	0	0
0	0	1	1
0	1	0	0
0	1	1	1
1	0	0	1
1	0	1	0
1	1	0	1
1	1	1	1

d.

A	B	C	F
0	0	0	0
0	0	1	0
0	1	0	1
0	1	1	1
1	0	0	0
1	0	1	0
1	1	0	1
1	1	1	1

d.

A	B	C	F
0	0	0	0
0	0	1	0
0	1	0	0
0	1	1	1
1	0	0	1
1	0	1	0
1	1	0	0
1	1	1	0

f.

A	B	C	F
0	0	0	1
0	0	1	1
0	1	0	0
0	1	1	1
1	0	0	1
1	0	1	1
1	1	0	0
1	1	1	1

13. Sketch the non-simplified versions of the logic circuits obtained from the truth tables in problem 12.

Section 7.6 – Simplification

---

14. Explain the benefits of simplifying a digital logic circuit.

15. Determine the simplified Boolean expressions for the minterm expressions found in problem 12.

16. Sketch the simplified circuit diagrams based on the expressions found in problem 15.

17. Simplify the following Boolean expressions.

  a. $A'B'C'+A'B'C + AB'C + ABC'+ABC$
  b. $A'BC'+A'BC + ABC'+ABC$
  c. $A'B'+A'BC'+A'BC + ABC$
  d. $A'B'C'D'+A'B'C'D + A'B'CD'+BC'D'+CD$

18. Simplify the following Boolean expressions.

  a. $A'B'C + A'BC + AB'C + ABC'$
  b. $A'B'+A'BC + ABC'+ABC$
  c. $A'B + A'BC + AB'C + ABC$
  d. $A'B'C'D'+A'B'C'D + A'B'CD'+A'B'CD + A'BC'D + A'BCD + ABC'D + ABCD$

Section 7.7 – Logic Word Problems

---

19. Assume there is a security door at a top-secret installation. The door has been designed with a three-stage electronic lock. Each stage requires that one or more agents provide the correct retina scan. Three agents have been granted access to the secure area. However, for security reasons, no one agent can pass all of the stages. Agent A's retina scan will open stages 1 and 2, Agent B's retina scan will open stages 2 and 3, and Agent C's retina scan will open stages 1 and 3.

  a. Create a truth table to determine the state of the door given all possible agent combinations. Use 1 to represent the door being open or the agent being present (assume that if he is present, he has provided the correct scan) and 0 to represent the door being locked or the agent being absent.
  b. Write the Minterm Expression for the truth table in part a.
  c. Simplify the expression found in part b.
  d. Sketch the resulting digital logic circuit.

20. You have been asked to design a digital logic system for a local chemical analysis company. This company tests for four different substances in the samples that they receive (Substances A, B, C, and D). Each device that tests for the presence of a substance returns a 0 if it does not detect an unusual amount of the substance and a 1 if the concentration of the substance is above a given level. Not all combinations of these substances are a problem. However, the following combinations are cause for concern:

> 1. Substances A and B if C is not also present
> 2. Substances B and D if A is not also present
> 3. Any one of the Substances alone
> 4. All four of the substances together

The company desires for an alarm to go off if the substances present in a sample meet any of the conditions for concern (Assume a logic output of 1 indicates the sounding of the alarm). It is your job to design a logic circuit to link the outputs of the three sensors to the alarm system.

a. Create a truth table that will determine in which cases the alarm must be sounded. Use a one for the presence of a substance and a 0 for its absence.
b. Write the Minterm expression for the truth table in part a.
c. Simplify the expression obtained in part b.
d. Sketch the resulting simplified logic circuit.

Chapter 8. Computer Organization and Architecture

Chapter Outline

8.1 Introduction

In the previous chapter, we learned that digital logic can be applied to a wide range of applications. One of the most important advances to come about through the application of digital logic is the digital computer. In fact, the digital computer, in some form, is present in many of the devices we use on a daily basis. In general, these computers fall into two broad categories. The first of these is the general computer. This type of computer is what most people think about when they refer to a computer. The second category is the embedded or single purpose computer. This type of computer often goes unnoticed because it operates in the background within a device, often without the user even knowing it is there. Thus, the main difference between a general computer, like the one likely sitting on your desktop at home, and an embedded computer is that the general computer is designed to perform a wide range of different operations from word processing and scientific computations to video gaming, but embedded computers are designed to perform one specific function. Because computers come in both general and embedded forms, they can be found not only on our desktops but also embedded in applications ranging from calculators to automobile engine control. If you really stop to think about it, computers are involved in many of the activities you perform every day. Since the computer is so critical to our lives, we will investigate its design and operation in this chapter. In order to truly understand the power and potential of the computer, one must have an understanding of its basic design and operation.

8.2   Basic Computer Organization

A typical computer, whether it is embedded or general-purpose, is comprised of several distinct parts. Figure 8.1 contains a block diagram that illustrates these parts. As you can see in the figure, far more than a simple computer processor is required for a computer to function properly. A basic computer requires a clock, a power supply, a central processing unit, memory, input and output devices, and a bus to connect it all together. In the following sections, we will take a closer look at each of these parts to gain a better understanding of how each part contributes to the overall operation of the computer.

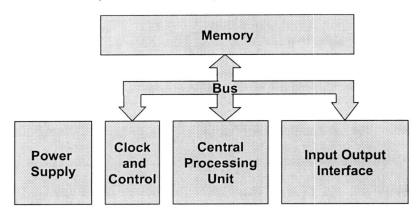

Figure 8.1 A block diagram of the components present in a typical computer.

8.2.1   The Power Supply

Regardless of the task the computer is designed to perform, it must have an electrical source in order to operate. The most readily available power source is a typical wall outlet. Unfortunately, wall outlets provide an alternating current at a high voltage. This type of power source is not suitable for powering sensitive digital equipment. Instead, the power from this source must be converted into a form that is suitable to operate the computer and its components. It is the task of the power supply to perform this conversion. Even if the power source is a battery, which provides a more suitable DC voltage, most computers require sources with multiple voltage levels. As a result, the power supply needs to provide power at a variety of voltages. For example, a typical power supply in a desktop personal computer provides voltage sources at +5V, -5V, and 12V for powering most of the input and output devices and an even lower voltage such

Figure 8.2 A photograph of a typical power supply from a desktop personal computer.

as 3.3V to supply the central processing unit. So, while the power supply is an often overlooked component in the structure of a computer, it has the vital job of converting an input source into one or more voltage sources that are suitable for use by the other components in the computer system. Figure 8.2 contains a photograph of a power supply taken from a typical desktop personal computer. The design of such a power supply follows methods that are similar to those discussed at the end of chapter 4.

8.2.2   The Clock

The digital logic that we studied in the previous chapter is a type of logic known as combinational logic. This means the output of the circuits that we looked at is a function of only their current input values. While computers do contain some combinational logic circuits, they also contain circuits from a second type of digital logic known as sequential logic. The output of a sequential logic circuit depends not only on the current input values, but also on previous input values. In other words, this type of logic has memory of previous input configurations and the output is influenced by this memory. In order for sequential logic circuits to operate properly, they must be provided with a periodic alternating digital signal known as a clock. This signal allows the sequential logic to proceed from stage to stage in an orderly fashion. As Figure 8.3 demonstrates, the clock signal for a sequential digital logic circuit is a square wave. This waveform is usually generated with the assistance of a clock crystal. In many cases, the clock speeds at which the various components of the computer operate will be different. For example, the logic circuits in the processor may operate at a high speed, perhaps several gigahertz, while the logic in the remainder of the system operates at a much slower speed, perhaps a few hundred megahertz. As a result, many computers will employ either multiple clock generators or one clock at the slowest required clock speed and a clock multiplier system to produce higher clock rates from the slower base clock.

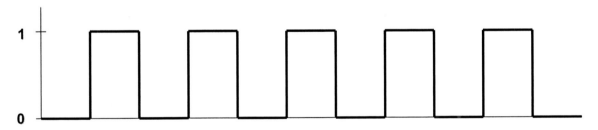

Figure 8.3 An illustration of a typical clock signal.

8.2.3   The Central Processing Unit (CPU)

The most important part of any computer is the central processing unit. The central processing unit, or CPU for short, is the brain of the computer system. The other components within the computer are dependent on and controlled by this special microprocessor. It is important to note that the central processing unit is not the only microprocessor in most computer systems. It is often assisted by additional microprocessors that perform specific dedicated tasks to allow the central processing unit to focus on its tasks without having to worry about dealing with unnecessary details. These assisting processors are often used to deal with input and output functions, such as, video and keyboard control. So, what exactly does the CPU do and what sub-units does it consist of? The CPU is a complex combination of digital logic circuits that not only

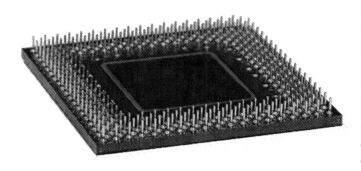

controls the operation of the rest of the systems in the computer, but is also responsible for performing calculations. A photograph of a CPU can be found in Figure 8.4.

A typical CPU can be broken down into a series of specialized units. This breakdown is illustrated in Figure 8.5. As you can see, a basic CPU is built from five basic components, including an arithmetic logic unit, a floating point unit, a control unit, an interface unit, and a series of temporary, high-speed storage

Figure 8.4 A photograph of a central processing unit.

locations known as registers. Since each of these sub-units has an impact on the overall functionality of the central processing unit, we will take a closer look at each one, beginning with the arithmetic logic unit.

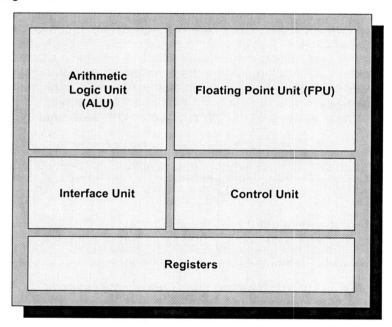

Figure 8.5 A breakdown of the components found in a typical central processing unit.

8.2.3.1   The Arithmetic and Logic Unit

The arithmetic and logic unit, or ALU for short, is the CPU component that has the responsibility of performing all integer and bitwise calculations. A basic ALU is able to perform binary computations such as addition, subtraction, and multiplication. In addition, it can perform many of the basic logic operations discussed in Chapter 7, such as AND, OR, NOT, XOR, and basic bit shifting. The ALU does not handle integer division or any kind of non-integer (floating-point) math.

A typical ALU operates by receiving two basic pieces of information. First, it requires the integer value or values that need to be operated on (in binary form), and second, it requires a code from the control unit that dictates what operation is to be performed. Once this information is received, the ALU will perform the requested calculation and provide the result as output. In addition, the ALU provides a few indicators or "flags" that can be used to determine whether or not any errors occurred during the calculation.

While all basic computers contain some form of arithmetic logic unit, many modern computers contain multiple ALUs. This allows the processor to perform multiple integer arithmetic operations at the same time and can increase the processing speed of the computer.

8.2.3.2   The Floating Point Unit

As we just discovered, the ALU is responsible for the integer arithmetic operations within the CPU. If the CPU requires the ability to perform more advanced mathematical operations, an additional specialized unit is often added. This unit is known as the floating point unit, or the FPU for short. The FPU is capable of performing non-integer mathematical operations, such as the addition, subtraction, multiplication, and division of numbers containing a decimal point. Some processors contain highly advanced FPUs that are capable of performing higher-level operations including trigonometric operations (sine, cosine, tangent, etc.) and exponential functions (square roots and powers).

Since floating-point numbers cannot be represented with binary in the same way that integers are, they are stored by the computer using a special notation. The most common method for representing a basic or single precision floating point value is the IEEE 754 format. This format specifies that the number is stored using 32 binary digits. The way in which these 32 digits are used is presented in Figure 8.6.

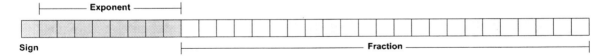

Figure 8.6 The layout of a single-precision floating point number.

As the figure shows, the first of the digits is a sign digit. If it has a value of zero, the number is positive, and if the value of this digit is one, the number is negative. The next eight binary digits of the number are used to represent the power of the normalized binary number that is being stored. The value in this position is offset by 127 so that negative exponents can be represented. The remaining digits within the 32 bit sequence are used to represent a fractional binary number. The easiest way to see how this storage method works is to look at an example.

Example 8.1

Problem: Represent the number -8245.0625 in single-precision floating point format.

Solution: In this problem we need to convert a decimal number to its 32 bit single-precision floating-point format. The first thing we notice is that the number is negative. This means that the first bit of our binary representation will be set to 1. Next, we need to convert the absolute value of the decimal number to its binary equivalent. First, we find the whole number portion by repeatedly dividing by two (see chapter one). If we do this, we will find that

$$(8245)_{10} = (10000000110101)_2 .$$

Next, we can find the binary representation of the fractional part of the decimal number by repeatedly multiplying by two (see chapter one). Doing this we will find that

$$(0.0625)_{10} = (0.0001)_2$$

Putting these two values together, we find that the binary equivalent of our number is

$$(8245.0625)_{10} = (10000000110101.0001)_2$$

Now we need to normalize the binary value. In other words, we need to represent the number in scientific notation. By representing the number in this way we find that

$$(10000000110101.0001)_2 = 1.00000001101010001 \times 2^{13} .$$

So, we see that the exponent of our normalized value is 13. Since the value in our 8 bit exponent field is offset by 127, we need to add 127 to the value. This gives us

$$13 + 127 = 140 ,$$

and in eight digit binary this can be written as

$$(140)_{10} = (10001100)_2 .$$

The fractional part of the number is simply obtained by taking the value to the right of the radix point of the normalized binary value and padding the end of it with zeros until it is a full 23 digits long. Note that if the fractional part is longer than 23 digits, it must be rounded off. This can be a source of error when dealing with floating point computations on a computer. In our case, however, we do not have to round. Thus, we have a fractional value of

$$00000001101010001000000 .$$

Putting all of this together, we can now write our decimal number in 32-bit format as

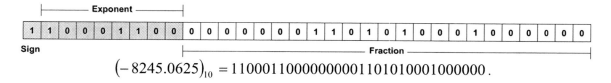

$$(-8245.0625)_{10} = 11000110000000001101010001000000 .$$

Thus, we have just seen how to represent a fractional decimal number in the single precision floating point format.

In the past, the floating point unit was not always included in personal computer processors. The floating point operations were performed through an emulation process using the ALU. While this meant that the computations took much longer to complete, it was far less expensive to design the CPU without the additional complex circuitry. Since only high-end scientific workstations justified the additional expense, many processor manufacturers designed the processor without the FPU and provided an additional FPU unit on another microchip that could be purchased separately. All recent personal computer processors, however, have the FPU integrated into the CPU design.

As was the case with the ALU, the FPU operates by taking both the values that are to be operated on and a control code as inputs. It then produces the result of the operation associated with the control code as output.

8.2.3.3   The Registers

When the ALU or FPU performs computations, the processor needs a temporary place to store the operands and results. In other words, before a computation can occur, the values involved in that computation must be fetched from the computer's memory. When these values are read from memory, they must be stored somewhere until the computation is complete. In addition, the result of the computation must be stored somewhere until it can be written to a memory location. The temporary storage locations in which these values are placed are called registers. Registers provide the processor with a small amount of memory to use for internal calculations. This memory operates at the same clock speed as the processor, as opposed to the system's main memory which often operates at a much slower speed. Since space on the processor is at a premium, the number of registers is often limited to just enough space to store a few values. The actual number of registers within a given processor depends on the design goals of the processor and the number of temporary memory locations that it is expected to require to perform its calculations.

Registers are often used within the processor for tasks other than just holding input and output values. The processor often contains several different types of registers dedicated to these diverse tasks. The registers found in most computers fall into one of three basic categories. These are general purpose registers, floating point registers, and special purpose registers. General purpose registers are used for storing values that the processor is currently using. These values may be basic data such as the operands and results of computations taking place in the ALU or FPU, or they could be memory addresses indicating where given values are to be retrieved from or stored to within the system memory. Floating point registers are much like general purpose registers. Their purpose is to store floating point values associated with operands required or results produced by the FPU. Finally, special purpose registers can be used for a wide range of purposes. These uses include the program counter register, which holds the memory location of the next instruction to be executed by the processor, the instruction register, which holds the instruction currently being executed, and the status register, which provides information about the current state of operations within the processor. Other optional special purpose registers such as constant registers that contain read-only constant values may also be present in a processor.

So, we see that registers are very important to the operation of the CPU because they provide the processor with locations to temporarily store critical values. Since the processor must access these values very quickly, they must be stored within the CPU itself instead of in the slower system memory.

8.2.3.4   The Interface Unit

Since the computation units within the central processing unit (the ALU and FPU) have direct access to only a limited number of internal registers. The processor would be unable to perform any useful operations if it could not access external resources such as, additional memory and input/output devices. However, the parts of the computer that are found outside the CPU often operate at slower speeds than the CPU itself. Because of this, it is difficult for the CPU to directly communicate with these devices. In practical terms, it would be like saying the CPU and external devices speak different languages. To overcome these communication challenges, CPUs possess an interface unit. This unit acts like an interpreter and governs the connections between the microprocessor and other parts of the computer system. The interface unit is responsible for sending and receiving all of the signals that pass between the CPU and the other devices within the computer system. So, if, for example, the ALU needs a piece of information from the computer system's memory, it is the responsibility of the interface unit to signal to the memory that a specific piece of information is required. The interface unit will then receive that information from the memory and store it in an internal register for use by the ALU. Thus, without an interface unit, the CPU would be isolated from the rest of the computer system and would be unable to perform many useful computations.

8.2.3.5   The Control Unit

Finally, just as the CPU is the central control or brain of the computer, the control unit is the brain of the CPU. The control unit makes sure that the CPU performs the tasks that it is instructed to perform in an orderly fashion. So, what exactly does a control unit do? Well, the responsibilities taken on by the control unit can vary significantly between different processor designs, but in its simplest form it directs the flow of information within the CPU.

Everything begins when the control unit receives a command in the form of a binary instruction code. The control unit interprets this command, and based on the desired outcome, it sends signals to the ALU, FPU, and interface unit to load data from memory into the registers, perform calculations, and store results to memory. Common instructions include loading a value from memory into a register, storing the value from a register into memory, and basic integer and floating point arithmetic.

There are two schools of thought surrounding the number and complexity of the operations that the CPU is expected to perform. The first of these two approaches is known as the CISC, or complex instruction set computer, approach. A CPU that is designed to follow the CISC approach is designed with the programmer in mind. A CISC CPU understands a large number of different instructions. Many of these instructions allow the programmer to carry out complex operations with a single command. Thus, the CISC processor allows the number of commands needed to carry out a series of computational operations (a program) to be minimized. This means that computer programs for CISC processors require less memory to store. The downside of this approach however, is that because some commands are simple and others are more complex, each instruction can take varying amounts of time to execute and must be decoded by the control unit into a series of steps required to perform the desired task. All of this takes time which means that while a command to a CISC processor can result in the performance of a complex operation, such an instruction will take more time to decode and execute than a simple instruction would.

This leads us to the second approach to processor design which is the RISC, or reduced instruction set computer, approach. The RISC approach came about when designers analyzed the way in which the CISC CPU was used and discovered that most programs could be written to use only a small subset of the available processor instructions. As a result, they decided that it would be more efficient to develop a processor that understood only this smaller subset of commands. This would eliminate much of the complexity of the CPU design and allow them to manufacture the CPU with fewer transistors. Not only did the reduction in transistors make the CPU cheaper to manufacture, but it also allowed for quicker instruction decoding and execution. So, while programs written for the RISC architecture require more commands to achieve the same result as a CISC program, the individual instructions are decoded and executed more quickly.

Today, both RISC and CISC architectures are implemented in the CPUs of personal computers. In fact, some of the more modern processors have taken the best aspects of both architectures to create a processor that could be considered a blend of the RISC and CISC approaches. Regardless of the instruction set used, however, the control unit continues to be a critical component of the CPU.

8.2.4   The Bus

As we found in the previous section, the central processing unit needs to be able to interact with other components in the system, particularly the memory and input/output devices. We also discovered that the processor communicates with these components through its interface unit. However, the interface unit can not communicate with these external devices until it is directly connected to them. The interface unit on the CPU connects to a series of pins or connectors on the outside of the

Figure 8.7 A photograph of a typical motherboard. Many of the traces on this circuit board make up the computer's bus.

microchip that holds the CPU. When a CPU is inserted into a computer system, these pins are connected to the memory and input/output components through a series of connectors called the bus. Usually, these connectors are simply metal traces on a large circuit board, commonly called the motherboard. Figure 8.7 contains the photograph of a common personal computer motherboard. As the bus runs along the motherboard, multiple devices connect to it.

While the bus, in general, can be thought of as the collection of connections joining the CPU to the other devices in the computer system, the bus can be divided into several smaller buses based upon the information that travels over each group of connections. Most common computer systems have a bus that can be divided into an address bus, a data bus, and a control bus. The diagram in Figure 8.8 demonstrates these three buses and the way in which data flows across them.

The first bus that we will discuss is the control bus. This bus deals with the connections that carry various control signals between the CPU and the other system devices. For example, when a component needs some attention from the CPU, it can send a signal on the control bus. This signal tells the processor that it needs to interrupt its current task to take care of the component's needs. Since these control signals interrupt the processor, they are commonly referred to as interrupt request lines. In addition, the control bus takes care of signals from the CPU to each of the remote devices in the computer system. For example, the CPU may need to send out signals indicating that it is ready to store a value to memory or that it needs to send some output to the screen. One of the most important signals that travel on the control bus is the clock. By placing the clock signal on the control bus, all of the components connected to the bus will be synchronized. This is essential for proper computer operation. Thus, the control bus contains the connections that are needed by the processor and external devices to send basic signals back and forth.

The data bus, as the name suggests, carries data. This data can be input from some device, such as the keyboard, or some value to be read from or written to memory. Regardless of the nature of the data, it is the purpose of the data bus to serve as a conduit between the CPU and the other devices in the system for the sole purpose of sending data back and forth. The width of this bus determines the amount of data that can be sent in a single memory access operation. For example, early computers had a data bus that was only eight wires wide. This meant that data had to be sent in 8 bit (1 byte) chunks. So, a 32 bit single precision floating point value would require four memory access cycles to transmit across the bus. Modern computers, on the other hand, have a data bus that is much wider, often 32 or 64 bits wide. In fact, some modern computers have even wider data paths.

The address bus provides a means for the CPU to specify where it intends to read data from or write data to within the system. For example, if the processor needs to read data from memory at a

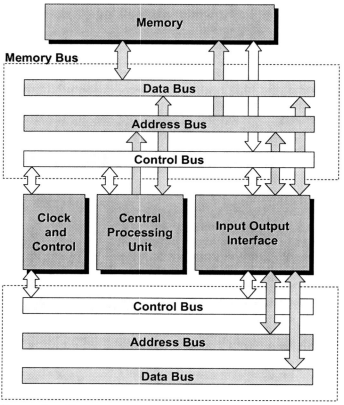

Figure 8.8 A diagram of a simplified computer system containing two buses, one for memory and one for the remaining devices. Each bus is broken down into three subsections, a control bus, an address bus, and a data bus.

particular location, it would place the address of that location on the address bus and then send a control signal to the memory controller that it wants to read data. The memory controller reads the address off of the address bus and places the requested data onto the data bus. It then signals the CPU that data is ready to read. The CPU can then read the data that it needs. A similar process is used in the writing of data. The value placed on the address bus determines the memory location of the data being dealt with. Many computer systems make no distinction between actual system memory locations and external devices such as video cards and keyboard interfaces. These devices are assigned memory locations and data is read from and written to these devices just as if they were system memory. The width (number of wires) of the address bus determines the total number of memory locations that the computer can access. Early personal computers had a 16 bit address bus which meant that they could access only 65,536 memory locations. Since most of these computers used an 8 bit or 1 byte memory location size, this meant that early computers could address only 64 kilobytes of memory. Note that when discussing sizes of storage related to computers, a kilobyte is not 1000 bytes, but rather the nearest base 2 equivalent, which in this case is 1024 (two to the tenth power) bytes. Thus, 64 kilobytes is actually equal to 65,536 bytes instead of 64,000. Modern computers use address buses with widths between 32 and 64 bits. This allows these computers to access more than 4 million address locations. Some computers with extreme memory requirements use even more than 64 bits for their memory bus.

As Figure 8.8 shows, many computer systems contain more than one bus. The configuration in the figure shows a computer system with two large buses, each composed of a control, address, and data bus. The common configuration shown above is used when the system designer wishes to operate certain parts of the system at different clock speeds. For example, since modern memory is far faster than the external input/output devices such as the hard drive and keyboard, it is desirable to run the memory bus clock at a higher rate of speed so that the transfer of data from memory can occur more quickly. At the same time, we do not want to require the input/output devices to respond at speeds above those that are practical. So, to allow these devices to use a slower clock, we simply create a bus with a slower clock rate and connect this to our higher-speed bus through an interface. This interface bridges the gap between the high speed bus connected to the memory and the processor and the slower speed bus that connects to the disk drives and external peripheral devices of the computer system. It passes all of the necessary data and communications between the two buses.

Another aspect of Figure 8.8 that is of interest is the fact that the input/output controller has both read and write access to the address bus. The only other device in the system that can write to the address bus is the CPU. The reason the input/output controller is allowed to write to the address bus is to allow data from a device such as the hard disk drive to be written directly into the system's memory without the need to involve the CPU. This approach called direct memory access (DMA), allows the input/output controller to take some of the work off of the CPU so that it is able to perform other tasks while the data is transferred. To do this, the input output controller simply requests temporary control of the bus from the CPU. When it is granted, the controller uses the address and data buses to move the required data either into the system memory or from the system memory to the peripheral device. Control of the bus is returned to the CPU whenever the CPU requests it or when the data transfer is complete.

Thus, we have found that the bus in a computer system is more complex that one would suspect at first glance. Providing an appropriate bus or series of buses while maintaining proper control of the flow of information between the CPU, the memory, and all of the peripheral devices is critical to the development of an efficient computer system.

8.2.5   Memory and Storage

The average computer has only a limited amount of storage space available to it within the CPU itself. This storage is provided in the form of the CPU's registers. This amount of storage only provides enough space for the processor to operate and perform the most trivial computations. In order to provide enough storage space to perform complicated and involved operations and to work with large amounts of data, a computer needs additional memory space. All memory is not equal, however. The memory within a computer can be broken down into several levels based upon how quickly the computer can access it. For example, since the registers are located within the CPU itself and operate at the same clock speed, they are considered the fastest type of memory in the computer system. The memory and data storage devices within the computer are often described using a pyramid model with the fastest memory at the top of the pyramid and the slowest at the bottom. Such a pyramid model is shown in Figure 8.9. Since we are already familiar with the registers at the top of the pyramid, let's begin by looking at the next memory class in the pyramid, the cache memory.

Cache memory is extremely fast memory that is situated as close to the processor as possible. In fact, many processors now integrate some amount of cache memory onto the same silicon chip as the CPU or at least into the same module as the CPU. The type of high-speed memory used in cache applications is expensive to manufacture, however. As a result, the amount of it that is included within a computer system is limited. Since this memory is significantly faster than the main system memory, it is assigned the task of storing a copy of information as it is retrieved from the main memory by the CPU. Since the CPU typically needs to access the same

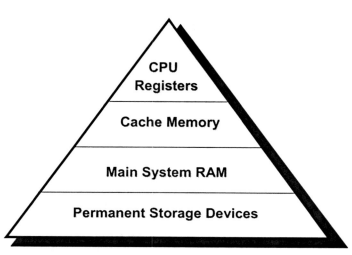

Figure 8.9 An illustration of the levels of memory present in most computer systems.

information from memory multiple times, the processor is slowed down by all of the memory accesses. However, if a copy of the information is stored in the high-speed cache memory, it can be accessed more quickly and the processor can operate more efficiently.

After the cache memory, we come to the main system RAM, or Random Access Memory. By random access, we simply mean that the memory locations within the memory can be accessed in any order. They do not have to be processed sequentially. The main system RAM, like that shown in Figure 8.10, is the memory that makes up the bulk of the computer's temporary storage. While most computers have only a few megabytes of cache memory, they will have hundreds of megabytes to gigabytes of system RAM. The computer system stores the bulk of its data and instructions in the RAM memory while it is operating. This memory, however, is volatile memory. This means that when the computer loses power, any information that was stored in the RAM memory is lost. The need to store information on a more permanent basis leads us to the final category of the memory and storage pyramid, the permanent storage devices.

There are many different methods of storing data in a more permanent fashion. The reason that these approaches are not used for the main system storage is that, as a rule, permanent storage takes more time to access. For example, it takes considerably more time to access data from a hard disk drive than it does to read it from the system RAM. Permanent storage can either be located on the system board or as part of a peripheral or input/output device.

On the system board, permanent storage usually takes on the form of ROM, or Read Only Memory. This type of memory is not erased when the power is turned off, but it cannot be written to like regular memory. Standard ROM is designed by creating a microchip such that the desired binary pattern is always present when the chip's memory addresses are accessed. Once the ROM chip is created, it can never be changed. This type of memory is often used to store important CPU instructions needed to

Figure 8.10 A photograph of typical computer system RAM modules.

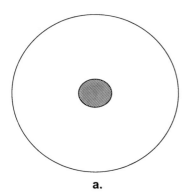

a.

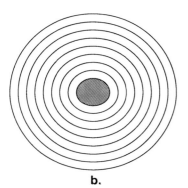

b.

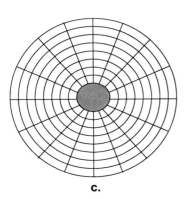

c.

Figure 8.11 An illustration of the disk formatting process. The blank disk (a) is divided into tracks (b) and then sectors (c).

initialize the computer system. Since this type of memory cannot be rewritten, it can only be used to hold information that will never change. For slightly less permanent information, an EPROM can be used. An EPROM, or erasable programmable read only memory, chip can be programmed with a desired set of memory patterns. Once programmed, the EPROM acts just like a ROM chip in the system. However, when exposed to the proper conditions, namely ultra-violet light, an EPROM can be erased and then reprogrammed. A more modern variant of the ERPOM is known as the EEPROM, or electronically erasable programmable read only memory. These devices function just like ROM chips after they are programmed, but they can be erased electronically and reprogrammed. The basic operating instructions for many devices are now written on EEPROM chips because it is easy to upgrade the device operation in the field without having to replace chips.

In addition to the use of the various types of ROM, data can be stored off of the main system board through the use of various peripheral devices. The most common of these devices is the hard disk drive. In this device, several metal disks, called platters, are coated with a magnetic material. These disks are spun at high speed and a magnetic read/write head is passed over the surface of the disks. As the disks spin, the read/write head can read information from and write information to the platter by sensing or creating magnetic fields.

Before a hard drive, or any storage disk for that matter, can be used, it must be prepared, or formatted. The formatting process involves the writing of some special pieces of information to the hard drive's platters. These pieces of information are used like signposts to allow for easy organization of the data to be stored there. The first step of the formatting process divides the platters up into a series of concentric circles called tracks. Once the tracks are established, the platters are divided by a series of radial lines, just like slicing a pizza. The portions of the disk formed where the concentric circles are divided by the radial lines are known as sectors. By combining a set of sectors, data units known as clusters are formed. The cluster is the smallest writable unit on a disk. This means that if the information to be written is only one byte, it will still require the space of a full cluster to store. Once this formatting process is completed, a basic file system directory is written to the drive and the drive is ready to store additional information. Figure 8.11 illustrates the concept of formatting a hard disk platter.

Other storage devices include optical disks like CDs and DVDs. These devices record data as a series of flat plateaus (lands) and depressions (pits) within the disk surface to represent the binary sequence of the data. This data is written in a track that spirals from the center of the disk to the outside edge. These pits and lands are read and converted into a binary data stream by using a laser to judge whether a pit or land is encountered in each data position along the track.

Regardless of the storage type or the technology used to record the data, a typical computer uses a combination of all of the memory types on the storage pyramid. Registers are used for holding values for immediate use by the processor, cache is used to speed up memory access, RAM is used to store the data currently being used by the system, ROM holds permanent instructions for the CPU, and the magnetic and optical disks store data for long periods of time.

8.2.6   Input / Output Devices

For a computer to be useful to its user there must be a pathway available for the user to communicate commands to the computer and for the computer to return results to the user. This pathway is established through the computer's input and output devices. Early computers had a limited number of devices that could be used. In fact, interaction with the early computer was often limited to just a simple keyboard and a display console. Data was stored on magnetic tape and punch cards. The modern computer, however, possesses a nearly limitless variety of interface components. Some of these components are listed in Table 8.1. As you can see, the table breaks down the current devices into four separate categories. These are devices whose primary function is to allow the user to communicate with the computer (Input devices), devices whose primary purpose is to provide the user with information (Output devices), devices that allow the computer to store information (Storage devices), and devices that allow computers to communicate with each other (Communication devices).

Table 8.1   A listing of some common input / output devices.

Input Devices	Output Devices	Storage Devices	Communication Devices
Keyboard	Monitors	Hard Drives	Network Card
Mouse	Printers	Optical Drives	Modem
Joystick	Speakers	Floppy Disk Drives	Serial Port
Video Camera	Plotters	Portable Keychain Drives	Parallel Port
Graphic Tablet		Tape Drives	USB
Touch Screens			
Microphone			

Input devices allow the computer's user to enter commands or provide external data for the computer to process. The two most common of these are the keyboard, which allows users to enter textual information, and the mouse, which allows users to select objects graphically and navigate graphical menus. Graphics tablets and touch screens are also used for graphical manipulation and selection. Joysticks can be used to provide the computer with directional information. Audio input can be provided through a microphone, and video input can be captured from a video camera. These and many other input devices are commonly found on personal computers.

Output devices provide the computer with a means of relaying information back to the user. This could be information about the computer's status, the results of computations, or graphical information. The most common form of output device is the monitor. The monitor allows the computer to supply the user with both visual textual information and graphical displays. If the user is in need of a physical copy of information, a printer or plotter can be used to put the information onto paper. Finally, audio feedback can be provided through speakers. There are many other output devices that are used on personal computers, but the ones described above are by far the most common.

Storage devices are both input and output devices. The distinction that separates storage devices from other peripherals is the fact that they are used to store data. By far the most common storage device is the hard drive. In fact, as we just saw, it is considered to be part of the memory and storage pyramid. Similarly important storage devices include optical drives, portable keychain drives, and floppy drives. Older computers used magnetic tape as a primary form of storage. However, due to its required sequential access and slow data retrieval speed, it is currently used mostly for data backups. Storage is a critical issue for most computers, and as a result, there are a wide range of storage devices available today.

Finally, communication devices allow the computer to communicate either with other computers or other remote devices. Computer to computer communication is most often performed through a network card or phone modem. These devices allow computers to send data back and forth to each other. The basic goal is to connect computers together into local or global information networks like an office network or the Internet. The computer can communicate with other devices by using devices like the serial port, parallel port, or USB. The serial port is an external connection that allows the computer to communicate with other devices by sending one bit of information at a time. Serial ports are often used to connect external modems and a wide variety of other external sensors and devices. Parallel ports are used to allow the computer to communicate with external devices by sending an entire byte of data at a time. Common parallel port devices are printers and scanners. Finally the USB or universal serial bus provides a more modern method of connecting external devices to the computer. The transfer speeds over USB are much faster than those available to the standard serial ports, and as a result, there have been a large number of devices designed to connect to the USB.

So, as we have seen, input / output devices are critical to getting the most out of a computer. These devices make interaction with the computer far easier. Whether that interaction is with the user, an external device, or another computer, input / output devices make it all possible.

8.3   Computer Operating Systems

It would not be proper to end our discussion of computers without some mention of the important role of software and the operating system. Typically, when an individual refers to "software" or "computer programs," they are simply talking about the collection of commands that must be sent to the CPU in order to get the computer system to perform a desired task. This task could be something as simple as adding two numbers and printing the result on the screen or as complex as graphically modeling the way in which electromagnetic waves interact with three dimensional surfaces. So, even the most complicated supercomputer cannot do anything useful without the proper set of instructions (program) to direct its actions.

The size and complexity of the set of CPU instructions is often highly dependent upon the task that the computer has been designed to accomplish. For example, embedded computers are usually designed to perform a very limited number of tasks, and as a result, they usually only run a single program. General purpose computers, on the other hand, are expected to perform a wide variety of tasks and must run a large number of different programs to accomplish these tasks. To simplify the process of loading the different programs into memory and executing the proper programs at the proper time, most general purpose computers rely on what is known as an operating system. The operating system is a special program that takes control of the computer's CPU. It then has the job of interacting with the user and determining what the user expects the computer to do. This may be accomplished textually through a series of commands typed on a keyboard or graphically through the selection of items and navigation of menus using a mouse and video display. Once the user makes known to the computer what program they wish to run, the operating system loads the series of instructions that make up the program into the computer's memory and direct the CPU to begin processing those instructions. Thus, the operating system's goal is to provide an interface to enable the user to easily control the operation of the computer and to execute various programs when they are needed.

So, when a basic personal computer is turned on, it follows a series of steps. When the power first comes on, all of the hardware in the computer is reset to an initial state. The processor then checks a hard-coded memory location for its first instruction. This instruction points the processor to a system initialization program contained in ROM. This program first performs a system test to make sure that the system is in proper working order. It then searches to see if any devices connected to the expansion bus have any special startup instructions. If so, these instructions are executed. Once this is complete, the system searches for a storage device that contains an operating system. Finally, the operating system is loaded into memory and control of the computer system is turned over to it. Thus, we can see that it is a delicate balance between hardware components and software that allow the computer to operate to its full potential.

8.4　　Chapter Review

In this chapter, we reviewed topics relating to the design and operation of computers.

- We began by learning that computers are one of the most important implementations of digital logic. In addition, computers can be grouped into two basic categories. These are embedded or dedicated computers and general purpose computers. Embedded computers work in the background and are developed to perform a specific task. General purpose computers are designed to be very versatile. They can perform a wide range of tasks.

- Next, we looked at the basic components that make up a typical computer. We found that a computer needs a power supply, clock, CPU, bus, memory bank, and input/output devices to operate properly.

- The first of these components that we looked at was the power supply. We found that all computers need power to operate and that it is the task of the power supply to reliably provide for these needs.

- We then learned that computers operate using a combination of combinational and sequential digital logic circuits. The sequential circuits require a periodic clock signal to operate correctly. Thus, the computer system must provide a clock to synchronize operations between the CPU and other devices.

- The "brain" of the computer system is the central processing unit (CPU). The CPU is itself composed of a series of sub-units. These units include an arithmetic and logic unit (ALU), a floating point unit (FPU), several high speed registers, a control unit, and an interface unit.

- The ALU is responsible for basic arithmetic and simple bitwise logic. The ALU is only designed to work with integer values.

- The FPU supports the ALU by assuming responsibility for mathematical operations that involve floating point inputs or results.

- In relation to the FPU, we also discovered the way in which the computer represents floating point values as sets of 32 binary digits.

- The registers are high speed memory locations located within the CPU. They are used to store values that are currently being used by the CPU.

- The interface unit allows the CPU to communicate with other system components such as the memory and input/output devices.

- The control unit is the brain of the CPU. It controls the other units of the CPU and maintains efficient order. It is the responsibility of the control unit to make sure that instructions issued to the computer are carried out properly.

- We discovered that CPUs are constructed with one of two strategies in mind. The first approach is to design a processor that understands a large number of complex instructions. Known as a complex instruction set computer, or CISC, this design allows programmers to write smaller more efficient programs. The other approach is to design a CPU that understands only a few simple commands. Known as a reduced instruction set computer, or RISC, this design makes the CPU cheaper to manufacture and allows for faster instruction execution, but it requires more instructions to accomplish the same result as a CISC processor.

- The computer's bus connects the processor to the other important components in the system. The typical bus can be broken down into an address, data, and control bus. The address bus contains the address of memory that the processor wants to access. The data bus carries data values between the memory, peripheral devices, and CPU, and the control bus carries control signals between the various system components.

- Memory, or data storage space, is very important to computer operation. The memory available to a CPU is often divided into categories based on the speed at which is can be accessed. This means that registers, which are located in the CPU, are the fastest type of memory. Cache memory, which holds recently accessed main memory data, is the second fastest, the main system random access memory (RAM) is next, and permanent storage such as read only memory (ROM) and the disk drive are the slowest types of storage.

- Input / Output devices provide a means for users to interact with the computer and for the computer to provide the user with feedback. These devices can be divided into four categories. These are input devices, which allow the user to provide the computer with data, output devices, which allow the computer to provide the user with feedback, storage devices, which offer long-term storage of data, and communication devices, which allow the computer to communicate with other computers and external devices.

- Finally, we looked at the importance of software and the operating system. We found that the hardware in the computer cannot do anything without an appropriate set of commands, called a program. In the case of general purpose computers, a special program known as an operating system is employed to supply a simple control interface for the user. The operating system can load other programs and execute them when the user requests it to do so.

8.5   Review Exercises

Section 8.2 – Basic Computer Organization

---

1. List the six basic parts that compose a computer. What is the purpose of each part?

2. Why is the power supply important to the computer?

3. List the two types of digital logic present in computers. Which of these types of logic requires a clock signal?

4. What does the abbreviation CPU stand for?

5. List the parts of a typical CPU.

6. What does ALU stand for? What is its purpose?

7. Which unit within the CPU takes care of floating point computations?

8. Describe the way in which the computer represents single precision floating point numbers.

9. Represent the following decimal values in the single precision floating point format.

   a. 245.0625   b. -385.425   c. 1036.825
   d. 0.1   e. -8075.5   f. 184732.825

10. What happens if the fractional part of the binary number is too large to fit in the fraction part of the single precision format? Does this impact the computer's computational accuracy?

11. What purpose do the registers serve? What information is stored within them?

12. Which sub-unit of the CPU deals with external communication?

13. Which sub-unit of the CPU oversees internal operations and makes sure that commands are executed properly?

14. Describe the two approaches to CPU design. What are their benefits and drawbacks?

15. What is the purpose of the bus?

16. Where is the bus found in most computer systems?

17. The system bus can be divided into three parts. What are these parts and what is their purpose?

18. Describe the memory and storage pyramid. Which attribute of the storage devices does it base its ranking on?

19. What is the purpose of cache memory?

20. What does RAM stand for? What is its purpose in a computer system?

21. What does ROM stand for? What is its purpose in a computer system?

22. What does EPROM stand for? What about EEPROM? How do these two devices differ?

23. Describe how a disk is formatted. What is a track? What is a sector?

24. What is the purpose of input / output devices?

25. Name four categories that input / output devices can be grouped into. Provide an example device for each category.

26. What is the difference between the way in which a parallel port and a serial port transmit data?

Section 8.3 – Computer Operating Systems

27. Why are programs important to the operation of a computer?

28. What is the purpose of an operating system?

29. List some common operating systems. What type of hardware do they run on?

30. Describe what happens when a computer is turned on.

UNIT 4

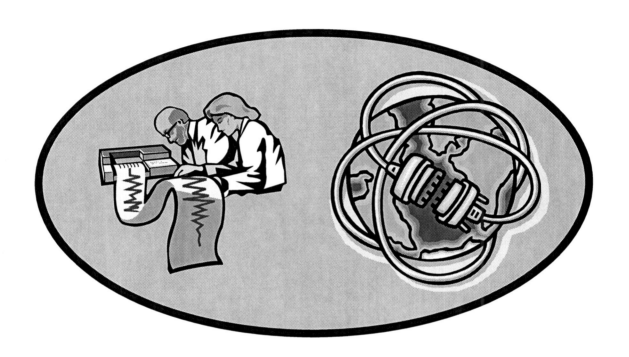

Digital Signal Processing
and
Communications

Chapter 9. Digital Signal Processing

Chapter Outline

9.1   Introduction and Overview

In everyday life, we are surrounded by all kinds of signals. Some of these signals can be heard. Others can be seen. But, most signals cannot be seen or heard. The study of these signals and the processes involved in extracting, storing, enhancing, and transmitting information embedded within them is the discipline that we know as signal processing. Signal processing is an important field within Electrical Engineering and there are many interesting applications of signal processing technology. In this chapter, we will provide you with some background in the area of signal processing and take a look at some of the basic theory involved in the processing of digital signals (analog signal processing is beyond the scope of this text). Some applications of digital signal processing in the areas of sound and image processing are reviewed in chapters 10 and 11 and the importance of signals in communication can be seen in chapter 12.

9.1.1   What is a Signal?

So, now that we know signal processing is an important area of study, let's answer that nagging question that I am sure that many of you have: What exactly is a signal? A signal can be thought of formally as a function of one or more variables that convey information. What exactly is this definition trying to tell us? Basically, a signal is any transmission that conveys some amount of information. For example, the siren and flashing lights of an ambulance convey information visually and audibly. In this case, the information can be extracted from the signal with our own eyes and ears. Other signals such as radio broadcasts, television, cell phone signals, and radar require special electrical devices to intercept and decode the signals so that they can be understood. So, basically any transmission, be it light, sound, electrical impulses, or electromagnetic radiation can be considered a signal as long as it contains information that can be extracted in some way. Now, let's think about this definition of a signal for a moment. Can you think of anything that you do that does not involve the processing of information taken from signals? Activities such as reading, writing, watching TV, listening to the radio, sports, and surfing the internet all involve information taken from signals. The human senses of sight, hearing, smell, taste, and touch are just special tools for extracting information from signals. In fact, your senses involve signals of their own. For example, your eyes send visual information to the brain in the form of electrical signals transmitted through your nervous system. Similar signals are sent by your ears, nose, tongue, and skin. These signals are then further processed and interpreted by your brain. So, as we can see, signals are everywhere and signals are important to our everyday lives. However, with so many different types of signals surrounding us, we need to come up with a way to organize these signals so that we can better understand them. Thus, we may ask: is there any way that we can classify the signals we encounter into more manageable categories? Yes, there are a wide variety of categories that we can break signals down into. In the next section, we will look at a few of the more common categories that we currently use to identify signals.

9.1.2   Classification of Signals

One way we can categorize signals is to break them down into the number of dimensions (how many variables) that they involve. An audio signal, for example, is one-dimensional. It varies in intensity with time ($x(t)$). Images, on the other hand, are two-dimensional. They consist of a plane of varying light intensities ($I(x,y)$). Video signals have three dimensions. Video signals arise from a series of two-dimensional images that change over time (about 30 images per second on a typical television) ($I(x,y,t)$).

Another type of classification similar to the dimensionality criteria is the number of channels involved in the signal. Not all signals involve a single stream of information. Instead, multiple streams may be embedded into a single signal transmission. For example, a radio station broadcasting music in stereo often uses a two-channel signal. This is due to the fact that information for both the right and left speaker must be included in the signal for it to be accurately reproduced at the receiver. So, a stereo radio broadcast can be considered a one dimensional, two-channel signal. Multi-channel signals can also arise from the collection of data from multiple sensors. For example, a 3 or 12 lead electrocardiogram will result in a multi-channel signal.

We can also group signals by whether they are deterministic or random in nature. A deterministic signal is one that can be uniquely determined by a well-defined process, for example, an equation or lookup table. The behavior of a random signal, on the other hand, cannot be predicted ahead of time. There is always a level of uncertainty as to how the signal will behave next. In signals of this type, future behavior is given in terms of probabilities instead of exact equations.

Signals can also be broken down into real and complex valued signals. For example, if we encounter a one-dimensional signal that consists of an intensity that varies in time, we can call it a real signal if the intensity value $x(t)$ at any given time t is a real value. However, if there exists a time, t, such that $x(t)$ is a complex number, we label the signal as complex. As we know, complex numbers can be broken down into the form $a+jb$, where a and b are real values. So, if we break our signal down into the form $x(t)=x_1(t)+jx_2(t)$, where $x_1(t)$ and $x_2(t)$ are real signals, we can think of a complex signal as a special form of a two-channel signal.

9.1.2.1 Continuous and Discrete-Time Signals

One of the most common distinctions between signal types, however, is the distinction between continuous and discrete-time signals. The difference between continuous, or analog and discrete-time signals is dependant on how the signal variables are represented. Let's look at the distinction for a one-dimensional signal that changes based upon time. If we are working with an analog signal of this type, we can find an intensity value for any value of time that we choose. For example, if we want to know what the amplitude of our signal is at 1.5354643 seconds, we can obtain that information. However, if we are working with a discrete-time signal, finding a value for that exact time may not be possible.

Discrete-time signals, as the name suggests, holds intensity values for the signal only at specific discrete points. Thus, the discrete-time signal, depending upon its resolution, may only have intensities for times 1, 2, 3, 4, etc. seconds. If the signal resolution is higher, a discrete-time signal may have a value every tenth of a second, or perhaps every hundredth, etc. It would be unlikely however, that a value would be available for a point as specific as 1.5354643. Another way to think about it is to look at the first second of the signal. In an analog signal, how many data points can we find in this time period? If you think about it, there are an infinite number of points. An analog signal will have a data point for any fractional time value between 0 and 1 second. On the other hand, if we have a discrete-time signal, the number of data points will be dependent on the resolution of the signal and will always be less than infinity. So, a discrete-time signal divides the time variable up into a series of discrete points at which the amplitude is known.

A digital signal is a special case of a discrete-time signal. In a digital signal, both the time and the amplitude are discrete and the amplitude value at each point in time is represented as a binary value. Thus, while digital signals have the drawback of not having values for every possible data point, they have one important benefit. Because they do not have an infinite number of points, they can be stored in a finite amount of space, for example on a computer hard drive. This makes the processing and storage of digital signals much easier than analog.   Figure 9.1 shows the difference between the plots of an analog and discrete-time signal.

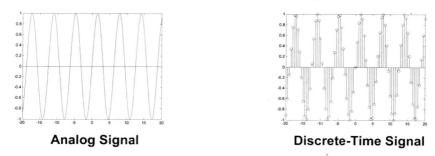

Analog Signal   **Discrete-Time Signal**

Figure 9.1 Examples of the plots of an analog and discrete-time signal.

The way in which analog and discrete-time signals are mathematically represented is also different. For example, a one dimensional signal that varies in time is represented as *x(t)* if it is an analog signal. In this representation, *t* can take on any real value. Thus, as we saw before, an analog signal is defined for all possible times. On the other hand, if our signal is a discrete- time signal, it is represented as *x(n)*. In this expression, *n* can hold only integer values (... -2, -1, 0 , 1, 2, ...). These integer values do not refer to an actual time, but rather they are an index number that refers to a specific discrete point in the sequence of values (sometimes called a sample number). Thus, to determine the actual time that a value in a discrete-time sequence occurred, you must know how many values the signal includes for each second of time (samples per second or sampling rate). We will learn more about the process of signal sampling (the process of converting an analog signal into a discrete-time signal) in chapter 10.

9.1.2.2   Even and Odd Signals

One last way we can classify a signal is based upon its symmetry. If a signal is symmetric about the *y* axis, it is said to be an even function. The mathematical result of this symmetry is that *x(t) = x(-t)* for all values of *t*. An example of such a signal would be the cosine function. If, on the other hand, the signal is anti-symmetric about the *y* axis, it is said to be an odd signal. Mathematically this means that *x(-t) = -x(t)*. The sine function could be considered an odd function. Figure 9.2 shows a plot of an even and an odd signal.

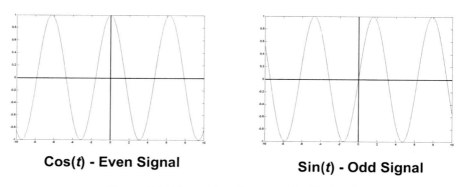

Cos(*t*) - Even Signal   **Sin(*t*) - Odd Signal**

Figure 9.2. Examples of even and odd signals.

Many functions, however, are not purely symmetrical or anti-symmetrical. Instead, they contain components of both even and odd signals. In other words, we can say that for any real valued signal x(t), we can write:

$$x(t) = x_o(t) + x_e(t)$$

Where x_o is the odd portion of the signal and x_e is the even part. Therefore, we should be able to break a given signal down into its even and odd components. Let's explore this in a bit more depth by developing some equations to represent the even and odd parts of a given signal. First, let's take our original expression from above and substitute $-t$ for t. By doing this we find that:

$$x(-t) = x_o(-t) + x_e(-t)$$

Now, let's see what happens when we add these two equations together.

$$x(t) + x(-t) = x_o(t) + x_e(t) + x_o(-t) + x_e(-t)$$

Next, we can take advantage of the behavior of even and odd signals. We know that since even signals are symmetric we can say:

$$x_e(-t) = x_e(t)$$

In addition, the anti-symmetric behavior of odd functions allows us to say:

$$x_o(-t) = -x_o(t)$$

Making these substitutions in our summation equation, we can write:

$$x(t) + x(-t) = x_o(t) + x_e(t) - x_o(t) + x_e(t)$$

$$x(t) + x(-t) = x_o(t) - x_o(t) + x_e(t) + x_e(t)$$

$$x(t) + x(-t) = 2x_e(t)$$

So, we see that the sum of x(t) and x(-t) is equal to twice the even part of the signal. Thus, if we divide both sides of the equation by 2, we will get:

$$x_e(t) = \frac{1}{2}(x(t) + x(-t))$$

The equation for the odd portion of the signal can be obtained by subtracting as follows:

$$x(t) - x(-t) = (x_o(t) + x_e(t)) - (x_o(-t) + x_e(-t))$$

Using the behavior of even and odd functions, we can rewrite this as:

$$x(t) - x(-t) = x_o(t) + x_e(t) + x_o(t) - x_e(t)$$

$$x(t) - x(-t) = 2x_o(t)$$

Once again, we can divide both sides by 2 to get the equation for the odd part of the signal.

$$x_o(t) = \frac{1}{2}\left(x(t) - x(-t)\right)$$

Now we have an equation for both the even and odd parts of a given signal.   Let's see what happens if we apply these signals to an actual signal.

Example 9.1

Problem: Find the even and odd parts of the following signal.

$$x(t) = 3(t+1)^3 + \cos(t) - \frac{1}{2}\sin(t)$$

Solution: In this example, we have been asked to find the even and odd parts of a signal.   To do this, all that we must do is apply the two equations that we just derived.   Before we do that, however, let's expand our equation to make the later steps easier.

$$x(t) = 3(t^3 + 3t^2 + 3t + 1) + \cos(t) - \frac{1}{2}\sin(t)$$

$$x(t) = 3t^3 + 9t^2 + 9t + 3 + \cos(t) - \frac{1}{2}\sin(t)$$

Now, we need to find *x(-t)* by replacing *t* with *–t*. This gives us:

$$x(-t) = 3(-t)^3 + 9(-t)^2 + 9(-t) + 3 + \cos(-t) - \frac{1}{2}\sin(-t)$$

If we factor a -1 out of each of the powers of *–t*, note that cos(t) is an even function so cos(-t) = cos(t), and note that sin(t) is an odd function, so sin(-t) = -sin(t), we can write:

$$x(-t) = 3(-1)^3 t^3 + 9(-1)^2 t^2 - 9t + 3 + \cos(t) + \frac{1}{2}\sin(t)$$

Now, simplifying the powers of -1 we write:

$$x(-t) = -3t^3 + 9t^2 - 9t + 3 + \cos(t) + \frac{1}{2}\sin(t)$$

We can now use this equation for –t in our equations for the even and odd parts of our signal. Let's start by finding the even part:

$$x_e(t) = \frac{1}{2}\big(x(t) + x(-t)\big)$$

$$x_e(t) = \frac{1}{2}\left(3t^3 + 9t^2 + 9t + 3 + \cos(t) - \frac{1}{2}\sin(t) - 3t^3 + 9t^2 - 9t + 3 + \cos(t) + \frac{1}{2}\sin(t)\right)$$

Combining like terms yields:

$$x_e(t) = \frac{1}{2}\big(18t^2 + 6 + 2\cos(t)\big)$$

$$x_e(t) = 9t^2 + 3 + \cos(t)$$

Now, let's look at the odd part:

$$x_o(t) = \frac{1}{2}\big(x(t) - x(-t)\big)$$

$$x_o(t) = \frac{1}{2}\left(3t^3 + 9t^2 + 9t + 3 + \cos(t) - \frac{1}{2}\sin(t) + 3t^3 - 9t^2 + 9t - 3 - \cos(t) - \frac{1}{2}\sin(t)\right)$$

Combining like terms yields:

$$x_o(t) = \frac{1}{2}\big(6t^3 + 18t - \sin(t)\big)$$

$$x_o(t) = 3t^3 + 9t - \frac{1}{2}\sin(t)$$

So, now we have an expression for both the even and odd parts of our signal. It is important to note at this time that the sum of the even and odd expressions will result in the original expression. If this were not the case, we would have made an error at some point in the process. Now, let's take a look at the plots of the original signal and its even and odd parts.

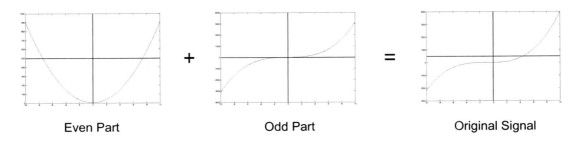

Even Part   Odd Part   Original Signal

Thus, we have found that even if a signal is not perfectly even or odd, we can divide it up into its even and odd parts.

___

We should now have a little better idea of some of the ways that we can organize the different types of signals that surround us. In many cases, the signal classifications can be combined to even further subdivide the signal types. For example, you might classify a typical stereo radio broadcast as a one-dimensional, two-channel, analog signal.

9.1.3   What is Digital Signal Processing

So far in this chapter, we have discovered that we are surrounded by signals. In fact, almost everything that we do involves dealing with signals in one way or another. In order to better study these signals, we tend to group them into categories based upon common characteristics. Some of these characteristics include signal dimension, the number of channels of information, whether the signal is random or deterministic, signal symmetry, and whether the signal is analog or discrete. All of these different ways of approaching signals help us to better understand and study the information that is embedded in signals. As we mentioned earlier, the field of signal processing is concerned with the study of these signals as well as the processes involved in extracting, storing, enhancing, and transmitting the information that is embedded in them.

This field of study can be further broken down into two major subfields. These subfields are built around one of the major signal classifications, analog vs. digital. While the basic theory involved is very similar regardless of whether the signal is analog or digital, the implementation of that theory can be very different. In addition, processing a signal in either analog or digital form can have certain drawbacks or benefits based upon the application and the information embedded in the signal. Let's take a quick simplified look at how each of the approaches work with signals and note some benefits and drawbacks. We will start by looking at analog signal processing (ASP).

Since most of the signals produced naturally are analog signals, analog signal processing was the first signal processing approach developed. In fact, until the late 1900's, nearly all signal processing was done on analog signals. Radio, television, and the telephone were all originally built on analog technology. So, how does analog signal processing work? As you would expect, analog signal processing starts with the acquisition of an analog signal. This signal may be in one of many forms, for example, a radio wave intercepted by an antenna or an electrical signal on a wire. Once the signal is acquired, it is fed into an analog signal processor. This processor is a device that is made up of electrical networks built from basic electrical components. These components include active components, such as, diodes, operational amplifiers, and transistors, as well as, passive components, including, resistors, capacitors, and inductors. The electrical networks in the analog signal processor act upon the signal to perform some desired transformation. This could include extracting information, removing noise, or injecting information into the signal. Regardless of the transformation that is performed, the output of an analog signal processor is another analog signal. This signal can then be fed into an additional analog signal processor for further processing, transmitted to another location, or displayed for the end user. For example, in a simple AM radio, the input analog signal is a radio wave that is received by an antenna. This signal consists of a high frequency sinusoidal waveform that changes in amplitude based upon the audio signal that it is carrying. This signal is fed into the radio receiver, which is an analog signal processor. The receiver extracts the lower frequency audio information from the high frequency carrier wave. This audio information, the signal processor's output, is sent to the speakers for the end user to listen to. A block diagram of an analog signal processing system is shown in Figure 9.3.

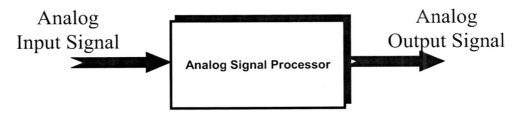

Figure 9.3 A block diagram of the operation of an analog signal processor

The approaches taken in analog signal processing are mature and well understood. In fact, they do a fairly good job. Analog signal processing also benefits from the fact that most naturally occurring signals are analog in nature. Thus, they can directly operate on these signals. Analog signal processors do, however, have some drawbacks. First, we know that the electrical networks used in analog signal processors are built from electrical components. These components do not hold exact values. Instead, they are manufactured to be within some tolerance of the value that they claim to have. Thus, circuits built from them will not yield precisely the designed output. While the difference is negligible and probably acceptable in most situations, applications that require precise operation may be difficult to implement with an analog design. In addition, designing and building an analog signal processing circuit can be both time consuming and costly. Once the system is designed, it lacks flexibility. Making changes to the operation of the processor requires changing the electrical circuit that makes it up. Thus, if you design a TV receiver, the functionality that you design into the original circuit is all that the receiver will have. It cannot be upgraded later without replacing circuit boards. Finally, the very nature of analog signals and analog signal processors make them susceptible to distortion from noise. If you have ever listened to AM radio during a thunderstorm, you have probably heard the pops and crackles introduced into the signal from electrical noise in the atmosphere. Analog signal processors can also pick up noise from nearby equipment and power lines.

Now, let's take a look at digital signal processing (DSP). Since most signals don't start out as digital, digital signal processing has the drawback that it is often necessary to convert an analog into a digital signal before any work can be done on it. In addition, most output devices, for example audio speakers, cannot handle digital signals. So, the digital signal needs to be converted to analog before it can be understood by the end user. Despite these drawbacks, digital signal processing is a powerful and important field of study. A digital signal processing system begins in much the same way as an analog one. It acquires a signal. In this case, the signal that it acquires may be analog or digital in nature. If the signal is already digital, no pre-processing is required and the signal is fed directly into the digital signal processor. If the signal is analog, it is first passed through a device called an analog to digital converter. This device converts the analog input signal into a digital signal by taking readings from the input signal at fixed points (a process called sampling). The readings taken at these points are then encoded into a digital signal (the process of converting an analog signal into a digital one is covered in more detail in chapter 10). Once the analog signal has been converted, it is fed into the digital signal processor. Unlike the analog signal processor, the digital signal processor consists of a computer or dedicated microprocessor chip. This device performs logic operations on the incoming signal to perform some desired transformation. This new signal is then fed out of the processor. This output signal is handled in one of two ways depending on the application. It can either be preserved in digital form for storage, future processing, or transmission, or it can be converted into an analog signal. This conversion is performed by passing the digital signal through a device called a digital to analog converter. This device takes the incoming digital signal and generates an approximated analog waveform as an output. This conversion to analog is necessary if the signal needs to be transmitted over older analog communication channels or, more commonly, if it is going to be played or displayed on an analog device such as a speaker or television screen. Figure 9.4 displays a block diagram of a digital signal processing system. It is important to note that the analog to digital and digital to analog conversion steps may be bypassed if conditions require it.

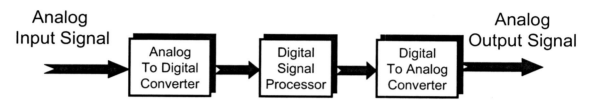

Figure 9.4 A block diagram of a digital signal processor.

So, with all the conversion steps required to process analog information on a digital signal processor, why would we want to bother with it? The reason we use digital signal processing is because of the large advantages that it holds over the analog approaches. Let's look at a few of these advantages:

1. **Flexibility:** As we noted earlier, an analog device, once developed, is fixed. In other words, it cannot be changed or improved without changing circuit boards. A digital system, on the other hand, can be changed simply by modifying the programming of the device. For example, if you have ever upgraded the "firmware" on an electronic product (digital camera, CD or DVD drive, or network switch for example), you have changed the programming of the microprocessor. The flexibility of digital signal processing devices is probably best illustrated by the computer. Based on the program that is running, a computer can do numerous things that involve processing signals from input and sending signals to output devices in a wide variety of ways. Compare this to your television set. It receives a television signal and decodes it to output the resulting signals to the screen and speakers. It can't be easily reconfigured to do anything else.

2. **Ease of Development:** While analog systems must be developed as electrical circuits, digital signal processing can be developed as computer software and later moved to a dedicated microprocessor system. Thus, development and testing of a digital signal processor is easier and more cost-effective than the equivalent processes for an analog signal processor.

3. **Accuracy/Stability:** Digital signal processors are based on digital logic. Thus, there are no sloppy tolerances involved. So the output of a digital signal processor will be far more accurate than that of an analog. In addition, digital signal processors operate on simple logic operations, such as addition and multiplication. This makes them easy to understand and stable to operate.

4. **Storage:** Unlike analog signals, digital signals consist of a finite number of data points. As a result digital signals can be easily stored and duplicated. A copy of a stored digital signal will be an exact duplicate with no loss in quality. Compare that with a copy of a tape or videocassette. Every time these stored signals are copied they lose quality. For example, digital audio signals stored on CDs or in digital computer files are of a much higher quality than those you will find on audio tapes or records.

5. **Error and Noise Handling:** It is a difficult task to remove noise from or detect errors in data in an analog signal. In a digital signal, however, it is far easier. The existence of extremely powerful error detection and correction algorithms for digital signals makes the use of digital approaches very appealing. Take a common audio CD for example. The music information is stored on the disk as a series of pits and rises or "lands" that represent zeros and ones when read by a laser. The data is encoded in such a way that minor scratches on the disk's surface that can make some portions of the data unreadable do not hurt the playback of the music. The encoding allows the device to recover from the loss of small amounts of data with no noticeable loss in quality. This is a very desirable ability. We will discuss error detection and correction in further detail in chapter 10.

From this discussion, we can see that the field of signal processing is divided into two large areas based upon whether the signal is processed in the analog or digital domain. We also understand some of the benefits involved in doing our processing in the digital domain. Throughout the rest of this chapter, we will examine some of the basics of digital signal processing.

9.2 Introduction to Some Special Signals

Before we take a look at some methods of working with signals, we first need to become familiar with a couple of commonly used signals. While most signals you will encounter will be far more complicated, it is important to understand the following basic signals because they are widely used in the representation and description of other signals. The two signals we will be investigating are the unit impulse function (sometimes called the unit sample function) and the unit step function.

9.2.1 The Unit Impulse Function

The first basic signal we will look at is the unit impulse function. This function, also referred to as the delta function, is a very simple function. In an analog domain, the delta function, written as $\delta(t)$, consists of a single spike located at $t = 0$. The analog delta function is defined as having a value of infinity at $t=0$ and zero everywhere else. The area under the analog delta function is exactly equal to one. Similarly, the discrete-time delta function, written as $\delta(n)$, holds a value only at the point $n=0$ and is equal to zero everywhere else. However, since the area under a discrete-time function is calculated by summing the values at each point in the signal, the discrete-time delta function is equal to one at $n=0$ so that the area under the discrete-time delta function is also equal to one. An example of the analog and discrete-time delta function is given in Figure 9.5.

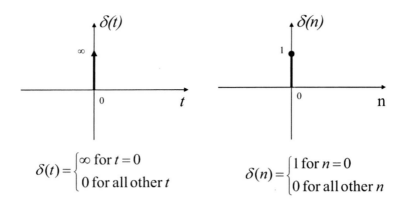

$$\delta(t)=\begin{cases}\infty\ \text{for}\ t=0\\0\ \text{for all other}\ t\end{cases} \qquad \delta(n)=\begin{cases}1\ \text{for}\ n=0\\0\ \text{for all other}\ n\end{cases}$$

Figure 9.5. The analog and discrete-time delta function.

9.2.2   The Unit Step Function

The second special function that we will investigate is the unit step function. The unit step function, sometimes referred to as the Heaviside unit function, looks just like a step, just as its name implies. As was the case with the delta function, the unit step function is defined slightly differently in the analog and discrete-time domains. In the analog domain, the unit step function, denoted $u(t)$, holds a value of 0 for any time less than 0 and a value of 1 for any time greater than 0. The unit step function is discontinuous at t=0 (i.e., it doe not have a defined value). However, to avoid confusion, it is typically assigned a value of ½. In its digital representation, the unit step function, denoted $u(n)$, holds a value of 0 for n<0 and a value of 1 for n≥0. Figure 9.6 demonstrates the appearance of the unit step function in both the analog and discrete-time domains.

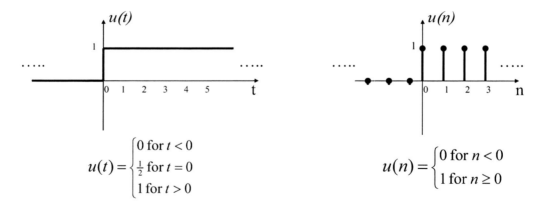

$$u(t)=\begin{cases}0\ \text{for}\ t<0\\ \frac{1}{2}\ \text{for}\ t=0\\1\ \text{for}\ t>0\end{cases} \qquad u(n)=\begin{cases}0\ \text{for}\ n<0\\1\ \text{for}\ n\geq0\end{cases}$$

Figure 9.6. The analog and discrete-time unit step function.

The two functions reviewed here will be encountered throughout your study of signal processing. Therefore, it is important to be comfortable with them. The delta function is typically used when a specific point within a signal must be isolated, and the unit step function is typically used to simulate the action of turning a signal on and off.

9.3 Plotting Discrete-Time Signals

In this section, we will be looking at some examples of discrete-time signals. We will pay particular attention to the way in which transformations of the independent variable (*n*) and dependent variable (the amplitude) change the plot of the signal.   Transformations of the independent variable include time shifting, time scaling and signal reversal. Dependent variable transformations include signal addition, signal multiplication, and scaling. We will briefly examine each of these transformations in the sections below and show how the plot of a common signal changes in response to each. We will begin by defining a discrete-time signal to work with. Let *x(n)* be:

$x(n)$ = {1, 2, 3, 4, 3, 2, 1} for *n* = 0 to 6, and 0 elsewhere.

Let's plot this signal out so we can compare it to the transformed signals later.

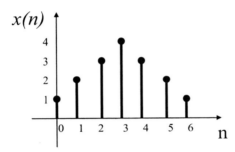

9.3.1   Time Shifting

One of the most common signal transformations that you will encounter is time shifting. A time shifted signal is one of the form: $x(n-k)$, where k is an integer value indicating the size of the shift. If *k* is a positive integer, the signal will be shifted to the right, or delayed, by *k* positions. Likewise, if *k* is a negative value, the signal is shifted to the left, or advanced, by *k* positions. Let's take a look at our sample signal shifted three to the left (*k*=-3) and right (*k*=3).

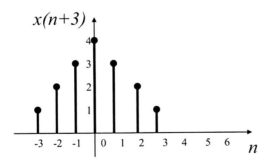

Time Advanced Signal (*k* = -3)

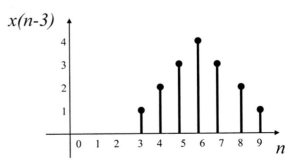

Time Delayed Signal (*k* = 3)

So, here we have seen how to shift a given signal by a given number of samples in time.

A Different Look at the Time Shifting Transformation

Performing transformations on the independent variable often leads to confusion, especially when dealing with more complicated transformations. Here we present a systematic tabular approach that works for any transformations on the independent variable. Let's look at the right shift transformation as an example.

Our original function is:

$x(n)$ = {1, 2, 3, 4, 3, 2, 1} for n = 0 to 6, and 0 elsewhere.

An equivalent statement is:

$x(n)$ = {1, 2, 3, 4, 3, 2, 1} $0 \leq n \leq 6$, where n is incremented by 1, and 0 elsewhere.

For the time shifted function $x(n-3)$ it follows $n-3$ = 0 to 6, and 0 elsewhere,

or

$$0 \leq n\text{-}3 \leq 6$$

or, if we add 3

$$3 \leq n \leq 9$$

Now that we have the new values of n we can create a table of n and our time shifted function $x(n-3)$ as follows;

n	$(n-3)$	$x(n-3)$
3	0	1
4	1	2
5	2	3
6	3	4
7	4	3
8	5	2
9	6	1

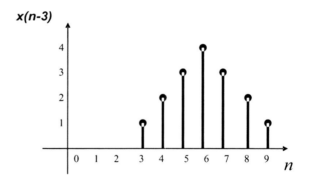

Time Delayed Signal (k = 3)

9.3.2   Time Scaling

Another common transformation used in signal processing is time scaling. In an analog signal, time scaling is performed by compressing or expanding the signal along the time axis. In the discrete-time domain, the time scaling of the signal is performed by processes known as up-sampling (expansion) and down-sampling (compression).

To understand the discrete-time approach to time scaling, it is important to first understand what happens when a continuous-time signal is expanded or compressed in time. For example, if we want to create a new continuous-time signal, $g(t)$, that is expanded to twice its normal size along the time axis, we would write $g(t) = x\left(\frac{1}{2}t\right)$, where $x(t)$ is the signal that we are expanding. From this example, we can see that a point in the expanded signal will possess an amplitude equivalent to the amplitude of the original signal at half the current time position. In other words, the expanded signal will have the same amplitude at t = 1 second that the original signal possessed at t = 0.5 seconds. Thus, this expression results in an effective doubling of the overall signal duration. So, what happens when we try to expand a discrete-time signal? In this case, our time expanded signal is written as $g(n) = x\left(\frac{1}{2}n\right)$. As you would expect, this expansion works in much the same way as the continuous-time version. The only difference is that our original signal only possesses amplitudes for integer values of n. So, we cannot find an amplitude for a value of n = 1 in the expanded signal because the amplitude for n = 0.5 is undefined in the original discrete-time signal. Because of this, a value of zero is usually stored for these points in the expanded signal.

The process of signal compression is similar to that of expansion. In this case, however, the independent variable is multiplied by a value larger than one. For example, we would compress a continuous-time signal by multiplying the independent variable by 2. Thus, we can write $g(t) = x(2t)$. This means that the amplitude of a point in the new, compressed signal is equivalent to that of a point occurring at twice the time interval in the original signal. This results in a signal that is compressed to half its normal duration. In the discrete-time equivalent, we write $g(n) = x(2n)$. In practical terms, this means that the compressed signal will contain every other value from the original signal. The other values are discarded. Thus, down-sampling a signal results in the loss of information related to the signal.

So, we have found that signal compression and expansion can be accomplished by multiplying or dividing the independent variable (n) by a fixed value k. We can achieve an up-sampled signal by dividing n by k ($x\left(\frac{n}{k}\right)$) and a down-sampled signal by multiplying n by k ($x(kn)$). Let's take a look at our sample signal up-sampled and down-sampled by two (k=2).

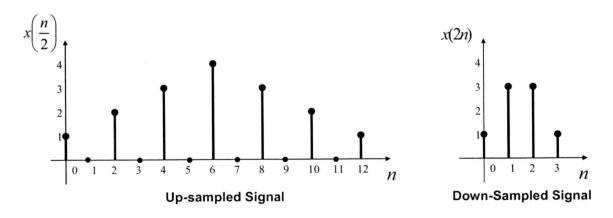

Up-sampled Signal   Down-Sampled Signal

So, up-sampling a signal expands the signal by inserting zero terms and down-sampling compresses a signal by removing existing terms.

A Different Look at the Time Scaling Transformation

Now let's look at an example of time scaling using the tabular approach.

$x(n)$ = {1, 2, 3, 4, 3, 2, 1} $0 \leq n \leq 6$, where n is incremented by 1, and 0 elsewhere.

For the up-sampled function $x\left(\dfrac{n}{2}\right)$ it follows $\left(\dfrac{n}{2}\right)$ = 0 to 6, and 0 else where,

or

$$0 \leq \left(\dfrac{n}{2}\right) \leq 6$$

or, if we multiply by 2

$$0 \leq n \leq 12$$

Now that we have the new values of n we can create a table of n, and our up-sampled function $x\left(\dfrac{n}{2}\right)$ as follows;

n	$\left(\dfrac{n}{2}\right)$	$x\left(\dfrac{n}{2}\right)$
0	0	1
1	1/2	0
2	1	2
3	3/2	0
4	2	3
5	5/2	0
6	3	4
7	7/2	0
8	4	3
9	9/2	0
10	5	2
11	11/2	0
12	6	1

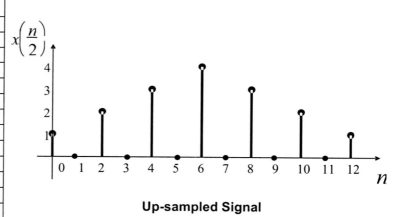

Up-sampled Signal

Recall our original function evaluates only for integer values, elsewhere it is 0.

For the down-Sampled case

$x(n)$ = {1, 2, 3, 4, 3, 2, 1} $0 \leq n \leq 6$, where n is incremented by 1, and 0 elsewhere.

For the down-sampled function $x(2n)$ it follows $2n$ = 0 to 6, and 0 elsewhere, or

$$0 \leq 2n \leq 6$$

or, if we divide by 2

$$0 \leq n \leq 3$$

Now that we have the new values of n we can create a table of n and our down-sampled function $x(2n)$ as follows;

n	$(2n)$	$x(2n)$
0	0	1
1	2	3
2	4	3
3	6	1

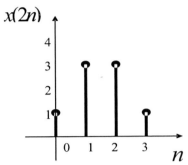

Down-Sampled Signal

9.3.3 Reversal

The final common transformation of the independent variable (n) that you will encounter is time reversal. In a time-reversed signal, the index order is reversed by replacing n with $-n$. This has the effect of flipping the signal around the y axis. The resulting signal will be a mirror image of the original. Let's apply time reversal to our sample signal.

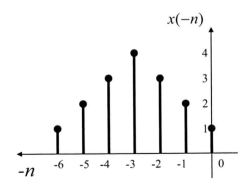

Time Reversed Signal

Here we see that the time-reversed version of the signal is simply its mirror image about the y axis.

A Different Look at the Time-reversed Transformation

Once again, we can apply the tabular method to time-reversal.

$x(n)$ = {1, 2, 3, 4, 3, 2, 1} $0 \leq n \leq 6$, where n is incremented by 1, and 0 elsewhere.

For the Time-reversed function $x(-n)$ it follows $-n$ = 0 to 6, and 0 elsewhere,

or

$$0 \leq -n \leq 6$$

or, if we multiply by -1

$$0 \geq n \geq -6$$

and rearranging things a bit, we find

$$-6 \leq n \leq 0$$

Now that we have the new values of n we can create a table of n and our time-reversed function $x(-n)$ as follows;

n	$(-n)$	$x(-n)$
0	0	1
-1	1	2
-2	2	3
-3	3	4
-4	4	3
-5	5	2
-6	6	1

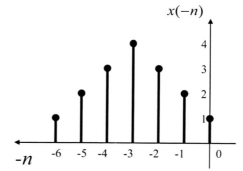

Time Reversed Signal

The previous three transformations have been transformations of the independent variable. In the next three sections we will examine some transformations that can be performed on the dependent variable (amplitude). We will begin with the most common amplitude transformation, scaling.

9.3.4   Scaling

The scaling of a signal involves a transformation of the dependent variable. In this case, we can manipulate the amplitude of the signal by multiplying the signal by a constant positive value, k, $(kx(n))$. If we use a value of k between zero and one, we will uniformly decrease the amplitude of the signal. This is a process known as signal attenuation. If we use a value greater than one, however, we will uniformly increase the amplitude. This process is known as amplification. Let's see what happens to our signal for values of k equal to 2 and 0.5.

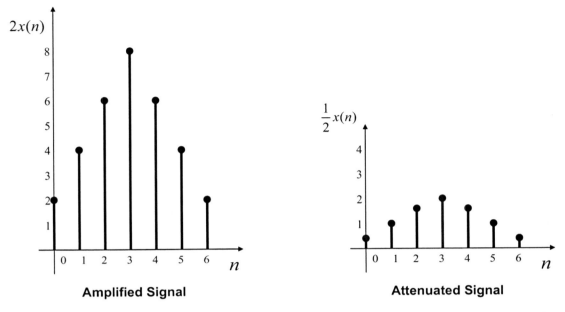

Amplified Signal   **Attenuated Signal**

Thus, we can see that by simply multiplying each value of our signal by a constant value, we can either amplify or attenuate the overall signal.

9.3.5   Signal Addition

Another way that a signal's amplitude can be transformed is by adding the amplitude of another signal to it. This is performed by adding the two signals point by point.   Let's add another signal to our example signal to illustrate the process.

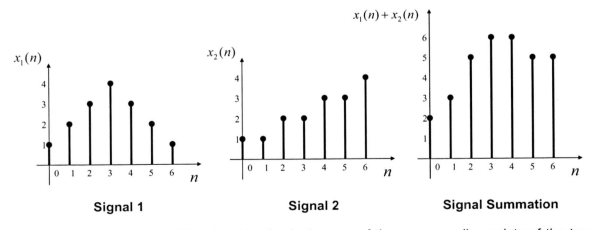

Signal 1   **Signal 2**   **Signal Summation**

Thus, we see that the resulting signal is simply the sum of the corresponding points of the two signals.

9.3.6   Signal Multiplication

The final amplitude transformation that we will examine is signal multiplication. As the name suggests, signal multiplication transforms a given signal through the multiplication of the signal with a second function. Signal multiplication is performed much like signal addition. In this case, however, the corresponding points of the signals are multiplied instead of added. Let's multiply our example signal with a second signal to examine the effects of signal multiplication.

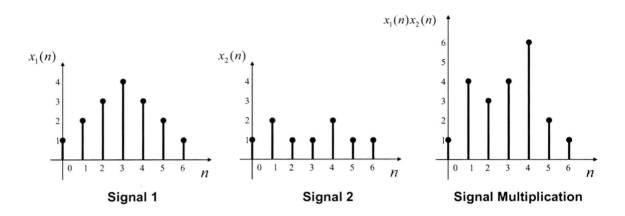

| Signal 1 | Signal 2 | Signal Multiplication |

So, the original signal has been transformed by multiplying it with another signal. It is important to note that signal scaling can be thought of as a signal multiplication where the second signal is a constant.

The previous six transformations illustrate the most common transformations that can be applied to a signal. To further illustrate the plotting of transformed discrete-time signals, let's look at an example.

Example 9.2
_____

Problem: Given the following two signals, $x_1(n)$ and $x_2(n)$, plot the following expression.

$$x_1(n) = \{1,1,2,2,1,1\} \text{ for } n = 0 \text{ to } 5 \qquad x_2(n) = \{1,2,3,3,2,1\} \text{ for } n = 0 \text{ to } 5$$

$$y(n) = 2x_1(n-2) + x_2(2(n-3))$$

Solution: In this example, we have been given two signals and asked to plot an expression that involves the application of several signal transformations. Let's start by plotting the two signals without any transformations.

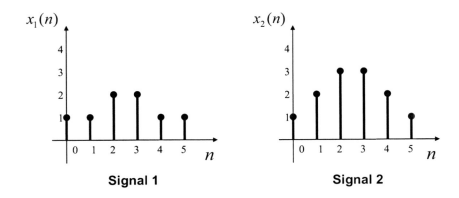

Signal 1   Signal 2

By looking at the expression that is to be plotted, we see that it is the sum of two transformed signals. Let's start by transforming the first signal. By looking at the expression $2x_1(n\text{-}2)$, we see that the signal is time delayed by 2 samples and amplified by 2. Applying these transforms to our plot yields the following.

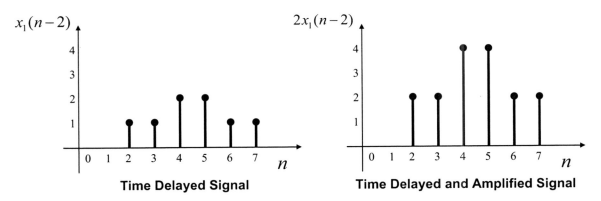

Time Delayed Signal   Time Delayed and Amplified Signal

Now, let's take a look at the second transformed signal. In this case, we have the transformed signal $x_2(2(n\text{-}3))$. Here we see that the original signal has been down-sampled by two and time delayed by three samples. Applying these transforms to our plot yields the following.

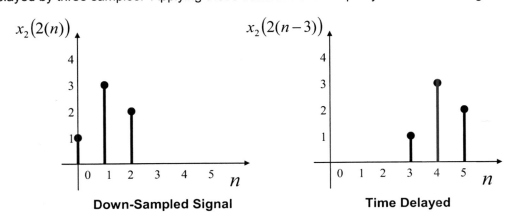

Down-Sampled Signal   Time Delayed

So, now we know what our two transformed signals look like, we can perform the final signal addition and plot the final result. Doing this yields the following plot.

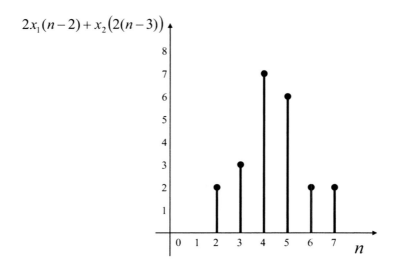

Thus, we have plotted the transformed signal and arrived at the solution.

Now that we have established some basics regarding the plotting and transformation of discrete-time signals, we can begin to look at some operations that we can perform on these signals to obtain and process useful information.

9.4 Working with Signals

In this section, we will begin looking at how we work with signals. We will do this by taking a look at some operations that we can perform to gain additional information about a discrete time signal. In particular, we will introduce the concept of signal summation and show how it can be used to give us additional information about our signal.

9.4.1 Signal Summation

In the previous section, we looked at a signal transformation called signal addition. This transformation required us to add a second signal to a given signal by performing a point-by-point addition of the two signals. It is important not to confuse this transformation with the concept of signal summation. Signal summation involves only one signal and results in a single value rather than a sequence of values. To perform a signal summation, we simply add a range of the terms in the signal together. To properly describe signal summation, we must introduce some new notation. The following equation represents a signal summation.

$$S = \sum_{n=-\infty}^{\infty} x(n)$$

This notation may be new to some of you. The large symbol is the capital Greek letter sigma. It is typically used in mathematics when you are describing the summation of a series of values. The expression under the sigma symbol tells us what variable we are changing in each term of the series and the value at which that variable starts. In this case, the variable is n, and the starting value is negative infinity. The value above the sigma is the final value that the variable will hold in the series, and the expression to the right of the sigma describes the terms that are being summed. So, the above equation tells us that we are summing the terms of $x(n)$ starting at n equal to negative infinity and ending with n equal to infinity. In other words, the result, S, will be equal to the total sum of all the terms in the discrete-time signal $x(n)$. So, let's say we want to represent the summation of the first ten terms of the signal, $x(n)$. We could do that by writing out the terms individually as follows:

$$S = x(0) + x(1) + x(2) + x(3) + x(4) + x(5) + x(6) + x(7) + x(8) + x(9)$$

This notation can easily become long and tiresome to write out, especially if we are summing a larger number of terms. So, instead we can represent the summation using the sigma notation as follows:

$$S = \sum_{n=0}^{9} x(n)$$

Both of the above expressions represent the summation of the first ten terms in the discrete-time signal $x(n)$, but the second representation is more compact and easier to write. Thus, much of the notation that you will encounter in digital signal processing will involve the summation of long series of values using the sigma notation. Therefore, it is important to completely understand what the notation means. Let's look at an example of calculating a signal summation.

Example 9.3

Problem: Given the following signal:

$X(n)$= {1, 4, 7, -2, 6, 2, 1, 9, 3, 5, 2} for n = 0 to 10

Calculate the following signal summations:

$$S_1 = \sum_{n=-\infty}^{\infty} x(n) \qquad S_2 = \sum_{n=0}^{5} x(n)$$

Solution: In this problem, we have been given a discrete-time signal consisting of 11 terms and asked to find two signal summations. This first summation ranges from negative infinity to positive infinity, and the second ranges from zero to five. Let's look at the first summation. It ranges from negative infinity to positive infinity. Thus, we are supposed to calculate the summation of the entire discrete-time signal. In this case, however, our signal does not have an infinite number of terms. It is limited to values of n ranging from zero to ten. All of the terms outside of this range are equal to zero. Thus, they will not contribute any value to the summation, and our summation can be rewritten as:

$$S_1 = \sum_{n=0}^{10} x(n)$$

So, we can obtain our answer by simply summing all of the values from *n=0* to 10. Doing this we get the following:

$$S_1 = x(0) + x(1) + x(2) + x(3) + ... + x(10)$$

$$S_1 = 1 + 4 + 7 - 2 + 6 + 2 + 1 + 9 + 3 + 5 + 2$$

$$S_1 = 38$$

Thus, we find that the total signal summation of *x(n)* is 38. Now, let's find the second sum. In this case, the summation ranges from zero to five. So, we just need to sum the first six terms in the signal to obtain our answer.

$$S_2 = \sum_{n=0}^{5} x(n) = x(0) + x(1) + x(2) + x(3) + x(4) + x(5)$$

$$S_2 = 1 + 4 + 7 - 2 + 6 + 2$$

$$S_2 = 18$$

As we can see, our second sum is equal to 18. So, we have now evaluated both of the required summations and have our answer.

_____

After working through this example, we should have a clear understanding of how the sigma notation works.

One interesting application of signal summation is that by coupling it with the delta function, we can decompose a discrete-time signal. In other words, if we multiply our signal with a time-shifted delta function, we can isolate any specific point in the signal. Likewise, by using the summation notation, we can describe the entire signal as a series of summations as follows:

$$y(n) = \sum_{k=-\infty}^{\infty} \delta(k - n)x(k)$$

This equation basically shows us that to obtain the value of a specific sample in x(n), all we need to do is multiply x(n) by an appropriately time-shifted delta function and sum up the values of the resulting function. Of course, since the delta function has a value for only one sample, the resulting signal will also have a value at only that position, and the sum will consist of the value at that point. So, if we wanted to describe a specific value in the signal, let's say x(8) for example, we could write the following:

$$y(8) = \sum_{k=-\infty}^{\infty} \delta(k-8)x(k)$$

In this case, the delta function is only defined for k=8 where it has a value of one. Thus, multiplying this delta function and the function x(k) together will yield a function with only one value at k=8 that will be equal to x(8). So, we have seen that the summation notation can be useful when trying to isolate specific samples within our signal. In the next sections, we will look at some additional signal operations that use the sigma, or summation notation. These operations can be used to give us more information about a given signal and will prove useful in future signal processing work.

9.4.2   Signal Energy

The total energy contained in a given signal is one piece of information that can be useful. Luckily, this information is easy to obtain with an understanding of signal summation and sigma notation. A given signal's total energy is given by the equation:

$$E = \sum_{n=-\infty}^{\infty} x(n)x^*(n) = \sum_{n=-\infty}^{\infty} |x(n)|^2$$

Where $x^*(n)$ is the complex conjugate of x(n). So, based upon this expression, we can obtain the total signal energy by first taking the square of the magnitude of each of the terms within the signal and then summing these values together. This simple concept is best illustrated by an example.

Example 9.4

Problem: Given the following signal, determine the signal energy.

x(n)={1, 2, 3, 4, 4, 4, 3, 3, 2, 2, 1, 1} for n= 0 to 11

Solution: In this problem, we have been given a discrete-time signal and asked to determine the energy present in the signal. From our previous discussion, we know that the energy contained in a given signal is described by the equation:

$$E = \sum_{n=-\infty}^{\infty} |x(n)|^2$$

In our case, the signal is only defined for values of n between 0 and 11. Since all the other values in the signal are zero, and thus, will not contribute to the summation, we can rewrite our expression as:

$$E = \sum_{n=0}^{11} |x(n)|^2$$

So, all we have to do to find the energy in our signal is to square the magnitude of each term for $n = 0$ to 11 and then sum the results. Thus, our summation expands to:

$$E = 1^2 + 2^2 + 3^2 + 4^2 + 4^2 + 4^2 + 3^2 + 3^2 + 2^2 + 2^2 + 1^2 + 1^2$$

$$E = 1 + 4 + 9 + 16 + 16 + 16 + 9 + 9 + 4 + 4 + 1 + 1$$

$$E = 90 \text{ Joules}$$

With the completion of the summation, we find that the given signal has a total energy of 90 joules.

This example has illustrated one of the useful pieces of information we can obtain by using signal summation. It is important to note that only time-limited signals will have an energy that is finite. Signals that are infinite in length, like the unit step function, for example, will have an infinite total energy. Similarly, any periodic function like the cosine or sine functions will have infinite energy. When dealing with periodic functions, however, there is another useful signal measurement that we can obtain through signal summation. This measurement is average signal power.

9.4.3   Signal Power

When dealing with periodic signals, finding the signal energy does little to help us describe the signal characteristics. This is due to the fact that non time-limited periodic functions all have infinite energy. However, we can turn to average power to help us describe the signal. The average power of a periodic signal is calculated over just one period. Thus, we are once again dealing with a more manageable number of samples. The method of calculation is also very similar to that of energy. If you think about it, power is simply the amount of energy delivered per unit time. Therefore, it should be no surprise that the average power of a periodic signal is simply the energy of one period divided by the number of samples in the period. Mathematically speaking, we can describe the average power as:

$$P = \frac{1}{N} \sum_{n=0}^{N-1} |x(n)|^2$$

Where N is the number of samples in one period of the signal. So, as we can see from the equation, we find the power by determining the energy within one period of the signal and then dividing that energy by the number of samples in the period. Let's look at an example to clarify the process.

Example 9.5

Problem: Find the average power of the following periodic signal.

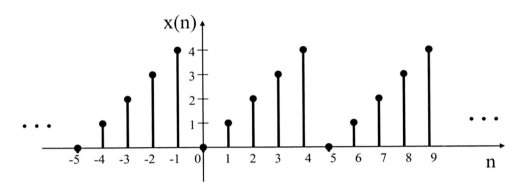

Solution: In this problem, we have been given the plot of a periodic discrete-time signal and asked to determine the average power of the signal. We know that to find the power, we must first find the energy for one period of the signal. To do that, we must first determine the length of one period of our signal. By looking at the plot, we can easily see that there are exactly five terms in one period. These terms are {0, 1, 2, 3, 4} occurring at positions n= 0 to 4. Thus, we know that the length of one period is N=5. Now, we can rewrite our expression for average power as the following:

$$P = \frac{1}{5}\sum_{n=0}^{4}|x(n)|^2$$

So, we need to calculate the energy in the first five terms of the signal and divide that value by five. Performing this operation gives us:

$$P = \frac{1}{5}\left(0^2 + 1^2 + 2^2 + 3^2 + 4^2\right)$$

$$P = \frac{1}{5}\left(0 + 1 + 4 + 9 + 16\right)$$

$$P = \frac{1}{5}\left(30\right)$$

$$P = 6 \text{ Watts}$$

Thus, the average power in the signal plotted above is 6 watts.

This example should provide us with a good understanding of the method used for the calculation of average power in a periodic signal. Thus, we have seen two pieces of information that we can determine through the use of signal summation. Next, we will examine a special type of signal that is often encountered in signal processing applications, the geometric series.

9.4.4   The Geometric Series

So far, we have been looking at summation applications that apply to determining information for general time-limited and periodic signals. However, the summations occur very frequently in signal processing applications. One special type of summation you will see frequently is the geometric series. The geometric series is a summation of the form:

$$S = 1 + \alpha + \alpha^2 + \alpha^3 + ... + \alpha^n$$

Where α is a constant and n is an arbitrary integer. As you can see, the geometric series is simply a series of values that are determined by raising some constant value to increasingly large powers. This expression can also be rewritten using sigma notation as follows:

$$S = \sum_{k=0}^{n} \alpha^k$$

Since sums like these come up often in signal processing applications, it is important to be familiar with them. As you can imagine, summing all of the terms of a geometric series individually would be a time consuming and tedious task. Luckily there is an easier closed-form solution to the geometric series. Let's see how it works. First, let's take the original expression given above for a geometric series:

$$S = 1 + \alpha + \alpha^2 + \alpha^3 + ... + \alpha^n$$

Now, let's multiply this sum by alpha. This will give us:

$$\alpha S = \alpha + \alpha^2 + \alpha^3 + ... + \alpha^n + \alpha^{n+1}$$

If we then subtract this sequence from the original sequence we will get:

$$S - \alpha S = \left(1 + \alpha + \alpha^2 + \alpha^3 + ... + \alpha^n\right) - \left(\alpha + \alpha^2 + \alpha^3 + ... + \alpha^n + \alpha^{n+1}\right)$$

In this subtraction, all but the first and last terms will cancel out. This leaves us with:

$$S - \alpha S = 1 - \alpha^{n+1}$$

Now, we can factor an S out of the two terms on the left, giving us:

$$S(1 - \alpha) = 1 - \alpha^{n+1}$$

And, finally, dividing both sides by (1-α) will yield:

$$S = \frac{1-\alpha^{n+1}}{1-\alpha}$$

Thus, by plugging in the appropriate values of n and α, we can easily determine the sum of a geometric series. Let's look at a quick example.

Example 9.6

Problem: Find the sum of the following geometric series.

$$S = 1 + 3 + 9 + 27 + ... + 3^{20}$$

Solution: In this example, we were given a geometric series and asked to find the sum. By looking at the given series we can easily see that it is a series in which three is being raised to successively larger powers. These exponents range from 0 to 20. Thus, we know that $\alpha=3$ and $n=20$. Plugging this into our closed-form solution, we get:

$$S = \frac{1-3^{21}}{1-3}$$

Simplifying this expression, we get:

$$S = \frac{-1.0460 \times 10^{10}}{-2} = 5.2302 \times 10^{9}$$

Thus, we have determined the sum of our series.

The closed form solution used in this example will work for any geometric series. However, we can simplify things slightly if the conditions allow. When we encounter a series that has a value of alpha that is between zero and one and has a relatively large value for n, we can make a simplifying assumption. Since a value between zero and one becomes smaller as it is raised to successively larger powers, we can assume that it will approach a value of zero for sufficiently large powers. Thus, if alpha lies in this range and n is relatively large, we can assume that the portion of our closed form solution that involves alpha raised to the $n+1$ power is going to be essentially zero. Thus, our closed-form solution will simplify to:

$$S = \frac{1}{1-\alpha} \quad \text{for } 0 < \alpha < 1 \text{ and large } n$$

So, as long as the conditions are met, we can use this simplified solution to determine the sum of a geometric series. Let's look at an example where we can use this simplification.

Example 9.7

Problem: Find the sum of the following geometric series.

$$S = 1 + \frac{1}{2} + \frac{1}{4} + \frac{1}{8} + \dots + \left(\frac{1}{2}\right)^{20}$$

Solution: In this example, we were given a geometric series and asked to find the sum. In this case, however, we can note that the value being raised to increasing powers, α, is between zero and one. In addition, the final power, n, is relatively large. Thus, we can use our simplified equation in this case. From the summation, we know that $\alpha=0.5$ and $n=20$. Plugging this into our simplified closed-form solution, we get:

$$S = \frac{1}{1 - 0.5}$$

Simplifying this expression, we get:

$$S = \frac{1}{0.5} = 2$$

Thus, we have determined the sum of our series.

We can verify our approximation by plugging into our exact form of the equation and seeing if the results agree. Doing this yields:

$$S = \frac{1 - 0.5^{21}}{1 - 0.5} = \frac{1 - 4.768x10^{-7}}{0.5} \approx \frac{1}{0.5} = 2$$

Thus, the result from the exact equation is approximately equal to the result of the simplified equation.

So, we have seen that the geometric series can be determined by a closed-form solution. We will wrap up this section by looking at one interesting application of the geometric series.

Example 9.8

Problem: Most of us are familiar with a chess board. It has 64 squares. If we begin by placing a single penny on the first square of the board, and then doubling the amount for each successive square (2 pennies on the second square, 4 pennies on the third square, etc.), how much money would be **required** to complete the operation?

Solution: In this example we have been given an interesting question. We must determine how much **money** would be required to complete the operation of placing increasing amounts of money on **each** **successive** square of a chess board. This particular process can be described as a geometric **series**. Each successive square of the board receives one power of two more pennies than the previous square. Thus, we can describe the total amount of money required to fill all 64 squares as:

$$S = 1 + 2 + 4 + 8 + ... + 2^{63}$$

So, we have a geometric series with $\alpha=2$ and $n=63$. If we plug these values into our closed-form solution for the geometric series we will get:

$$S = \frac{1 - 2^{64}}{1 - 2}$$

$$S = \frac{1 - 1.8447x10^{19}}{1 - 2} = 1.8447x10^{19} \text{ pennies}$$

Thus, in order to finish the activity, we would need $1.8447x10^{17}$ dollars. That's 184,470 trillion dollars. So, it is unlikely we could ever completely fill the chess board.

This example illustrates one practical problem that can be solved using the geometric series. When working in the field of signal processing, you will encounter many other situations where you will find the geometric series a useful tool.

9.5   An Introduction to Systems

Up to this point, we have been talking primarily about signals. We have looked at some different kinds of signals, investigated some of the transforms we can apply to signals, and learned how to use summation notation to help us process and gain useful information from signals. However, without hardware and software to help us process and extract information, signals would be of little use to us. This is where systems come into the picture. Systems are the physical tools that we use when we work with or process signals. The rest of this chapter will be dedicated to the study of signal processing systems.

9.5.1   What is a System?

So, what exactly is a system? We know that it is a physical tool that helps us work with signals. That definition is a good start. We can think of a system in its most basic sense as a physical device that takes one or more signals as input, performs some kind of operation on these input signals, and then provides one or more signals as output. So, essentially, systems are what we use when we want to perform a given operation on a signal. Whether that operation is extracting audio from a radio wave or removing the hiss from an audio tape recording, systems exist to work with signals. If we want to approach systems in the mathematical sense, we can model a system as an operator $T[\bullet]$ that takes an input signal, for example $x(t)$ and transforms it into a different signal, let's call it $y(t)$, the output signal. Thus, we can describe the output of a system as:

$$y(t) = T[x(t)]$$

This can also be described by using a block diagram as follows:

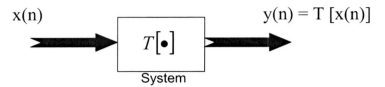

$$x(n) \qquad\qquad\qquad\qquad y(n) = T\,[x(n)]$$

$$T[\bullet]$$

System

So, now we know that when we talk about a system, we are simply referring to a physical device, or mathematical model of a device, that takes input in the form of signals and provides transformed signal as output. However, as was the case with signals, there are many different types of systems. In many cases, these systems classifications are based on the type of signals that the system processes, but systems can also be broken down based on their basic design. A few of the more common system classifications are discussed in the next section.

9.5.2 Classification of Systems

When we discussed signals earlier in the chapter, we noted that it was important to classify signals into various groups in order to make them easier to understand and process. This practice of classification is also useful when dealing with systems. Since there are so many different kinds of signals, you would expect that there would be a wide array of approaches that could be taken to process them. As a result, there exists a wide variety of different system classifications.

9.5.2.1 Analog and Discrete-Time Systems

One of the most intuitive ways to classify systems is to divide them up based upon the types of signals that they process. One of the most obvious classifications that can be made based upon signal type is analog vs. discrete-time systems. As you might guess, an analog system accepts analog signals as input and provides analog signals as output. Thus, analog systems are used exclusively in analog signal processing (ASP) applications. Analog systems would include any device that works with analog signals like televisions, radios, and cassette tape players. A discrete-time system, on the other hand, is one that accepts discrete-time signals as input and provides discrete-time signals as output. So, as we would expect, discrete-time systems see wide use in the digital signal processing (DSP) arena. Discrete-time systems work exclusively with discrete-time signals, and would include devices like digital cellular phones, computers, and CD players. So, one major difference between system classes is the type of signal that they work with.

9.5.2.2 Linear Discrete-Time Systems

Another important system classification is linearity. Basically, a system is said to be linear if it obeys the principle of superposition. To understand superposition, let's begin by considering the example of a system that takes a single discrete-time signal as input and provides a single discrete-time signal as output. Now, let's pass two separate signals through the system, call them $x_1(n)$ and $x_2(n)$, and record the two resulting outputs, $y_1(n)$ and $y_2(n)$. Now, we run a third signal through the system that consists of the sum of the first two signals, $x_1(n)+x_2(n)$. If the system satisfies the principle of superposition, the output of this third signal, $y_3(n)$, will be equal to the sum of the two previous outputs, $y_1(n)+y_2(n)$. In other words:

$$T\big[x_1(n) + x_2(n)\big] = y_3(n) = y_1(n) + y_2(n) = T\big[x_1(n)\big] + T\big[x_2(n)\big]$$

So, for a system to be linear, the transformation it provides for a signal that is composed of the sum of two other signals must be the same as the sum of the two signals transformed independently. This is illustrated graphically below.

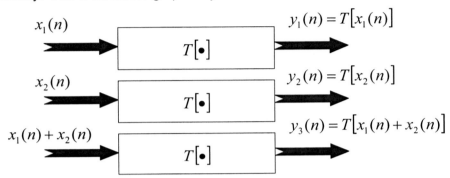

$$T[\bullet] \text{ is linear if and only if } T\big[x_1(n) + x_2(n)\big] = T\big[x_1(n)\big] + T\big[x_2(n)\big]$$

Thus, we can see what is required to satisfy superposition and make a system linear. Let's look at an example.

Example 9.9

Problem: Determine whether the following systems are linear or not.

   a. $T\big[x(n)\big] = 3x(n)$

   b. $T\big[x(n)\big] = 2x(n) + 1$

Use the following signals as system inputs.

$$x_1(n) = \{1,2,3,4\}$$

$$x_2(n) = \{2,2,3,3\}$$

Solution: In this example, we have been asked to determine whether two systems are linear or not. We have also been given two input signals to use to test linearity. All that needs to be done to solve this problem is to test to see if either of the systems satisfies the principle of superposition. Let's start by testing the first system.

Passing the first signal through the system yields:

$$T[x_1(n)] = 3x_1(n) = \{(3)(1),(3)(2),(3)(3),(3)(4)\} = \{3,6,9,12\}$$

Likewise, passing the second signal through the system will give us:

$$T[x_2(n)] = 3x_2(n) = \{(3)(2),(3)(2),(3)(3),(3)(3)\} = \{6,6,9,9\}$$

Now, let's process the sum of the two inputs:

$$x_1(n) + x_2(n) = \{1,2,3,4\} + \{2,2,3,3\} = \{3,4,6,7\}$$

$$T[x_1(n) + x_2(n)] = \{(3)(3),(3)(4),(3)(6),(3)(7)\} = \{9,12,18,21\}$$

If our system is linear, this output will be equal to the sum of our previous two outputs.

$$T[x_1(n) + x_2(n)] \overset{?}{=} T[x_1(n)] + T[x_2(n)]$$

$$\{9,12,18,21\} \overset{?}{=} \{3,6,9,12\} + \{6,6,9,9\}$$

$$\{9,12,18,21\} \overset{?}{=} \{3+6,6+6,9+9,12+9\} = \{9,12,18,21\}$$

So, as we can see, the two sides of this expression are equal. Thus, superposition is satisfied and our system is linear.

Now, we can investigate the second system.

Passing the first signal through the system yields:

$$T[x_1(n)] = 2x_1(n) + 1 = \{(2)(1)+1,(2)(2)+1,(2)(3)+1,(2)(4)+1\} = \{3,5,7,9\}$$

Likewise, passing the second signal through the system will give us:

$$T[x_2(n)] = 2x_2(n) + 1 = \{(2)(2)+1,(2)(2)+1,(2)(3)+1,(2)(3)+1\} = \{5,5,7,7\}$$

Now, let's process the sum of the two inputs:

$$x_1(n) + x_2(n) = \{1,2,3,4\} + \{2,2,3,3\} = \{3,4,6,7\}$$

$$T[x_1(n) + x_2(n)] = \{(2)(3)+1,(2)(4)+1,(2)(6)+1,(2)(7)+1\} = \{7,9,13,15\}$$

If our system is linear, this output will be equal to the sum of our previous two outputs.

$$T[x_1(n) + x_2(n)] \overset{?}{=} T[x_1(n)] + T[x_2(n)]$$

$${7,9,13,15} \overset{?}{=} {3,5,7,9} + {5,5,7,7}$$

$${7,9,13,15} \overset{?}{=} {3+5,5+5,7+7,9+7} = {8,10,14,16}$$

In this case, the two sides of our expression are not equal. This means that superposition is not satisfied by this system. Therefore, the second system is non-linear.

So, here we see how we can determine whether a given system is linear or not by using superposition.

9.5.2.3   Linear, Time-Invariant, Discrete-Time Systems

Time-invariant systems form another important system class. So, what is a time-invariant system? Let's suppose we get an output $y(n)$ from a system when it is provided with an input of $x(n)$. If we then provide the same input, $x(n)$, at some later time, our system input would look like $x(n-k)$, where k is some time delay. Now, if the output provided by the system for this delayed input is the same as the output it gave for the non-delayed input except for a time delay in the output, the system is called a time-invariant system. Basically, what we mean is that if the system behaves in exactly the same manner, irrespective of the instant at which the input is applied, it can be called a time invariant system. We can show this graphically as follows:

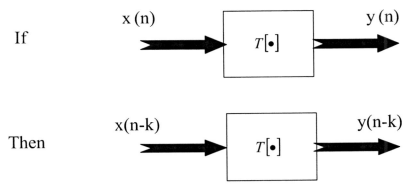

So, we can see that if our system is time invariant, the output that the system provided will be the same whether we provide the input now or three weeks from now. This is obviously an important trait for a system to have. If our system behaved differently depending on when we gave it input it would not be very useful.

Systems that are linear and time invariant are referred to as linear time-invariant or LTI systems. These systems, as we will see in the next section, form an important class of systems.

9.5.2.4   Causal Systems

A system is said to be causal if it does not anticipate its input. In other words, if the output of a system at some arbitrary time t_0 depends only on signals that have occurred at time instants before t_0, then the system is causal. So, the output of a causal system at the present time depends only on the present and past values of the input and not its future values. All of the systems we have seen so far are causal systems. The following system would be an example of a non-causal system.

$$T[x(n)] = x(n+1)$$

In this example, we can clearly see that the output at a given value of n depends upon an input sample that has not yet entered the system, $n+1$. Thus, the system is said to be not-causal.

9.5.2.5   Stable Systems

A system is termed *bounded-input, bounded-output* stable, or BIBO stable, if for some input signal, $x(n)$, such that the magnitude of $x(n) < k_1$ (some constant), it produces an output signal, $y(n)$, such that the magnitude of $y(n) < k_2$ (some constant). Basically, what this means is that if a system receives an input signal with an amplitude that is not infinity and produces an output signal with an amplitude less than infinity, it is said to be a BIBO stable system. BIBO stability is usually a desirable quality in a system.

9.5.2.6   Feedback Systems

The concept of a feedback system is one that is of great importance in engineering. Feedback control systems are used in many electrical engineering applications from guided missiles to thermostats. In a feedback system, the output signal is fed back and added to the input to the system. Thus, the input to the system will change depending upon the system's output. At some point, equilibrium is reached. At that point, small changes from equilibrium can be resisted by the feedback loop and the system will be resistant to small variations. The following diagram illustrates the concept of a feedback system.

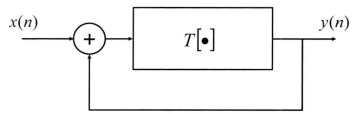

9.5.2.7   Systems With and Without Memory

The final system classification that we will discuss is system memory. A system is said to be memoryless if its output at any instant of time depends on only the input at that instant of time. Otherwise the system is said to have memory. In other words, a system with memory will "remember" some of its previous inputs. These input values will have some effect of the system's output values. In the analog realm, an example of a memoryless system is a resistor, R. If we take the current through the resistor to be the input, $x(t,)$ and the voltage dropped across that resistor to be the output, *y(t)*, then by Ohms law, y(t) = Rx(t). Thus, the output depends only upon the instantaneous input. In the case of a capacitor, however, the output is not so simple. It depends upon the previous charge stored in the capacitor, making the capacitor a system with memory.

9.6 Working with Systems

In the previous section, we learned what a system was, and we discovered that there are a lot of different kinds of systems. In this section we are going to focus on one specific group of systems, namely those that are linear and time invariant (LTI Systems). Systems in this group are widely used in signal processing applications. This is, in part, because there are convenient methods for determining the output for systems of this type. In the following sections we will look at how we can model systems of this type and investigate the methods used for determining their output.

9.6.1   Determining a System's Output – Convolution

Since we just mentioned that there is a convenient method for finding the output for LTI systems, you are probably wondering what this convenient approach is. The output for an LTI system can be determined by performing an operation that is known as convolution. The output, *y(n)*, is calculated by convolving the input signal sequence, *x(n)*, with the system's impulse response, *h(n)*. This process is illustrated below.

$$x(n) \quad \boxed{\quad h(n) \quad} \quad y(n) = x(n) * h(n)$$

Now, since we have just introduced some new terminology here, let's make sure we understand what is going on. We know that convolution is an operation that will help us obtain the output for our system, but what is a system's impulse response? You may remember that earlier in this chapter we discussed two special functions. One of these was the unit impulse, or delta function ($\delta(n)$). The system's impulse response is simply the output (or response) that the system provides when its input is a delta (impulse) function. Thus, if we feed a delta function into our LTI system, we would expect to find that the system's output was exactly equal to its impulse response. Thinking of the impulse response from a slightly different perspective, we can say that the impulse response simply provides an abstract characterization of a given system in the form of a discrete sequence of values.

Now that we understand what an impulse response is, how exactly does this convolution operation work? The convolution of two sequences usually denoted with an asterisk (*), is written as follows:

$$y(n) = x(n) * h(n) = \sum_{k=-\infty}^{\infty} x(k)h(n-k)$$

So, once again we have encountered the use of sigma notation to indicate signal summation. In this case, we are summing the product of two signals. The first signal is the input sequence, and the second signal is a time-reversed and shifted version of the impulse response. We will look at several ways to evaluate this expression in a moment, but first, let's take a look at some important properties of the convolution operation.

9.6.1.1   Properties of Convolution

Convolution is a linear operation, and as such, it possesses several useful properties. Much like any other linear operation, convolution satisfies the commutative, associative and distributive properties. Let's look at how these properties are applied to the convolution operation and how we can use them to simplify convolution expressions.

We will start with the simplest property of a linear system, the commutative property. This property basically states that changing the order of the terms in the convolution expression will not change the system output. This can be stated mathematically as follows:

$$x(n) * h(n) = h(n) * x(n)$$

This means that we can interchange the order of the impulse response and the input signal in cases where doing so will make the computations easier. We can also describe this property graphically in the following block diagram.

Another useful property of the convolution operator is that it is associative. In other words, if there are multiple convolutions in an expression, the order in which they are evaluated does not change the results. Mathematically speaking, this means:

$$\left[x(n) * h_1(n)\right] * h_2(n) = x(n) * \left[h_1(n) * h_2(n)\right]$$

This relationship is particularly useful in situations where a signal is fed into one system and the output of this system is fed into a second system. Determining the output of the whole configuration would involve convolving the input with the first impulse response and then the second. However, since the convolution operator is associative, we can find the convolution of the two impulse responses first and then convolve that new response with the input signal. This allows us to simplify the system configuration and develop an impulse response that describes the two systems. We can then use this new impulse response for any later input signals instead of having to recalculate both convolutions for every new input signal. This simplification is illustrated in the block diagram below.

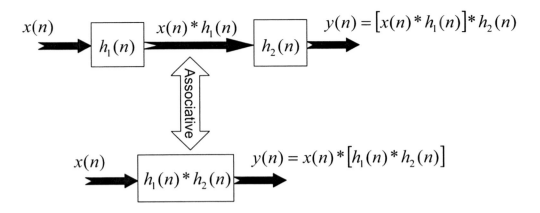

Finally, the convolution operator possesses the distributive property. This means that a convolution operation can be distributed across the terms in a summation. Thus, we can write the following:

$$x(n) * [h_1(n) + h_2(n)] = [x(n) * h_1(n)] + [x(n) * h_2(n)]$$

This property can prove useful when you encounter a group of systems that process an input signal in parallel. Normally, you would have to take the convolution of the input signal and the impulse response of the first system and then add that result to the convolution of the input signal and the second system's impulse response. Thanks to the distributive property, however, we can develop an impulse response for the entire parallel arrangement of systems by adding the individual impulse responses together. This overall impulse response can then be convolved with any input signal to determine the configurations' overall output. This simplification is illustrated in the block diagram below.

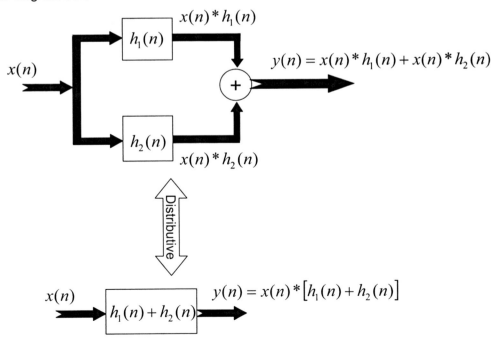

　　　The properties that we have discussed here are not all of the properties that the convolution operation possesses, but they are some of the most important. As the diagrams illustrated, these properties can be used to simplify the calculation of the final output when multiple systems are involved. So, now that we understand the basics, let's take a look at how we can actually calculate the output of a system using convolution.

9.6.1.2　Methods of Computation

　　　There are many methods that have been developed to evaluate the convolution. The most obvious approach is to simply evaluate the convolution equation directly. While this is possible, and in some cases the only way to reliably compute the convolution, it is often a more complicated approach than is needed. To properly evaluate the expression, we would need to calculate the summation for all possible values of n. This can become a difficult task. In most cases, however, we can use slightly more intuitive methods to arrive at the result. This is particularly true when working with short, time-limited sequences like we will encounter in this text. Thus, we will examine several of these approaches in this section. Let's begin by taking a graphical look at what the convolution equation is actually describing. To do that, we will define two sample signals. We will let $x(n) = \{1, 2, 3, 4\}$ and $h(n) = \{4, 3, 2, 1\}$. Thus, we have an input signal and an impulse response for the system. Let's look at the plots of these signals.

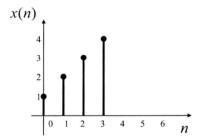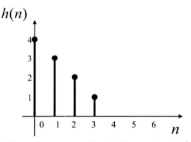

　　　Now, by looking at the equation for the convolution, we see that the output for a particular instance in time, n, is determined by summing up the product of the original input signal and a time-reversed and shifted copy of the impulse response from negative to positive infinity. So, if we time-reverse the impulse response of the system and plot it below the input, we will see how the signals line up.

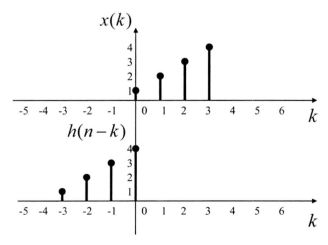

If we were to perform a point-by-point multiplication on these signals we can see that we will only get non-zero results where the two functions overlap. Thus, we only need to worry about finding output values of n that shift the impulse response in such a way as to make it overlap with the input sequence. Note that negative values of n will shift the impulse response to the left and positive values will shift it to the right. We can also see from our two equations that the first point at which they will overlap is at n=0. Thus, we can begin calculating the convolution output for y(0). We can also see that the last point at which the signals will have terms in common will be when n=6. So, our convolution output will consist of seven terms. As it turns out, we can always determine the length of our output signal from the lengths of the input and impulse response sequences. The output sequence length will be equal to the sum of the lengths of the input and impulse response sequences minus one. So, if our input is of length L_x and our impulse response is of length L_h, then the output length, L_y can be defined as:

$$L_y = (L_x + L_h) - 1$$

In this particular case, both our input signal and impulse response have a length of four, so we would expect our output to be of length:

$$(4 + 4) - 1 = 7$$

This length agrees with what we just observed on the graph of the two signals. So, now that we know what to expect, let's begin calculating the outputs. We will begin by calculating the output for n=0. In this case our signals will overlap in only one point. Thus, the product of these two signals will yield only one non-zero term. So, the sum of all the values in the product will be equal to that one term. In this case, we find that:

$$y(0) = (4)(1) = 4$$

The first term in our system output is 4.

We can now shift the impulse response to the right and calculate the value for y(1). This shift has two points that overlap. Thus, the output will be the sum of the products between these two points. We will continue to shift, multiply and add the results until we have calculated the value for y(6). At this point we will have all of the value in the output sequence. The following plots illustrate the calculation of the terms in this convolution.

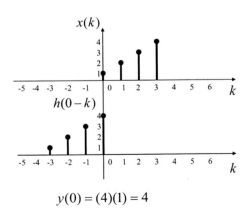

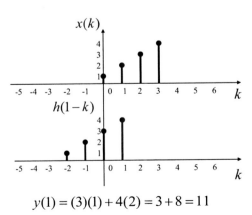

$$y(0) = (4)(1) = 4$$

$$y(1) = (3)(1) + 4(2) = 3 + 8 = 11$$

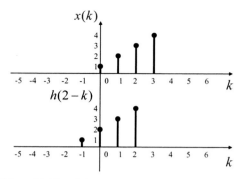

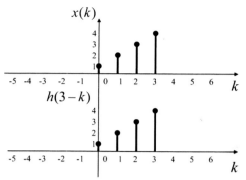

$$y(2) = (2)(1) + 3(2) + (4)(3) = 2 + 6 + 12 = 20$$

$$y(3) = (1)(1) + (2)(2) + (3)(3) + (4)(4) = 1 + 4 + 9 + 16 = 30$$

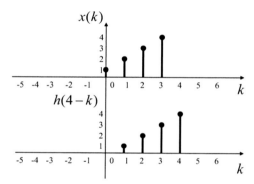

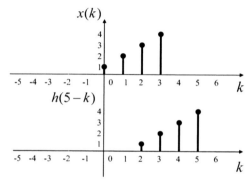

$$y(4) = (1)(2) + (2)(3) + (3)(4) = 2 + 6 + 12 = 20$$

$$y(5) = (1)(3) + (2)(4) = 3 + 8 = 11$$

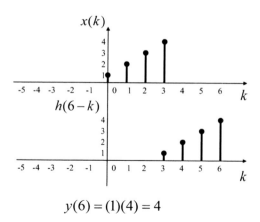

$$y(6) = (1)(4) = 4$$

With the completion of the calculation for the seventh term, *y(6)*, we now have the complete output sequence.

$$y(n) = \{4, 11, 20, 30, 20, 11, 4\}$$

Finally, we can plot our output sequence to get an idea of what it looks like.

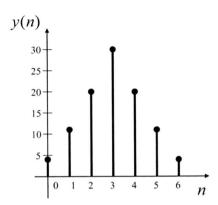

So, we have graphically determined what the output of our system would be. By examining a convolution graphically, it should now be easier to see what is going on. While the graphical approach will give us the correct solution, it works better as a teaching tool to illustrate the process. There are more efficient ways to calculate a convolution. Let's take a look at a few non-graphical ways to calculate the convolution.

The first approach we will look at is called the sliding sequence method (sometimes also referred to as the slide rule method). In this approach, we write the first sequence down along the edge of a sheet of paper. We then write the time-reversed version of the second signal down on a second sheet of paper. By overlapping the two pieces of paper and sliding them past each other you can see the corresponding terms for each shift. The process is best illustrated by an example.

Example 9.10

Problem: Using the sliding sequence method, compute the output of the following system.

$$x(n) = \{1,2,3,4\} \qquad \boxed{h(n) = \{4,3,2,1\}} \qquad y(n)$$

Solution: In this example, we have been given an input and impulse response sequence and asked to find the output. Obviously we need to use a convolution to obtain the output. We have been asked to use the sliding sequence method for the computation in this case. Thus, we begin by writing our input sequence vertically on the edge of one sheet of paper and the time-reversed impulse response along the edge of another sheet of paper as illustrated here:

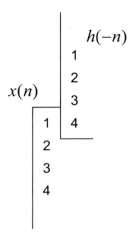

Now, we simply multiply the corresponding terms and sum the results. Once we calculate one term, we shift the right-hand sheet of paper down by one number and repeat the multiplication and summation. We will continue shifting until there are no more overlapping numbers. Doing this we get the following:

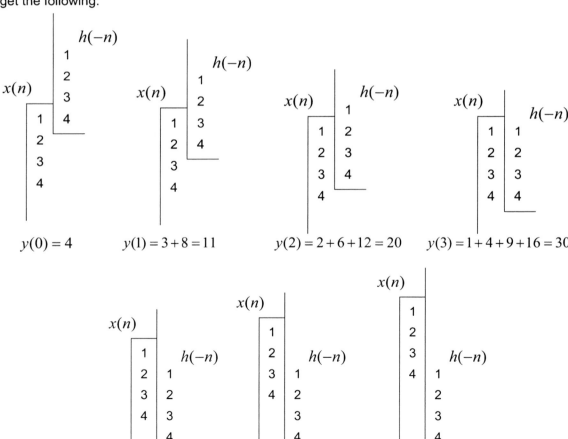

$$y(0) = 4 \qquad y(1) = 3 + 8 = 11 \qquad y(2) = 2 + 6 + 12 = 20 \qquad y(3) = 1 + 4 + 9 + 16 = 30$$

$$y(4) = 2 + 6 + 12 = 20 \qquad y(5) = 3 + 8 = 11 \qquad y(6) = 4$$

So, we find that the output of this system is:

$$y(n) = \{4,11,20,30,20,11,4\}$$

This is the same output that we calculated previously by the graphical method.

Another method for the calculation of a convolution uses matrix multiplication. In this approach, we create a matrix where each column is a shifted version of the input sequence. We create as many columns in the matrix as there are elements in the impulse response. Once we create this matrix, we can determine the system output by multiplying the matrix by a column vector containing the impulse response. Let's look at how this approach works.

Example 9.11

Problem: Using the matrix method, compute the output of the following system.

$$x(n) = \{1,2,3,4\} \quad\longrightarrow\quad \boxed{h(n) = \{4,3,2,1\}} \quad\longrightarrow\quad y(n)$$

Solution: In this example, we have been given the same system we worked with in example 9.10. This time, however, we need to use the matrix approach to calculate the convolution. Let's begin by constructing our matrix columns from shifted versions of the input sequence. Since our impulse response is four elements long, we will need four columns in our matrix. Thus our matrix will look like this:

$$\begin{bmatrix} 1 & 0 & 0 & 0 \\ 2 & 1 & 0 & 0 \\ 3 & 2 & 1 & 0 \\ 4 & 3 & 2 & 1 \\ 0 & 4 & 3 & 2 \\ 0 & 0 & 4 & 3 \\ 0 & 0 & 0 & 4 \end{bmatrix}$$

Now, we can get our output sequence simply by multiplying this matrix by the impulse response vector. Doing this yields the following:

$$\begin{bmatrix} 1 & 0 & 0 & 0 \\ 2 & 1 & 0 & 0 \\ 3 & 2 & 1 & 0 \\ 4 & 3 & 2 & 1 \\ 0 & 4 & 3 & 2 \\ 0 & 0 & 4 & 3 \\ 0 & 0 & 0 & 4 \end{bmatrix} \begin{bmatrix} 4 \\ 3 \\ 2 \\ 1 \end{bmatrix} = \begin{bmatrix} 4 \\ 8+3 \\ 12+6+2 \\ 16+9+4+1 \\ 12+6+2 \\ 8+3 \\ 4 \end{bmatrix} = \begin{bmatrix} 4 \\ 11 \\ 20 \\ 30 \\ 20 \\ 11 \\ 4 \end{bmatrix}$$

So, our output sequence is:

$$y(n) = \{4,11,20,30,20,11,4\}$$

This is exactly the same answer we obtained using both of the previous methods.

The final method that we will discuss for the calculation of the convolution is a method that we call the long multiplication method. We call it this because it works much like the basic method for long multiplication that you learned in grade school. If you recall the way in which two multiple-digit numbers are multiplied you will remember that we first multiply of all the top digits with the right-most bottom digit. We write each product in a row moving from right to left. When the first digit is done, we multiply all the upper digits with the next lower digit. The product values are stored in the next row down. This row is shifted one position to the left and padded with a zero. As it turns out, we can compute a convolution in much the same way. Let's look at an example to see exactly how it works.

Example 9.12

Problem: Using the multiplication method, compute the output of the following system.

$$x(n) = \{1,2,3,4\} \qquad h(n) = \{4,3,2,1\} \qquad y(n)$$

Solution: In this example we are once again dealing with the system from example 9.10. In this case, we need to use the multiplication approach to get our solution. So, let's set up our sequences.

$$\begin{array}{cccc} 1 & 2 & 3 & 4 \\ 4 & 3 & 2 & 1 \\ \hline \end{array}$$

Now, beginning with the right-most lower term, we will multiply the upper terms moving from the right-most to the left-most. We will record each of the products in the first row below the line. The first row of products will be:

```
1  2  3  4
4  3  2  1
─────────
1  2  3  4
```

Now, we place a zero in the right-most position of the second row under the line and get the products for the second bottom term. This gives us:

```
   1  2  3  4
   4  3  2  1
   ─────────
   1  2  3  4
2  4  6  8  0
```

Processing the products with the third term gives us:

```
      1  2  3  4
      4  3  2  1
   ─────────────
      1  2  3  4
   2  4  6  8  0
3  6  9  12 0  0
```

Finally, we find the products for the final term:

```
         1  2  3  4
         4  3  2  1
      ──────────────
         1  2  3  4
      2  4  6  8  0
   3  6  8  12 0  0
4  8  12 16 0  0  0
```

The output sequence is simply the sum of the rows that we have created. Thus, we find that the system output is:

$$
\begin{array}{rrrrrrr}
 & & & 1 & 2 & 3 & 4 \\
 & & & 4 & 3 & 2 & 1 \\
\hline
 & & & 1 & 2 & 3 & 4 \\
 & & 2 & 4 & 6 & 8 & 0 \\
 & 3 & 6 & 8 & 12 & 0 & 0 \\
4 & 8 & 12 & 16 & 0 & 0 & 0 \\
\hline
4 & 11 & 20 & 30 & 20 & 11 & 4
\end{array}
$$

And:

$$y(n) = \{4,11,20,30,20,11,4\}$$

So, we have discovered that this method also produces the same results as our previous approaches.

As you can see, we have now found five ways to evaluate a convolution and find the output of an LTI system. First, we could evaluate the convolution equation directly. If we opt not to do that, we could try a graphical approach by plotting out the signals and shifting them past each other. Or, finally, we could try one of the three shortcut approaches that were outlined in this section. The approach that should be used is determined by both the signals being processed and the method you are most comfortable using.

9.6.2   Sequence Correlation

Another signal processing application you will encounter often is correlation. It is used in a wide array of applications ranging from communication systems to radar. The correlation of two different signals (sometimes called cross-correlation) is performed by using the following equation:

$$R_{xy}(n) = \sum_{k=-\infty}^{\infty} x(k)y(k-n)$$

One of the first things we should notice when we look at this equation is that it is nearly identical to the equation for convolution. In fact, the only difference is that the second term in the summation involves $k-n$ instead of $n-k$. This basically means that the cross-correlation of two signals is performed exactly like the convolution of two signals with the one exception being that the second signal is shifted but not time-reversed. Thus, when two signals are correlated, they are simply shifted past one another. At each shift, the signals are multiplied and the products are summed together to yield one term in the output sequence. Therefore, all of the methods that we learned previously for the purposes of calculating the results of a convolution can be adapted to work with correlation by simply reversing the order of the second signal in the calculation (This undoes the reversal that was already built into the approach). So, let's look at some examples of cross-correlation.   We will adapt the three shortcut approaches we previously learned for convolution to calculate the correlation of our signals.

Example 9.13

Problem: Given the following two signals, compute the cross-correlation using the sequence sliding technique.

$$x(n) = \{1,2,3,2\}$$
$$y(n) = \{1,2,3\}$$

Solution: In this example, we are going to adapt the sequence sliding method to correlation. As you recall, to perform the sequence sliding method, we write one sequence along one sheet of paper and then a time-reversed version of the second signal on the second sheet of paper. In the case of correlation, we simply do not time reverse the second signal. Doing this gives us the following arrangement:

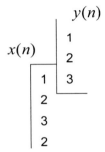

It is important to note that in this configuration, both of the sequences begin at n = 0. Thus, the second sequence must be shifted back by two samples to get the first overlapping pair. As a result, the output sequence from this correlation will begin at n = -2. So, if we start with n = -2 and slide our sequences past each other just as we did in the convolution example, we will get the following results:

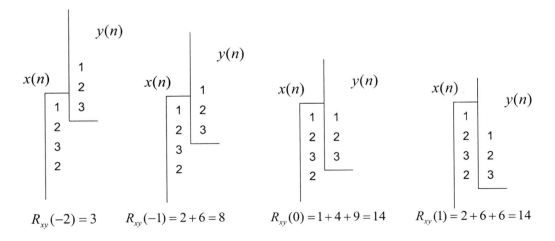

$$R_{xy}(-2) = 3 \qquad R_{xy}(-1) = 2+6 = 8 \qquad R_{xy}(0) = 1+4+9 = 14 \qquad R_{xy}(1) = 2+6+6 = 14$$

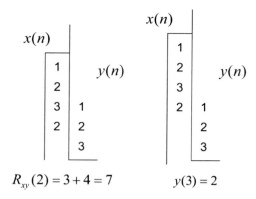

$$R_{xy}(2) = 3 + 4 = 7 \qquad y(3) = 2$$

So, we find that the results of this cross-correlation will be:

$$R_{xy}(n) = \{3,8,14,14,7,2\} \text{ for } n = -2 \text{ to } 3$$

Example 9.14

Problem: Given the following two signals, compute the cross-correlation using the sequence sliding technique.

$$x(n) = \{1,2,3,2\}$$
$$y(n) = \{1,2,3\}$$

Solution: In this example, we will adapt the matrix multiplication method for use in determining correlations. If we look back at the matrix multiplication method used to find the convolution in example 9.11, we can see that we first construct a matrix consisting of shifted versions of the first signal and then multiply that matrix by a vector containing the second signal. To make this approach work for correlations, all we have to do is reverse the order of the vector containing the second signal. So, setting up the matrix multiplication for the signals given will give us the following:

$$R_{xy} = \begin{bmatrix} 1 & 0 & 0 \\ 2 & 1 & 0 \\ 3 & 2 & 1 \\ 2 & 3 & 2 \\ 0 & 2 & 3 \\ 0 & 0 & 2 \end{bmatrix} \begin{bmatrix} 3 \\ 2 \\ 1 \end{bmatrix}$$

Now, we simply multiply and get:

$$R_{xy} = \begin{bmatrix} 1 & 0 & 0 \\ 2 & 1 & 0 \\ 3 & 2 & 1 \\ 2 & 3 & 2 \\ 0 & 2 & 3 \\ 0 & 0 & 2 \end{bmatrix} \begin{bmatrix} 3 \\ 2 \\ 1 \end{bmatrix} = \begin{bmatrix} 3 \\ 6+2 \\ 9+4+1 \\ 6+6+2 \\ 4+3 \\ 2 \end{bmatrix} = \begin{bmatrix} 3 \\ 8 \\ 14 \\ 14 \\ 7 \\ 2 \end{bmatrix}$$

Thus, our output is:

$$R_{xy}(n) = \{3,8,14,14,7,2\} \text{ for } n = -2 \text{ to } 3$$

Example 9.15

Problem: Given the following two signals, compute the cross-correlation using the long multiplication technique.

$$x(n) = \{1,2,3,2\}$$
$$y(n) = \{1,2,3\}$$

Solution: In this example, we will adapt the long multiplication method for use in the calculation of a correlation. You will remember that the long multiplication method for convolution required that we line up the two signals and then multiply the values in much the same way as we would multiply two multi-digit numbers. The basic layout is the same for correlation, with the exception that we need to reverse the order of the lower sequence. Doing this with our signals will give us the following:

```
1  2  3  2
   3  2  1
_____
```

Now, if we perform the multiplications and sum the resulting columns, we will get:

```
          1   2   3   2
              3   2   1
      _____
      0   0   1   2   3   2
      0   2   4   6   4   0
      3   6   9   6   0   0
      _____
      3   8  14  14   7   2
```

So, once again, our solution is:

$$R_{xy}(n) = \{3,8,14,14,7,2\} \text{ for } n = -2 \text{ to } 3$$

Thus, we have seen that we can use the same methods to calculate correlations that we used for convolutions by making some minor modifications. It is important to note that, unlike convolution, the correlation does not satisfy the commutative property. In other words, the order of the sequences matters in correlation. So, in most cases,

$$R_{xy} \neq R_{yx}.$$

If the sequences are correlated in the reverse order, the output sequence will not be the same. Instead, the resulting sequence will be a reversed version of the output given when the signals were correlated in the other order.

An additional variant of the correlation equation arises when the signal is correlated with itself rather than another signal. This is called the autocorrelation function and is represented as follows:

$$R_{xx}(n) = \sum_{k=-\infty}^{\infty} x(k)x(k-n).$$

So, we can see that if we correlate a signal with another signal, we call it a cross-correlation, and if we correlate a signal with itself, we call the operation an autocorrelation.

We have now seen how to calculate a correlation, but why would we need to use it? We already know how to calculate the output of a system by using the convolution, so what it the correlation's purpose? One of the main pieces of information that the correlation of two signals provides is an estimate of signal similarity. For example, if we want to find out if there are any portions of a given signal that are similar to a signal that we are looking for, we can simply correlate the signals. The correlation will give us an idea of what parts of the signal are similar to our reference and for how long they are similar. One important application of the correlation is in radar. Radar works by broadcasting a signal and then listening for the signal to echo back. The returning echoes indicate that the signal was reflected by an object. Thus, if we can detect these echoes in the return signal, we can use the time delay between the signal transmission and reception to calculate a distance to the object. To see how correlation can help us in this kind of application, lets' say we transmit a signal similar to the one below.

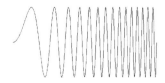

Now, we listen for any return echoes of our signal. However, the returning echoes are often weak in amplitude and buried in signal noise. So, our received signal might look like the following.

It is difficult to see if there are any echoes of our reference signal in this received signal. However, if we correlate our reference signal with the noisy received signal, we will get the following results.

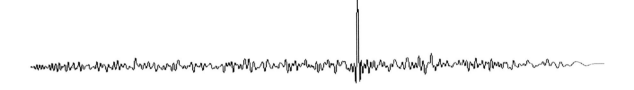

In this signal, we can easily see that there is a large peak. This peak indicates that there is a portion of the received signal that is highly similar to our transmitted signal. So, we have found an echo of our original signal within the received signal. Now, we can use the timing of this echo to determine the distance to our target. If we transmit a chain of signals, we can track the object over time and determine the object's distance, speed, and direction of movement. So, as we have just seen, the correlation function can be very useful in an actual application.

9.6.3   Describing a System – Difference Equations

We have now spent some time discussing signals and systems. In addition, we have looked at methods of working with both signals and systems. In our studies, we have found that by using systems we can make use of and modify the information contained in signals. Now, we need to investigate ways in which we can describe these systems. One way that we already looked at is the impulse response. We found that the impulse response could be used to describe the behavior of a system. Since it gives the output from an input impulse, it can be used both to describe a system and, through convolution, to determine the output given other input sequences. While this approach is of some use, there is another method that is often used to describe and model discrete-time systems. It is based on a method that we commonly use to describe analog systems, differential equations.

Differential equations are built on the principles of calculus and make use of integrals and derivatives. For example, assume we are studying the decay of a radioactive element. As you know, radioactive elements break down over time. This breakdown is quantified by a measure called a half-life, the amount of time required for half of the existing amount to decay. So, if we have some initial amount, R_0, of our radioactive substance, then we would expect that after some period of time, t_{hl}, we will have only half the original amount and after an additional period of time, only half of the remaining amount would be left, and so on. The differential equation describing such an exponential decay is written as:

$$\frac{dR}{dt} = -\tau R$$

Where R is the amount of radioactive substance and τ is the rate of decay. Basically, this equation tells us that the change in the amount of radioactive material over time is going to be equal to the existing amount of material multiplied by a decay rate. To be able to determine the actual amount of material that will be present at a given time, t, we must solve this differential equation. It turns out that the solution to this equation is:

$$R(t) = R_0 e^{-\tau t}$$

Where R_0 is the initial amount of radioactive material, τ is the decay constant, and t is time. So, differential equations can be used to describe systems in continuous time, but what do we use to describe discrete-time systems? Discrete-time systems are described by a discrete approximation of differential equations called difference equations. Difference equations use addition instead of integration and subtraction instead of differentiation (thus the name difference equations). For a given linear, time-invariant, discrete-time system we write the difference equation describing the system as summation of the terms of the input and output signals as follows:

$$a_0 y(n) + a_1 y(n-1) + ... + a_N y(n-N) = b_0 x(n) + b_1 x(n-1) + ... + b_M x(n-M)$$

Where the coefficients a_i and b_j are constants, y(n) indicates system outputs, x(n) indicates system inputs, and $N \geq M$. This equation can be written more compactly using summation notation as follows:

$$\sum_{i=0}^{N} a_i y(n-i) = \sum_{j=0}^{M} b_j x(n-j)$$

Thus, we can describe a system with a difference equation. To see how difference equations work, let's return to our radioactive decay example. Assume that we begin with a sample of 200 grams of a radioactive substance. If we take a measurement of the amount that remains each day for a period of time, we can determine the rate at which the element is decaying. Let's say that we find that five percent of the material decays each day. Using this information we can write the following difference equation to describe the operation.

$$R(n) = R(n-1) - 0.05R(n-1)$$

Combining like terms, we can write:

$$R(n) = 0.95R(n-1)$$

So, we see that at a given measuring instance $R(n)$, we will find ninety-five percent of the amount of material we had at the previous measuring instance. Since in this case each measuring instance represents one day, we could determine the amount of material left after a given number of days by plugging into the difference equation for each successive day until we reach the target day. For example, if we want to know how much of our 200 gram sample $(R(0)=200)$ will remain after 5 days we can plug into the equation as follows:

$$R(1) = 0.95R(0) = (0.95)(200) = 190 \text{ grams}$$

$$R(2) = 0.95R(1) = (0.95)(190) = 180.5 \text{ grams}$$

$$R(3) = 0.95R(2) = (0.95)(180.5) = 171.475 \text{ grams}$$

$$R(4) = 0.95R(3) = (0.95)(171.475) = 162.901 \text{ grams}$$

$$R(5) = 0.95R(4) = (0.95)(162.901) = 154.756 \text{ grams}$$

Thus, after five days we would expect 154.756 grams of our material to remain. This example shows us that difference equations can be useful. Instances of exponential growth and decay similar to the example shown here occur in population studies, charging and discharging of capacitors, bank interest calculations and many other applications. In the remainder of this section, we will explore two additional application examples in which difference equations will play a role in both describing the systems and arriving at a solution.

Example 9.16

Problem: Assume you have been given a pair of newborn rabbits. These rabbits are great breeders and will reproduce every month yielding an additional pair of rabbits. Assuming that newborn rabbits cannot reproduce until their second month and that each newborn pair contains one male and one female rabbit, how many rabbits will you have after a year (12 months).

Solution: We begin our solution by analyzing the problem. We know at our initial time $(n=0)$ we have two newborn rabbits. Since they cannot reproduce until the second month, there will be no change in the number of rabbits for the first month. After this initial period, we can expect the first pair to produce a new pair every month. Each of these pairs will produce an additional pair when they reach their second month and so forth. So, for a given month, $y(n)$, we will have, at a minimum the same number of rabbits as we had the previous month, $y(n-1)$. In addition, any rabbit that existed two months ago, $y(n-2)$, will produce a new pair. So, our system can be modeled by the following difference equation:

$$y(n) = y(n-1) + y(n-2)$$

Thus, the number of rabbits in a given month is simply the sum of the number of rabbits that existed in the previous two months. If we assume the initial conditions of $y(0)=1$ pair and $y(-1)=0$ pairs, we can plug into our difference equation and determine the number of rabbits that we will have after 12 months. Doing this yields the following.

$$y(1) = y(0) + y(-1) = 1 + 0 = 1 \text{ pair}$$

$$y(2) = y(1) + y(0) = 1 + 1 = 2 \text{ pairs}$$

$$y(3) = y(2) + y(1) = 2 + 1 = 3 \text{ pairs}$$

$$y(4) = y(3) + y(2) = 3 + 2 = 5 \text{ pairs}$$
$$y(5) = y(4) + y(3) = 5 + 3 = 8 \text{ pairs}$$
$$y(6) = y(5) + y(4) = 8 + 5 = 13 \text{pairs}$$
$$y(7) = y(6) + y(5) = 13 + 8 = 21 \text{ pairs}$$
$$y(8) = y(7) + y(6) = 21 + 13 = 34 \text{ pairs}$$
$$y(9) = y(8) + y(7) = 34 + 21 = 55 \text{ pairs}$$
$$y(10) = y(9) + y(8) = 55 + 34 = 89 \text{ pairs}$$
$$y(11) = y(10) + y(9) = 89 + 55 = 144 \text{ pairs}$$
$$y(12) = y(11) + y(10) = 144 + 89 = 233 \text{ pairs}$$

So, after only one year, our single prolific pair of rabbits has grown to 233 pairs or 466 rabbits. On a side note, the sequence described by this system in which each term is the sum of the two previous terms is referred to as the Fibonacci series.

Example 9.17

Problem: In signal processing many occasions will arise in which you need to smooth out a signal. One method of doing this is through the application of a moving average, or smoothing filter. A moving average filter is a system that achieves a smoothing operation by replacing each signal value with the average value of itself and a given number of its neighbors. By doing so, sudden changes in the signal amplitude, often caused by high frequency noise, can be minimized to reduce their impact on the signal. A typical moving average filter is defined as follows:

$$y(n) = \frac{1}{M} \sum_{k=0}^{M-1} x(n-k)$$

Where M is the width of the filter (number of neighbors to average). This value is usually a relatively small odd number, such as 3, 5 or 7. Using this equation for the moving average filter, find the result of applying a moving average filter of width 3 to the following input sequence.

$$x(n) = \{1,2,4,6,8,10,12\} \text{ for } n = 0 \text{ to } 6$$

Solution: In this example, we are dealing with a common signal processing application, the filtering of an input signal. In this case, we are applying a smoothing filter to our input. We have been given the equation for a moving average filter and told we will be using a filter of width 3. With this information, we can expand the filter notation and write out a difference equation for our specific filter as follows:

$$y(n) = \frac{1}{3}\left[x(n) + x(n-1) + x(n-2) \right]$$

Since our input signal ranges from 0 to 6, we can assume any signal before and after the defined values are zero. Thus, x(-1)=0 and x(-2)=0. Using these initial conditions and our input signal, we can calculate our system output by plugging into the difference equation until there aren't any more terms in our input involved in the output. Thus, we begin at y(0).

$$y(0) = \frac{1}{3}[x(0) + x(-1) + x(-2)] = \frac{1}{3}[1 + 0 + 0] = \frac{1}{3}$$

$$y(1) = \frac{1}{3}[x(1) + x(0) + x(-1)] = \frac{1}{3}[2 + 1 + 0] = 1$$

$$y(2) = \frac{1}{3}[x(2) + x(1) + x(0)] = \frac{1}{3}[4 + 2 + 1] = \frac{7}{3}$$

$$y(3) = \frac{1}{3}[x(3) + x(2) + x(1)] = \frac{1}{3}[6 + 4 + 2] = 4$$

$$y(4) = \frac{1}{3}[x(4) + x(3) + x(2)] = \frac{1}{3}[8 + 6 + 4] = 6$$

$$y(5) = \frac{1}{3}[x(5) + x(4) + x(3)] = \frac{1}{3}[10 + 8 + 6] = 8$$

$$y(6) = \frac{1}{3}[x(6) + x(5) + x(4)] = \frac{1}{3}[12 + 10 + 8] = 10$$

$$y(7) = \frac{1}{3}[x(7) + x(6) + x(5)] = \frac{1}{3}[0 + 12 + 10] = \frac{22}{3}$$

$$y(8) = \frac{1}{3}[x(8) + x(7) + x(6)] = \frac{1}{3}[0 + 0 + 12] = 4$$

So, the output of a width 3 moving average filter when applied to the given input signal is:

$$y(n) = \left\{ \frac{1}{3}, 1, \frac{7}{3}, 4, 6, 8, 10, \frac{22}{3}, 4 \right\}$$

These examples serve to illustrate the usefulness of the difference equation in the description, modeling, and solution of signal processing problems. The second example illustrates one of the most important aspects of signal processing, the filter. The application of various filters to an input signal is a routine task in signal processing. Many of the devices you work with on a daily basis make use of filters.

9.7 Filters and Filtering

In general, we can think of filters as specialized systems that transform an input signal into a desired format, which serves as an output signal. While filtering can apply to almost any transformation, thus making basically any system a filter, the most commonly used filters can be classified based on how they modify the frequency content of a signal. As we have seen, filters often operate on the signal in the time-domain. However, we often think of filters in terms of frequency. In other words, when we design a filter, we may be building a system that acts upon the signal's individual samples, but we are actually attempting to manipulate the frequency content of the signal. In this section we will look at some of the basic ways a filter may be designed to manipulate the frequency of a signal. To do this, we will show the filter's response in terms of frequency. A value of one in the frequency response of a filter indicates that that frequency is passed through the filter unmodified. Values greater than one indicate amplifications and values less than one indicate the attenuation of the corresponding frequency. As you recall, filters are applied in the time-domain by performing a convolution. However, in the frequency spectrum, the frequency spectrum of the signal is multiplied by the response of the filter. There are five basic categories that frequency manipulation filters can be grouped into. These are

   1. All-Pass Filters
   2. Low-Pass Filters
   3. High-Pass Filters
   4. Band-Pass Filters
   5. Band-Stop Filters

We will examine each of these categories in more detail below. Let's begin by looking at the simplest category, the all-pass filter. An all-pass filter, as the name suggests, allows all of the frequencies in the input signal to pass through. However, all of the frequencies may be amplified or attenuated. The defining characteristic of an all-pass filter is that its frequency response is a constant value. The phase of the signal, however, can be changed by such a filter. The following figure illustrates the frequency response of an allpass filter.

In this case, the all-pass filter passes all of the frequencies without any attenuation or amplification. All-pass filters are typically used in systems that require some phase changes within the signal but all of the frequencies must be preserved. One purpose of an all-pass filter would be signal delay. In many cases we may need to delay one or more of our input signals. This delay can be accomplished by using an all-pass filter. Many audio effects require such a phase modification or delay, and thus, you would find all-pass filters in audio effects generators among numerous other systems.

   The second category of filters is low-pass filters. Filters in the low-pass category block all frequencies above a given cutoff frequency while allowing the low frequencies in the signal to pass through. The following figure depicts the frequency response of an ideal low-pass filter.

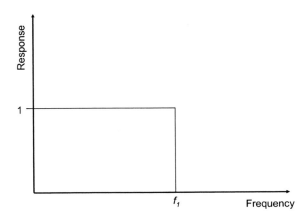

   As you can see from the figure, the frequencies above the cutoff frequency of f_1 are blocked and the frequencies lower than the cutoff are passed without modification. Low-pass filters have numerous uses in electronics systems. When applied to a signal, they have a smoothing effect. Sharp changes in a signal are representative of high frequency components in the signal. By removing these high frequency components, the signal is smoothed. The moving average filter that you worked with in example 9.17 is one example of a low-pass filter. Low-pass filters are used in applications ranging from power supplies to radio demodulation to audio processing and are one of the most common filter types that you will encounter.
   The compliment of a low-pass filter is a high-pass filter. It performs the exact opposite task of its low-pass counterpart by allowing high frequencies to pass while blocking the lower frequencies. The frequency response of a high-pass filter is shown below.

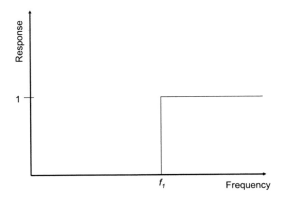

   So, any frequency above f_1 is passed through the filter with no modification while those lower are blocked. High-pass filters also see wide use in various applications. One major use is in audio. The speaker boxes on most home stereos consist of at a minimum, two speakers. One of these speakers is designed to reproduce high frequencies and the other is better suited to lower frequencies. In order to drive these speakers, the input signal must be split into a low frequency and high frequency component. This is done using a set of filters. Typically this is one high-pass and one low-pass filter. The output of the high-pass filter feeds into the high frequency speaker (tweeter) and the output of the low-pass filter feeds into the low frequency speaker (woofer). This is just one of many applications in which a high-pass frequency is employed.

If you have ever tuned in a radio or television station, you have used a band-pass filter. When it is necessary to only allow a small range of frequencies to pass through a system, a filter of the band-pass variety is needed. The frequency response of a band-pass filter is shown below.

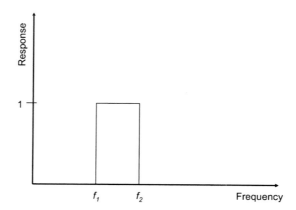

As you can see, the frequencies between f_1 and f_2 are passed through while all others are blocked. There are many applications for such a filter. As we just mentioned, a band-pass filter is particularly useful in frequency selection circuits. In a radio receiver, for example, the antenna picks up the transmissions of all available radio stations. This signal is fed into the receiver circuitry. In order to send only one station's audio signal to the speakers, we must remove the other information from the input signal. This is done using a user-tunable band-pass filter. When you tune your radio to a selected frequency, you are actually selecting the center frequency of a band-pass filter. It is this filter that allows you to hear the station that you want and not all of them at once.

In some instances, there is a specific frequency that causes problems. This is particularly the case when communication lines run too close to electric power lines. The electric field in the power lines induces a voltage in the communications lines. This results in a hum that is at the same frequency as the electric power in the transmission line. Thus, in many cases we find an undesirable 60 Hz hum in our signal. This can be removed by our final class of frequency filter, the band-stop filter. The response of this filter is shown below.

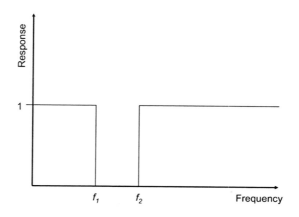

In this case, the frequencies between f1 and f2 are blocked and all other frequencies are allowed through. It is easy to see how this type of filter would be useful to remove our unwanted hum as well as any other frequency that might be causing a problem.

The five types of frequency filters that we have just discussed are seen in countless signal processing applications. All of these filters can be useful, and in fact, most systems utilize multiple types of these filters. It should be noted that the filter frequency responses shown above are ideal cases. In reality the sharpness of frequency selection shown are not achievable. Instead, we develop approximations to these filters that achieve the level of selectivity needed for the desired application. An example of the response of an actual low-pass filter is shown below.

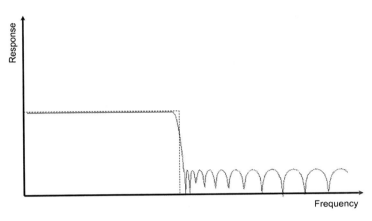

As you can see, the response is not as clean as the ideal case, but it achieves the desired result.

9.8 Chapter Summary

In this chapter, we introduced the basics of signals, systems, and signal processing.

- We began by explaining that signals can be defined as a function or transmission that conveys information. We found that we are surrounded by signals and make use of them in virtually every aspect of our lives.

- Next, we looked at some ways that we could classify the many signals we encounter. We looked at classifications based on number of dimensions and channels, whether the signal was real or complex, random or deterministic or analog or discrete-time.

- We then found that signal processing is the branch of electrical engineering that is dedicated to working with the information contained in signals.

- Next, we introduced two special signals, the delta function and the unit step function.

- Then we looked at discrete-time signals and some of the transforms that we could apply to them. We saw that we can transform the independent variable by time shifting, time reversal, and time scaling and the dependent variable by amplitude scaling, signal multiplication, and signal addition.

- Once we understood the basics of signals, we investigated some applications of signals. We looked at signal summation and how we could use it to find the energy and average power of a signal.

- In addition, we looked at the geometric series, and learned how to find a closed-form solution to such a series. We found that this series is encountered in many signal processing applications.

- With a clear understanding of signals, we began to investigate systems. We found that systems are the means by which we manipulate signals. We looked at the classifications of systems. They can be classified based upon the types of signals that they process and the way in which they are designed.

- We found that one way we could find the output to a system is through convolution. By convolving the systems impulse response (the output when the system receives an impulse as input) with its input, we will obtain the system output.

- Next, we investigated an operation similar to the convolution in which the second signal was not time-reversed. We found that this operation was called correlation and that it could be used to find areas of similarity between two signals. We also looked at how this could be useful in the application of radar.

- The difference equation was discussed next. We found it could be used to describe a system in terms of its inputs and outputs. We saw how we could use difference equations to describe, model, and solve signal processing problems.

- We wrapped up the chapter by discussing the concept of filters. We learned about the five basic categories of frequency modifying filters, the all-pass, low-pass, high-pass, band-pass, and band-stop filters. The frequency response for each filter class was investigated and areas of application were noted.

9.9 Review Exercises

Section 9.1 – Introduction and Overview

1. What is a signal?

2. Name at least three of the categories that we use to classify signals.

3. What is the difference between a continuous and discrete-time signal?

4. Find the even and odd parts of the following signals

   a. $4x^3 - 2x^2 + 3$

   b. $(x-1)^3 + \cos(3x)$

   c. $(2x^2 - 1)^2 + 3x + 1$

   d. $3\sin(x) + 4\cos(x) - (x+1)^2$

   e. $(x^3 - 4)^3 - (x+1)^2$

   f. $3x^5 - 2x^2 + 17$

   g. $(x^3 - 2x^2 + 17)^2$

   h. $4(x^3 - 2x)^2 + \cos(4x) - \dfrac{1}{2}\sin(x)$

5. Give three examples of analog signals.

6. Give three examples of digital signals.

7. Describe the advantages of using digital signal processing as apposed to analog signal processing.

Section 9.2 – Introduction to Some Special Signals

8. What does a delta function look like as an analog signal? What about as discrete-time signal?

9. What does a unit step function look like as an analog signal? What about as discrete-time signal?

10. Plot the following delta and step functions.

   a. $\delta(n)$

   b. $\delta(n-3)$

   c. $u(n)$

   d. $u(n-4)$

   e. $3u(n+2)$

   f. $2\delta(n+1)$

Section 9.3 – Plotting Discrete-Time Signals

---

11. Given the following discrete-time signal

$$x(n) = \{1,2,3,3,3,2,1\} \text{ for } n = 0 \text{ to } 6,$$

    a. Sketch the result of up-sampling the signal by 2.

$$y(n) = x\left(\frac{n}{2}\right)$$

    b. Sketch the result of down-sampling the signal by 2.

$$y(n) = x(2n)$$

    c. Sketch $y(n) = x(n-4)$

    d. Sketch $y(n) = 3x(n+2)$

12. Given the discrete-time signals

$$x(n) = \{1,2,3,4,3,2,1\} \text{ for } n = 0 \text{ to } 6$$

$$g(n) = \{4,3,2,1,1,2,3,4\} \text{ for } n = 0 \text{ to } 7$$

Sketch the following from n = 0 to 10.

    a. $y(n) = x(n) + g(n)$
    b. $y(n) = 2g(n) - x(n)$
    c. $y(n) = x(n) - u(n)$
    d. $y(n) = x(n)u(n-2) + g(n)u(2-n)$
    e. $y(n) = x(2n) - g\left(\frac{n}{2}\right) + \delta(n-3)$
    f. $y(n) = \delta(n) + 3\delta(n-1) + 2\delta(n-2) - 4\delta(n-5) + \delta(n-6)g(n)$
    g. $y(n) = \delta(n) + 2\delta(n-1) + 3u(n-4)$
    h. $y(n) = x(n) + 4u(4-n)g(n)$

Section 9.4 – Working With Signals

___

13. Given the signal $x(n) = \{1,2,0,4,-2,1,7,6,5,4,3,2,1,0,0,1,4,8,-2,3,-1,-1,-3,4\}$, find the following signal summations:

a. $S = \displaystyle\sum_{n=-\infty}^{\infty} x(n)$

b. $S = \displaystyle\sum_{n=0}^{23} x(n)$

c. $S = \displaystyle\sum_{n=4}^{8} x(n)$

d. $S = \displaystyle\sum_{n=0}^{10} x(n)$

e. $S = \displaystyle\sum_{n=3}^{5} x(n)$

f. $S = \displaystyle\sum_{n=7}^{15} x(n)$

14. Determine the signal energy for the following signals:

a. $x(n) = \{1,2,3,4,3,2,1\}$
b. $x(n) = \{-1,2,-2,3,-3,4\}$
c. $x(n) = \{2,4,6,8,6,-7,1\}$
d. $x(n) = \{-7,-6,-5,-4,-3,-2,-1,0,1,2,3,4,5,6,7\}$
e. $x(n) = \{4,-1,2,-1,0,3,7,9\}$
f. $x(n) = u(n) - u(n-7)$

15. Find the average power for the following periodic signals:

a. $x(n) = \{\ldots,-1,2,4,3,6,-1,2,4,\ldots\}$
b. $x(n) = \{\ldots,0,1,2,3,4,3,2,1,0,1,2,\ldots\}$
c. $x(n) = \{\ldots,-5,2,3,5,6,-5,2,3,\ldots\}$
d. $x(n) = \{\ldots,-2,1,4,5,6,-2,-1,\ldots\}$
e. $x(n) = \{\ldots,3,4,5,6,1,2,3,4,5,6,\ldots\}$
f. $x(n) = 3j^n$
g. $x(n) = \sin(n\pi)$
h. $x(n) = \cos\left(n\frac{\pi}{2}\right)$

16. Find the sum of the following geometric series:

a. $S = 1 + 5 + 25 + \ldots + (5)^{12}$

b. $S = 1 + \frac{1}{2} + \frac{1}{4} + \ldots + \left(\frac{1}{2}\right)^{40}$

c. $S = 1 + 7 + 49 + \ldots + (7)^{8}$

d. $S = 1 + \frac{1}{5} + \frac{1}{25} + \ldots + \left(\frac{1}{5}\right)^{50}$

e. $S = 1 + (-2) + 4 + \ldots + (-2)^{15}$

f. $S = 1 + \left(-\frac{1}{2}\right) + \frac{1}{4} + \ldots + \left(-\frac{1}{2}\right)^{75}$

Section 9.5 – An Introduction to Systems

17. What is a system?

18. Use the input sequences $x_1(n) = \{1,2,3,4,5\}$ and $x_2(n) = \{1,1,2,3,3\}$ to determine whether the following discrete-time systems are linear or not:

a. $y(n) = 5x(n)$
b. $y(n) = 3x(2n)$
c. $y(n) = 4x(n) + 7$
d. $y(n) = 2x(n) + 3x(n-1)$
e. $y(n) = x(n)x(n-1)$
f. $y(n) = 3x(n) + 1$
g. $y(n) = 4x(n-2)$
h. $y(n) = 3x\left(\frac{n}{2}\right)$

19. What is required for a system to be causal?

20. Describe the construction of a feedback system.

21. What is meant by the term memory when it is applied to a system?

Section 9.6 – Working With Systems

22. Given the linear, time-invariant system shown below,

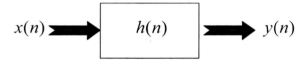

$$x(n) \quad\longrightarrow\quad \boxed{h(n)} \quad\longrightarrow\quad y(n)$$

determine the output signal when:

a. $x(n) = \{1,2,3\}$ and $h(n) = \{2,4,-1\}$
b. $x(n) = \{1,2,1,-1\}$ and $h(n) = \{5,0,-2\}$
c. $x(n) = \{-1,2,4,-3,7\}$ and $h(n) = \{4,3,2,1\}$
d. $x(n) = \{3,7,-2,1\}$ and $h(n) = \{-2,4,-6\}$
e. $x(n) = \{2,4,6,8,6,4,2\}$ and $h(n) = \{1,2,3,2\}$
f. $x(n) = \{\frac{1}{2},2,\frac{1}{3},7\}$ and $h(n) = \{\frac{1}{2},-\frac{1}{2},1,-1\}$

23. Given the input signal $x(n) = \{1,2,3,4,3,2,1\}$, and the impulse responses
$h_1(n) = \{-1,-2,-1,0,1\}$, $h_2(n) = \{1,-1,2\}$, and $h_3(n) = \{-4,2,-3,1\}$,
compute the following:

    a. $x(n) * h_1(n)$ b. $x(n) * h_2(n)$

    c. $x(n) * h_1(n) * h_2(n)$ d. $x(n) * [h_1(n) + h_2(n)]$

    e. $x(n) * h_3(n) * h_2(n) * h_1(n)$ f. $x(n) * h_1(n) + x(n) * h_2(n) + x(n) * h_3(n)$

24. Perform the correlation of the following signal pairs

    a. $x(n) = \{1,2,3,4,3,2,1\}$ and $y(n) = \{1,2,3,4\}$
    b. $x(n) = \{1,2,3,1,2,3,1,2,3\}$ and $y(n) = \{1,2,3\}$
    c. $x(n) = \{4,-1,2,3,4\}$ and $y(n) = \{1,2,3\}$
    d. $x(n) = \{-1,-2,-3,-2,4,6,8\}$ and $y(n) = \{2,-1,3\}$

25. Given the input signal $x(n) = \{1,2,3,4,5,4,3,2,1\}$, compute the requested output term for each difference equation.

    a. $y(n) = x(n) + 2x(n-1) + 3x(n-2)$ Find $y(7)$
    b. $y(n) = 2y(n-1) + 2y(n-2)$ given $y(0) = 1$ and $y(-1) = 3$ Find $y(5)$
    c. $y(n) = 3x(n) + 2x(n-1) - y(n-1)$, given $y(0) = 5$ Find $y(7)$
    d. $y(n) = -2x(n) + 3x(n-1) + y(n-1) - y(n-2)$, given $y(0) = y(-1) = 0$ Find $y(3)$

Section 9.6 – Working With Systems

26. List the 5 basic types of filters and sketch their frequency responses.

Chapter 10. Applied DSP – Sound Processing

Chapter Outline

10.1 Introduction

In the previous chapter, we introduced some of the basic concepts that relate to digital signal processing. We outlined the information that you need to begin looking at and processing many of the signals that surround you. In this chapter and the next, we will look at two types of signals that you encounter on a daily basis. The first of these, which will be covered in this chapter, is sound (images will be covered in chapter 11). On any given day, we are bombarded with sound. When we hear a car horn, the sirens on an emergency vehicle, or the doorbell ring, we gain important information about our surroundings. The car horn may indicate some impending danger on the road, the sirens tell you that an emergency vehicle needs to get through traffic, and the doorbell lets you know that someone is waiting at your door. In each of these situations and the countless others that occur throughout the day, sounds provide us with an enormous amount of information about our environment. Sound signals go far beyond simply providing us with basic environmental information, however. If you think about it, sound signals also form the basis of human communication. We convey huge amounts of information to each other every day simply by producing specific sounds in recognized patterns (language). In fact, we have developed specialized equipment such as the telephone and radio to help us send these signals from one place to another over long distances. Since sound plays such a large role in our lives, it should be no surprise that the processing of sound is an important field of digital signal processing. Before we can investigate some of the many signal processing techniques that we can apply to sound signals, however, we must first get a better understanding of the signal that we are working with. So, before we can talk about the signal processing techniques that can be applied to sound signals, we must answer questions like: What exactly is sound? How is sound produced? What are some of the basic characteristics of a given sound signal? By answering these questions we will gain a better understanding of the signals we are dealing with, and thus, be better equipped to understand the reasoning behind some of the signal processing techniques applied to sound signals.

10.2 The Basics of Sound

In order to work with sound signals effectively, we must understand a few fundamental things. First, we need to know what a sound wave is. Once we understand that, we can look at the numerous ways in which these audio signals are created. Finally, we must understand some of the distinguishing characteristics of sound signals. This will include basic measurements like frequency and sound pressure (amplitude). This section will provide some of this important background information. So, let's begin by looking at just what a sound wave is.

10.2.1 What is sound?

We already know one important thing about sound. We know that sound is a signal and that this type of signal is very important in our day-to-day lives. More specifically, however, sound is a mechanical wave. This means that unlike radio waves and other electromagnetic waveforms that travel through electrical and magnetic fields, sound is propagated by mechanical interaction. So,

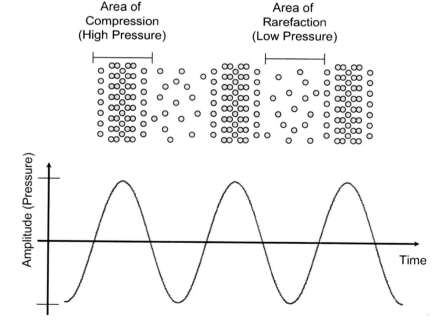

Figure 10.1 An illustration of the air compressions making up a typical sound wave.

in order for a sound wave to travel, it requires some sort of medium (substance) through which to move. In most cases, this medium is air (sound can also travel through other substances, such as water, or any other gaseous, liquid, or solid material). Sound starts out as a vibration. Whether that vibration is in your vocal chords or the string of a guitar, it is these vibrations that produce sound. When a sound is produced, it interacts with particles in the medium. The particles are displaced and in turn, displace other particles. In this way, the sound propagates from its origin. Let's look at a sound wave in air to see exactly how this works. Let's say that you pluck a string on a guitar. When you do this, the string undergoes a periodic vibration. These vibrations in the string in turn cause movements in the surrounding air. As the string moves in one direction, it pushes the air molecules in front of it together creating an area of compression. Then, as the string moves in the other direction, it leaves a void that the air can expand into. This creates an area of lower pressure or rarefaction. These areas of compression and rarefaction move out from the string in much the same way the ripples radiate in a pond after a rock is thrown in. This train of air compressions is what we refer to as a sound wave. So, when we hear a sound, our ear is simply detecting vibrations or minute changes in the pressure of the air surrounding us. Figure 10.1 demonstrates a train of compressions and rarefactions in air.

As we can see in the figure, the air molecules undergo periods in which they are compressed together and then allowed to expand apart. The sinusoidal wave displayed below the compression train in the figure shows how this waveform would be displayed if plotted on a time vs. pressure scale. Based on our study and the illustration in Figure 10.1, there are now several things we can say about sound.

1. Since sound conveys information, it is a signal.
2. Sound is a mechanical wave. Thus, it requires a transmission medium such as air.
3. A sound wave consists of alternating areas of high and low pressure within the medium.
4. Sound radiates in all directions from its source.
5. The larger the pressure differences in the sound, the higher the perceived amplitude.
6. Sound Waves are periodic in nature.

Of these observations, the first four should be fairly obvious based upon our study. The final two observations are based on what we see in Figure 10.1. First, the amplitude, or perceived intensity of a sound can be seen to be a result of the degree of compression within the medium. The more compressed the molecules in the medium are, the more displacement is required to compress them. The sinusoidal plot in the figure helps to illustrate this. The more compressed the molecules become, the higher the air pressure, and the larger the amplitude of the resulting sinusoidal waveform. Thus, we can conclude that larger compressions in the medium result in greater sound intensity. Think of it this way. If you lightly pluck a guitar string, the distance that the string moves when it vibrates is small. As a result, it does not displace, or compress, as much air and the resulting sound will have a small amplitude. However, if you pluck the string with more force, it will move a greater distance as it vibrates. Thus it will displace, or compress, more air. This results in a louder (larger amplitude) sound.

So, how do we compare the intensity of one sound to another? To do this, we need some kind of measurement unit. When comparing the intensity of one sound to another we turn to the unit called the Bel. However, this unit of measure, named after Alexander Graham Bell, is a fairly large unit. One Bel describes a sound that is twice as loud as a reference sound. This tends to be too coarse of a measurement to use in practical applications. As a result, we describe differences in sound intensity using decibels (tenths of Bels). So, when describing the intensity of a sound, we usually describe it relative to the softest sound that we can hear (This sound is often called the threshold of hearing and is considered a 0 dB sound). The resulting relative intensity difference is expressed in decibels. As the power of the signal increases, so does its decibel

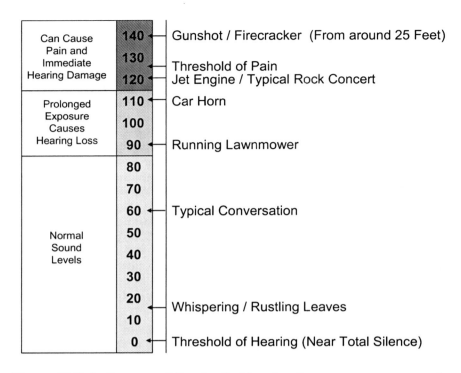

Figure 10.2 A diagram of the decibel levels of some common sounds

level. Thus, a signal that is twice as loud as a sound at the threshold of hearing is 10 dB louder. A signal that is four times louder than the threshold of hearing is considered to be a 20 dB sound, etc. Figure 10.2 displays the decibel levels of some common sounds and environments.

Another important piece of information that we can take from Figure 10.1 is the observation that the train of compressions and rarefactions occurs at regularly spaced intervals. From this observation, we would expect sound waves to be periodic. This is in fact, the case. This periodicity is a direct result of the way in which sound waves are created. Since sound is the result of vibrations, which are naturally periodic, we should obviously expect the resulting sound wave to have the same period as the vibrating object. Since sound waves are periodic, we can classify sounds based upon their frequency. So, how do we find the frequency of a sound wave? Frequency in sound waves is determined in the same way as the frequency of any waveform. First, one period of the signal is identified. The period is the amount of time it takes for one full cycle of the waveform to pass a given point. Once the period is determined, we can calculate the frequency as the number of cycles will pass a point in one second. Thus, we can write:

$$f = \frac{1}{T}$$

Where f is the frequency and T is the period. Figure 10.3 shows one period of the sound wave from Figure 10.1.

Every sound signal consists of one or more frequencies. These frequencies may all originate from one source or come from multiple sources. In addition, these frequencies may be either audible or inaudible. While the human ear is very sensitive, some frequencies cannot be heard by humans (animals such as dogs are another matter). At its best, the human ear is capable of

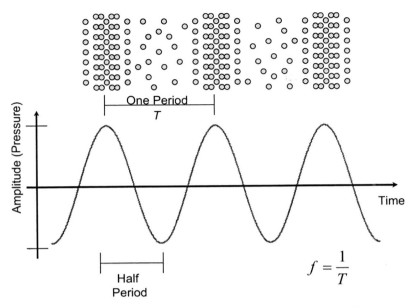

Figure 10.3 An illustration of one period of a sound wave

detecting sounds with frequencies in the range of 20 to 20,000 Hz. Sounds with frequencies below our detectable level are often felt as distinct pulses rather than being detected by the ear. We call sounds at these frequencies infrasound. Frequencies above 20,000 Hz are higher than even the most sensitive human ear can detect. Sound in this frequency range is called ultrasound. While it is inaudible, ultrasound has found significant use in the medical field, particularly for medical imaging. Figure 10.4 shows the spectrum of sound and illustrates the three ranges of sound as it relates to human hearing.

Infrasound	Range of Human Hearing	Ultrasound

0 Hz   20 Hz   20000 Hz

Figure 10.4 The audio spectrum.

So, while we are not capable of hearing all frequencies of sound, it does not mean that we cannot make use if it in signal processing applications. Ultrasound imaging is only one small example of this. All this talk about what frequencies we can and cannot hear brings up another interesting question. How does the human ear process the sound signals that it receives?

10.2.2　How do we hear sounds?

The way in which the human ear captures sound waves and converts them into electrical signals to send to the brain is truly amazing. The ear can differentiate an intensity difference of as little as 1 dB between signals. In addition, it can handle signals varying over a large range of intensities (0 dB to over 120 dB) and is capable of distinguishing frequencies from 20 to 20,000 Hz. So, let's take a quick look at how this is accomplished. Unlike many of the other human senses, the ear processes sound through purely mechanical means. The ear can be divided into three sections: the outer ear, the middle ear, and the inner ear. Each of these sections plays an important role in how we perceive sound. The outer portion of the ear, sometimes called the ear flap or pinna, is responsible for collecting sound signals from the environment and funneling them into the auditory canal where the middle ear takes over. The middle ear consists of a thin membrane called the tympanic membrane or eardrum and three small bones, the hammer, anvil, and stirrup. These bones are collectively referred to as the ossicles. As the sound signals travel down the auditory canal, the pressure differences in the air cause the tympanic membrane to vibrate. The more intense the pressure variation of the sound signal, the more pronounced the vibrations in the membrane. As the eardrum vibrates, it causes the ossicles to move. This motion transfers and amplifies the eardrum vibrations. These amplified vibrations are passed to the cochlea in the inner ear. In the cochlea, the vibrations are picked up by tiny hair-like nerve cells.

So, as these nerve endings vibrate, they produce nerve impulses that are sent to the auditory section of the brain by way of the auditory nerve. It is these nerve impulses that we perceive as sound. The diagram in Figure 10.5 illustrates the basic operation of the human ear. The outer ear collects the sound, the middle ear amplifies the sound, and the inner ear detects the sound. The detected sound leaves the ear as electrical impulses that travel on the auditory nerve. Damage to any portion of the ear can result in hearing loss. The inner ear is particularly sensitive. As you age and as you encounter situations in which sound is too intense, you can damage the sensitive nerve endings in the inner ear. This results in an inability to perceive some frequencies of sound. The damage usually starts in the higher frequencies. This is why children often have far more sensitive hearing than adults, particularly in the highest frequency ranges.

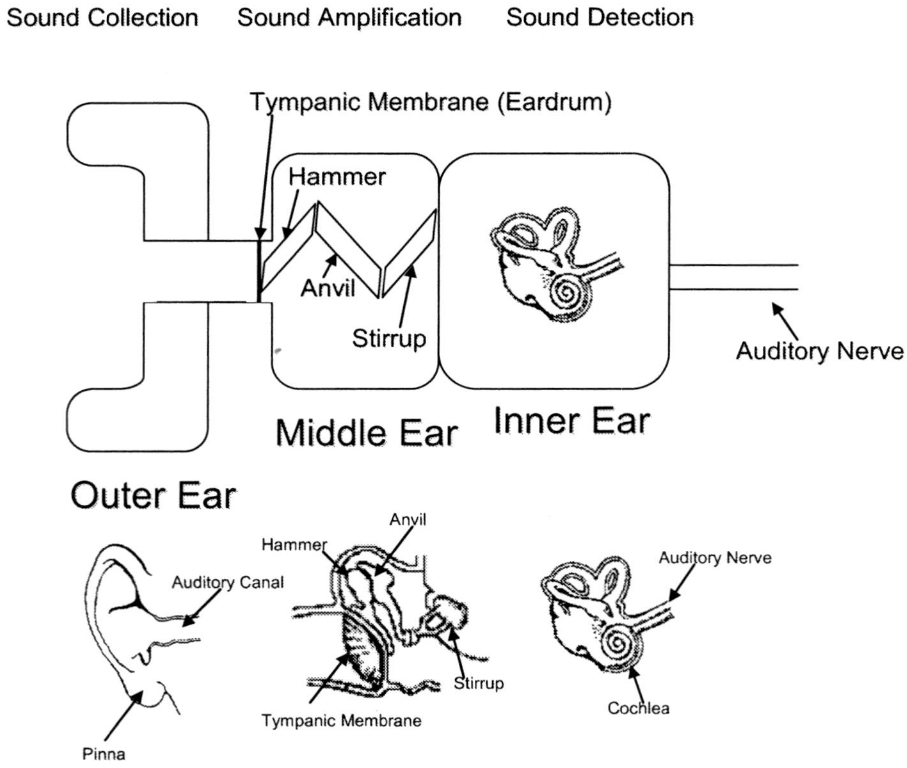

Figure 10.5 The inner workings of the human ear.

While the human ear has amazing capabilities, there are some animals that possess even greater auditory perception. For example, the dog is able to hear frequencies far above those that people can here. This is how a dog whistle works. It emits an ultrasonic sound that humans cannot hear, but that is within the audible range for dogs. Of all the animals, it is believed that the dolphin could have the best sense of hearing. A dolphin's sense of hearing is roughly 14 times better than ours.

10.2.3 Sources of Sound

At this point, we are familiar with what sound waves are, how they are created, and how we are able to hear them. Another interesting question to ask is: what are some common sources of sound? Throughout a given day, we encounter numerous sounds from countless sources. Many of these sources can be grouped into a few broad categories. For example, most sounds we encounter can be grouped into one of the following four categories.

1. Environmental Sources
2. Vocal Sources
3. Instrumental Sources
4. Artificial Sources

Environmental sounds, as the name suggests are produced in nature. These consist of things like rustling leaves, whistling wind, and crashing waves. The human ear is well equipped to receive and process these sounds. Vocal sources tend to be more complex than natural sources. Whether the source of the voice is animal or human, the general purpose of vocal sound sources is communication. Humans communicate through spoken language, and while we do not know the total extent to which animals communicate, we know that they use vocal expressions for a variety of purposes. As with any sound, vocal sounds are produced from vibrations. In the case of human voice, the vibration occurs in the vocal cords. While it is the vocal cords that are vibrating, they are not solely responsible for the human voice. The main components of the human voice production system are the lungs, trachea (windpipe), larynx (the organ of voice production containing the vocal cords), the throat, the mouth, and the nose. The system that we refer to as the vocal tract consists of the throat and the mouth. This vocal tract receives the output of the larynx and modifies the sound up until it exits the mouth at the lips. There are some finer anatomical details that are essential to the voice production process. These include the vocal cords, soft palate or velum, tongue, teeth and lips. We call these parts articulators because they help us modify the basic sound formed in the vocal cords into understandable speech. Figure 10.6 displays a general model of the different parts of the vocal tract involved in human speech. As you can see, the sounds produced by the vocal chords are modified and amplified by the various cavities and openings that are encountered along the vocal tract. It is the combination of all of these parts that gives you your distinct voice and allows you to speak.

So, what makes a male voice different from a female voice? Why are children's voices higher pitched than those of adults? The answer to these questions lies in the anatomical characteristics of the individual. For example, the average adult male has an average vocal tract length of about 17 centimeters. Compare this to an adult female with a length of 14 cm or a child which on average has a vocal tract length of only 10 cm. The length of your vocal tract governs the general range of vocal frequencies that you are able to produce. So, one reason that adult men have deeper voices is because they have a longer vocal tract (Adult men also have a larger larynx than women. Thus, they have longer vocal cords which produce lower frequencies).

As we now know, the larynx, or voice box, contains the vocal cords. The vocal cords play a central role in human speech. It is the vocal cords in the larynx that generate the basic sound of

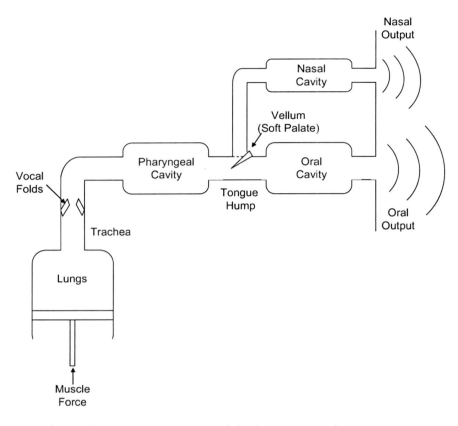

Figure 10.6 A model of the human vocal system.

your voice. When examining the sounds involved in human speech, we can classify them as either voiced or unvoiced. So, what is the difference between a voiced and an unvoiced sound? Whether or not a sound is voiced or unvoiced is simply a matter of whether or not the vocal cords play a role in its production. The periodic vibration of the vocal cords is responsible for the voiced speech. The unvoiced sounds, however, are produced closer to the end of the vocal tract. They are typically produced in the mouth and with the vocal cords at rest. In the English language, vowels, such as a, e, i, o, and u are all voiced sounds. They are all produced somewhere in the middle of the throat and the vocal cords are fully involved in their production. On the other hand, most consonants do not have a great deal of contribution from the vocal cords and are hence weaker in amplitude. Some consonants are voiced and others are not. Examples of voiced consonants are B, D, L, M, N and R. Unvoiced consonants include the letters S and F, which are produced almost at the end of the vocal tract. So, as we can see, there is a lot involved in the production of the human voice.

The third category of sound sources is instrumental sounds. This includes the wide range of sounds we produce using musical instruments. These sounds can be produced in a wide variety of ways. One thing that remains the same, however, is that all the sounds are produced by some type of vibration. This could be the vibration of a column of air like that in a pipe organ, a vibrating reed like the one in a clarinet or saxophone, or vibrating strings like those on guitars, harps, and pianos. Let's look at a few different types of instruments and the ways in which they produce sound.

In the case of a stringed instrument like the violin or cello, the part of the instrument that vibrates is the strings. The strings are made out of cat gut (or some equivalent substance) and, as the bow is slid across the strings, the strings vibrate. These vibrations resonate and are magnified in the hollow wooden cavity below the strings. This gives the sound enough energy to travel over a longer distance and be heard. Other acoustic stringed instruments like the acoustic guitar and banjo work in much the same way. In this case, however, the strings are strummed or plucked by hand to generate the vibrations. These strings, typically made of steel are stretched across a nut and a bridge and a specified tension is maintained on each string. This enables the string when plucked to emit a note of a specified frequency and hence a particular sound. The thickness of each string varies based upon the notes it is designed to play. Strings designed to play bass (or notes with less frequency and hence lower pitched sounds) are much thicker than the strings that are designed to produce a higher pitch.

Another important feature of stringed instruments is that many have a fret board, as in the case of the guitar. The basic idea is that the length of the string may be adjusted during playing time by pressing the finger on a fret. The fret is just a raised metallic piece on the fret board. By varying the length of the string, the frequency of the note produced is different. Thus, a wider range of frequencies can be produced. In the end analysis however, it is the string that vibrates and makes sound – we simply control the dynamics of the string to create different sounds and make music.

In the case of the woodwind and brass instruments, music is created either by the vibrations caused by air moving over reeds or by the vibration of the player's lips. Reeds are generally thin pieces of wood (or reed) that are placed in the path of air blown from the mouth. The vibrating reeds set the air around them in motion, creating a sound wave. Most woodwind and brass instruments are shaped in such a way that the sound created by the lips or reed resonates inside them, and thus is amplified. An example of the shape of an instrument aiding in the magnification of sound would be the tuba. The large volume of the tuba allows the vibration from the player's lips to resonate in the chamber. The frequency of the sound produced is then varied by adjusting the length of the path of the sound wave in the chamber (by pressing down the various valves).

Instruments that produce sound when they are struck, such as drums, also have a vibrating element that produces sound. In the case of the drum, the material that is stretched over the frame vibrates when struck and creates ripples in the air. These ripples are amplified in the space below the membrane and a loud sound can be obtained. Drums can also be tuned by adjusting the tension on the stretched material to get a sound of a desired frequency.

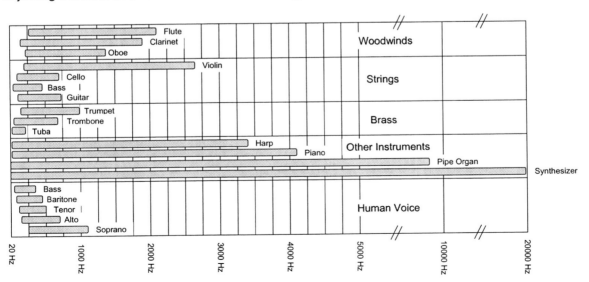

Figure 10.7 Frequency ranges of some common musical instruments.

One interesting aspect of all the different varieties of musical instruments that we have just discussed is that they are designed to produce signals at specific frequencies (musical notes). So, for a given instrument, we can determine a range of frequencies in which it operates based upon the notes that it is capable of playing. Figure 10.7 displays the frequency ranges of some common instruments as well as the frequency ranges of the human singing voice.

One of the things we can see immediately by looking at Figure 10.7 is that many of the musical instruments overlap in their frequency range. If that is the case, why do different types of instruments sound so different when playing the same note? There are actually two reasons for this. The first has to do with the way that the note is played. When a musical instrument plays a note, each type of instrument uses a slightly different method of producing the sound. If you were to play just one note on the instrument, you would find that the resulting sound wave had three distinct phases, an attack phase, a steady state phase, and a decay phase. The waveform shown in Figure 10.8 displays these three phases.

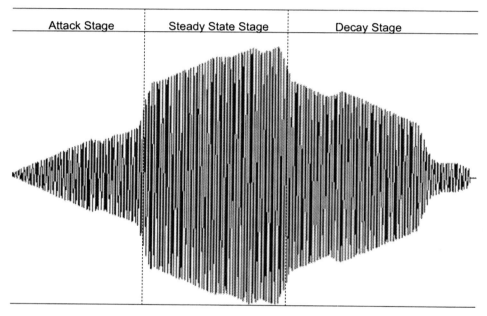

Figure 10.8 The three stages of a musical sound.

As you can see, the attack stage occurs when the instrument begins to play the note, for example, when the hammer strikes the string on a piano or when you pluck the string on a guitar. The instrument then reaches a point where it produces nearly constant amplitude and frequency. This is known as the steady state of the note. Finally, the note dies out. This is the decay stage. Even if the instruments produce similar steady state sounds, each instrument has slightly different attack and decay stages. These stages are determined by the way in which the instrument is played. Thus, the differences in the way the note begins and dies out are one reason why different instruments sound different even when playing the same note.

Another reason that instruments sound different is the fact that they produce additional frequencies that are higher than their base frequency. These frequencies, called harmonics, are different for each type of instrument. These higher frequencies add definition to the lower base or fundamental frequency. So, while the fundamental frequency determines the basic shape of the waveform, the higher frequency harmonics add a distinct color or flavor to the sound. Therefore, by producing distinct attack and decay stages and different harmonics, an instrument is able to achieve a unique sound.

The final class of sound sources contains artificially created sound. These are sounds that are not naturally occurring, are not produced vocally, and are not created by musical instruments. Sounds of this type are usually created electronically using signal processing techniques and then converted into sound by a speaker. By producing the signals electronically, it is possible to create nearly any sound imaginable. This can range from simple single tones to complex mixes. A common musical instrument that uses this technique is the synthesizer. It is capable of producing signals at frequencies that cover the whole range of human hearing. Thus, while it is difficult to perfectly reproduce the harmonics of other instruments, the synthesizer is capable of using signal processing techniques to both approximate other instruments and produce sounds that would otherwise be impossible.

10.3 Preparing Audio Signals for Processing – Digitizing Sound

Now that we have a basic understanding of sound waves and how they are produced, we can begin looking at what we can do with them. The first step in working with sound signals is getting them into a form that we can easily process. In our case that means we need to find a way to convert a mechanical analog waveform into a digital signal. Doing this will require several steps. The first of these is to convert our analog mechanical sound wave into an electronic waveform that we can process further. This is typically achieved through the use of a microphone. A microphone is a simple device that can take a mechanical sound wave as input and convert it to an electrical wave. There are several different types of microphone. Each of these types uses a slightly different method to perform the conversion. The two most common types of microphones are the dynamic and condenser varieties.

The dynamic microphone is basically a reverse application of the technology used in speakers. A coil of wire is attached to a membrane. As the membrane vibrates back and forth, the coil of wire is moved through a magnetic field. This movement causes a current to flow through the coil of wire. Thus, when a sound wave comes in contact with the microphone's membrane, it causes it to vibrate. These vibrations cause the coil to move resulting in a current flow in the wire. This electronic signal, which represents the sound wave, can then be amplified and processed further.

Condenser microphones are slightly more complicated than their dynamic counterparts. These microphones operate based on capacitance. The microphone itself is basically a capacitor with one fixed plate and one plate that moves in response to sound. So, the second plate of the capacitor in a condenser microphone will usually be a thin conductive membrane that will vibrate in the presence of sound. When a voltage is placed across this capacitor, the changes in plate distances caused by the vibration of the membrane will result in small fluctuations in the capacitor's capacitance value. These minute changes can be amplified into a usable electronic representation of the sound wave.

So, as we have seen, we can convert the sound wave from a mechanical to an electronic signal by using a microphone. Thus, we now have an analog electrical signal to work with. Since we need to process the sound digitally, we need to go through three more steps to convert the analog signal into a digital one. These three steps are sampling, quantitization, and binarization. Let's start by looking at sampling. When we sample an analog signal, we basically read its value at evenly spaced instants in time. The amount of time that occurs between the times that we read the analog signal is called the sampling period, usually denoted as T_s. The total number of times that we read the analog signal in one second is called the sampling frequency. This is usually denoted as f_s. So, as you can see, the process of sampling converts an analog signal into a discrete-time signal. Figure 10.9 demonstrates the results of sampling an analog signal with a sampling period of 0.05 seconds resulting in a sampling frequency of 20 samples per second.

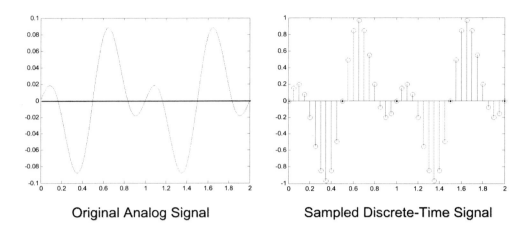

Original Analog Signal   Sampled Discrete-Time Signal

Figure 10.9 An analog signal and its sampled discrete-time counterpart.

As you can see, the discrete-time signal provides a fairly good approximation of the original signal. In reality, however, most signals will need to be sampled at rates much higher than the twenty samples per second that was used in the figure. In fact to accurately capture a given signal you must use a sampling frequency that is equal to or greater than twice the highest frequency that is present in the signal. Thus, if you wanted to accurately represent everything in the range of sound that a human can hear, you would need to sample the signal with a sampling frequency of at least 40,000 samples per second (2*20,000 Hz). This rule regarding sampling is known as the Nyquist sampling rate and will be encountered throughout your studies of signal processing. So, what happens if we ignore the Nyquist rate and sample our signal at less than twice its highest frequency? When we sample at too low of a sampling rate, we cannot accurately reconstruct the original analog waveform from the resulting discrete-time approximation. Instead the reconstructed signal will be a distorted version of the original. This is an effect known as aliasing. Basically what happens is that when we sample at too low of a frequency, we do not have enough information about the higher frequencies to realize that they are present in the signal. Therefore, the components of the higher frequencies in the signal are incorrectly interpreted to be components of lower frequency within the signal. For example, we may only detect every third peak of a high frequency component of the signal. The resulting captured peaks will then look like a periodic signal component with a frequency that is one third that of what the original was. Thus, the reconstructed analog signal is polluted by the presence of these higher frequency components. Take the music on your CDs for example. The recording engineers are well aware of the fact that humans can only hear frequencies up to 20,000 Hz. As a result, they use the Nyquist rate and determine that to accurately represent every frequency we can hear in the signal will require 40,000 samples per second. However, a given analog audio signal can contain harmonics that are higher than 20,000 Hz. If the analog signal is sampled with these harmonics present, they will cause aliasing effects that will distort the music when it is played back from the CD. To solve this problem, analog band-pass filters are employed to strip out the frequencies above the range of the human ear. The resulting signal is then sampled at around 44,000 Hz (actually 44,100 Hz) to account for some of the frequencies close to 20,000 Hz that the filter may not have completely suppressed. The result is a sampled signal that can be used to accurately reproduce the analog signal within the range of human hearing. So, when digitizing sound, the sampling frequency is an extremely important factor to consider.

Now that the signal has been sampled, we have a signal that has values at discrete instances in time. Thus, we have drastically reduced the amount of space required to represent the signal.

Unfortunately, the signal can still hold any value at each sample point. This means that there are an infinite number of possible amplitudes for each sample. If we want to represent our signal digitally and in a finite amount of space, we must also make the signal discrete in amplitude. This is done through a process known as quantitization. When a signal is quantitized, its samples are assigned one of a pre-selected set of values. The number of these values that are available depends on how much space we are willing to allow each sample to consume. For example, if we allow each sample to use 8 bits (1 byte) of digital memory, we can only represent 256 possible values in this space. Since our signal can have both positive and negative values, we will typically assign one of these values to be zero and then split the remaining values across the positive and negative amplitude ranges. To illustrate the concept of quantitization, let's look at a slightly simpler case. Let's say we are willing to use only 4 bits to represent each sample value in our signal from Figure 10.9. This means that we must assign the amplitude of each sample one of sixteen possible levels. Thus, we spread our sixteen levels evenly across the range of amplitudes taken by the signal and then assign each sample the value of the level to which it comes closest. Doing this would result in the quantitized signal shown in Figure 10.10.

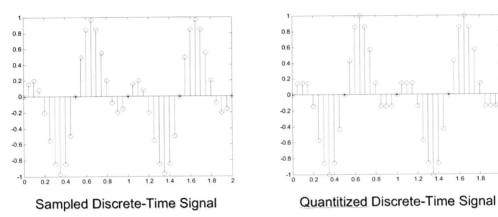

Sampled Discrete-Time Signal   Quantitized Discrete-Time Signal

Figure 10.10 An example of quantitization using 4 bits per sample.

Now, since the quantitization levels are evenly spaced across the signal's amplitude range, we do not need to store the actual value of the amplitude. We only need to store the number of the level that the sample's amplitude is closest to. Thus, we can represent the sample amplitudes as a series of integer numbers. These numbers can then be converted into binary digits and stored in memory. This conversion to binary digits, called binarization, is the final step of the sound digitization process. So, our original analog signal from Figure 10.9 has been reduced to the final binary digital signal shown in Figure 10.11.

1000, 1001, 1001, 1001, 0111, 0100, 0010,
0001, 0010, 0101, 1000, 1011, 1110, 1111,
1110, 1100, 1001, 0111, 0111, 0111, 1000,
1001, 1001, 1001, 0111, 0100, 0010, 0001,
0010, 0101, 1000, 1011, 1110, 1111, 1110,
1100, 1001, 0111, 0111, 0111, 1000

Binary Digital Signal

Figure 10.11 The final binary digital signal produced by digitizing the analog signal in Figure 10,9 using f_s=20 samples per second and 4 bits per sample.

We have just seen that, in order to begin working with an audio signal, we first must transform the analog mechanical sound wave into a stream of digital bits. This process is summarized in the block diagram shown in Figure 10.12.

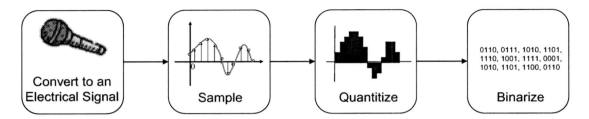

Figure 10.12 A block diagram of the process involved in converting a sound wave into a digital signal

Our process for digitizing sound allows us to work with sound signals using DSP techniques. It also allows us to store the signals in digital form. The amount of storage that is required for a given signal is a function of the length of the signal and the choices that were made when digitizing it. The choices that we are referring to are the sampling rate at which the signal was captured (time resolution) and the number of bits that were used to represent each sample (amplitude resolution). If we know these values, we can use basic mathematics to determine the amount of storage space that is needed to represent it. Let's look at an example.

Example 10.1

Problem: Assume that you are digitizing an analog signal at 44,000 samples per second and with a sample size of 16 bits per sample. How much computer memory is required to store 30 seconds of audio at these settings? How long can you record before you fill 650 MB?

Solution: This problem requires us to make use of our knowledge of signal sampling to calculate some simple storage size values. First, we know we are taking 44,000 samples each second. Secondly, we know that each of these samples requires 16 bits (2 bytes) of space to store. This means that for one second of audio at these settings we will need:

$$44,000 \frac{Samples}{Second} * 2 \frac{Bytes}{Sample} = 88,000 \frac{Bytes}{Second}$$

So, to find out how much space we need for 30 seconds we simply multiply our requirement for 1 second by 30 to yield:

$$88,000 \frac{Bytes}{Second} * 30 \ Seconds = 2,640,000 \ Bytes \approx 2.52 MB$$

Thus, digitizing 30 seconds of audio at the specified quality will require a little more than two and a half megabytes of storage space. Now, we need to find out how long we can record before filling up 650 MB. To do this, we can return to the amount of space that one second of audio requires. Next we simply divide the available space by the amount of space per second. Thus, we arrive at the number of seconds that we can record. First, however, let's find the number of bytes in 650 MB.

$$650 MB * 1024 \frac{KB}{MB} * 1024 \frac{B}{KB} = 681,574,400 \; Bytes$$

Now that we know how many bytes there are in 650 MB, we can do our division.

$$\frac{681,574,400 \; Bytes}{88,000 \dfrac{Bytes}{Second}} \approx 7745 \; Seconds \approx 2.15 \; Hours$$

So, we can fit around 2.15 hours of sound captured at our specified quality in 650 MB of storage space.

10.4 Working with Sound - Applications

Now that we can convert sound waves into a format that we can process using DSP, we can look at a few of the things we can do with sound signals. In this section we will look at four examples of sound processing applications. We will begin with a simple example of extracting useful information from a signal. Then we will look at modeling sound systems and transforming sound to add special effects. The third application deals with creating sound signals using DSP, and finally, we will look at an application that is still an important area of sound processing research. This area is sound identification.

10.4.1 Finding the Fundamental Frequency of an Audio Signal

When we discussed sources of sound, we talked about the basic or fundamental frequency of a sound. We noted that it is the fundamental frequency of a sound that gives it its basic form and the higher frequency harmonics that add color and definition to the signal. So, how do we go about finding this fundamental frequency? Finding this frequency is not difficult if we have an understanding of the signal we are working with. For example, there are two ways of plotting a digital sound signal. We can plot its amplitude against either time or sample number. Figure 10.13 demonstrates this. The top signal is plotted using the sample number on the horizontal axis and the lower signal is plotted vs. time.

When we first encounter a sound in the form of a digital signal, it will be expressed in terms of sample, just as the top signal in Figure 10.13. So, how do we find the time that a given sample occurs? This is an easy task as long as we know the sampling frequency that was used when the sound was captured. By dividing the sample number by the sampling frequency, we obtain the time at which the sampled amplitude occurred. Thus, by dividing each sample number by the sampling frequency, we can plot our signal vs. time just as we see in Figure 10.13. Now that we see how to relate an actual time to each sample in our signal we can look at what is required to calculate the signal's fundamental frequency.

A signal's fundamental frequency is the frequency of its base waveform. Thus, if we can determine the frequency of the basic sound waveform, we will have the signal's fundamental frequency. We can find this frequency in much the same way that we determine the frequency of any periodic signal. First, we need to find the period of the waveform. This is the distance

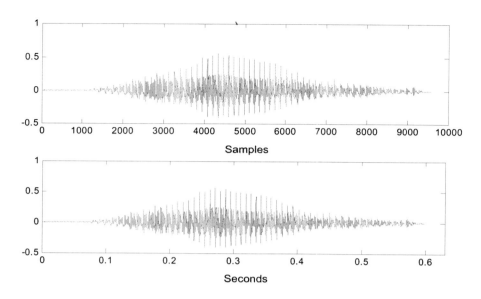

Figure 10.13 An example of a sound signal plotted vs. sample number and time.

between two similar points in the signal. In order to find these points in our signal, we need to focus in on a part of the signal that has fairly large amplitude and has a regular waveform. We demonstrate this in Figure 10.14.

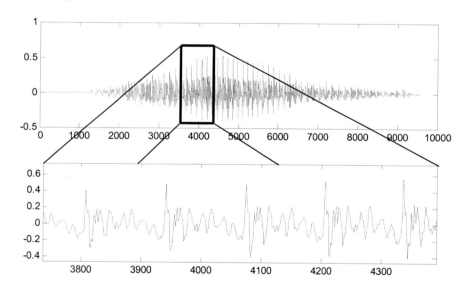

Figure 10.14 An illustration of the periodic nature of a sound wave.

In this figure, we can easily find two similar points between which to measure the period of the waveform. You could pick any two similar points in the signal, but it is usually easiest to measure between two similar peaks as shown in Figure 10.15.

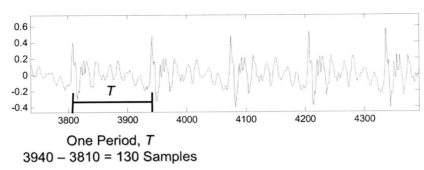

One Period, T
3940 – 3810 = 130 Samples

Figure 10.15 One period of the sound waveform

In Figure 10.15 we found two similar points. One at around sample number 3810 and one around sample 3940. The distance between these points is our signal period. If we subtract the smaller sample number from the larger, we find that our period consists of 130 samples. To proceed further, we need to know the sampling frequency of this sound. The sound in Figure 10.15 was sampled at 16,000 samples per second. Thus, the period can be expressed in terms of time by dividing the number of samples in the period by the sampling frequency. Doing this will give us the following.

$$T = \frac{130\,\text{Samples}}{16{,}000\,\dfrac{\text{Samples}}{\text{Second}}} = 0.008125\,\text{Seconds}$$

So, our period is 0.008125 seconds. Now all we need to do to find the fundamental frequency of our sound is to take the inverse of the period. Doing this will give us:

$$f = \frac{1}{T} = \frac{1}{0.008125} \approx 123\,\text{Hz}$$

Thus, the fundamental frequency of the sound in Figure 10.15 is approximately 123 Hz. As we can see, the determination of a signal's fundamental frequency is a simple task. This example demonstrates that we can obtain useful information from our signal by using DSP techniques. There are many applications in which it is useful to know the frequency of a sound. For example, we may have a system where commands are transmitted as a series of tones. In order for the remote system to determine what command was sent, it is necessary to determine the frequencies that were sent.

10.4.2 Modeling Echo

Another interesting application is the modeling of physical environments. Let's take the simple example of an echo. The echo is a phenomenon that we have all experienced. So, how does an echo occur naturally and how can we reproduce this effect using signal processing techniques? The basic explanation of an echo is fairly simple. In large caves or empty halls sound waves reflect off wall surfaces. These reflected waves all travel slightly different distances before they are picked up by the ear. Since all sound waves travel at a given speed, these waves will arrive at the ear with slightly different delays. In addition, as a sound wave travels and is reflected off of objects, some of its power is dissipated. As a result, the delayed waves will become softer as time goes on. If echoes are simply caused by reflected sound waves, then why do we not hear an echo in a furnished hall or a hall filled with people? This is because as the reflected waves move through the room, they are absorbed by the material in their path – that is, their energy is dissipated into the objects in the hall. In the case of the empty hall, however, the only place for the sound waves to dissipate their power is the walls themselves. So, if the walls of the room do not absorb much of the power from the sound waves, the room will produce clear echoes. This is why recording studios cover their walls with sound absorbing foam. If echoes were to occur in the recording studio, they would reduce the quality of the sound being recorded. Sometimes, however, the addition of echo to a sound is a desired effect. This is why it is possible to find echo generators for use with recording systems and musical instruments. If used correctly, an echo can add fullness and power to the perceived sound of a voice or piece of music. Thus, the addition of one or more echo signals can be a useful signal processing task.

Let's reflect back on our description of how an echo naturally occurs and use that information to first build a mathematical model of an echo using a difference equation and then look at how we can employ this equation to add an echo through the use of DSP. As you recall, an echo begins with a sound source. This source radiates sound waves in all directions. Some of these sound waves will follow the most direct route to the listener, and thus, arrive before any others. Let's refer to these sound waves as $x(n)$. In other words, these sound waves represent the original un-delayed version of the sound. Now, some of the other sound waves radiating from the sound source will arrive at the user by an indirect route. Thus, they will have some delay, D_i, associated with them where D_i is the delay associated with the ith sound wave to arrive. Let's refer to these delayed copies of the original sound wave as $x(n-D_i)$. Since these sound waves have traveled further and reflected off of other objects, they will have lost some of their initial intensity and thus be perceived as softer when compared to the sound wave that arrived first. Because of this, we must associate an attenuation factor, k_i, with these delayed copies of the sound where k_i is the attenuation factor associated with the ith copy of the signal to arrive. The attenuation factor is a value between zero and one that indicates the degree to which the sound's intensity has been decreased. For example, an attenuation value of 0.5 would indicate that the associated copy of the original signal would be half the intensity of the original (a 10 dB decrease). Figure 10.16 illustrates this process graphically.

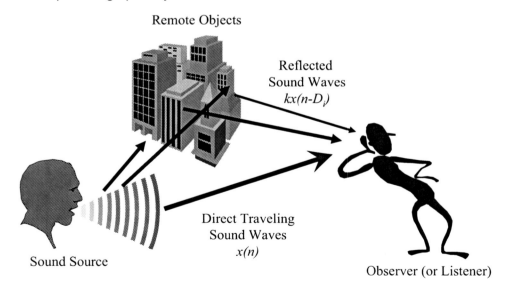

Figure 10.16 A graphical illustration of the process involved in creating an echo.

Now, we can put all of this information together to describe what the listener hears. As each sound wave arrives it interacts additively with the other sound waves that the listener hears. Thus, the overall sound that will be heard by the listener can be modeled as follows.

$$y(n) = x(n) + k_1 x(n - D_1) + k_2 x(n - D_2) + ... + k_i x(n - D_i)$$

If we use summation notation, we can rewrite this in a more compact form.

$$y(n) = x(n) + \sum_{m=1}^{i} k_m x(n - D_m)$$

It is important to note that the delay value, D, is expressed in samples and not time. In order to account for a specific time delay, t_d, we must multiply the time delay by the signal's sampling frequency and round the result to the nearest integer value. For example, if the sound was sampled at 44,100 samples per second, a 0.1 second delay would amount to:

$$Round[(f_s)(t_d)] = Round[(44,100)(0.1)] = 4410 \, Samples$$

So, let's finish up this application by taking a look at a simple example involving the addition of a single echo signal.

Example 10.2

Problem: Develop a difference equation that describes a system that adds a single echo signal to its input signal. The resulting signal should contain en echo that arrives 0.1 seconds after the initial signal and is attenuated to 75% of the original sound's intensity. Assume that the input signal has a sampling frequency of 16,000 samples per second.

Solution: In this example, we simply need to make use of our existing model for a signal with echoes and plug the appropriate values in for attenuation and delay. We know that the attenuation must be 75% of the intensity of the original. Thus, our attenuation factor will be 0.75. Our next step is to calculate the number of samples in a 0.1 second delay. Plugging our sampling rate of 16,000 samples per second and our delay of 0.1 seconds into the equation developed above gives us:

$$Round[(f_s)(t_d)] = Round[(16,000)(0.1)] = 1600 \, Samples$$

Now, we can plug this information into our difference equation model to get the following model for a system providng a single echo with the provided delay and attenuation.

$$y(n) = x(n) + 0.75x(n - 1600)$$

This can be shown in block diagram form as follows:

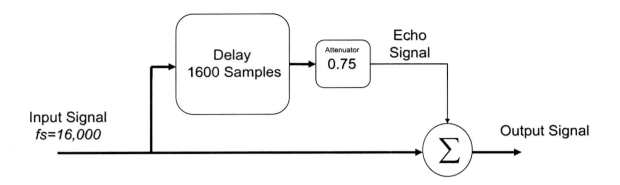

In a mathematical sense, we are simply doing the following:

$$y(n) = \{[x(n)][1600 \text{ zeros}]\} + 0.75\{[1600 \text{ zeros }][x(n)]\}$$

Basically this means that we are adding the original sound signal with 1600 samples of silence following it to a second signal that consists of 1600 samples of silence followed by the original sound multiplied by an attenuation factor of 0.75. We can illustrate this process graphically by running an actual sound signal through the system that we have just modeled. Let's try this with the following signal.

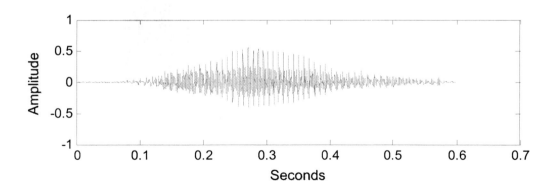

Using this signal, the resulting echo can be found as is shown.

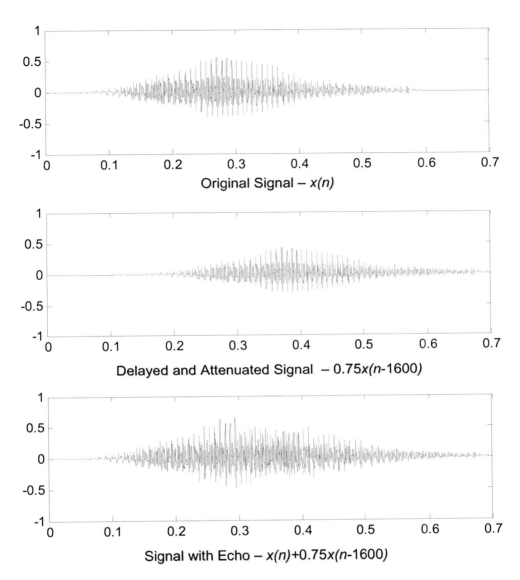

Original Signal – *x(n)*

Delayed and Attenuated Signal – 0.75*x(n*-1600)

Signal with Echo – *x(n)*+0.75*x(n*-1600)

So, we have derived and examined the use of a difference to model the addition of a single echo to an input signal.

10.4.3 Generating Audio Tones – Touch Tone Phones

While the processing, modeling, and modification of sound signals are important areas of signal processing, another key application of signal processing is the generation of sound signals. One such application is the generation of signaling tones for push-button telephone dialing. When you press the buttons on a typical telephone in order to place a call, a series of tones are generated for each button that you press. These signals are part of a special communication coding system called DTMF (dual tone multi frequency). This system is used by phone switches to route your call to the desired destination. This frequency-based information encoding system is also employed by many voice-mail and automated answering systems to provide the user with interactive options and to internally route calls. So, how does it all work? When you dial a number, the phone employs the DTMF encoding scheme by generating two tones of specific frequencies for each button that you press. So that these control signals cannot be accidentally generated by the human voice during normal conversation, one tone is generated from a high-frequency group of tones and the other from a low frequency group. The following table lists the two frequencies associated with each of the buttons on your phone.

Button	Low Frequency (Hz)	High Frequency (Hz)
1	697	1209
2	697	1336
3	697	1477
4	770	1209
5	770	1336
6	770	1477
7	852	1209
8	852	1336
9	852	1477
0	941	1336
*	941	1209
#	941	1477
A	697	1633
B	770	1633
C	852	1633
D	941	1633

Seven frequencies are used to code the 10 decimal digits and the two special buttons marked '*' and '#' that you see on typical telephone dialing pads. A special eighth frequency is used to denote the digits 'A', 'B', 'C', and 'D'. These buttons are not present on normal phones, but are instead used for internal phone signaling and are reserved for possible future use on regular phones. They were originally used on military phone systems to indicate a level of call priority, but now they are most often found in amateur radio applications.

Now, let's take a look at what happens when you dial a number. As we just mentioned, when you press a button, one low frequency signal and one high frequency signal will be transmitted to uniquely identify the button that you pressed. The low band frequencies are 697 Hz, 770 Hz, 852 Hz and 941 Hz, and the high band frequencies are 1209 Hz, 1336 Hz, 1477 Hz, and 1633 Hz. Simple signal addition is used to combine the two signals. Therefore, in the case of digitally generated tones, the two signals will simply be added sample by sample. So, is there any pattern to the way the two frequencies are chosen? Yes, the dialing pad on the phone is laid out in a grid pattern as shown in Figure 10.17. Each row represents one of the four low frequency signals and each column represents one of the four (three on most standard phones) high frequency signals. So, when you press a button on the phone dialing pad, the low frequency associated with its row will be added to the high frequency associated with its column, and the resulting signal will be sent out. The receiving equipment then analyzes the incoming tones for their frequency content. By finding the two distinct frequencies in each tone, the remote equipment can determine exactly which buttons you pressed. This information is then used to route your call to the appropriate destination. It is interesting to note that since this dialing scheme consists simply of broadcasting specific frequencies on the phone line, it is possible to dial a number without using the pad on the telephone. Any device that is capable of generating the appropriate tones could be placed near the microphone on the telephone handset. This device could then dial a number by producing the tones for the desired number.

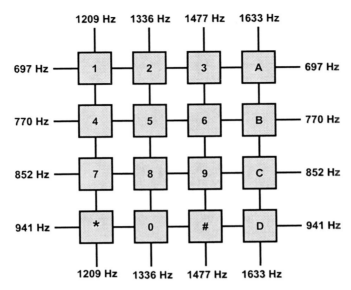

Figure 10.17 The grid layout of the telephone dialing pad and the frequencies associated with each row and column of the grid.

So, as we have seen, audio signal processing techniques play an important roll in something as simple as dialing the telephone.

10.4.4 Identifying Sounds

As we near the end of this chapter, we will discuss one final application of audio signal processing. This application involves the use of signal processing techniques to identify sounds. This may be something as simple as searching for a known sound within a larger signal or as complicated as trying to recognize human speech and convert it to text. In either case, we will find that signal processing techniques play a vital role. We will begin this section by investigating a simple application, the identification of a given sound through a library comparison. After this, we will briefly turn our attention to an application that is the subject of ongoing research, speech recognition. This far more complicated application has many potential uses from text dictation to language translation and, thus, is an exciting application of audio signal processing.

The identification of sounds is one important audio signal processing application. Let's say for example that you are researching bird songs and have amassed a vast library of recordings from identified birds. Now, let's say you are out in the field and record the song of a bird that you do not recognize. You cannot find the bird for visual identification, but you manage to make a recording of the song. The song seems somewhat familiar, but you cannot remember what bird it is from. Chances are high that you have a sample of this bird's song in your library of digitally recorded bird songs. The question is, how can you use the digital audio information in your library to identify this particular song without having to manually listen to every recording until you find a match? The answer is quite simple. You can make use of the correlation technique that we discussed in chapter 9 to compare the signals. If you take a small sample from the unknown recording and correlate it with all of the samples in your library, you can look at the results that return large spikes to find a matching recording. Before doing this, however, there is one additional step you can take to reduce the number of samples that you need to convolve your sample with. You can examine the fundamental frequencies of both your unknown sound and those in your library. If the library samples do not have a similar fundamental frequency they could not be matches and thus do not need to be compared. This is sometimes known as the form and frequency approach. First you check for similar frequency, and then you can check to see if the waveforms match. An example of the comparison between an unknown signal and a library of known signals is shown in Figure 10.18. In the figure, an unknown bird song signal has been correlated with a library of four known bird songs.

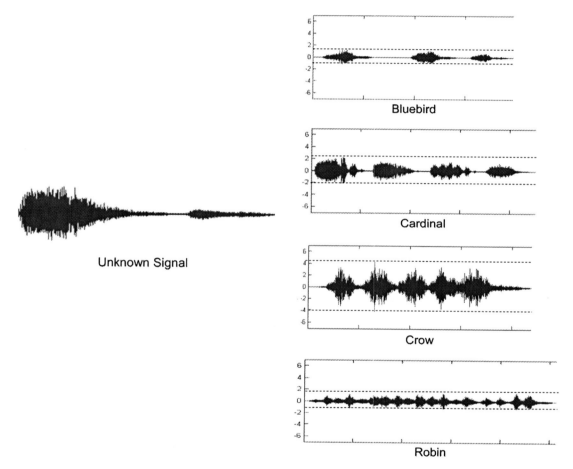

Figure 10.18 An illustration of sound signal identification. An unknown bird song is correlated with four known examples.

As you can see in the figure, the correlation between the unknown sample and the recorded songs of the bluebird cardinal, and robin do not contain any large peaks. In fact, they obtain responses with amplitudes no larger than two. The correlation with the crow song however shows a large response of slightly higher than four. From this information, we can make the assumption that our unknown signal is most likely some variety of crow (which in fact, it is). Thus, we can see that it is possible to identify an unknown signal by using signal processing techniques.

A much more practical and also more complicated application of sound signal identification is speech processing, particularly speech recognition. There are several areas in which speech recognition can be useful. Some of these include voice to text translation, identity verification based on voice, and vocal language translation. As you might imagine, all of these applications of speech processing are very complex and are currently areas of ongoing research. For example, before we can begin to implement spoken language translation, we must first be able to master voice to text translation. It is no easy task to get a computer to recognize the content encoded in spoken language. The many variations that are introduced into the voice over the course of time serve to further complicate the problem. For example, different people's voices sound different, they have different frequencies and two different people may pronounce the same word differently. In addition, a person's emotional state and physical wellbeing can affect the pitch and harmonics in their voice as well as the speed at which they speak. All of these variables make a universal speech recognition system incredibly complicated and difficult to implement.

A typical voice recognition system operates in two basic stages. First, the input signal is processed in order to extract some key elements called phonemes. Phonemes are the basic sounds that make up human language. A given language constructs words by arranging its base phonemes in varying orders. The English language, for example, consists of between 40 and 45 phonemes. A list of 44 of these phonemes and their international phonetic symbols are shown in Figure 10.19.

Symbol	Examples
ʌ	Cup, Truck
ɑː	Harm, Father
æ	Fat, Track
e	Fed, Bed
ə	Away, Cinema
ɜ˞	Burn, Turn
ɪ	Hitting, Kit
iː	See, Meat
ɒ	Cot, Sock
ɔː	Call, Four
ʊ	Good, Put
aɪ	Five, Eye
uː	Blue, Mood
aʊ	Cow, Out
eə˞	Where, Pair
oʊ	Tome, Go
eɪ	Say, Neigh
ɪə˞	Fear, Here
ɔɪ	Toy, Join
ʊə˞	Sure, Tour

Symbol	Examples
b	Bar, Slab
d	Kid, Lady
f	Stiff, First
g	Rag, Good
h	Hop, Hike
j	Yellow, Yes
k	Black, Cat
l	Light, Little
m	Man, Clam
n	Ten, Not
ŋ	Finger, Singing
p	Sap, Pet
r	Rest, Trap
s	Sand, Miss
ʃ	Ship, Trash
t	Tea, Setting
tʃ	Cheer, Church
θ	With, Thin
ð	Mother, This
v	Vice, Rave
w	Wish, Widow
z	Zip, Lazy
ʒ	Vision, Pleasure
dʒ	Jet, Huge

Figure 10.19 Some of the phonemes associated with the English language.

Once the Phonemes have been detected, the speech recognition system moves to its second stage. It uses large dictionaries and complex statistical algorithms to attempt to guess what was said. The system tries to make use of both the phonemes and the context of other words and sentences around the current word to make a more educated guess. So, as you can see, the task of determining the content of speech signals, let alone their actual meaning is a complicated and difficult task. Thus, speech recognition and identification systems remain an area of extensive research.

The previous sections have shown some of the many applications of signal processing in the area of sound. These included examples to show the application of signal processing in several aspects of sound processing. We saw that we could extract useful information from sound signals in the form of the fundamental frequency. We learned that we could model acoustic phenomena by examining the echo. We saw that we could generate sound signals artificially and use these signals to convey information. As an example, we examined the DTMF encoding system used in telephone systems. Finally, we examined the application of signal processing in the area of sound recognition and saw that there is a great deal of work that remains to be done. There are many other additional areas of application in the area of sound processing. In fact sound processing is one of the most common areas of signal processing involving singe dimensional signals and many of the techniques used in sound processing can be applied to general signal processing applications as well.

10.5 Chapter Summary

In this chapter we discussed sound and the application of signal processing techniques in the field of sound.

- We began by discussing the basics of sound. We learned that sound is a mechanical wave that is the result of vibrations. These vibrations propagate through the surrounding medium as a serried of compressions and rarefactions.

- Next, we learned what was involved in the human perception of sound. We discovered that the human ear consists of three parts, the outer, middle, and inner ear. The parts of the ear are each responsible for different aspects of our hearing. The outer ear gathers sound waves from the environment and funnels them into the ear canal. In the middle ear, the eardrum vibrates as a result of the compressions and rarefaction in the sound waves. These vibrations are amplified by the ossicles, the three bones in the middle ear. These amplified vibrations are then detected by the cochlea and translated into electrical nerve impulses which are sent to the brain for processing.

- Following this, we learned about some of the sources of sound. We found that there are four sources of sound. These are natural sounds, vocal sounds, instrumental sounds, and artificially generated sounds.

- In order to use our digital signal processing techniques on sound waves we first must convert them to digital signals through a process known a digitization. For sound waves, this process involves four steps.

- First, the sound is converted from a mechanical wave to an analog electrical wave. This waveform is then sampled to arrive at a discrete-time signal. The discrete-time signal must be quantitized to make it discrete in amplitude as well. The quantitized signal is then represented in binary to arrive at the final digital signal.

- Once we learned how to convert sound into a manageable digital signal, we looked at some applications.

- First we found that we could extract information form a sound wave. We looked specifically at obtaining the fundamental frequency.

- Next, we learned how to model a physical phenomenon by using a difference equation to model an echo. We saw how this echo could be implemented mathematically and looked at an example.

- After that we saw that signal processing could also be used to create new sound signals. To illustrate this, we looked at how the phone system uses DTMF encoding to represent the buttons that are being pressed on the phone keypad.

- Finally, we looked at a more complicated application of sound processing. We looked at how sound processing could be used to identify unknown sound signals. This was illustrated by looking at the identification of an unknown bird song. We also examined how this could be extended to identify the phonemes in human speech in order to perform speech recognition. We looked at how such a speech recognition system would operate and noted that these systems are still an area of important research and development.

10.6 Review Exercises

Section 10.2 – The basics of Sound
___

1. Explain how sounds are created. How do sound waves propagate?

2. Describe the system used to compare sound intensities.

3. Explain the process involved in human hearing. What is the range of frequencies that the human ear can hear?

4. List the four categories of sound sources. Give an example of a sound in each category.

5. Explain how human speech works.

Section 10.3 – Preparing Audio Signals for Processing
___

6. After a sound is captured and converted into an electrical signal, there are three steps involved in converting the analog electrical signal into a digital one. List these three steps and explain what happens in each step.

7. Assume you are recording digital sound. You are using a sampling frequency of 44,100 samples per second and representing each sample with 16 bits. How long can you record before you fill the following storage allocations? (Assume that 1 KB = 1024 Bytes, 1 MB = 1024 KB, and 1 GB = 1024 MB)

   a. 1.44 MB (Floppy Disk)
   b. 700 MB (CD-ROM)
   c. 100 MB (Zip Disk)
   d. 4.7 GB (DVD)
   e. 1 GB
   f. 40 GB

8. Assume that you are digitizing sound at a sampling rate of 11025 samples per second with a sample size of 8 bits. How much space is required to record for the following durations?

   a. 5 minutes
   b. half an hour
   c. 30 seconds
   d. 4 hours
   e. 12 hours
   f. 45 minutes

9. How often must we sample a signal to accurately reproduce it? What is this rate called?

10. When stereo sound is digitized, both the left and right channels are digitized separately. With this in mind, how much storage space would it take so store a 5 minute song in stereo if we sample it at 41,000 Hz and use 16 bits to store each sample (CD Quality)? How many minutes of stereo sound can be recorded onto a 700 MB CD?

Section 10.2 – The Basics of Sound

11. Explain how the pitch of a sound can be determined.

12. The following figure displays a waveform that was digitized with a sampling rate of 22050 samples per second. Using this figure estimate

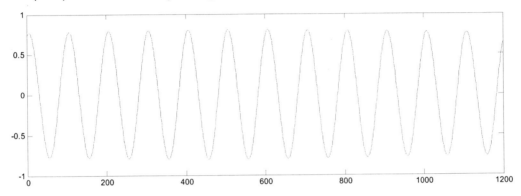

   a. The period of the waveform
   b. The fundamental frequency of the waveform

13. The following figure displays part of a waveform that was digitized with a sampling rate of 11025 samples per second. Using this figure estimate

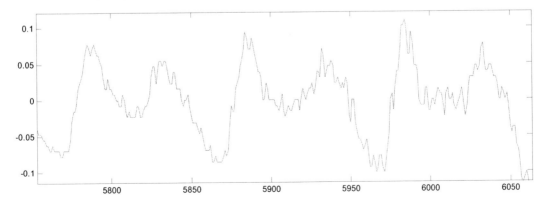

   a. The period of the waveform
   b. The fundamental frequency of the waveform

14. Explain how an echo occurs.

15. Using the echo model, write a difference equation for an echo with the following delay and attenuation properties. (Assume a sampling rate of 44,100 samples per second.)

    a. A delay of 0.5 seconds and an attenuation factor of 50%
    b. A delay of 0.1 seconds and an attenuation factor of 75%
    c. A delay of 0.2 seconds and an attenuation factor of 25%
    d. A delay of 0.01 seconds and an attenuation factor of 10%.

16. Expand upon the echo model to write a difference equation for a double echo with the following properties. (Assume a sampling rate of 41,000 samples per second.)

a. $d_1 = 0.2s$, $d_2 = 0.3s$, $k_1 = 50\%$, $k_2 = 75\%$
b. $d_1 = 0.1s$, $d_2 = 0.4s$, $k_1 = 75\%$, $k_2 = 25\%$
c. $d_1 = 0.5s$, $d_2 = 0.7s$, $k_1 = 60\%$, $k_2 = 40\%$
d. $d_1 = 0.05s$, $d_2 = 0.3s$, $k_1 = 90\%$, $k_2 = 75\%$

17. Explain how DTMF (Touch tone Phone) signals are generated.

18. What frequencies will be sent when the following phone buttons are pressed?

    a. 1   b. 5
    c. 3   d. 9
    e. 2   f. #
    g. 0   h. 7

19. Explain how correlation can be used to identify sounds.

20. What applications might there be for automated sound identification?

Chapter 11. Applied DSP – Image Processing

Chapter Outline

11.1 Introduction

There is an old saying stating that 'a picture is worth a thousand words.' To someone specializing in image processing, however, a picture is worth even more than that. If you think about it, from the moment that you open your eyes in the morning until you go to sleep at night, you are processing a nearly constant stream of images. On top of that, the widespread use of computers has resulted in high performance multimedia in which high-resolution pictures and sound are transferred back and forth on high-speed networks. Image processing in the modern world plays a part in almost every high technology area. Its applications range from medical image processing to data transmission and storage. In the area of medical image processing for example, images of different parts of the body generated through imaging technologies like x-ray imaging, the CAT scan, and ultrasound imaging enable physicians to diagnose tumors, fractures, and a whole host of other physical ailments without having to operate on the patient. This is just one example of the countless applications of image processing.

As we begin our study of image processing, we should begin by asking ourselves one simple question. What exactly is an image? Basically, an image is a special type of two-dimensional signal. In other words, an image contains intensity values that, instead of being arranged linearly in time, as was the case with sound signals, are arranged along an x and y spatial axis. When you first think about what an image is, the concept that first comes to mind is most likely that of a photograph. We typically think of an image as a picture, and although there are other kinds of images, we will use the terms image and picture interchangeably in this chapter.

11.2 Analog Image Processing

The photograph that most of us think of when an image is mentioned is an example of an analog image. Photographs, at least black and white ones, are usually formed through a chemical process in which silver halide is converted into silver in the presence of light. The actual quantity of silver resulting from the conversion process can be any amount within a specific range of values (ranging from no silver halide converted into silver to all the silver halide converted into silver). As a result of this chemical process, an analog grayscale image can have an infinite number of gray shades ranging from black to white. Since photographs are what we most often think of when thinking about images, let's take a look at how basic analog photography works (we will mention digital photography later).

11.2.1 Analog Photography

Analog photography begins with the camera. A camera is essentially a light-proof box with a small opening, or aperture, to allow light to enter. Inside this box, light-sensitive film is placed so that light passing through the aperture will encounter it. The length of time that light is allowed to enter the camera is controlled by a shutter covering the aperture. This shutter opens and closes very quickly so that the film is exposed to light for only a short period of time. In most cases, the camera also includes a series of lenses to ensure that the light entering the camera body properly focuses on the film. So, the way this all works in a typical SLR (single lens reflex) camera is as follows. First, the desired scene is focused using the lens. The camera user determines whether the scene is in focus by adjusting the lens position while looking through the viewfinder. A series of mirrors and prisms redirect the light passing through the lens to the viewfinder until the picture is taken. When the user is satisfied with what they see through the viewfinder, they press a shutter release button. When this button is pressed, the redirection mirror flips up and out of the light path and the shutter briefly opens. This allows the incoming light to come in contact with the film and set the chemical reaction in motion. After a preset amount of time, the shutter closes, the mirror flips back down, and the film is advanced to the next frame. The path of light through a typical SLR camera is shown in Figure 11.1.

So, what happens to the film in the camera? Black and white photographic film is made up of a number of layers, one of these is known as the emulsion layer. The emulsion layer is covered with tiny crystals of silver halide. When the film is exposed to light, the silver halide grains absorb optical energy and undergo a complex change. The grains that have absorbed sufficient energy contain small flecks of silver. So, the incoming light causes a chemical reaction to take place in the film causing some of the silver halide crystals to be transformed into silver. The degree to which the silver halide is converted is dependent on the amount of energy that is absorbed from the incoming light. Thus, the more light that is absorbed, the more of the silver halide will be transformed into silver. Now that the film has been exposed to light, it must be developed before the picture can be viewed.

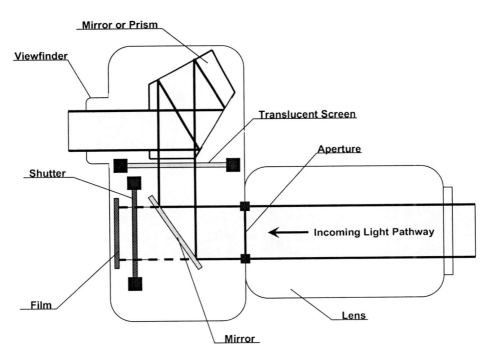

Figure 11.1 An illustration of the basic inner workings of a film camera.

In order to develop the film, it is placed into a developing agent. This chemical will act on all of the silver halide grains that absorbed enough photonic energy when exposed within the camera. These silver halide crystals, which contain silver ions due to their earlier photochemical reaction, are converted into opaque silver crystals. The longer the film is exposed to the developing agent, the more of these exposed silver halide crystals will be converted into silver crystals. Once the silver crystallizes and the photographer is happy with the contrast, the development reaction must be stopped. This is done by rinsing the developing chemical off of the film. Now, in order to prevent the film from absorbing additional light, it must be 'fixed.' Fixing is the process by which the remaining silver halide crystals are removed from the film, thus preventing any further changes due to light exposure. Thus, the picture is permanently 'fixed' at its current gray shade levels. So, we now have a negative image, in other words, those parts that absorbed more light will be more opaque or darker and those that absorbed little or no light will be clear. Once the negative is formed it can be enlarged by projection. The negative is projected onto another piece of photographic paper resulting in a positive image. Thus, a black and white photograph is produced from the light originally captured by the camera.

The process is slightly more complicated when dealing with color film. Color film contains three layers of silver halide crystals. Each of these layers is sensitive to a different color of light, red, green, and blue. When the film is developed, different chemicals are used for each of the three layers. Instead of depositing opaque silver crystals, the silver is removed from color negatives to leave special color dyes in each of the film layers. The colors of these dyes are obtained by subtracting the desired color from white. Thus, the red layer is represented with cyan dye, the green layer with magenta dye, and the blue with yellow dye (we will talk more about additive and subtractive color models later in the chapter). This negative color image is then projected onto color photographic paper to yield the desired positive color print. So, as we have seen, analog photography is capable of producing images with both an infinite range of colors and an infinite range of gray shades.

11.2.2 Human Sight

While we may initially think of an image in terms of a photograph, we would not be able to see images at all without our eyes. The human eye is an amazing image processing tool. So, how do our eyes work? To understand that, we need to take a look at how the eye is constructed. As is shown in Figure 11.2, the human eye is composed of several major structures. Let's look at some of these.

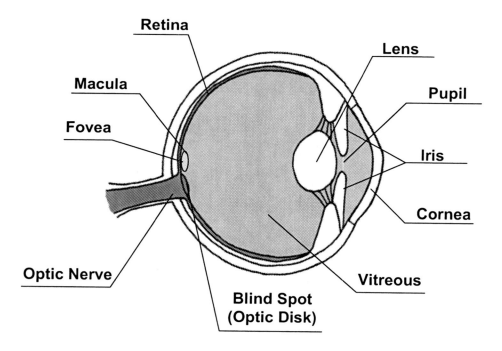

Figure 11.2 The anatomy of the human eye.

Starting at the front of the eye, we encounter the cornea. The cornea is the clear front window of the eye. It serves the purpose of transmitting and focusing light into the eye. In fact, the cornea is responsible for much of the job of focusing what you see. Next, the amount of light that is allowed to enter the eye is regulated by the iris, the colored part of the eye. It contracts and expands to change the size of the pupil, the hole in the center of the iris that allows light to enter the inner eye. Upon passing through the pupil, light is focused by the lens. The lens is the second part of the eye's focusing system. Incoming light is first focused by the cornea which takes care of the rough focusing and then the lens performs the task of fine tuning the focus. Muscles attached to the lens modify its shape in order to change its optical properties. The light focused by the lens is projected on to the retina, a layer of nerve cells that line the back of the eye. These cells are light sensitive and convert the light that they receive into electrical impulses that travel down the optic nerve to the brain where they are decoded into what we see.

The photosensitive nerve cells in the retina combine to form two basic types of structures, rods and cones. Most of the retina is composed of rods. These structures are responsible for the gross detection of movement, shapes, light and dark, peripheral vision, and vision in low light. The vision provided by rods is black and white. Each human eye contains over 100 million rods. The second photoreceptive structure is the cone. Each eye contains around 7 million cones. While there are cones scattered throughout the eye, most of the cones are located in the portion of the retina that deals with the center of vision. This area of the retina is called the macula and the center of the macula is called the fovea. In this area, there are around 150,000 cones per square millimeter. Unlike rods, cones are capable of sensing color and can determine fine visual details. Although the cones provide us with color vision, they are not as light sensitive as the rods. In dim lighting, we must rely on our rods for vision. This is one reason why it is hard to tell the color of an object in a darkened area. So, both rods and cones contribute to our vision differently, but while they serve different purposes, they are both necessary for our eyes to function properly.

The final structures labeled in Figure 11.2 are the **vitreous**, the clear, jelly-like substance that fills the eye and the **optic disk**, which is the point at which the optic nerve connects to the retina. This portion of the retina is incapable of sensing light and, thus, is sometimes referred to as the eye's blind spot. Thus, we now have a general understanding of the basic construction of the human eye. With this information we can begin to explore how all of these structures work together to allow us to see.

Our vision is dependent on light being reflected off the objects around us. After this light has bounced off an object, it enters the eye through the cornea. This transparent membrane provides the initial focusing of light as it enters the eye. Next, the iris contracts or relaxes in order to adjust

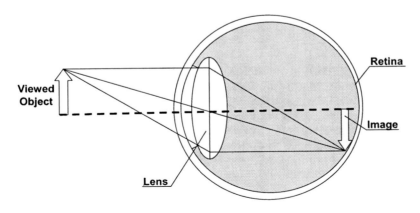

Figure 11.3 The path of light in the eye.

the amount of light reaching the lens. The lens in turn changes shape to fine tune the focus of the incoming light on the retina. The path of light through the eye is shown in Figure 11.3. As the figure demonstrates, the image that falls on the retina is actually upside-down. The brain actually flips the image back over after the image is sensed by the retina. So, how does the retina detect the light that lands on it?

Over a century ago, an English physicist named Thomas Young predicted that the human eye detects color by the relative amounts of red, yellow, and blue light, which bounces off an object and into the eye. He chose this three-color combination because he knew that painters could produce practically every other color simply by combining a mixture of these three. Thus, he theorized that inside the eye, there existed three types of receptors, one for red, one for yellow, and one for blue. He believed that the color of an object was determined by the relative amounts of each of these colors. This supposition regarding light became known as the tri-chromatic theory. This theory has since been revised to use the colors: red, green, and blue.

The first time that the eye actually senses light energy is when light hits the rods and cones in the retina. However, this light must first travel through several layers of nerve cells before it reaches the rods and cones. As mentioned before, rods cannot sense color but are capable of peripheral vision and night vision. Cones, on the other hand, are less sensitive to light but can sense color and are capable of detecting more detail. The ratio of rods to cones in the eye is twenty to one. Thus, a large portion of the eye is dedicated to simply sensing motion and light while a smaller portion is reserved to the resolution of color and fine detail. Rods contain the photoreceptor rhodopsin, and are one thousand times more sensitive to light than cones are. This is because rods contain many more membranous disks and thus a larger total number of receptors. However, since rods only have one type of photoreceptor, they can not distinguish between different colors. In contrast, cones contain any one of three different photoreceptors, one optimally activated at the 570nm wavelength of light (red), one optimally activated at 535nm (green) and lastly one optimally activated at 445 nm (blue). So, color is distinguished by the relative activation of these three types of receptors in the retina. Once the photoreceptors in the rods and cones are activated, they send nerve impulses via the optic nerve to the central visual system of the brain. So, we see things because light energy causes a signal to be sent from nerve cells in the retina to the brain where it is processed into a visual image.

Now that we understand what analog images are, how we capture these images as photographs, and how the human eye senses image signals, we can begin to explore the world of digital images. Digital images form an important subclass of the overall set of images. Since most of the image processing that engineers perform is done on computers, an understanding of the basics of digital images is essential.

11.3 Digital Images

So, what is a digital image and how is it different from an analog one? Good digital images can often be difficult to visually distinguish from their analog counterparts. This is because the human eye is not capable of sensing extremely small details. Think about images printed in newspapers. The pictures are not continuous analog images, but are rather built up of a series of tiny colored dots of ink. These dots only become visible on very close inspection, typically with some magnification. Even photographs taken on regular film are not completely analog images. Their resolution is limited by the number of silver halide crystals that can be placed on the film. If you enlarge them beyond a certain point, the image quality will degrade and the image will exhibit a noticeable graininess. Photographs are still considered analog images, however, because they do exhibit the capability to produce an infinite number of shades of gray (or an infinite range of colors). A true digital image, on the other hand, is restricted in both its spatial coordinates and in its allowed intensities. As you can see in Figure 11.4, if you zoom in on a digital image, at some point you will begin to see a series of varying intensities arranged in a grid pattern. Each element of the grid can hold only one color or shade of gray. These discrete elements that make up a digital image are called picture elements, or **pixels** for short.

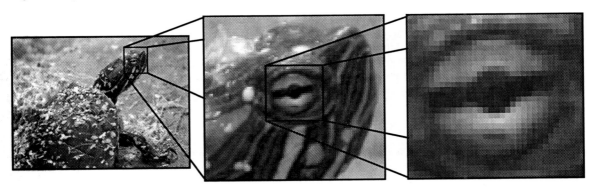

Figure 11.4 An illustration of the individual elements or pixels present in a digital image.

Each pixel in a digital image can hold one of a preset range of values. The size of this range is determined by the amount of space that we are willing to use to store our image. The image as a whole is made up of a large number of these pixels arranged in a grid. Thus, a digital image is composed of a set of evenly spaced points that hold one of a predefined set of values. As you have no doubt seen, this is very similar to the way that digital sound signals are constructed. The only difference is that images have intensities that vary based on spatial location in two dimensions and sounds have amplitudes that vary in a single dimension (time).

Now that we know that an image is constructed from pixels arranged in a grid, how do we go about describing these images? How can we talk about a single element in the image's grid? We work with images in a form that, at this point, we should all be familiar with. Images are typically worked with in the form of matrices. If you think back to our discussions of matrices in chapter three, you will recall that they are a perfect fit for describing digital image. We can easily map an image into a matrix by recording the intensity value for a pixel into the matrix element that corresponds to the pixel's row and column location in the image. Thus, an image that is 800 pixels wide and 600 pixels high can be represented as a 600 x 800 matrix (600 rows and 800 columns). By mapping the pixels intensities into the elements of a matrix, we can also describe the intensity of an individual pixel within the image simply by using the row and column of its

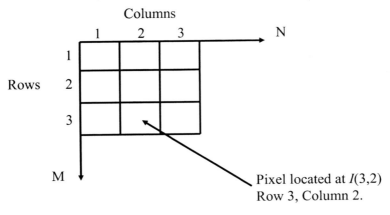

Figure 11.5 An example of the matrix notation used to describe a digital image.

position in the matrix. So, if we want to describe the intensity of a pixel located on a specific row and column of an image, we can refer to it as $I(i,j)$. Where I is a matrix representing the image, i represents the desired row of the element and j represents the desired column. Therefore, the intensity of a pixel located in the third row and second column of an image could be referred to as $I(3,2)$. This is illustrated in Figure 11.5.

So, now we know that digital images are simply two-dimensional signals that are discrete both in amplitude and spatial location. These discrete elements are known as pixels and the intensity of a given pixel can be referred to as a location in a matrix. This answers our question of what a digital image is. Therefore, the next logical question we should ask is how do we obtain these digital images?

11.3.1 Obtaining Digital Images

When we looked at sound in the last chapter, we found that digital sound signals were approximations of their analog counterparts. These signals were obtained by first converting the analog signal to a discrete-time signal by sampling it, and then creating a signal that was discrete in both time and amplitude by quantizing the signal. This resulting signal could then be easily encoded into binary as a digital signal. Now, we must expand our digitizing process to handle signals of more than one dimension. In the case of images, we have a series of intensities that are arranged in a grid pattern. So, when we sample our analog image, we must take intensity readings at evenly spaced locations in both the x and y directions. This basically amounts to placing an evenly spaced grid over the analog image and taking intensity samples in all the locations that the grid lines intersect. The spacing between samples in our image determines the amount of detail that we are able to capture. Thus, it is very important to make sure that the sampling rate is high enough to capture the information that is needed. Once samples of the analog image have been taken, we have an image that is discrete in its spatial coordinates, but not in its intensities. Thus, we must next quantize the sampled values to arrive at a signal that is discrete in position and amplitude. Figure 11.6 demonstrates the processes of sampling and quantization on a typical grayscale image.

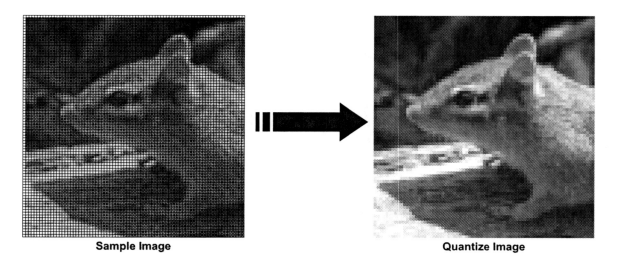

Sample Image   **Quantize Image**

Figure 11.6 An illustration of the processes of sampling and quantization

The range of colors or shades of gray that can be represented in the image depend on the amount of space we are willing to use for each sample. In most grayscale images, 1 byte is allotted for each sample. This means that we can store one of up to 256 different intensity values for a given pixel. These values range from 0, which represents black to 255, which represents white. The quantized image in Figure 11.6 was quantized with 1 byte per sample or 256 gray

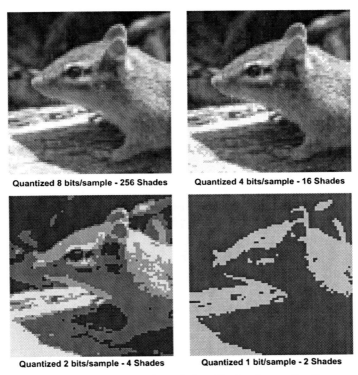

Quantized 8 bits/sample - 256 Shades   Quantized 4 bits/sample - 16 Shades

Quantized 2 bits/sample - 4 Shades   Quantized 1 bit/sample - 2 Shades

Figure 11.7 An illustration of the effects of quantization.

levels. If we were to decide we wanted to use less space, we could quantize the image using a smaller set of values. Figure 11.7 illustrates the results of quantizing an image with 256, 16, 4, and 2 allowed values. As you can see, the use of fewer values allows us to use less space to store each pixel, but this comes at the price of sacrificing the visual quality of the image.

Finally, once we have sampled and quantized our analog image, we can complete the process of analog to digital signal conversion by encoding the quantized sample values into a binary sequence. This sequence of binary bits can then easily be transferred, processed, or stored for future reference.

Our discussion up to this point has dealt with the process involved in transforming an analog image into a digital one. However, we have neglected to look at how we sense the analog image in the first place. Obviously, our signal needs to be in a form that allows us to perform the actions of sampling and quantization. Our image typically originates from a series of light intensities. So, how do we get these varying intensities of light into an electronic format that allows us to perform sampling? The answer to this involves specialized hardware. We saw that when we process sound waves, we first had to convert them into an electrical signal using a sensor, a microphone. When we work with images, we need to use special sensors that react to different wavelengths of light. Typically these sensing elements are very small devices. An example of what one of these sensing elements might look like is shown in Figure 11.8. Each of these elements is capable of sensing one sample point of incoming light. Thus, we must use a large number of these sensors configured in various ways to achieve a useful conversion from light into electrical signals.

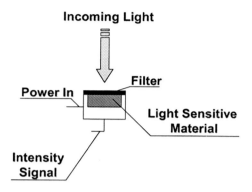

Figure 11.8 An example of the layout of a typical light sensitive element.

If we are trying to convert a stationary image, such as a photograph or a page in a book, we can use a linear arrangement of these sensors and pass this line of sensors across the image. An example of a linear arrangement such as this is shown in Figure 11.9.

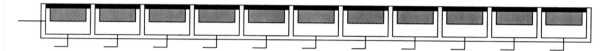

Figure 11.9 An example of a linear sensing array similar to those used in flat-bed scanners.

As you might have guessed, arrays like this one are used in scanners. The image is placed on a transparent surface above the scanning array. A light source then illuminates the object we wish to scan while the linear sensor array is moved across the object. Each time the array moves we get a new row or column of pixel values. Thus, by moving across the entire image, we can sense the light intensities at all of the required sample points and construct a digital copy of the original analog image. In this type of conversion application, our sampling frequency in one direction is determined by how closely spaced the sensors are within the array and in the other direction by the distance traveled by the array between column or row captures.

If, on the other hand, we need to capture a digital image from a live scene, we do not have the time move a linear array of sensors around. We must capture our entire image all at once. To do this, we must arrange our sensing elements in a two dimensional structure. Now, when light hits this two dimensional sensor array, we can read an intensity value from each element. Thus, we can capture an image with a resolution based upon how tightly packed the elements are within the array. This is the method used by digital cameras. Instead of using film, we place an array of sensors behind the shutter, as shown in Figure 11.10. When the shutter release is pressed, the shutter will open just as it did when we were using film. In this case, however, the light encounters our light sensitive array. Each element in this array is responsible for capturing the intensity information for one pixel in the resulting image. Thus, when the shutter closes, we quantize the intensities sensed by each sensor element and encode these values into a digital image. This image can then be displayed or saved to some form of storage media.

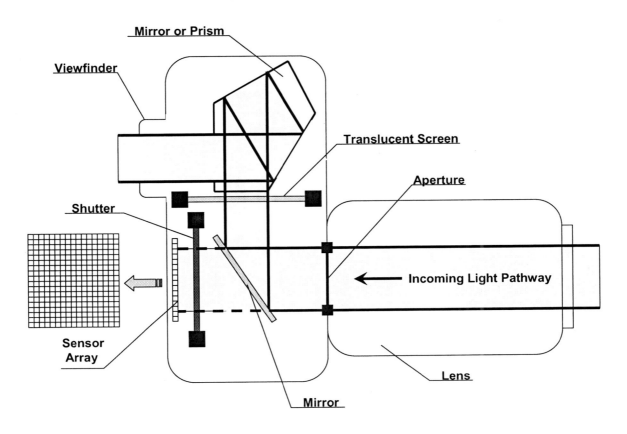

Figure 11.10 An example of the light path in a digital camera. Note that the only difference between this camera and the one in Figure 11.1 is that the film is replaced by a light sensitive array.

So, we have seen how analog images are converted into digital images. We have also examined some of the specialized hardware that is used to create these digital images. Now let's look in a little more detail at how these images are represented in memory once they are captured.

11.3.2 Grayscale Images

We will begin by looking at grayscale images. Most of the discussion so far has dealt the capture and conversion of images of this type. This is due to the fact that the processes of conversion, storage, and manipulation of grayscale images are somewhat easier to understand than the similar processes that deal with color images. As we mentioned above, when a black and white image is captured using a scanner or digital camera, the image is first sampled. In most cases, this sampling is somewhat restricted by the hardware that is being used. You cannot sample in higher detail, or resolution, than the hardware allows. Where you do have some degree of choice is in the quantization process. You can choose the number of distinct levels that you wish to represent based on the space you allow per sample. In most cases, however, grayscale images are always stored with a sample size of one byte (8 bits) per pixel. This is because the human eye has difficulty noticing any difference if additional levels are used. Thus, these levels are usually considered an unneeded waste of storage space. If however, the image is intended for a use other than viewing, it may be beneficial to store a more detailed value for each sample. So, while you may find that most systems allow for only 256 gray levels, you will occasionally encounter some images with more levels than this.

Once a grayscale image has been captured and converted into a digital image, it is stored as a two-dimensional array (a matrix) in computer memory. This configuration is illustrated in Figure 11.11 below.

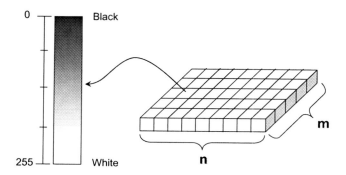

Figure 11.11 An illustration of the way in which grayscale digital images are represented and stored.

Each element in this two-dimensional array contains the quantized intensity that was associated with that portion of the image during the digitization process. In most cases, this element will contain a value ranging from 0 to 255. A value of zero corresponds to black (no intensity) and a value of 255 corresponds to the brightest white (maximum intensity). Any value between these two extremes will be displayed as a shade of gray. Thus, the representation of grayscale images is a simple process, but what about color images?

11.3.3 Dealing with Color

Color images are a bit more difficult to deal with. We just learned that to digitize a grayscale image, we look at the overall intensity level of the sensed light. We must take a few additional steps to capture color information. You no doubt noticed that the sensing elements shown in Figure 11.8 included a filter. The purpose of this filter component is to restrict the light sensed by the element to a specific range of light frequencies. Thus, if we use this filter to restrict the light sensed by the sensor so that it lies within the wavelength range of a specific color of light, we can detect the intensity of that specific color for that sample location. The wavelength range of the visible spectrum of light is shown in Figure 11.12.

Now, if we model the cones in the human eye and create three sets of sensors, we can detect the intensities of the same three colors that the cones sense, red, green, and blue. Since we know from physics that any color of light consists of a combination of frequencies contained in the three "primary" color ranges (red, green, and blue occupy a range of wavelengths in the spectrum), we can digitally represent these colors as the combination of the intensities sensed by our red, green, and blue frequency tuned sensors. Thus, the detection of a color image comes down to sensing the intensities of each of the three primary colors instead of just the overall intensity of the incoming light. Once we have sensed the intensity of the red, green, and blue content of the light for each pixel in the image, we quantize the values for each color. Here we must choose how much space each sample should use. Since we have to store values for red, green, and blue for each pixel, the required space will three times what would be required to store a similar grayscale image.

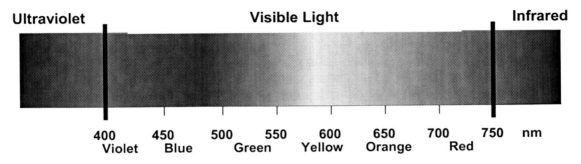

Figure 11.12 The wavelengths of visible light (in nanometers)

Most computer imaging systems utilize what is known as 24 bit color. In this quantization method, each of the three primary color intensities is allowed one byte of storage per pixel for a total of three bytes per pixel. In this arrangement, each color ranges from a value of 0 representing none of that color present in the pixel color to 255 representing the maximum amount of that color present in the pixel. The combinations that can be made with 256 levels for each of the three primary colors amounts to over 16 million distinct colors ranging from black (0,0,0) to white (255,255,255) and including everything in-between. This is more colors than the human eye can usually distinguish. So, a color image constructed in this way yields a result that looks natural and exhibits smooth transitions between colors. The easiest way to visualize this color configuration is as three distinct color layers or "planes." When these three planes are sandwiched together, a full-color image is produced. Most computers store color digital image information in three dimensional arrays. The first two indexes in this array specify the row and column of the pixel and the third index specifies the color "plane" where 1 is red, 2 is green, and 3 is blue. Figure 11.13 demonstrates this method of storing color image information.

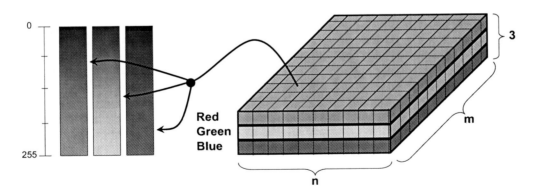

Figure 11.13 An example of the way in which color digital images are represented and stored.

Using the red, green, and blue content of the sensed light to digitally represent the color also works well for displaying the image on a monitor or television. A typical monitor consists of a series of scan lines (rows). Along each of these lines you will find a series of triangular "pixel dots." Each of these triangular structures contains three types of phosphorus, one that glows red, one that glows green, and one that glows blue. When the different types of phosphorus in these pixel dots are excited to various extents, a specific color is produced. Thus, by storing the color in this format, the computer already knows to what extent to excite each of the phosphorous colors when displaying the image on screen, but what about when we want to print an image on paper?

11.3.4 Ways of Describing Color – Color Models

The description of colors using pigments, such as printer inks, does not work the same way that describing colors in terms of light does. When mixing colors of light, we do so additively. For example, if we wish to add more blue to a color, we simply increase the intensity of the blue component of the color. Pigment colors, on the other hand work through subtraction. When you look at a printed color, the color that you see is based upon the colors of light that are reflected from the printed surface. Thus, if you look at a pure blue dot on paper, you see the blue light reflected from the dot. All other colors of light are absorbed. If you wish to increase the blue content of a color using a subtractive color method, you must remove yellow. The yellow pigment reflects red and green light, but absorbs blue. Thus, if we remove some of the yellow pigment, less blue light will be absorbed and the resulting color will contain more blue light. So, when you mix all the different colors of light, you will get white light, but if you mix all the different colors of pigment, you will get black. As you have just seen colors can be described in different ways. Because of this, we use different models to represent them. Let's take a look at some of these models now.

11.3.4.1 The RGB Model

By now, we should be fairly comfortable with the RGB model. As we have already mentioned, the RGB model is one of the most common ways of describing the color content of an image. This is probably because of the model's simplicity and the fact that the human eye uses a similar approach. Since many displays (monitors for example) create colors by mixing red, green, and blue light, this model is a natural fit to a large range of applications.

When looking at light from a practical perspective, the quantity that differentiates one color from another is its frequency. When describing light waves, however, we usually talk about wavelengths rather than frequencies. Since wavelength is simply the inverse of the frequency we are essentially describing the same thing, just in different terms. We know that when white light falls on any object, the object absorbs certain frequencies, or wavelengths, and reflects some others. These reflected wavelengths fall on the retina and stimulate the photoreceptors (particularly the cones when dealing with color), which then tell the brain that it "sees" a particular color. A body that reflects light that is relatively balanced in all visible wavelengths, that is it absorbs no frequency and reflects all frequencies looks white in color. A leaf on the other hand, predominantly reflects frequencies in the 500 to 570 nm (10^{-9} meter) range and absorbs all other frequencies. The 500 to 570 nm range of wavelengths are the wavelengths of light that we perceive as the various shades of green. So a natural classification of colors is to separate the red, green and blue and this is what the RGB model does. Red, green and blue are called the primary colors.

When describing the RGB model we can start by visualizing a three-dimensional coordinate axis. Next, we represent the primary colors red, green and blue on these three axes. Now, we can describe all of the colors that can be represented in the RGB model by providing their coordinates in this system using the Cartesian coordinate system. An example of the RGB color cube can be seen in Figure 11.14. In this system, the range for each color is between 0 and 1 (The color intensity value divided by the maximum possible intensity value). So, to represent pure red, for example, we would choose (1,0,0). The first digit represents red, and the maximum value that it can take is 1. The other places represent green and blue respectively. Another color such as cyan would be represented as (0,1,1) since it is a mixture of blue (0,0,1) and green (0,1,0). Black is represented at the origin (0,0,0) and white is a mixture of all three primaries, and hence it is represented as (1,1,1). Thus, we can see how the RGB system is additive. To find the position in the cube for a mixture of two other colors we simply need to add their color components together. Thus any color can be formed within this one unit cube. It is important to note that the range of grayscale colors used in black and white images lies along a diagonal line running from the origin (0,0,0) to the color white (1,1,1). Thus, the grayscale values are the points in the RGB model space where all of the color intensities are equal.

RGB Color Cube

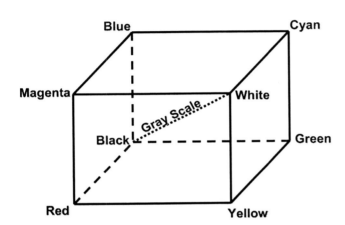 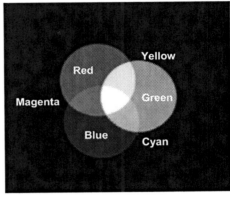

Figure 11.14 An illustration of the RBG color space and a demonstration of the results of mixing the various RGB primaries.

11.3.4.2 The CMY Model

While the RGB model uses an additive approach, the CMY (Cyan, Magenta, Yellow) color model is based on a subtractive primaries scheme. Basically, this means that while you define a color in RGB based upon the amounts or red green and blue that it contains, a color in the CMY system is defined based upon the colors of light that it absorbs. For example, the color yellow absorbs the primary color blue and reflects the other two. Likewise, Magenta absorbs green and cyan absorbs red. A subtractive scheme such as this is widely used in cases where color pigments have to be handled. For example, when you mix two colors of paint, the new color is not determined by adding the colors as it would be in the RGB model. Mixing two colors of paint increases the range of wavelengths that the resulting pigment can absorb. So, if we add cyan, which absorbs red light and magenta which absorbs green light, the resulting pigment will absorb both red and green light. This means that the color you will see is the blue light that will be reflected. Thus, if we mix all of the subtractive primaries (cyan, magenta, and yellow) together, the resulting pigment will absorb all wavelengths of light resulting in the color black. You can see some of the results of mixing the subtractive primaries as well as the color cube for the CMY model in Figure 11.15. So, as we have just seen, the subtractive method used by the CMY model is good at describing the effect of mixing various pigments. This makes it perfectly suited for use with devices such as printers. This leads to an interesting question. If we have our image colors defined in the RGB model for use on the computer, how do we convert them into the CMY color so that we can print the image? If you think about it this is not a difficult problem. Each subtractive primary has an additive primary that it absorbs. This means that it can be thought of in the RGB model as the sum of the remaining two colors. Using this information we can come up with a matrix equation to convert a color from the RGB scheme into the CMY scheme. This equation is written as follows.

$$\begin{bmatrix} C \\ M \\ Y \end{bmatrix} = \begin{bmatrix} 1 \\ 1 \\ 1 \end{bmatrix} - \begin{bmatrix} R \\ G \\ B \end{bmatrix}$$

So, we see that we can arrive at the correct values of cyan, magenta, and yellow by subtracting the values for red, green, and blue from one.

CMY Color Cube

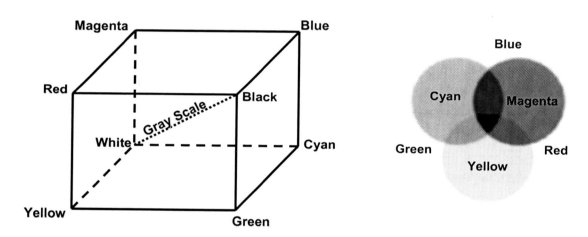

Figure 11.15 An illustration of the CMY color space and a demonstration of the results of mixing some of the various subtractive primaries.

11.3.4.3 The YIQ Model

Another color model you are likely to encounter is the YIQ model. In this model, Y represents luminance (the brightness factor – proportional to the amount of light perceived by the eye) and I and Q represent the chromaticity (the color information). In this case, I stands for in-phase color component and Q stands for quadrature phase color component.

So, you may ask where would you ever encounter this color model. The YIQ model is the default model used in the transmission of color television broadcasts. One major advantage that the YIQ model has over other models where television is concerned is that it is compatible with black and white television sets as well as color. The idea is that the chromaticity (color information) is processed separately from the brightness (luminance). Thus a black and white television set can disregard the chromaticity information and use only the information on brightness. This also works well because one of the properties of the eye is that it is more sensitive to changes in brightness than it is to changes in hue or saturation (chromaticity). Thus, by decoupling the color component from the brightness, this model allows us to use more signal space to code the luminance and less space to code the chromaticity. Another benefit of the YIQ model is that we can easily convert between the RGB and YIQ model by using the following matrix equation.

$$\begin{bmatrix} Y \\ I \\ Q \end{bmatrix} = \begin{bmatrix} 0.299 & 0.587 & 0.114 \\ 0.596 & -0.275 & -0.321 \\ 0.212 & -0.523 & 0.311 \end{bmatrix} \begin{bmatrix} R \\ G \\ B \end{bmatrix}$$

11.3.4.4 The HSI Model

In this model, H stands for hue, S for saturation, and I for intensity. Artists that work with color coordination use this model frequently as it comes closest to describing human perception. The benefit that this model has over the previously described YIQ model is that H (hue) and S (saturation) components are more intuitive when describing the color properties of light.

The best way to describe the HSI model is to imagine a two-ended hexagonal pyramid. The middle hexagon has corners at the three additive and the three subtractive primaries. The middle is white or achromatic light (grayscale). Thus, the angle around the hexagon is the hue, and distance from the center axis is saturation. As you go up or down the pyramid, you add to or subtract from the intensity of the light. When you get to an intensity of zero, everything fades to black. As you increase the intensity (at a fixed distance from the center axis), everything becomes less saturated and eventually becomes pure white. Thus, the values of black and white lie at the top and bottom tips of the double-tipped pyramid. This model is illustrated in Figure 11.16.

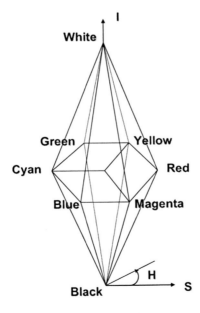

Figure 11.16 An illustration of the HSI color model.

11.3.4.5 The HSL Model

The HSL model is very similar to the HSI model mentioned above, except that it uses a cylinder instead of a hexagonal pyramid. In this model, the L stands for luminosity and the H and S represent hue and saturation just as in the HSI model. So, the difference between the two models is that luminosity is used here instead of intensity. Luminance is similar to intensity in that it is proportional to the physical power of the light source. Luminance was created to describe scientifically why we experience different sensations towards objects with different levels of brightness. For example, when we watch television, the sun in the picture appears to be brighter than a building or tree. It appears to be emitting more light so it has a greater luminance. But in reality, each pixel in the screen is emitting the same amount of energy, thus everything on the screen has an equal intensity. So, what does this model look like?

The best way to visualize this model is to imagine a cylinder that contains all the colors in the spectrum. The hue is described as an angle. In other words, if red is located at 0 degrees then cyan, the opposite of red, can be found at 180 degrees or exactly across from red. So as you move around the perimeter of the circle, the hue changes. The saturation of the color is based upon the radial distance from the center of the cylinder. Thus, the color circle shown in Figure 11.17 below represents a cross section of the middle of our HSL cylinder.

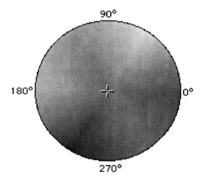

Figure 11.17 An example of the color circle obtained by taking a cross section of the HSL cylinder.

The definition of hue and saturation alone cannot fully describe a color. So, we also make use of luminance. In our HSL cylinder model, luminance is the vertical component ranging from 0% luminance at the bottom of the cylinder to 100% at the top. The color circles in Figure 11.18 represent cross sections of the cylinder taken from bottom to top starting with 0% luminance, then 25%, 50%, 75% and finally 100% luminance. Notice that the color disks for 0% and 100% luminance consist entirely of black and white respectively. The colors wash out to white as luminance increases and colors darken to black as luminance decreases.

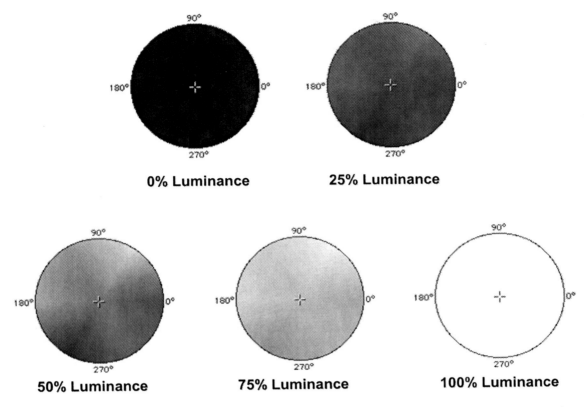

Figure 11.18. Examples of cross sectional color disks taken at different luminance levels in the HSL cylinder.

As we have seen in the previous sections, there are many models for describing color. In fact, the models listed here are just some of the more common ones and are by no means a complete list. Even though you are not likely to encounter all of these models on a regular basis, it is important to be aware of them in case you encounter an application that could benefit from their use.

11.4 Basic Digital Image Manipulation

Have you ever wondered how your basic image processing and paint programs perform some of their processing? Now that we have a basic idea of how grayscale and color images can be represented, we can begin examining some of the approaches used by these applications. Our examples from this point on will show the application of our techniques on grayscale images. This is done primarily because the techniques are simpler to illustrate on images of this type. The same techniques can be used on color images as well, but would require processing on all three of the RGB color levels. In the next sections, we will examine some of the most basic operations that can be performed on an image. Once we understand these, we can move on to more complicated topics such as filtering, noise removal, and compression. Let's begin by looking at the simple operations involved in horizontally and vertically flipping images.

11.4.1 Horizontal and Vertical Flipping

One of the most basic transformations that we can perform on an image is to flip it horizontally or vertically. The operation requires only that you change the order of the rows (vertical flipping) or columns (horizontal flipping). Let's start by looking at horizontal image flipping, sometimes called mirroring. To perform this particular transformation, as shown in Figure 11.19, we need to reverse the order of the image's columns. In other words, the last column becomes the first column and the first column becomes the last column of the new image. We can describe this mathematically as follows.

$$F(i, j) = I(i, (N+1) - j)$$

Where F is the flipped image, I is the original image, N is the number of columns in the original image, and i and j are indexes for the rows and columns such that:

$$i = 1,2,3,...,M$$
$$j = 1,2,3,...,N$$

Recall that M and N are the number of rows and columns in the image respectively.

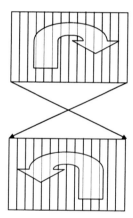

Figure 11.19. An illustration of the process of horizontally flipping (mirroring) an image.

So, if all we have to do to flip an image horizontally is reverse the order of its columns, it would follow logically that we would have to reverse the order of the image rows to flip it vertically. This is exactly the case. To vertically flip an image, as shown in Figure 11.20, we reverse the row order. This can be described mathematically in much the same way as the horizontal flip.

$$F(i, j) = I((M+1) - i, j)$$

Where F is the flipped image, I is the original image, M is the number of rows in the original image, and i and j are indexes for the rows and columns such that:

$$i = 1,2,3,..., M$$

$$j = 1,2,3,..., N$$

Once again, *M* and *N* are the number of rows and columns in the image respectively.

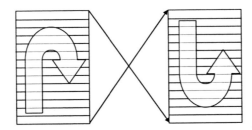

Figure 11.20. An illustration of the process of vertically flipping an image.

Thus, we find that the process of flipping an image either horizontally or vertically amounts only to an interchanging of the orders of the rows or columns of the images. Keeping this in mind, how would we perform the following transformation?

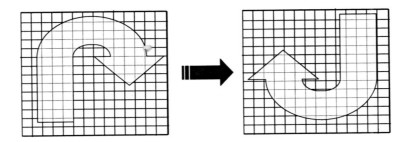

As you might have guessed, this is a transformation in which the image is flipped both horizontally and vertically. So, all that we need to do to implement it is modify our math so that we interchange the order of both the rows and columns at once. Such a transformation would look something like this:

$$F(i, j) = I((M + 1) - i, (N + 1) - j)$$

Where *F* is the flipped image, *I* is the original image, *M* is the number of rows in the original image, *N* is the number of columns in the original image, and, as usual, *i* and *j* are indexes for the rows and columns such that:

$$i = 1,2,3,..., M$$

$$j = 1,2,3,..., N$$

This transform can also be viewed as a 180 degree rotation of the image. We will talk more about image rotation in a moment. The application of each of the image flipping transforms that we just discussed is displayed in Figure 11.21.

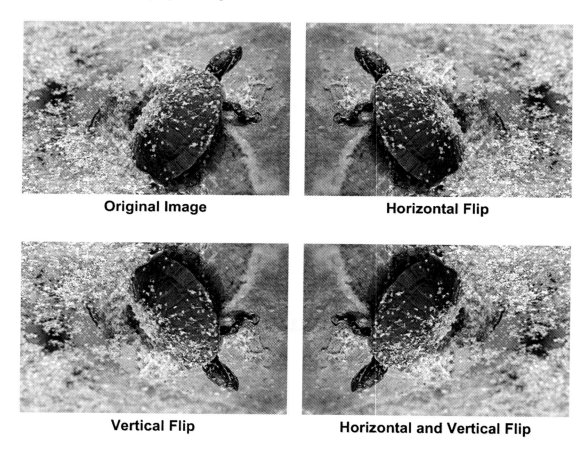

Original Image   **Horizontal Flip**

Vertical Flip   **Horizontal and Vertical Flip**

Figure 11.21 An illustration of each of the image flipping techniques.

11.4.2 Cut and Paste

Another simple but useful image transformation involves the copying of the information contained in one area of the image into another area. This is often called "cut and paste in image processing applications. So, how does it work? It is a simple operation, but requires several steps to carry out. Let's look at these steps along with a set of example images to illustrate the concept.

1.   Select an area in the original image to copy.

142 Row x 148 Column Selection within a 266 row x 400 column image

2.   Copy the contents of this area into a new, temporary memory array.

142 Row x 148 Column temporary image

3.   Select an additional area of equal size to copy the first selection into.

A new 142 Row x 148 Column selection

4. Determine row and column offsets for the top left corner of the new selection. These offsets are defined as the number of rows and columns above and to the left of the top left pixel in the selection.

Row Offset O$_R$=100, Column Offset O$_C$=230

5. Copy the contents of the temporary array into the newly selected area pixel by pixel. This can be mathematically expressed as follows:

$$I(O_R + i, O_C + j) = T(i, j)$$

Where *I* is the image we are pasting into, *T* is the temporary image created in step 2, O$_R$ and O$_C$ are the row and column offsets of the region we are pasting into, and *i* and *j* are indexes such that:

$$i = 1,2,3,...,r$$
$$j = 1,2,3,...,c$$

Where r and c are the number of rows and columns in the selected region.

The original image *I*, with the copied section, *T* pasted in at an offset of 100 rows and 230 columns.

So, as you can see, the process of copying one part of an image and pasting it over another area is a simple task of finding the appropriate offsets and copying the sections pixel by pixel.

11.4.3 Rotation

The rotation of images is an important transformation. There are essentially two different types of rotation that we can perform. The first type is a simple rotation. Rotations of this kind are taken in 90 degree increments. Thus, rotating and image by 90, 180, or 270 degrees can be done by simple pixel repositioning. We saw one of these rotations when we examined image flipping. When an image is flipped both horizontally and vertically, it is the same as rotating that image by 180 degrees. The other two angles of rotation, 90 and 270 degrees can be achieved by rotating the image 90 degrees clockwise and counterclockwise. Let's look at how this is accomplished.

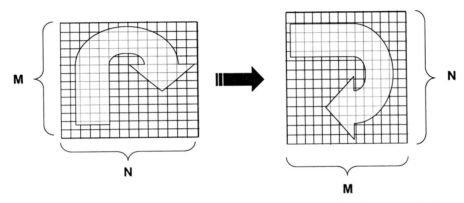

Figure 11.22. An illustration of a 90 degree clockwise image rotation.

As we can see in Figure 11.22, a 90 degree clockwise image rotation transforms the rows of the original image into the columns of the rotated image. We can also see that M elements in each row will consists of the elements of the M elements in each column of the original image, but in reversed order. Thus, we can describe the transformation as follows.

$$R(i, j) = I((M+1) - j, i)$$

Where R is the rotated image, I is the original image, M is the number of rows in the original image, and i and j are row and column indexes such that:

$$i = 1,2,3,...,N$$
$$j = 1,2,3,...,M$$

The rotation in the counterclockwise direction is done in a similar way. This rotation is shown in Figure 11.23.

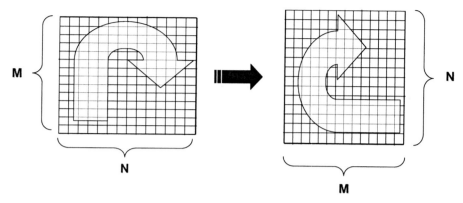

Figure 11.23. An illustration of a counterclockwise image rotation.

In this case, we see that, once again, the rows of the original image have become the columns of the rotated image. This time, however, it is the column order that must be reversed when the columns become the rows of the rotated image. In other words, the N elements in each row of the original image become the N elements in each column of the rotated matrix, but in reversed order. Thus, we can write this transformation in the following way.

$$R(i, j) = I(j, (N+1) - i)$$

Where R is the rotated image, I is the original image, N is the number of rows in the original image, and i and j are row and column indexes such that:

$$i = 1,2,3,..., N$$
$$j = 1,2,3,..., M$$

So, now we know how to perform all of the 90 degree rotations.

The second type of image rotations involves rotating the image by some arbitrary angle, θ. In order to perform this kind of an operation, we need to individually map each pixel in the original image to a location in a new image. This mapping is performed through the use of a rotator matrix multiplication as follows.

$$\begin{bmatrix} j' \\ i' \end{bmatrix} = \begin{bmatrix} \cos(\theta) & -\sin(\theta) \\ \sin(\theta) & \cos(\theta) \end{bmatrix} \begin{bmatrix} j \\ i \end{bmatrix}$$

Where i' and j' are the new row and column, i and j are the original row and column, and θ is the angle that we want to rotate our image. Several difficulties arise in this format beyond just having to perform a more complicated calculation. First, the coordinate values that result form the rotator matrix multiplication are not always integer values. We can resolve this problem by rounding the resulting values to the nearest integer values. However, doing this causes an additional problem. There will be some pixels in the new image that do not get values assigned to them because none of the mappings will be close enough to be rounded to those indexes. This means that there will be some undefined points within the new image. The result of this mapping problem is demonstrated in Figure 11.24.

Original Image   **Rotated Image**

Figure 11.24. An illustration of the undefined pixels that occur in a rotated image because the rotated pixel locations do not always map to integer values and must be rounded.

As you can see, there are a series of black dots scattered throughout the image. These dots are there because no pixel in the original image mapped to a pixel in those regions. To correct this problem, special processing steps must be taken. These steps perform what is known as "interpolation." Interpolation is a process in which each of the undefined pixel's values are determined by the values of the pixels immediately surrounding them. There are several methods for doing this, but the end result will be a smooth-looking rotation like the ones shown in Figure 11.25.

Rotated 45 degrees clockwise   **Rotated 45 degrees counterclockwise**

Figure 11.25. Examples of images that were rotated with interpolation.

So, now you should have a better idea how imaging programs perform image rotations.

11.4.4 Resizing Images

The final basic transformation that we will explore is basic image resizing. The process used to resize images is simply a two-dimensional extension of the processes of up-sampling and down-sampling which we saw during our study of signal processing back in chapter 9.

The easiest size adjustment we can make is to shrink our original image. This process simply involves the discarding of some of the images pixel information. For example, if we want to shrink our image to half of its original size, we just need to discard half of the image's pixel information. The easiest way to do this is to throw out every other row and column in the image. Thus, if the image began with 256 rows and columns, the resized image would have 128 rows and columns. We can write this operation as:

$$S(i, j) = I(xi, xj)$$

Where S is the new image of reduced size, I is the original image, i and j are indices (ranging from 1 to their original length divided by x), and x is the down-sampling factor. In the case where we wish to reduce our image size by half we would use a down-sampling factor of 2. Some examples of down-sampled images are shown in Figure 11.26.

256 x 256   **128 x 128**   **64 x 64**   **32 x 32**

Figure 11.26. Some examples of down-sampled images. Each image is half the size of the previous.

Thus, we can see that the process of shrinking an image is relatively simple. Enlarging an image, on the other hand, is slightly (but not much) more difficult. In this case, we need to add new information to our image. One way of doing this is based on the up-sampling method that we learned in chapter 9. Basically, what we need to do is increase the size of the image by making additional copies of each row and column in the image. For example, if we want to double the size of our image, all we need to do is double the width of each row and column in the image. So, each pixel in the image will become a 2x2 block of pixels, all with the same intensity. We can describe this mathematically as follows.

$$S(i, j) = I\left(ceil\left(\frac{i}{x}\right), ceil\left(\frac{j}{x}\right)\right)$$

Where S is the new image of expanded size, I is the original image, i and j are indices, x is the up-sampling factor. The "ceil" operation indicates that the results of the division of the index by the up-sampling factor should always be rounded up to the nearest integer. In this equation the values for i and j will range from 1 to x times the length and width of the original image. In other words:

$$i = 1...xM$$
$$j = 1...xN$$

Where *M* is the number of rows in the original image and *N* is the number of columns in the original image. The images in Figure 11.27 demonstrate the results of up-sampling an image.

32x32   **64x64**   **128x128**   **256x256**

Figure 11.27. Examples of enlarging images by up-sampling.

11.5 Image Filtering and Enhancement

Basic image transformations are only a small subset of the image processing techniques that we can employ. The transformations that we have examined so far are only capable of modifying an image's size and orientation. To obtain any truly useful information from our images we need to go beyond these simple transformations. One way of doing this is through the application of filtering. So, what exactly is filtering and what can it do for us?

Image filtering is a technique for modifying or enhancing an image. The job of a filter is to remove or isolate certain features within the image in order to extract specific information or make the image more pleasing to look at. For example, filters can be applied to an image to emphasize the edges of objects in the image, to remove noise, or to smooth or sharpen the transitions between the pixel values. As you can see, a filter can be used for many purposes. In the following sections, we will examine filters that can be used for several of these purposes. We will begin by looking at filters that can be used for image enhancement. By enhancement, we basically mean the process of manipulating pixel values for the purpose of making the image more pleasing to the viewer. The techniques that we will look at in this area are blurring and sharpening. Next, we will look at filters that can be used to isolate important features. The best example of this is edge detection. Finally, we will investigate filters that can be employed to remove unwanted information, namely noise, from an image.

11.5.1 How does filtering work?

Before we can investigate all these applications of filtering, however, we need to know how filtering works. If you think back to our discussion of convolution and correlation in chapter 9, you will recall that we mentioned that basic filtering was performed by convolution. So, if we want to apply a filter to a one-dimensional signal, we can do it by convolving the signal with a special impulse response. This impulse response is referred to as a filter. The process of filtering a two-dimensional signal, such as an image, is done in a similar way. In many cases, however, we perform image filtering by using correlation. You will recall that correlation was essentially the same as convolution with one small exception. When two signals are correlated, we do not perform a time-reversal on one of the signals as we did in convolution. We will look at how this correlation process works on a two-dimensional signal in a moment, but first, we must look at what we mean by a filter. In image processing, a filter, also known as a mask, is a two-dimensional signal. A filter can best be thought of as a small, usually square, matrix containing a series of values or weights. Common sizes for filters are 3x3, 5x5, and 9x9, but other sizes can also be used. Two filters are shown in Figure 11.28, the first is a 5x5 blurring filter, and the second is a 3x3 sharpening filter.

$\frac{1}{25}$	$\frac{1}{25}$	$\frac{1}{25}$	$\frac{1}{25}$	$\frac{1}{25}$
$\frac{1}{25}$	$\frac{1}{25}$	$\frac{1}{25}$	$\frac{1}{25}$	$\frac{1}{25}$
$\frac{1}{25}$	$\frac{1}{25}$	$\frac{1}{25}$	$\frac{1}{25}$	$\frac{1}{25}$
$\frac{1}{25}$	$\frac{1}{25}$	$\frac{1}{25}$	$\frac{1}{25}$	$\frac{1}{25}$
$\frac{1}{25}$	$\frac{1}{25}$	$\frac{1}{25}$	$\frac{1}{25}$	$\frac{1}{25}$

-1	-1	-1
-1	9	-1
-1	-1	-1

5x5 Blurring Filter   **3x3 Sharpening Filter**

Figure 11.28 Examples of some commonly used image filters.

When these filters are applied to an actual image, the first filter will produce a softer, blurred image, and the second filter will produce a sharper, more defined image. So, how do we apply filters like these to images? As we mentioned before, images are usually filtered by performing a two-dimensional correlation between the image and the filter. Practically speaking, this is a seven step process. Let's look at these steps now. Each step will be illustrated with a small example. To begin our example, we need to define a filter and image.

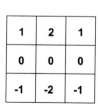

0	0	0	0	0	0	0	0	0
0	0	0	0	0	0	0	0	0
0	0	0	0	0	0	0	0	0
0	0	1	1	1	1	1	0	0
0	0	1	1	1	1	1	0	0
0	0	1	1	1	1	1	0	0
0	0	1	1	1	1	1	0	0
0	0	0	0	0	0	0	0	0
0	0	0	0	0	0	0	0	0

3 x 3 Filter   **9x9 Image**

Now, let's begin the filtering process.

1. Create a new empty image that is the same size as the original.

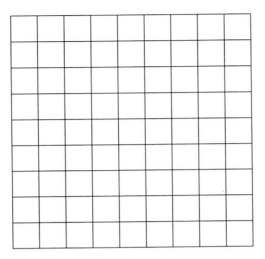

2. Copy the border pixels from the original image into the new image. By border pixels, we mean the first and last rows and columns. The number of rows and columns that must be copied depends on the size of the filter you are using. A 3x3 filter requires only one row and column. So, we would copy the first row and column and the last row and column. A 5x5 image would require 2 rows and columns from the edges. The number of rows and columns can be determined by dividing the filter size by two and rounding down to the nearest integer value. Thus, to find the number of rows and columns for a 9x9 filter we simply divide 9 by 2 to get 4.5 and round this down to 4. So, we would copy the first and last four rows and columns from our original image to the new image. Since our example filter is 3x3, we will copy single pixel edges. The reason that we need to do this will become apparent in a moment.

0	0	0	0	0	0	0	0	0
0								0
0								0
0								0
0								0
0								0
0								0
0								0
0	0	0	0	0	0	0	0	0

3. Superimpose the filter on the top left corner of the original image.

1	2	1	0	0	0	0	0	0
0	0	0	0	0	0	0	0	0
-1	-2	-1	0	0	0	0	0	0
0	0	1	1	1	1	1	0	0
0	0	1	1	1	1	1	0	0
0	0	1	1	1	1	1	0	0
0	0	1	1	1	1	1	0	0
0	0	0	0	0	0	0	0	0
0	0	0	0	0	0	0	0	0

4. Multiply each value in the mask with the value of the pixel over which it lies and add the resulting products together.

$$(1)(0) + (2)(0) + (1)(0) + (0)(0) + (0)(0) + (0)(0) + (-1)(0) + (-2)(0) + (-1)(0) = 0$$

5. Place the resulting sum in the new image. This value is placed in the pixel location that has the same coordinates as the pixel in the original image over which the filter is currently centered. You can now see why we had co copy the border pixels earlier. Since the center of the mask will never be centered over any of the edge pixels, they must be copied over so that our resulting image will have the appropriate size. Some techniques other than copying can be used. One of these involves padding the original image with zeros so that all the pixels in the original will be processed, but no matter what technique is used, the edge pixels in a filtered image are unreliable. This unreliability is typically referred to as edge effects.

0	0	0	0	0	0	0	0	0
0	0							0
0								0
0								0
0								0
0								0
0								0
0								0
0	0	0	0	0	0	0	0	0

6. Move the filter one pixel to the right and repeat steps 4 and 5. This will continue until the last non-edge pixel in the current row of the new image has been calculated.

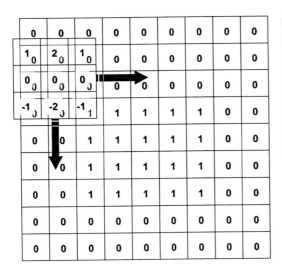 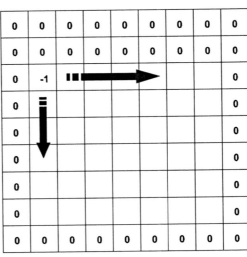

7. Move the filter down one row and back to the left side of the image. Repeat steps 5 through 6 until all of the non-edge rows have been processed.

These seven steps are all that is required to perform basic image filtering. The concept itself is not really all that difficult. It is, however, a time-consuming and tedious process to perform this type of operation by hand. This is why computers are almost always used to do the actual filtering. So, while you will probably never need to perform image filtering by hand, it is important to understand what happens when images are filtered. An awareness of this process makes it easier to understand how filters are designed and allows you to get a general idea of what a filter will do simply by looking at it. If we were to continue the filtering operation that we began on our example 9x9 image, we would arrive at the following filtered result.

0	0	0	0	0	0	0	0	0
0	0	0	0	0	0	0	0	0
0	-1	-3	-4	-4	-4	-3	-1	0
0	-1	-3	-4	-4	-4	-3	-1	0
0	0	0	0	0	0	0	0	0
0	0	0	0	0	0	0	0	0
0	1	3	4	4	4	3	1	0
0	1	3	4	4	4	3	1	0
0	0	0	0	0	0	0	0	0

The particular filter that was used in this example happens to be a filter that detects horizontal edges within an image. This is why the filter only returned non-zero values along the horizontal edges of the block of ones. We will talk about edge detection in more detail in a later section. Now that we are well acquainted with the way in which filtering works, we can begin to explore some of the applications of image filtering. Let's begin by examining a filter that we can apply to achieve image blurring.

11.5.2 Low Pass Filters

One common image processing technique that can be applied through filtering is blurring or softening. In some cases, an image may contain too much sharp detail. Such an image appears harsh and cold when it is viewed. In cases like this it may be useful to reduce the sharpness of the image to some degree. As you may recall, the definition or detail in a signal is determined by its high frequency components and its basic form is defined by its low frequency components. Thus, if we want to reduce the sharpness of an image, we need to remove some of the high frequency information from the image. This is done with a low-pass filter. We looked at a simple example of a low pass filter in chapter 9 when we studied the moving average filter. In this one-dimensional case, we simply averaged the values within a certain distance window from the current sample and replaced the sample with the result of that average. We should be able to extend this example to work for a two dimensional case without much difficulty. Keeping the way we perform image filtering in mind, how can we achieve a two-dimensional moving averaging filter? All we need to do is define a two dimensional window that we wish to average over. Since the filtering operation multiplies the value in each filter position by the corresponding image pixel and then sums the result, we can easily implement an averaging operation. First we define a averaging window size, let's say N x N. Now, we create a filter of the same size as our window and place the value of 1 in each of the filter's elements. Finally, we divide the value in each of the filters elements by the total number of elements in the filter, N^2. By doing this, the sum of the products that are taken on each filter step will represent the average of the values under the filter. So, we can write this out mathematically as:

$$B(i, j) = \frac{1}{N^2} \sum_{k=0}^{N-1} \sum_{l=0}^{N-1} I(i + (k - n), j + (l - n))$$

Where B is the blurred image, N is the length of one side of the filter, i and j are indices of the original image, k and l are offsets from the center of the filter, and n is half of N rounded down to the nearest integer. Thus, for a 3x3 blurring filter, we can write:

$$B(i, j) = \frac{1}{9} \sum_{k=0}^{2} \sum_{l=0}^{2} I(i + (k - 1), j + (l - 1))$$

Essentially, this equation simply states that for each element in the blurred image, the take the average value of the same point in the original image and its eight neighboring pixels. Figure 11.29 illustrates two commonly used blurring filters and Figure 11.30 demonstrates the effects of applying blurring filters of varying sizes to an example image.

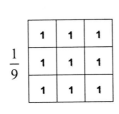

3 x 3 Averaging Filter

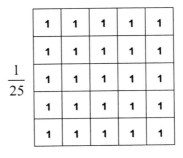

5 x 5 Averaging Filter

Figure 11.29 Examples of some common blurring filters.

As you can see in Figure 11.30, the larger the averaging window size the more extreme the blurring effect will be. A 3x3 filter produces minor softening, while the much larger 55x55 filter makes the flower in the image almost impossible to distinguish from the background. So, we can assume that the larger the window, the more high frequency information is removed from the image.

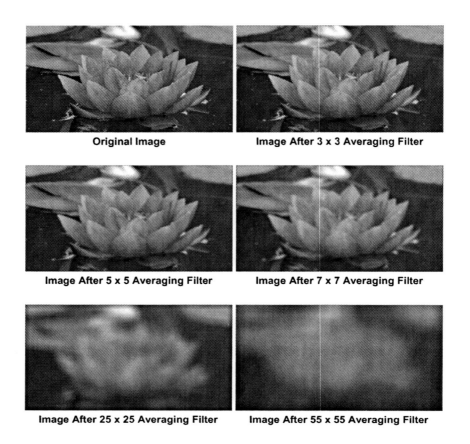

Original Image   **Image After 3 x 3 Averaging Filter**

Image After 5 x 5 Averaging Filter   **Image After 7 x 7 Averaging Filter**

Image After 25 x 25 Averaging Filter   **Image After 55 x 55 Averaging Filter**

Figure 11.30 Examples of the effects of applying blurring filters to an image.

11.5.3 High Pass Filters

We have just seen how we can soften an image by using a low-pass filter to remove some of the high frequency information from the image, but what happens to the image if we remove the low frequency information. Since we know that the low frequency information provides the image with its basic form, we should expect a high-pass filtering operation to remove all but the sharp transitions between color shades. Before we see what a high-pass filter will do, let's find out what one looks like. The following is a typical 3x3 high-pass filter.

-1	-1	-1
-1	8	-1
-1	-1	-1

Now that we have a filter defined, we can see what happens when we apply it to our image. If we take a look at the second image in Figure 11.31 we see that a high pass filter does indeed remove all but the high frequency edge transition information. While this information by itself can be useful, it can often be even more helpful if we add this information back into the original image. This operation will enhance the high frequency information and add additional definition to the image. Because it makes the image appear with more crispness or sharpness, the practice of adding the high frequency information back into the original image is often called sharpening. The result of adding the high frequency component back into the original image is illustrated in the third image in Figure 11.31.

Original Image   **Result of High-Pass Filtering**   **Sharpened Image**
                                                                      (Original Image + High-Pass)

Figure 11.31 Examples of high-pass filtering and sharpening.

As you can see, the petals of the flower and the leaves around it have much more definition around their edges than they do in the original image. Thus, we have enhanced the original image using high-pass filtering techniques.

11.5.4 Edge Detection

Another area of image processing that employs filters is edge detection. The ability to locate the edges of an object within an image can be helpful in many situations. Let's say, for example, that we are designing an automated assembly line system for a circuit board manufacturing company. At the end of the manufacturing line, we wish to implement an automated quality control system. To do this, we place a digital camera on the line. As each completed circuit board passes the camera a picture is taken and sent to a computer where a program has been written to examine the traces on the circuit board to make sure there are no broken traces. Writing a program to analyze a standard grayscale image by itself is a difficult task. Small variations in the intensities of the image might cause the program to think there is a problem with the circuit board when none exists. If we could enhance the edges in the image, however, it would be much easier for our program to make sure there are no breaks. But, how do we find the edges within the image? With a little bit of investigation, we soon discover that an edge within the image occurs when there is a sudden change in intensity values. So, if we trace along a row of the image and look for places that the intensity suddenly increases or decreases, we can find vertical edges, and if we trace along the columns of an image, any sudden increases or decreases in intensity would indicate horizontal edges. Now, can you think of any way that we can systematically look for these changes? What if we look at the slope of the signal representing a given row or column of the image? If the slope is small, there is little to no change in the neighboring values, but if the slope is large, we must be near an edge. Now, how do we find the slope of this signal? If the signal was analog, we could take a derivative, but since we are dealing with a digital signal, we only need to find the difference between two neighboring points. Several edge detection filters have been developed based upon this basic idea. Let's take a look at one of them.

1	2	1
0	0	0
-1	-2	-1

1	0	-1
2	0	-2
1	0	-1

Horizontal Edge Detector   **Vertical Edge Detector**

As you can see, the horizontal edge detector simply takes the difference along the columns under the mask. The central column is emphasized by doubling its contribution to the result. So, for each pixel in the image, the horizontal edge detection filter calculates the differences along the column containing the pixel and its two neighbors. The overall change along all three columns is then summed to determine a value for that particular point. Likewise, the vertical edge detection filter takes the difference along the row containing the current pixel and its two neighbors. So, after applying these two filters to an image, we would expect to have two separate images, one indicating the horizontal edges and another indicating the vertical edges. An example of these two images can be found in Figure 11.32. Now that we have the horizontal and vertical edges we can add the two images together to get a single image containing all of our edges (see Figure 11.32). The resulting image contains all of the edge information for the image, but it is still not ready to be used in an application such as our assembly line example. We have the information for the edges, but the lines denoting the edges are too wide and there are often light lines indicating weak edges. To solve this problem, we use a thresholding operation in which we keep only the values above a certain limit. This insures that we only look at edges of sufficient strength. In most cases, we are not interested in finding weak edges that can often be caused by minor changes in the grayscale image. We are only concerned with strong, well defined edges. Once we have removed the weak edges from the image, we need to apply a thinning operation to the remaining edge information to reduce the width of the indicating lines to a single pixel. Once this is complete the edges in our image are marked by single pixel wide lines as shown in the final image of Figure 3.32.

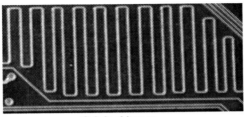

Original Image

Result of Horizontal Edge Detection

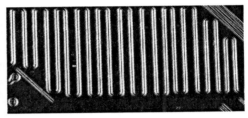

Result of Vertical Edge Detection

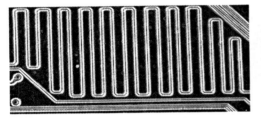

Result of Horizontal and Vertical Detection

Filtered Image After Thresholding and Thinning

Figure 11.32. Examples of different stages of the edge detection process.

With our edge information reduced to nice thin lines, a computer application can easily follow the lines to search for breaks or compare the edge information to a previously stored template to make sure everything is as it should be. Thus, we have found yet another useful application of image filtering.

11.6 Noise and Image Restoration

The process by which a digital image is created can often have limitations. These limitations can lead to images that are less than acceptable. For instance lenses, film or the digitizer may not perform adequately, resulting in a less than perfect image. In the analog domain, we have little choice but to accept these limitations. However, in the case of digital images, we have considerable liberty to manipulate the image to remove these deficiencies. In addition older analog pictures that have degraded over time can now be digitized. Once in digital form, the degradation can be reduced if not eliminated completely. The process used to correct the errors in and remove the noise from image is known as image restoration. The main goal of image restoration is to bring the image closer to what it would have been if it had been recorded without any degradation.

To understand the processes involved in restoring an image, we need to understand what kind of noise (degradation) our image might be exposed to. So, let us first take a look at noise, its characteristics, and the different types of noise that we might encounter. Once we understand this we can further investigate methods used to remove this noise.

11.6.1 Some Common Types of Noise

Noise can be found everywhere in nature. In the world of science and technology, it is often an unwanted phenomenon. As a result, a great deal of time has been spent researching and attempting to artificially model noise. Noise is present in all analog devices, but digital signals have a built-in tolerance to noise. Digital signals are formed from combinations of 1 and 0 only, everything else is considered to be an error. Thus, much of the noise that would cause problems in an analog signal can be easily removed from digital signals. The human senses are accustomed to processing information containing noise and minor errors. In fact, the brain is capable of correcting and compensating for such errors in a near transparent manner. As a result, the unusually clean signals produced by digital systems can often seem too sterile to the human senses. Images that are too crisp and clean, for example, tend to appear unnatural.

There are many different kinds of noise. Noise can be coarse or soft. In some cases, we describe it with color (for example pink noise and white noise), and in other cases, it is described based upon the way in which it looks when viewed in an image. The most important thing to remember when dealing with noise, however, is that it is always unpredictable. Real, naturally occurring noise cannot be fully predicted, but it can be simulated with various mathematical models. Let's look at some of the types of noise we might encounter.

Noise with a flat spectrum (meaning that it contains an equal amount of all frequencies) is called "white noise". White noise sounds like the rush of a waterfall when present in a sound signal, and it looks like the "snow" you see on a television screen that is tuned to a channel with no broadcast signal. When present in an image, white noise cannot be predicted at all and is said to be completely uncorrelated from point to point. In other words, if you were to correlate two signals containing "white" noise, you would get no large responses indicating any similarities between the two signals.

Several other types of noise are described with colors. Noise that fits the form $1/f$ (1/frequency), for example, is called "pink noise." It has equal energy in each octave, whereas white noise has equal energy for any given bandwidth. In other words, white noise has an equal amount of all frequency components (flat frequency response) while the frequency response of pink noise changes with the frequency. In pink noise, lower frequencies have more energy than higher frequencies. Thus, pink noise sounds softer and less harsh than white noise, a little bit like a low hiss. In addition to pink noise, we might encounter other colors of noise such as brown or Brownian noise which has a frequency spectrum fitting the form of $1/f^2$. There are other "colors" of noise as well, but they are less commonly encountered.

Aside from color descriptions, noise is also described by the way it looks. For example, we may encounter "salt and pepper noise." This type of noise gets its name from the way in which it impacts images. It is essentially a "spike" or impulse noise that drives the intensity values of random pixels to either their maximum or minimum values. The resulting black and white flecks in the image resemble salt and pepper.

Another kind of noise that gets its name from the way it affects images is speckle noise. Speckle noise is a form of multiplicative noise in which the intensity values of the pixels in the image are multiplied by random values, and the resulting image takes on a "speckled" look.

So, we now see that noise is essentially a random occurrence which can not be predicted exactly. Thus, to model noise we must use a probabilistic approach. Various probability density functions have been used to model noise over the years. One of the most commonly used is the

Gaussian function. This is why you might hear noise described as Gaussian and white or Gaussian and pink and so on. The Gaussian term simply describes the way in which the noise was modeled. The three most common types of random noise you are likely to encounter in images are white noise, salt and pepper noise, and speckle noise.

In addition to these random noises, you are likely to encounter noises that are less random in nature, for example periodic noise. Periodic noises are the result of electronic or mechanical interference with your signal, and thus, are not typically naturally occurring noises. Periodic noises tend to occur at specific frequencies, and thus, can be removed by filtering out those frequencies. Since random noises are less predictable, they tend to be more difficult to remove and, thus, are somewhat more troublesome. Examples of the effects of three most commonly encountered types of image noise are illustrated in Figure 11.33.

Original Image   **Image + Gaussian White Noise**

Image + Speckle Noise   **Image + Salt & Pepper Noise**

Figure 11.33. Examples of images that have been affected by three common types of random noise.

11.6.2 Removing Noise

So, now that we know about some of the types of noise that are out there and how noise can affect an image, we can begin to look at methods we can use to reduce or remove noise from images. While removing most types of noise is difficult, some types are easier to deal with than others. Of the three major types of noise shown in Figure 11.33, two of them are difficult to remove. These two are white noise and speckle noise. Because they modify the pixels within the image in a more subtle way as compared to salt and pepper noise, it is difficult to determine what the value of a given pixel should be. The simplest approach used in an attempt to deal with noise of this type is to use an averaging or blurring filter. The results of using such a filter on an image affected by white noise are shown in Figure 11.34. As you can see in the figure, the noise has been reduced, but so has the overall quality of the image. The averaging filter removes the most noticeable high frequency elements of the white noise, but in doing so, it also blurs the image. In general, the blurring is less troubling to a viewer than the high frequency noise, but the fact still remains that some of the fine detail in the image will be lost. So, we can now see that removing some types of noise can be difficult. White noise is particularly difficult to deal with because it has frequency components across the entire spectrum. Thus, to completely remove them using a simple filtering approach is impossible without significantly degrading the frequency content of the image itself. More complicated approaches can achieve better results, but these approaches use advanced statistical properties of the image and their description is beyond the scope of this text.

Image with White Noise   **Image After 5x5 Averaging Filter**

Figure 11.34. An example of a simple noise removal technique for white noise.

At this point, let's turn our attention to the one type of noise that is a little easier to deal with, salt and pepper noise. As we learned earlier, salt and pepper noise affects an image by driving some of its pixel's intensities either to their maximum or minimum values. Since the values of the pixels that have been disturbed by the noise tend to be significantly different from the values of the pixels around them, it is a little easier to locate and correct them. To do this, we need to learn about a statistical value known as the median. So, what is the median and what does it have to do with the removal of salt and pepper noise?

The median is a number within a set of numbers such that half of the numbers in the set are less than or equal to it and half of the numbers in the set are greater than or equal to it. Thus, the easiest way to find the median value for a set of numbers is to first sort the set in ascending order and then pick the middle element if the number of elements is odd. If there are an even number of elements, the median value is defined as the average of the two middle numbers. Let's look at an example to see how this works.

Example 11.1

Problem: Find the median of the following sets of numbers

a) $\{1,3,4,10,4,2,6,7,3\}$

b) $\{9,6,1,8,4,13,7,3,45,21\}$

Solution: In this problem, we have been given two sets of numbers and asked to find the median value of each set. Starting with the first set, we need to put it in ascending order. Doing this we get:

$$\{1,2,3,3,4,4,6,7,10\}$$

Now we simply pick the element in this set such that half the elements are in front of it (less than) and half of the elements are behind it (greater than). In this case, since we have 9 elements in the set, we will pick the 5th element from our sorted set. Thus, the median is 4.

Turning our attention to the second set, we place it in ascending order as follows:

$$\{1,3,4,6,7,8,9,13,21,45\}$$

This set of values has an even number of elements. Thus, we must set the median equal to the average of the two central elements. Since there are ten elements in the set, we need the average of the 5th and 6th elements. Thus, the median value is computed as:

$$\frac{7+8}{2} = \frac{15}{2} = 7.5$$

So, the median of this set of values is 7.5.

From our example, we see that finding a median value is a simple task, but how will this help us clean salt and pepper noise from an image? Well, what if look at a small group of pixels in the image, say, for example, a 3x3 neighborhood around a pixel? This will give us a set of intensity values. The median of these values will give us an idea of what a typical intensity value for that neighborhood should be. Thus, the median value will give us an idea of a typical value without being influenced by any extreme high or low value in the neighborhood. At this point, we simply replace the value at the center of the neighborhood with the median value. After processing the neighborhoods for all the pixels in the image, we arrive at a final filtered result in which any extreme pixels have been replaced by the median value of the pixels around them. Figure 11.35 demonstrates how this kind of processing can improve an image that has been affected by salt and pepper noise.

Noisy Image   **Image Cleaned by Median Filtering**

Figure 11.35. An example of an image that has been processed with a median filter.

As we can see, the image quality is greatly improved by the median filtering process. However, some image content can be removed by median filtering. Thus, median filtering can degrade the image, but the degradation is not as noticeable as that caused by an averaging or blurring filter like the one used in the example in Figure 11.34.

In the previous examples, you have seen some of the simple approaches to noise removal in images. Thus, we have found that we can manipulate the content of images in a wide variety of useful ways through the use of filters. These filters can perform basic enhancements like blurring or sharpening. They can extract important information for images, such as the edges of object in the image, and they can help remove unwanted noise. So, filtering is an extremely useful technique in image processing.

11.7 Image Compression

We have seen the usefulness of processing digital images. Now we will look at some of the issues that are encountered when storing and transmitting digital images, particularly the processes involved in reducing the size requirements. Most of the images that we deal with are represented either as grayscale images using one byte per pixel to represent the intensity or as RGB color image which require 3 bytes per pixel to represent the red, green, and blue color intensities. So, if we want to store a large number of images in a fixed amount of space, it is in our best interest to find ways to compress the information. To see why we might want to do this, lets look at an example.

Example 11.2

Problem: Assume that you are taking pictures with a digital camera. Each picture is 3072 pixels wide and 2048 pixels high. How much storage space does one image require if it is stored as a grayscale image? What if it is stored as a color image? If you are using a 128 MB memory card, how many grayscale photos can you take? How many full color images can you store?

Solution: In this example, we are asked to examine the limitations of storing images taken on a digital camera. Our image consists of a matrix that is 3072 pixels wide and 2048 pixels high. Thus, we can calculate the total number of pixels in our image as:

$$3072 * 2048 = 6291456 \text{ Pixels}$$

Now, we know that a grayscale image uses one byte per pixel to store intensity information so, our grayscale image will require:

$$6291456 \text{ Pixels} * 1 \frac{\text{byte}}{\text{pixel}} = 6,291,456 \text{ Bytes} = 6 \text{ MB}$$

Likewise, a full color image requires 3 bytes per pixel. Thus, we can determine its storage requirements as follows:

$$6291456 \text{ Pixels} * 3 \frac{bytes}{pixel} = 18,874,368 \text{ Bytes} = 18 \text{ MB}$$

Now, if we are using a 128 MB memory card we can find the number of grayscale images we can store:

$$\frac{128MB}{6MB} = 21.3 \text{ Images}$$

So, we can only fit 21 full images on our memory card. Now, if we extend our calculation to full color images we find:

$$\frac{128MB}{18MB} = 7.1 \text{ Images}$$

Thus, we can fit only 7 full color images on the memory card.

_____

So, as we have just seen, without some method of reducing the amount of space that our images take up, we will require huge amounts of storage to make practical use of imaging devices and systems. Thus, a large amount of study has been done relating to methods that can be used to store and transmit data with reduced space requirements. So, in a nutshell, data or image compression is concerned with removing redundancy from the image data or encoding the data in images more efficiently.

Studies of human vision have shown that in most cases, it is not required to have very great detail in the images. Beyond a certain point, a higher resolution does not make any perceivable difference to the human eye. So, some of the data in the image may be excluded and the picture will still convey the information without any loss in meaning. The crucial step here is the separation of the data that is important to the image and data that is redundant. An example of a case in which we could discard some of the image's information is digital photography. Let's say, for example, that you are on a seaside vacation and you have taken some pictures on your digital camera that you wanted to email back home. In this case, it is more important to have a small image that can be sent quickly than a high-quality image that may take several minutes to send.

On the other hand, there are cases, especially in biomedical and forensic image processing, where no compromise can be made in the accuracy of data. Take for example an X-ray image of a hairline fracture in a bone. The hairline crack would be visible only on the clearest image rendered with the highest resolution possible. If some data is lost, then the physician might not be able to detect the hairline crack and the X-ray image becomes useless to the doctor. Thus, we need to preserve all of the image detail. However, even in this case, it becomes necessary to employ some form of compression to keep the size of the image comparatively small. This is particularly important if there is a need to preserve the digital images for long periods of time. Take, for example, forensic fingerprint images. The FBI needs to keep long-term records of all the fingerprint images that it acquires. This amounts to millions and millions of images that need to be stored. If we did not employ some kind of compression, the storage requirements would become prohibitive. So, we need to encode all of the image data so that it occupies the least amount of space, without losing any information. In other words, we want to be able to represent the data in the least amount of space while still being able to recreate the image exactly as it was originally recorded. By using this form of compression, we can preserve all of the image information while reducing the overall storage requirements required to store the images.

Based on the types of compression required in the previous two examples, we can divide image compression techniques into two groups:

- Lossy compression
- Lossless compression

A lossy compression algorithm discards information that it cannot later restore. As a result, it does not promise that the uncompressed data is the same as the data before compression. It only says that it is a close approximation. It is hoped that the lost information will not be significantly noticable. Lossy algorithms are used to compress digital imagery including video. This makes sense because such data often contains more information than the human eye can perceive and, for that matter, may already contain errors and imperfections that the human eye is able to compensate for. In addition, a lossy compression algorithm can achieve a higher compression ratio than its lossless counterpart.

A lossless compression algorithm, on the other hand, can faithfully reproduce the original data verbatim, without any change. This kind of algorithm is used to compress file data such as executable code, text files and numeric data because programs that process such file data cannot tolerate mistakes in the data. For the same reason, this form of compression is used in biomedical signal processing as well.

So, based upon the application, both lossy and lossless image compression can play an important roll. In the next two sections, we will explore some of the techniques, used for both lossless and lossy image compression.

11.7.1 Lossless Compression

Lossless compression is very important for a wide variety of applications ranging from forensics to biomedical. Most lossless image compression techniques take advantage of repetition or redundancy within the image. Let's take a look at a few of these.

11.7.1.1 Run Length Encoding

Run length encoding or RLE has a very simple underlying concept. The idea is to replace consecutive occurrences of a given symbol with only one copy of the symbol and a count of how many times that symbol occurs, thus the name "run length". For example, "AAABBCDDDD" could be encoded as 3A2B1C4D. Run length encoding can be used to compress digital imagery by comparing adjacent pixel values and then encoding only the changes. For images that have large homogeneous regions, such as scanned documents, this technique is quite effective. It can achieve compression ratios on the order of 8 to 1 for scanned text images. This form of encoding works well on text images because they often contain a large amount of white space (the background paper) that can be removed. RLE is the key compression algorithm used to transmit faxes. So, if the image contains large regions of identical color or uses very few colors, run length encoding can be a good choice for performing lossless compression. The drawback to run length encoding is that it does little good if there are not large identical regions within the image.

Thus, it should be avoided in cases where images contain a large amount of variation. In fact, in cases such as this, run length compression might actually increase the storage requirements rather than reducing them.

11.7.1.2 Differential Pulse Code Modulation (DPCM)

Another simple lossless image compression technique is differential pulse code modulation or DPCM. In this approach, we begin by selecting a reference symbol. Once we pick this symbol we represent all of the following symbols as the difference between the current value of the data and the reference symbol. So, if we start with the string "AAABBCDDDD" and choose 'A' as our reference symbol, the resulting encoded string would be "A0001123333." As you can see, the length of the string is actually longer if you count the reference symbol. However, if we choose the reference symbol carefully so that the differences will be small, we can encode the differences with a smaller number of bits than are required to encode the entire character. Thus, over a large image, compression can be achieved. DPCM is particularly effective on images that are not very vibrant and don't change much – this keeps the difference from the reference symbol small.

11.7.1.3 Colormap and Dictionary-Based Methods

The dictionary-based compression methods all use a substitution string as part of their approach. The GIF or Graphical Interface Format is an example of one of these dictionary-based compression methods. This format is an important standard algorithm for image storage, and since it is essentially a lossless technique, the picture quality is very good.

As we saw earlier, most color images are represented digitally using three bytes of information of each pixel in the image. This means that there over 16 million different distinguishable colors possible. Now, the first step of the GIF algorithm is to reduce this number of colors to 256. It does this by picking the 256 colors that most closely approximate the colors in the picture. The colors in the image that do not match these colors are set to the nearest of the 256 selected color values. The 256 colors in the image are then stored in a table. The columns of this table hold the red, green, and blue component values for the selected colors. Thus, each row in the color table, or "colormap" represents one of the 256 colors. Since there are exactly 256 rows in the table, we can refer to them using a single 8-bit binary number. As a result, the value for each pixel in the image can be replaced by the appropriate colormap index number. Thus, the image can be thought of as a paint-by-number image since each pixel describes a specific color in the colormap table. This image representation format is illustrated in Figure 11.36. As you may have noticed, this method is only lossless if the original image contains less than 256 distinct colors to begin with. Otherwise there will be some information loss involved.

The next stage of the GIF algorithm is where the dictionary-based substitution comes in. Common sequences of pixels are selected and given a short form notation. For instance, if red, green, blue, yellow, or RGBY was a sequence that was repeated often within the image, we could label RGBY as Z and Z could be used in place of RGBY. Of course, the challenge in any dictionary-based compression method comes in defining an appropriate dictionary. We can have two kinds of dictionaries: static dictionaries – dictionaries that are defined and remain the same irrespective of the image being compressed and adaptive dictionaries – those that are dynamically developed specifically for each image. There are many statically defined dictionaries that can be used based on the type of image that you are tying to compress. Building an effective adaptive dictionary, on the other hand, is an area of extensive research.

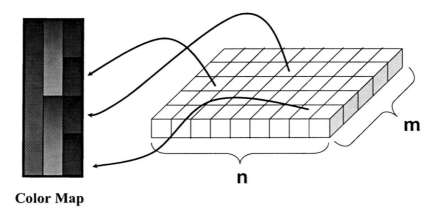

Color Map

Figure 11.36. An illustration of the colormap representation of a color image. Space is saved by reducing the total number of colors and storing an index number in each pixel location that points to that pixel's color in an associated colormap table.

11.7.2 Lossy Compression

In the previous section, we looked at a few methods that we could use to compress images without throwing away any of their information. While this is an important approach, especially in areas where perfect reproductions are required, lossless compression cannot achieve the level of compression that can be achieved by lossy compression. The key to performing good lossy compression is determining what information in the image can be discarded. There are several approaches that can be taken to make this determination, and thus, there are several different types of lossy image compression. The most common of these approaches is the Joint Photographic Experts Group standard, also known as the JPEG standard. The JPEG format, named after the committee that came up with the algorithm is a lossy compression algorithm that discards some of the information within the image, but in such a way that the general information conveyed by the picture is not lost. The benefit of this trade off in information accuracy is that the image will require a significantly smaller amount of space to store. In other words, it will have less data in it than the original image. Since the JPEG standard is so widely used, let's take a look at how it works. There are three stages in the JPEG process:

- The DCT (Discrete Cosine Transform) phase
- The Quantizing Phase
- The Encoding Phase

Let's look at each of these stages now.

The DCT Phase:

DCT stands for Discrete Cosine Transform and it is a mathematical operation that takes an 8x8 matrix of pixel values as input and outputs an 8x8 matrix of frequency coefficients. We say that the pixel values have been transformed into the frequency domain. As we know, the low frequency components of an image make up the overall gross features of the picture, while the high frequency components correspond to the fine detail in the picture and the overall sharpness of the picture. So, representing the image in frequency domain allows us to separate the picture in terms of the essential gross components and the less important detail features.

The first frequency coefficient at location (0,0) in the output matrix is called the DC coefficient. We can see that the DC coefficient is a measure of the average value of the 64 input pixels. The other 63 elements of the output matrix are called the AC coefficients. They add the higher-spatial-frequency information to this average value. Thus, as we move from the first frequency coefficient toward the 64th frequency coefficient, we are moving from low frequency information to high frequency information – from the broad strokes of the image to the finer detail. The higher frequency coefficients become increasingly unimportant to the perceived quality of the image as you move further from the DC coefficient. This leads us to the second phase of the JPEG algorithm in which it is decided which portion of which coefficients to throw away.

Quantization Phase:

The second phase of the JPEG algorithm is where the compression becomes lossy. The discrete cosine transform merely transforms the image into a form that makes it easier to determine what information to remove. Quantization is easy to understand. It is simply the dropping of the insignificant bits of the frequency coefficients.

To see how the quantization phase works, imagine that we need to compress some whole numbers less that 100, say 45, 98, 23, 66 and 7. If we decide that knowing these numbers truncated to the nearest multiple of 10 is sufficient, then we can divide each number by the "quantum" 10 using integer arithmetic. This would yield the numbers 4, 9, 2 and 0. Now, we can encode these numbers in 4 bits instead of the 7 bits needed to encode the original numbers. Thus, by determining that we could approximate the numbers allowed us to reduce the amount of space needed to store them.

Rather than using the same quantum for all 64 coefficients, JPEG uses a quantization table that provides a quantum to use for each of the coefficients. This quantum table can be thought of as a parameter that can be set to control how much information is lost and correspondingly how much compression is achieved. If we require a higher level of compression, we can change these values so that more information will be discarded.

In the example of the quantization table given in Figure 11.37, the coefficients for the low frequency information have a larger value. This means that little low frequency information is lost. On the other hand, high frequency coefficients have larger values meaning that more of the high frequency coefficients end up being set to 0 after the quantization phase. The more frequencies that are set to zero in this phase, the more compression we can achieve in the final phase.

$$
Quantum = \begin{bmatrix}
3 & 5 & 7 & 9 & 11 & 13 & 15 & 17 \\
5 & 7 & 9 & 11 & 13 & 15 & 17 & 19 \\
7 & 9 & 11 & 13 & 15 & 17 & 19 & 21 \\
9 & 11 & 13 & 15 & 17 & 19 & 21 & 23 \\
11 & 13 & 15 & 17 & 19 & 21 & 23 & 25 \\
13 & 15 & 17 & 19 & 21 & 23 & 25 & 27 \\
15 & 17 & 19 & 21 & 23 & 25 & 27 & 29 \\
17 & 19 & 21 & 23 & 25 & 27 & 29 & 31
\end{bmatrix}
$$

Figure 11.37. An example quantum table used in the JPEG quantization phase.

Encoding Phase:

The final phase of the JPEG algorithm encodes the quantized frequency coefficients in a compact form. This results in additional compression, but this compression is lossless. Starting with the DC coefficient in position (0,0), the coefficients are processed in the zigzag sequence shown in Figure 11.38 below. Along this zigzag, a special form of run length encoding is used. The run length encoding is applied to only the 0 coefficients. This is significant because many of the later coefficients have been reduced to 0 by the quantization phase. In addition, since the DC coefficient contains a large percentage of the information about the 8x8 block from the source image and images typically change slowly from block to block, each DC coefficient is encoded as the difference from the previous DC coefficient. This is similar to the strategy used in differential pulse code modulation. Thus, the 64 quantized values are compressed using lossless techniques.

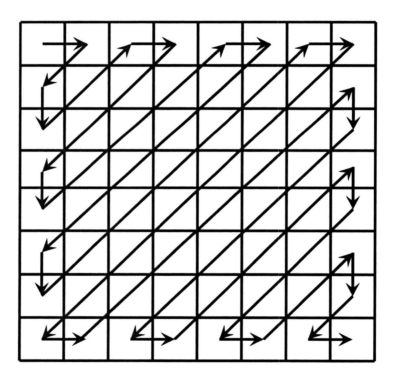

Figure 11.38 The zigzag path used when encoding the 8x8 matrix of quantized values.

So, as we have just seen, lossy compression can be a fairly complex operation, but if the application in which it is used can tolerate minor image degradations, it has many benefits when compared to lossless compression. The foremost of these is its high compression ratios. In fact, the more information that we discard in the quantization phase, the higher these compression ratios become. Figure 11.39 illustrates this. You can see that as the compression level increases, the quality of the image decreases.

Original Image 253 KB   **Low Compression 46.1 KB**

Medium Compression 17.7 KB   **High Compression 7.3 KB**

Figure 11.39 An illustration of some different compression levels in JPEG images.

11.8 Images in Motion - Video

So far, we have limited our discussion of image signals to single stationary images, such as photographs. Now, however, we will briefly turn our attention to moving images, or video. Video signals are composed of a series of two-dimensional images. These images are presented one after the other in quick succession. Thus, we can think of video as a three-dimensional signal in which the two-dimensional images vary with time. So, now when we talk about the value held in a pixel, we need to specify three indexes to uniquely identify it, its horizontal and vertical position, and the time. There are many practical applications of digital signal processing in the area of video. These range from tracking the motion of objects to noise removal, but one of the most valuable areas of study is in video compression. In this section, we will briefly investigate this interesting area. First, however, we need to know how video works.

11.8.1 How Video Works

As we mentioned above, a video signal consists of a series of rapidly changing images, also known as frames (a term carried over from the film industry). Since the human eye cannot react to these changing images fast enough to distinguish them as individual pictures, it merges the information in these images into a constant stream. Thus, the objects in the image can appear to move. This is a trait of the human eye known as persistence of vision and it is the reason that we can enjoy both movies and television. Take television for example. During each second of time that you watch the television screen, approximately 30 separate images flash by. Each of these images is slightly different from the previous one. When viewed by the eye, these small differences between images are perceived as motion. The way in which the images in a typical television broadcast are arranged is shown in Figure 11.40. If you think of time zero as being what is currently displayed on the screen, by the time one second has elapsed, all 30 of the images shown will be displayed.

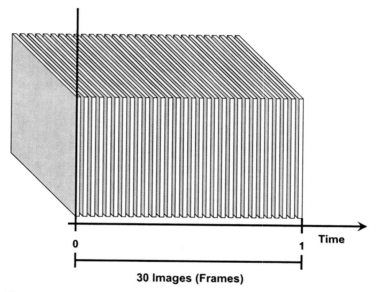

Figure 11.40 An illustration of the composition of a typical television video signal. Each rectangle is one image or frame in the overall video signal.

So, the only difference between an image signal and a video signal is the fact that the intensity values of the pixels will change over time. Thus, a video signal is digitized in exactly the same way as a still image except that we must capture an entire image a given number of times per second. In other words, we are taking samples of the video signal at a discrete time interval, known as a frame rate, and each of these samples consists of a two-dimensional digital image.

As we capture all of these digital images from our video signal, we need to store them. As you might imagine, this can consume a great deal of storage space. Let's take a look at an example to get an idea of how much.

Example 11.3

Problem: Assume you are capturing high quality video at a resolution of 720 rows by 480 columns and at a frame rate of 30 frames per second (DVD Quality). How much storage space will be required for each digital image if the image is grayscale? What if the image is color? Now, assume that you want to capture a two hour movie. How much storage space is required if the images are stored in grayscale format? How much if they are stored as RGB color images?

Solution: This problem begins by asking us to determine the storage space requirements for a single frame of video. We can determine this in the same way we did for a single image. First we find the number of pixels in the frame:

$$720 * 480 = 345,600 \text{ Pixels}$$

Now, we can answer the storage requirement questions. For a grayscale image we need one byte per pixel. So the storage requirement for a grayscale image is:

$$345,600 \text{ Pixels} * 1 \frac{\text{Byte}}{\text{Pixel}} = 345,600 \text{ Bytes} = 337.5 \text{ KB}$$

Now, the color image will require three times this amount since we store one byte each for the red, green, and blue, color intensities. Thus, our color image will require:

$$345,600 \text{ Pixels} * 3 \frac{\text{Bytes}}{\text{Pixel}} = 1,036,800 \text{ Bytes} \approx 0.99 \text{ MB}$$

So, now we know how much space we need to store one frame of video, but we want to record for two hours. This means we need to store:

$$2 \text{ Hours} * 60 \frac{\text{Minutes}}{\text{Hour}} * 60 \frac{\text{Seconds}}{\text{Minute}} * 30 \frac{\text{Frames}}{\text{Second}} = 216,000 \text{ Frames}$$

If we store our frames as grayscale images, 2 hours of video will require:

$$216,000 \text{ Frames} * 345,600 \frac{\text{Bytes}}{\text{Frame}} = 74,649,600,000 \text{ Bytes} \approx 69.5 \text{ GB}$$

If we opt for RGB color frames, we will need:

$$216,000 \text{ Frames} * 1,036,800 \frac{\text{Bytes}}{\text{Frame}} = 223,948,800,000 \text{ Bytes} \approx 208.57 \text{ GB}$$

So, as we can see, digitizing video signals can quickly consume huge amounts of storage space. Given that a typical DVD can only hold around 9 GB of data, we need to come up with some way of reducing the size of the data to make it fit. This is where video compression enters the scene. In order to get a two hour movie to fit on a DVD, we need to employ complex methods to drastically reduce the amount of redundant material within the video stream. In the next section we will give a brief overview of two approaches to doing just that.

11.8.2 Video Compression

As we just saw, uncompressed video signals can be enormous. Our study of image compression showed us that it is possible to greatly reduce the storage requirement for images, so there is hope that we can do something similar for video. In most cases video signals are simply too large to try to compress using any form of lossless compression. These compression methods simply do not achieve high enough compression levels to sufficiently reduce the size of the signal. Thus, we are almost always forced to turn to lossy methods for image compression. Whether using lossless or lossy compression, however, there are two general approaches to dealing with a video signal.

- Intra-frame Compression
- Inter-frame Compression

Intra-Frame Compression

Intra-frame compression, as the name suggests, applies compression to each of the frames independently. In other words, we simply treat each frame of the video as an independent image and compress it using a standard image compression method. One common image compression approach that does just that is known as MJPEG, or motion JPEG. This compression approach, as you might have guessed from the name, applies JPEG compression to each frame of the video independently. Thus, it provides spatial compression by removing the redundant information from each individual frame of the video. However, Intra-frame compression techniques, such as this one, do not attempt to remove redundant information along the time axis.

Inter-Frame Compression

Inter-frame compression picks up where intra-frame compression stops. Inter-frame compression techniques not only compress the information in the individual frames, but also take advantage of the large amount of redundant information between frames. If you think about it, how much information really changes between any two frames of a television signal? You are viewing 30 frames every second, and most likely most of those frames are nearly identical over that period of time. For example, if you are viewing a scene that involves two people talking to each other. The only changes between frames will most likely be the movements of those two people. The rest of the scene is background and is most likely not moving or changing very much. As a result, if we only record the information that changes between two frames of a video signal, we can often greatly reduce the amount of space needed to represent the signal. Of course, if we encode a video signal in this way and then want to view some frame mid-way through the signal, we first must reconstruct the first frame and then apply all the changes up to the point that we want to view. To minimize this problem, many inter-frame compression methods occasionally include a new complete frame so that the string of change frames does not go on for too long. Thus, when we want to look at a specific frame, we simply locate the nearest one of these complete frames and apply the changes from there. These embedded complete frames are often called key frames. A popular example of an inter-frame compression technique is MPEG.

MPEG, not to be confused with MJPEG, stands for Motion Picture Experts Group, the group that developed the method. The first version of the MPEG technique, called MPEG1 was developed in the late 1980's. Later, in the 1990's the MPEG2 standard was developed for use on DVDs. Even more recently a new standard called MPEG4 was released. In this text, we will describe the video encoding of MPEG1 since it is the easiest to understand.

A typical MPEG encoded video can contains three different types of encoded frames. These are:

- I-Frames, also known as intracoded frames. These frames consist of complete images from the video signal that are encoded using the JPEG compression method.
- P-Frames, also known as predictive frames. These frames contain information about what has changed since the previous frame.
- B-Frames, also called bidirectional frames. These frames contain information about the changes between the current frame and the frames before and after it.

By using these three types of frames in specific sequences, we can remove a great deal of the redundant information, both spatial and temporal, from the video signal. Most video applications use the following sequence of frames when compressing a video stream:

IBBPBBPBBPBBPBBPBB

This sequence is repeated over and over until the end of the video signal is reached. The sequence can be broken by the insertion of additional I-frames, typically due to drastic image content changes like scene changes in movies. Thus, we can achieve a high level of both spatial (intra-frame) and temporal (inter-frame) compression using the MPEG compression format.

So, as we have just seen, we can use an intra-frame (MJPEG) or inter-frame (MPEG) compression technique based upon what we plan to do with the video signal. If we need to perform future processing on the signal that may require us to deal with individual frames, we are better off using an intra-frame compression method to avoid the difficult task of reconstructing a random frame from a series of changes applied to a key frame. On the other hand if we are simply compressing the video signal to view later, we can use an inter-frame compression approach to fully take advantage of the increased compression capabilities gained from the removal of temporal redundancies.

11.9 Chapter Summary

In this chapter, we examined images and explored techniques for manipulating, enhancing and compressing them.

- We began with a basic overview of what makes up an image.

- Next, we looked at some practical examples of analog signal processing by illustrating the way in which cameras work and how the human eye senses light.

- The human eye operates by using the cornea and lens to focus light onto the retina at the back of the eye. In the retina, special light sensitive structures, rods and cones, sense various aspects of the incoming light. Rods sense information about motion and intensity while cones are responsible for detecting details and color. These structures convert the information they sense into nerve impulses that are sent to the brain for processing via the optic nerve.

- Once we learned about analog images, we explored the processes involved in converting an analog image into a digital one. We found that images are first sampled horizontally and vertically to form a grid of pixel locations. Then the values at these locations are made discrete in amplitude through the process of quantization. Finally, the values for each pixel are stored as a digital signal.

- With the image in digital form, we turned our attention to some of the ways that the pixel information of the image can be represented, particularly in the case of color images. We examined several of the commonly used color models.

- Next, we looked at some of the most basic transformations that we could apply to images. In particular, we looked at methods for flipping, rotating, and resizing images.

- We found that some of the most useful image transformations require the use of filters. Through the use of filters we found that we could enhance images, or make them more pleasing to look at by blurring or sharpening them. We also saw that we could find useful information such as the edges of objects within the image by applying the correct filters and processing the output.

- Next, we learned that images can be influenced by noise. This noise could enter the image in several ways, for example, imperfections in the original image, problems with the digitizing hardware, or errors in transmission could all lead to noise in the digital image.

- We looked at several types of noise that could degrade a digital image. This noise could be described with color, such as white or pink noise, or it could be described by the way it looks, for example speckle or salt and pepper.

- Some types of noise are easier to remove than others. We found that white noise is extremely difficult to remove. A blurring filter can be used, but it will reduce the quality of the image. Salt and pepper noise is easier to deal with. We found that it can be removed by using a median filter.

- Digital images are extremely useful, but can require a large amount of space to store in their raw format. To address this problem a great deal of work has been done it the field of image compression.

- There are two different types of image compression. Lossless compression in which the original image can be reconstructed exactly from the compressed information, and lossy compression in which some of the image information is thrown out resulting in reconstructions that are only approximations of the original. The type of compression that should be used is heavily influenced by the application requirements. Biomedical and forensic applications require exact reproductions while photographs and illustrations can still convey sufficient information even after some information is discarded.

- Finally, we looked at the concept of video. We discovered that a video signal is simply a three dimensional signal in which images change over time. We learned that video requires even larger amounts of storage than images, and thus, is an even better candidate for compression.

- There are two ways to compress video, spatially and temporally. Spatial compression works through the use of intra-frame compression in which each frame of the video is compressed independently. Temporal compression works by taking advantage of the redundant data between successive frames of the video. Since very few pixels change between frames, all that we need to encode is the changes between frames. We noted that the MJPEG format was a spatial compression method while MPEG was a temporal compression scheme. Image compression continues to be an area of important research especially with video intensive applications such as DVD.

11.10 Review Exercises

Section 11.2 – Analog Image Processing

1. Describe the process of analog photography.

2. How is an image represented on film?

3. Describe how black and white film is developed. What about color film?

4. Name the two light sensitive structures in the human eye. Which of these is responsible for color vision, and which can only sense intensity? Which of the two is more numerous in the human eye? Which colors are the color-sensitive structures to?

Section 11.3 – Digital Images

5. Describe the structure of a digital image.

6. What is a single element of a digital image called?

7. How are digital images obtained?

8. How is a grayscale image stored? What do the stored values for each element represent?

9. Describe the RGB method of image storage.

10. Compare the RGB and CMY color models. How are they different? How are they similar? How do you convert from one to the other?

11. Convert the following RGB values to their CMY equivalents.

    a. [255, 255, 255]　　　　　　b. [255, 0 , 0]
    c. [0, 255, 0]　　　　　　　　d. [0, 0, 255]
    e. [128, 250, 11]　　　　　　f. [103, 219, 65]
    g. [35, 12, 88]　　　　　　　h. [102, 35, 192]

12. Convert the following RGB values to their YIQ equivalents.

    a. [255, 255, 255]　　　　　　b. [255, 0 , 0]
    c. [0, 255, 0]　　　　　　　　d. [0, 0, 255]
    e. [150, 210, 50]　　　　　　f. [45, 89, 105]
    g. [202, 115, 70]　　　　　　h. [190, 15, 83]

13. Describe the HSI model.

14. Describe the HSL model.

Section 11.4 – Basic Digital Image Manipulation

15. Describe how an image is flipped horizontally.
16. Describe how an image is flipped vertically.

17. What would be necessary to flip an image both horizontally and vertically?

18. Describe the cut and paste process in image processing.

19. How can an image be rotated by 90 degrees clockwise? How about counter-clockwise?

20. What is required to rotate an image by an arbitrary number of degrees?

21. How are images up-sampled and down-sampled?

Section 11.5 – Image Filtering and Enhancement

22. Describe the process involved in applying a filter to an image.

23. Given the following image matrix, apply the specified filters.

$$
\begin{matrix}
0 & 0 & 0 & 0 & 0 & 0 & 0 & 0 & 0 & 0 & 0 & 0 \\
0 & 0 & 0 & 0 & 0 & 0 & 0 & 0 & 0 & 0 & 0 & 0 \\
0 & 1 & 1 & 1 & 0 & 0 & 0 & 1 & 1 & 1 & 0 & 0 \\
0 & 1 & 1 & 1 & 0 & 0 & 0 & 1 & 1 & 1 & 0 & 0 \\
0 & 1 & 0 & 1 & 0 & 0 & 0 & 1 & 0 & 1 & 0 & 0 \\
0 & 1 & 1 & 1 & 1 & 1 & 1 & 1 & 1 & 1 & 0 & 0 \\
0 & 1 & 1 & 1 & 0 & 0 & 0 & 1 & 1 & 1 & 0 & 0 \\
0 & 1 & 1 & 1 & 0 & 0 & 0 & 1 & 1 & 1 & 0 & 0 \\
0 & 1 & 1 & 1 & 1 & 1 & 1 & 1 & 1 & 1 & 0 & 0 \\
0 & 0 & 0 & 0 & 0 & 0 & 0 & 0 & 0 & 0 & 0 & 0 \\
0 & 0 & 0 & 0 & 0 & 0 & 0 & 0 & 0 & 0 & 0 & 0 \\
\end{matrix}
$$

a. 3 x 3 blurring filter (low pass)
b. 3 x 3 high-pass filter
c. Horizontal edge detection (Sobel)
d. Vertical edge detection (Sobel)

24. Describe some practical uses for image filtering.

Section 11.6 – Noise and Image Restoration

25. What is image noise? What are some common image noise models?

26. How can salt and pepper noise be corrected?

27. Find the median of the following sets of numbers.

   a. $\{1,4,3,8,16,72,105,96,12\}$ b. $\{8,3,2,4,7,5,7,2,10\}$
   c. $\{4,2,8,7,5,16,88,25,15,50,13,29\}$ d. $\{1,4,3,7,2,9,6,8,13,1,5,9,2,6\}$
   e. $\{9,1,3,10,23,1,4,9,18,8,7,6\}$ f. $\{100,255,203,98,0,52,105,96\}$

28. Describe how the median filter is applied to images to remove noise.

Section 11.7 – Image Compression

29. Why would we want to use image compression?

30. What are the two main categories of image compression? What is different about these two approaches?

31. How much storage space would be required to store grayscale images of the following dimensions?

   a. 640 x 480   b. 800 x 600   c. 1024 x 768
   d. 512 x 512   e. 6000 x 3000   f. 2800 x 765

32. How much space would be required to store images of the dimensions in problem 31 if they are stored using RGB color?

33. How does run-length encoding work? Which compression category does it belong to?

34. How does DPCM encoding work? Which compression category does it belong to?

35. Describe the colormapping approach to image compression.

36. Briefly describe how JPEG compression works. What compression category does it belong to?

Section 11.8 – Images in Motion – Video

37. Why would we need to compress video?

38. Determine the storage requirements for the following grayscale video clips.

     a. 640 x 480, 30 frames per second for 1 hour
     b. 720 x 480, 30 frames per second for 30 minutes
     c. 320 x 280, 30 frames per second for 2 hours
     d. 720 x 480, 12 frames per second for 1 hour

39. What would the storage requirements be if the video clips in problem 38 were stored in RGB color?

40. What are the two approaches to the compression of video frames? How does each method achieve a reduction in storage requirements?

Chapter 12. Communication and Coding

Chapter Outline

12.1 Introduction to Communication

The ability to communicate with each other has become a vitally important part of our daily lives. In fact, the way in which we are able to effortlessly communicate with others over great distances is often something we take for granted. Advanced technologies like computers, telephones, fax machines, radios, and televisions all help us to communicate large amounts of information quickly and effectively.

Since before recorded history, the ability for people to communicate with each other has been important. This began as established spoken language and then came written languages. Soon, people discovered that they could send information to each other by using drum beats, smoke signals, and by other visual means. With the later advent of the telegraph, people were able to send information as a series of electrical pulses that traveled down a wire. This was expanded upon with the eventual development of the telephone. Next, it was discovered that electromagnetic waves could be used to convey information. This led to the development of the wireless telegraph and then the radio and television. In recent years great strides in communication technology have been made. Today, it is possible to make phone calls from anywhere on the planet through the use of satellite phones or to retrieve information over the Internet from computers all over the world.

So, as we can see, communication is an important area of study. As electrical engineers, we focus primarily on the development of new and innovative ways to transport information from place to place in the form of various types of signals. This could include the development of new hardware that makes communication easier or more secure, or new ways of processing or encoding signals. Thus, there are many aspects of communication that are impacted by electrical engineering. Since this is the case, a basic understanding of communication theory is very important. Let's start our study by taking a look at what makes up a typical communication system.

12.2 The Parts of a Communication System

The purpose of a communication system is to allow the movement of information from one place to another. In order for this to occur, there are a series of steps that must be taken. From these steps, we can formulate five basic parts that compose a basic communication system. If we think about it logically, in order to convey information, we must first have some kind of information to communicate. Thus, the first part of our communication system is an information source. Next, this information must be formatted in a way that makes it possible to transmit. The way in which it is formatted depends upon the way in which you plan to transmit it. For example, if you are going to communicate your idea by speaking, the idea must be translated into a spoken language which is essentially a series of sounds. However, if you plan to convey your idea by writing it down, you must transform the idea into written language, essentially an image. In either case, this translation stage is performed by the second part of a communication system, the transmitter. The medium over which the transmitter conveys the information, whether it is air, paper, or copper wire, is known as the channel. Thus, we can think of the channel as the third part of a basic communication system. Once the signal is placed on the channel by the transmitter, we need a way to get it back off of the channel and back into a useful form. For this, we need a receiver. Thus, the purpose of the fourth part of the system, the receiver, is to take the communicated information from the channel and translate it into useful information for the final part of the communication system, the destination. So, if we follow our example of communication by speech all the way through, we can see that an idea is formed in the brain (the source). This idea is translated into speech by the vocal tract (the transmitter). The resulting sound waves travel through the air (the channel) and eventually arrive at someone's ears (the receiver) where they are translated into nerve impulses and sent to that individual's brain (the destination). Thus, we see that no matter what method the communication system utilizes, the resulting process will fit into this five step model, shown in Figure 12.1. In the following sections, we will look at each of the five stages in a communication system. In each section we will look at what is involved in each phase of the communication system model.

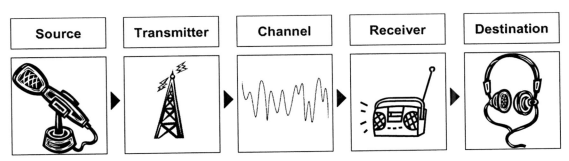

Figure 12.1 The five parts of a typical communication system.

12.3 The Source

The information that flows through a communication system must originate somewhere. This origin is what we know as the source. The purpose of the source in a communication system is to generate the signal that we wish to communicate. There are a wide variety of different sources and several ways in which they can be categorized. One common broad classification is to group sources by the type of signal that they generate. We saw in chapter 9 that one of the primary distinctions that we use to categorize signals is whether the signal is digital or analog. Thus, we often think of communication system signal sources as either analog or digital. Analog sources would include signals like the human voice captured by a microphone or a continuous stream of data from a remote sensor. On the digital side of things, we will also encounter a wide variety of signals. The primary source of digital signals, however, is computers. So, whether it originates from a digital device or the human voice, we must have a signal to transmit in order for our communication system to be useful.

In reality, there are three major sources of information that communications systems are designed to work with. These are:

1. Audio Signals – This primarily consists of the human voice, but can also include music or any other sound that is captured with a microphone. These audio signals may be captured in either analog or digital signals. Radio and the telephone are both communication systems that are designed to transmit audio signals.

2. Video Signals – This includes full motion video signals and static images. The two main communication applications that use video signals as sources are the television and the fax machine. The television transmits true video signals using electromagnetic waves and the fax machine transmits still images (usually scans of sheets of paper) over telephone lines.

3. Computer Data – Computer data consists of information in digital form that can be sent over a wide variety of communication systems ranging from telephone to satellite.

While these signal sources are the most commonly encountered in communications systems, they are by no means the only types of sources. If you spend any amount of time working in the field of communications, you will most likely encounter some of these others, but the three source types listed here will be by far, the most common.

12.4 The Transmitter

Once we have a signal to transmit, it must be converted into a form that can be transmitted over some specific physical medium. This, often complicated, task is undertaken by the second part of our communication system, the transmitter. When most people hear the word transmitter, they usually envision a radio transmission tower. While this is indeed a transmitter, it is by no means the only way to transmit a signal. Many devices you deal with every day have transmitters within them. For example; the telephone, hand-held two-way radios, wireless microphones, and networked computers all transmit information, and thus utilize a transmitter. Each transmitter, based upon the type of channel that it is required to transmit on and the type of information that it needs to transmit, will utilize different techniques. One of the techniques used by a wide variety of transmitters, particularly radio transmitters, is modulation. So, what exactly is modulation and how does it work? Let's take a look.

12.4.1 Modulation

Modulation is a technique that modifies the source signal, also known as the message signal, so that it is more suited for transmission over the channel. This can range from being a fairly simple process to something that is rather complicated, depending on the type of modulation that is used. There are several different approaches to modulation used throughout the field of communication. The type of modulation used in a specific situation is usually chosen based upon how well it is suited for the application that is being developed. For example, one type of modulation may require less complex hardware to implement while another may be more complex but, will offer more resistance to signal distortions and noise. Thus, the modulation technique is picked from those available based on a variety of factors. The key to all of these modulation schemes, however, is the use of higher frequency base, or carrier, signals. Modulation basically works like this: The message that we want to transmit (the modulating signal) interacts with, or "modulates" the higher frequency signal (the modulated signal) by inducing a change in some particular aspect of the carrier wave, for example its amplitude or frequency. The resulting waveform is a high frequency signal that has the lower frequency message signal embedded in it. This signal is then placed onto the channel.

12.4.2 Why We Modulate Signals

At this point, you are no doubt wondering why we even bother to modulate signals at all. Why not just transmit the message signal directly? There are actually two main reasons that we modulate signals. These are:

1. Effective use of the available channel
2. Reduction of antenna length for wireless systems

The first of these reasons should make sense from a practical perspective. On any given channel that is shared by several entities, there is a need for each of these entities to use the channel without causing problems for the others. Think of a room full of people as an example of a channel. If only one person is talking (using the channel), everyone can hear and understand what they are saying. If everyone in the room tries to talk at once, however, it is difficult and often impossible to communicate effectively. This is the same problem we run into on any channel. A perfect example of this is radio. If all of the radio stations were to simply broadcast their source, or message, signal, they would all overlap and you could not make any sense out of the resulting mess. So, without any modulation, only one radio station could broadcast information at a time. To solve this problem, we divide up the frequency range that is allocated for radio stations into a series of frequency bands. Each radio station is assigned one of these bands by the Federal Communications Commission (The FCC). The station then uses a carrier wave of the assigned frequency to modulate its message signal. As a result, each message signal resides in a different frequency range and can be listened to by tuning a receiver to a specific frequency band. This approach is not limited to wireless transmissions. The transmission of signals over almost any medium that, is being used by several groups at once, requires some kind of sharing method. Thus, modulation is a popular technique used by many transmitters to allow for sharing of the channel.

In addition to the ability to share the channel, there is another benefit to modulation that is particularly important to wireless systems. When a signal is broadcast as an electromagnetic wave, you must have an antenna that is one fourth of the signal's wavelength in order to be able to receive it. Many low frequency signals have wavelengths that are simply too long to be picked up by an antenna of practical length. In fact, we can use a simple calculation to determine what length of antenna we need for a given frequency. First, we need to find the wavelength of the signal for the frequency we are interested in. Since electromagnetic waves travel at the speed of light, we can calculate the wavelength as follows:

$$\lambda = \frac{c}{f}$$

Where λ is the wavelength, c is the speed of light (approximately 3×10^8 meters / second), and f is the frequency of the signal we are looking at. Once we know the signal's wavelength the length of the required antenna can be calculated as follows:

$$h = \frac{\lambda}{4}$$

Where h is the height of the required antenna and λ is the wavelength of the signal. Now, to see how the length of the antenna is affected by the frequency of what we transmit, let's look at an example.

Example 12.1

Problem: Calculate the antenna lengths required for the following signal frequencies.

   a) 10,000 Hz
   b) 87 MHz
   c) 1 GHz

Solution: In this problem, we have three frequencies for which we must determine the required antenna length. Thus, our first step will be to calculate the wavelength of our three signals. Using our equation for wavelength yields:

$$\text{a) } \lambda = \frac{3x10^8}{10,000} = 30,000 \text{ meters}$$

$$\text{b) } \lambda = \frac{3x10^8}{87x10^6} \approx 3.45 \text{ meters}$$

$$\text{c) } \lambda = \frac{3x10^8}{1x10^9} = 0.3 \text{ meters}$$

So, we see that the wavelengths for our three signals are 30,000 meters, 3.45 meters and 30 centimeters respectively. Now, to find our required antenna lengths, we must divide these values by four. Thus, the antenna lengths required for each of the above frequencies is:

$$\text{a) } h = \frac{30,000}{4} = 7,500 \text{ meters} \approx 4.66 \text{ Miles}$$

$$\text{b) } h = \frac{3.54}{4} = 0.885 \text{ meters} \approx 2.9 \text{ Feet}$$

$$\text{c) } h = \frac{0.3}{4} = 0.075 \text{ meters} = 7.5 \text{ cm} \text{ or } 2.95 \text{ Inches}$$

So, here we have found that to receive a signal at 10,000 Hz requires an antenna of 7,500 meters, a signal of 87 MHz requires an antenna 0.885 meters long, and a signal of 1 GHz requires an antenna length of 7.5 cm.

As we can see in the example, the higher frequency that we use, the shorter the antenna our receiver will require. Trying to receive a signal broadcast at 10,000 Hz would require an antenna length of over four and a half miles. This is obviously not practical for a mobile device. 87 MHz is actually the lowest frequency at which FM radio is broadcast. This is why your typical car radio antenna is about three feet long. Finally, cellular and cordless phones operate around the 1 GHz frequency. Thus we see much shorter antennas on these devices. So, we can clearly see the benefits of modulation in this case. Now that we see why modulation is important, let's look at some of the more popular modulation methods.

12.4.3 Types of Modulation

As we know, modulation works by changing some aspect of a higher frequency carrier wave. If we look at the equation for a sinusoidal waveform, we can get an idea of how these different approaches come about.

$$A\cos(\omega t + \phi)$$

Where A is the amplitude of the waveform, ω is the frequency, t is time, and ϕ is the phase. So, as we can see, there are three basic aspects of a given sinusoidal wave that we can modify. These are the amplitude, frequency, and phase. It is no coincidence then that, the three common analog and three common digital approaches to modulation take advantage of these variable aspects of sinusoids. Let's start by looking at analog modulation and then we will discuss the digital approaches.

12.4.3.1 Analog Modulation

You have probably heard of two of the three analog modulation approaches since you are familiar with radio broadcasts. These are, of course, amplitude modulation and frequency modulation. The simpler of these two is amplitude modulation. In this approach, the amplitude of the carrier signal is adjusted based upon the amplitude of the message signal. So, let's consider the following two signals, the message signal, $m(t)$, and the carrier signal, $\cos(\omega_0 t)$. In a simple amplitude modulation approach, these two signals are simply multiplied yielding the following:

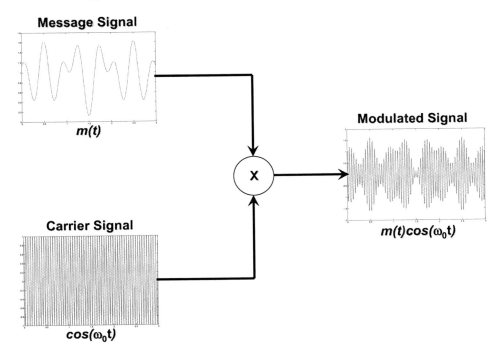

Figure 12.2 An illustration of the amplitude modulation process.

$$m(t)\cos(\omega_0 t)$$

Where *m(t)* is the message signal and ω_0 is the frequency of the carrier wave. This simple multiplication results in a high frequency signal with an amplitude that mirrors that of the message signal. This process is illustrated in Figure 12.2. As you can see in the figure, the product of the two signals does indeed yield a high frequency signal of varying amplitude. Amplitude modulation is one of the simplest forms of modulation. Since it requires only a simple signal multiplication, it is easily implemented in hardware. This is why the first radio broadcasts were modulated using the AM approach. Unfortunately, amplitude modulation has a few disadvantages. One of the largest of these is that it is easily distorted by noise. You have probably heard this distortion if, you have ever listened to AM radio during a thunderstorm. The electrical discharges in the atmosphere introduce crackling and popping into the signal. This is because the electrical interference changes the amplitude of the signal and as a result distorts the message that it was carrying. This is one reason why many radio stations switched to the second type of analog modulation, frequency modulation.

Frequency modulation was designed to overcome some of the limitations of amplitude modulation. In this approach, the frequency of the carrier wave is modified based on the message content. Since we are modifying the frequency, any noise that impacts the amplitude has minimal effect on the message contained within the signal. This makes this approach much more resistant to the effects of atmospheric noise. As a result, it is the modulation method of choice for high quality radio broadcasts. An illustration of frequency modulation is shown in Figure 12.3.

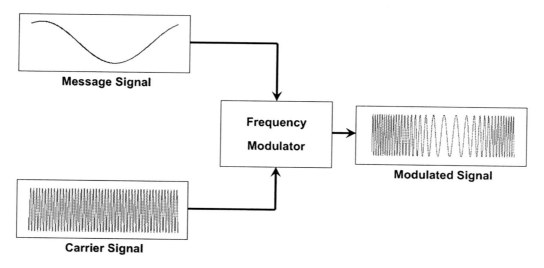

Figure 12.3 An illustration of frequency modulation.

The final analog modulation approach is probably one that you are less familiar with. Phase modulation has not been as widely used as amplitude and frequency modulation. Both phase and frequency modulations are known as angle modulation techniques. The basic idea behind phase modulation is that, the phase of the carrier wave is shifted based upon the message signal. An illustration of this technique is shown in Figure 12.4.

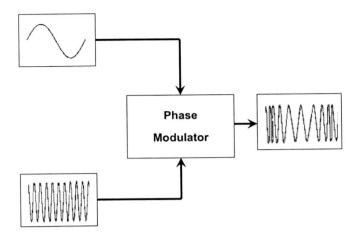

Figure 12.4 An illustration of phase modulation.

So, we have seen that by modifying one of the three aspects of the carrier sinusoid based upon out message signal, we can transmit our message at any frequency that we need.

12.4.3.2 Digital Modulation

Now that we have seen the analog modulation techniques, let's see how digital signals are modulated. Digital signals, unlike analog signals only require us to transmit two different values, zeros and ones. Thus, we can use simplified versions of the analog methods when dealing with digital signals. The three basic digital methods based on their analog counterparts are:

1. Amplitude Shift Keying (ASK)
2. Frequency Shift Keying (FSK)
3. Phase Shift Keying (PSK)

As was the case with the analog modulation techniques, each of these methods work by adjusting one of the aspects of the carrier sinusoid. The difference is that, since there are only two values to transmit, we can choose two distinct versions of the base sinusoid and chain them together to form our resulting transmission. For example, in ASK we choose two sinusoids with the same frequency and phase, but different amplitudes. We will associate one of these sinusoids with a digital zero and one with a digital one. Then we transmit a short duration of the respective sinusoid for each bit in our message signal.   An example of ASK is shown in Figure 12.5.

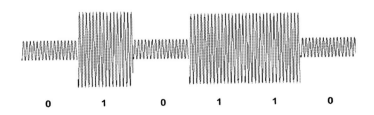

         0   1   0   1   1   0

Figure 12.5 An example of amplitude shift keying

Frequency shift keying works very similarly to amplitude shift keying except that, in this case, the two sinusoids have the same amplitude, but different frequencies. This technique is illustrated in Figure 12.6.

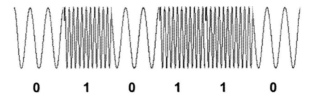

Figure 12.6 An illustration of frequency shift keying.

Finally, phase shift keying works by using two sinusoids of identical amplitude and frequency, but different phases. As was the case with the previous two approaches, we transmit one version of the sinusoid for a binary one and the other for a binary zero. This process is illustrated in Figure 12.7.

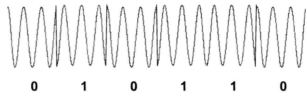

Figure 12.7 An illustration of phase shift keying.

As we can see, digital modulation is just a simplified form of analog modulation.

12.4.4 Classification of Communication Systems by Modulation Scheme

One way we can describe a communication system is based upon the type of modulation that it uses. The determination of whether a communication system is analog or digital is not dependent on the data, but rather on the modulation technique. If the system uses analog modulation, even if it transmits digital data, it is called an analog communication system.

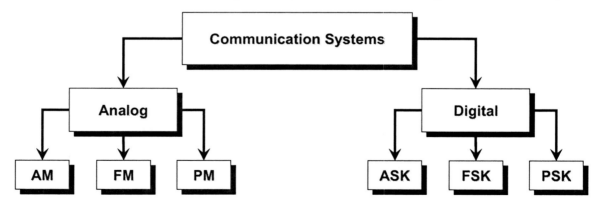

Figure 12.8 An illustration of the different classifications of a communication system based on the modulation approach.

Likewise, if a digital modulation scheme is used, the system is called a digital system. Figure 12.8 displays the different system classifications we can make based on modulation scheme. Under each broad system classification, we have listed the modulation schemes that fit into that classification.

12.5 The Channel

Once the transmitter has prepared the signal, it places it onto the channel. The channel is basically, just the medium over which the transmitted signal travels. It connects the transmitter to the receiver and can consist of any material that the signal can travel through. This can range from basic copper wire to empty space.

12.5.1 Common Channel Types

There are essentially two broad classifications we can make when dealing with communication channels. The distinction between the two classifications is whether the channel requires a physical connection to the receiver or not. If a physical connection, such as a wire, is required, the channel is called a wired channel. If, on the other hand, the channel utilizes the medium surrounding both the transmitter and receiver, such as air, the channel is called a wireless channel. There is a wide variety of materials that can be used to construct both wired and wireless channels. Each of these has its own strengths and weaknesses. Let's investigate a few of the most common of these now.

12.5.1.1 Wireless Channels

In general, a wireless channel utilizes the material that surrounds both the transmitter and receiver to carry the signal from one to the other. As a result, the transmitter and receiver can move around without breaking the link between them. Thus, an obvious benefit to using a wireless communication channel would be the ability to create mobile devices. This benefit is balanced by several drawbacks. First, since wireless channels require that the transmitter broadcast the signal, any other devices in the vicinity of the transmitter could potentially intercept the signal. This can be a serious concern if you are transmitting sensitive information. In addition, wireless channels are more susceptible to interference and noise. Thus, when deciding to use a wireless channel, it is necessary to balance the benefit of portability with the drawbacks of noise and information security concerns.

As you might imagine, a wireless channel can consist of any material that the transmitter and receiver can move around in. Thus, the most obvious wireless channel materials would be air and free space. This is the material used for the transmission of radio signals. Whether we are communicating with a mobile unit in the next room or a deep space probe millions of miles away, the wireless channel that carries our radio waves is made up of either air or empty space. Another less common channel material is water. This is the primary method for underwater communications and sonar systems. So, we can see that the one of the main requirements for a wireless channel material is that, it must allow relatively unrestricted movement of the transmitter or receiver while communication takes place.

12.5.1.2 Wired Channels

Wired channels on the other hand, require that the transmitter and receiver be physically connected. Thus, the receiver and transmitter must remain stationary. However, since the information is sent through a physical connection, access to the information can be controlled to a greater degree than would be the case with a wireless channel. Wired channels also tend to be more resistant to noise. This is because the channel can be shielded from external interference. The degree to which both of these benefits can be realized is dependent on the type of material used for the wired channel. For example, unshielded copper wire like that found in the phone wiring in your home can be affected by noise. Running the phone cable too close to an electrical wire can result in noise within the phone signals. In addition, transmissions on this type of cable can still be intercepted if the right equipment is used. Optical fibers, on the other hand, provide a much more noise resistant and secure path for transmissions. Since fiber optic cable carries the signal as a series of light pulses on a shielded glass fiber thinner than a human hair, there is little chance of external signals interfering with the transmission. External electronic and radio signals have no impact on an optical transmission. In addition, it is extremely difficult to tap fiber optic cables without being detected, and there is no way to intercept the transmission without physically splicing into the fiber. As a result, fiber optic cable is the medium of choice for secure transmissions. Both copper cables and optical fiber can carry a great deal more information than wireless channels.

Common wired channels include shielded and unshielded copper cable. A good example of unshielded copper cable is the cable used in Ethernet computer networks and your household phone systems. An example of shielded copper cable would be the coaxial cable used for cable television. Copper cable is a preferred channel for wired networks due to its low cost and ease of implementation. The other commonly used wired channel material is fiber optic cable. This cable which is thinner than a human hair can carry small pulses of light over great distances. It is almost completely immune to external noise since it carries the signal as light instead of electrical impulses and is extremely secure. In addition, the amount of information that can be transmitted on a fiber optic cable is enormous. Hundreds of thousands of phone conversations can be carried on a single fiber optic cable. So, as we have found, wired channels restrict the free motion of the transmitter and receiver, but in return, they offer better security, higher transmission rates, and fewer problems with noise. Thus, whether a wired or wireless channel is used it is often the result of a matching between the strengths and weaknesses of the channel and the desired requirements of the application.

12.5.2 Classification of Communication Systems by Channel Type

We saw earlier that, we could classify communication systems based upon the type of modulation that is used by the transmitter. Now, we can further classify a system based upon the

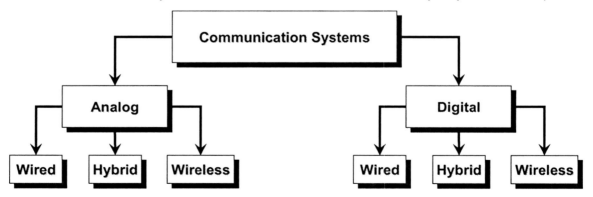

Figure 12.9 An example of the ways in which a communication system can be classified when using modulation method and channel type.

type of channel that it uses. There are essentially, three possible classes of systems based on channel type. A system can use a wireless channel, like commercial radio for example, a wired channel, like cable television, and a hybrid channel in which both wired and wireless channels are used. An example of a hybrid system would be the telephone network. In this network, there is an extensive wired network connecting a large number of phones. At the same time, there are a large number of wireless cellular phones. These phones communicate to a cellular tower. These towers then connect to the wired network and relay the signal to other areas. Thus, both wired and wireless channels are used within the system. Figure 12.9 demonstrates some of the classifications that we can make by using both modulation method and channel type.

12.5.3 Allocating the Channel

In many cases, a given communications application does not have complete control over the channel that it is transmitting on. A good example of this is radio. A given radio station does not have complete control over the entire electromagnetic spectrum. Instead, this channel must be shared by a large number of other radio stations as well as, other types of wireless transmitters. Even when one entity does control the entire channel, it is often desirable to allocate certain amounts of the channels transmission potential to a variety of different tasks or applications. As a result, several strategies have been developed to allow a given channel to be used by a variety of entities and in a variety of different ways. One of the simplest ways to split up the transmission potential of the channel is to divide its bandwidth into a series of smaller units.

12.5.3.1 Bandwidth

The definition of bandwidth is somewhat different depending on whether you approach it from an analog or digital perspective. From the analog point of view, bandwidth is defined as the band or range of frequencies on a channel over which data can be transmitted (measured in Hertz). On the other hand, bandwidth from a digital perspective is defined as the amount of data that can be transmitted over the channel during a given period of time (measured in bits per second). From these definitions, we can assume that there are two good ways to divide up the channel bandwidth, either by time or by frequency. The act of sending multiple data streams over the same channel is a concept known as multiplexing. Thus, we can perform either frequency-based multiplexing or time-based multiplexing (or in some cases both).

Frequency-based multiplexing (often called frequency division multiplexing or FDM) is achieved by dividing up the available frequencies and allowing different data streams to use a subset of the overall frequency space. This is exactly what we see in the commercial radio industry. The use of the entire electromagnetic spectrum is controlled by a government organization known as the federal communications commission. If tells each entity what frequencies can be used for a given application. For example, an FM radio station would be given a small frequency range centered at a frequency somewhere between 88 and 108 MHz and an AM station would receive a frequency somewhere in between 535 and 1605 KHz. Since each individual station uses a part of the overall spectrum, the total bandwidth of the wireless channel is efficiently utilized. Similar approaches are taken when broadcasting over wired channels. For example, by dividing up the frequencies that can be carried on a piece of copper wire or fiber optic cable, large numbers of simultaneous phone conversations can be carried on the same cable at one time without interfering with each other. Figure 12.10 illustrates the concept of frequency division multiplexing.

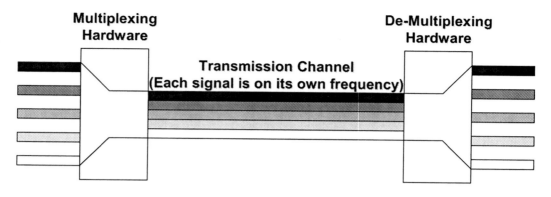

Figure 12.10 An illustration of frequency division multiplexing. Each input signal is modulated to its own carrier frequency and then transmitted on the channel. This allows multiple data streams to occupy the same physical channel. At the remote end, the frequencies are separated and the data streams are reconstructed.

Time-based multiplexing (often called time division multiplexing or TDM) operates by allowing each data stream a fixed amount of time to transmit on the channel. When its time is up, the next data stream is allowed to transmit and so on. Thus, all of the data streams can transmit information in short bursts after which they must wait for their next turn. By doing this, multiple data streams can use the same channel without overlapping. On the receiver side of the channel, the data streams are separated back out based on the time interval during which they were transmitted. Figure 12.11 illustrates the way in which a time division multiplexing system would work.

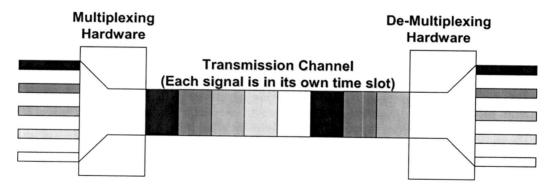

Figure 12.11 An illustration of time division multiplexing. Each input signal is transmitted in its own specific time slot. This allows multiple data streams to occupy the same physical channel. At the remote end, the data streams are reconstructed by piecing the time slots back together.

So, as we have seen, we can divide up the bandwidth of the channel, in order to accommodate multiple data streams, by dividing the overall bandwidth of the channel into either frequency or time slots. It is important to note that frequency division multiplexing works best for signals that require constant communication (like phone conversations) and time division multiplexing works best for data that does not require real-time transmission (like computer networks). Thus, to some degree the application that is communicating over the channel dictates the type of channel allocation that can be performed.

12.5.3.2 Switching

Another way to deal with channel allocation is to connect a data source to the channel only when it needs to transmit information. By doing this, you can accommodate more input sources than you have channel bandwidth to transmit, as long as, they do not all need to send data at once. There are two ways to achieve this task, either by circuit switching or packet switching.

The concept of circuit switching is best explained by looking at the telephone system. There can be millions of telephones in a large city. The telephone transmission lines could never transmit calls from all these phones at once, since there is a limited amount of bandwidth available. As a result, when your phone is not actively participating in a telephone call, it is not using any of the system's bandwidth. When you want to make a call, you dial a number on your telephone. The number you enter is processed by the phone company's switching network and a dedicated channel is set up between you and the person you call by allocating a portion of the available transmission resources on the network. While you talk on the phone the resources allocated to your connection are reserved for your use only. When you hang up, the connection is torn down and the resources that you were using are placed back into the pool of available resources so that someone else can use them. Thus, the phone company is able to service a large number of individual phones, without having to provide dedicated lines for them all of the time. This is also why you will sometimes get a message that all circuits are busy when you try to make a call. This simply means that, there are not enough resources available to construct a connection between you and the person you wish to contact.

Packet switching, on the other hand, is often found in computer networks. When you wish to send a file or piece of information to another computer, the data is divided up into a series of small units called packets. Each of these packets contain some of the overall data along with other information such as error detection codes, a source and destination address, and a sequence number. These packets are then released out onto the network. As they arrive at each node in the network, they can be routed down different paths based on network congestion and a number of other factors. As a result, parts of the same data file may travel different paths between the source computer and the destination. Some later packets may even arrive before some of the earlier packets. When a packet does arrive at the destination, it is checked for errors and then stored based on the sequence number it was sent with. Thus, regardless of the order in which the packets arrive, the original data can be reconstructed.

Packet switching works well for computer data, because it is not critical that the communication occurs in real time or that all the information arrive in the proper order. This would not work well at all for applications that required real-time dedicated resources. For those types of applications, circuit switched networks are required. So, once again we find that the method of channel allocation is dependent on the type of information being sent and the application that is sending it.

12.5.4 Dealing with Channel Noise

One of the main problems faced when transmitting information on any channel is noise. Noise can be thought of as, any kind of undesired signal aside from the message signal that we are trying to send. This could be caused by natural sources such as sunspots, cosmic rays in the earth's atmosphere, or lightning. It can also result from man-made sources like power lines and other signals improperly transmitted on the medium. Since noise in the channel distorts the message signal, we must make an effort to reduce its effect if we want to reliably send information to the receiver. There are several ways in which this can be done. Some of these include:

1. Choose a different modulation scheme.
2. Use a different channel.
3. Increase the power of the signal
4. Employ filters to remove the noise.
5. Use codes to detect errors in the data.

The choice of a new modulation scheme can often help reduce noise. For example, the FM technique was developed (in part) to address the susceptibility of AM to atmospheric noise. Thus, using FM instead of AM to transmit a signal will yield a reduction in noise.

If that is not a possibility, we can try to use a different channel. This could simply mean sending the data over a frequency range within the current channel medium that is less affected by noise or, it could mean using a different channel medium altogether. For example, if we are currently using copper wire to transmit our signal but, are encountering a great deal of signal noise from surrounding wiring, we could switch to a shielded copper wire or even a fiber optic cable to reduce the impact of that noise.

Another possible solution is to increase the power of the signal. This will help us to distinguish the signal from the noise more easily. In fact, we measure the power of the signal relative to the power of the noise in what we call a signal to noise ratio. This measure helps us determine how reliable our channel is and how much information we can reliably transmit for a given signal power. Thus, a channel with a large signal to noise ratio is better than one with a lower ratio. However, the power of a signal can only be increased by a certain amount,. The channel can often only support a certain amount of power. In addition, the increase in power could cause your signal to interfere with someone else's signal. The more power in the signal, the further it will travel. For example, if a radio station in one city boosts its power, it may carry further and interfere with a station that is using the same frequency in another city some distance away. Thus, particularly in the area of wireless communication, signal power is restricted by regulatory groups. When a radio station is allocated a frequency by the FCC it is also given a maximum signal power that it can use. This way, the same frequencies can be reused in different geographic areas.

If the noise that is distorting our transmitted signal lies within a specific frequency range, we can use filters to remove it from the received signal. However, since noise often varies in type and frequency depending on what is generating it, different filters may be required in different areas.

Finally, we could represent our data using a series of codes. These codes, some of which we will look at later, can often help us recognize when an error has occurred in our data stream and in some cases even correct that error.

So, as we have seen, noise can cause problems in a communication system, but we have a range of different approaches we can take to reduce or eliminate that noise. The reduction of signal noise is an area of ongoing research. We are always looking for a better solution to the ever-present problem of channel noise.

12.6 The Receiver

Once our message has made it across the channel, it is intercepted by the receiver. The task of the receiver is to take the message that is delivered by the channel and convert it into a form that will be useful to the end user. Thus, the receiver performs two basic tasks. First it isolates the signal that it is supposed to receive. In a radio receiver, for example, filters are used to lock in on the carrier frequency of the station that we want to listen to. The signals for all of the other stations are rejected. Once the signal is isolated, the receiver performs its second task, demodulation.

12.6.1 Demodulation

As you might expect, demodulation is the reverse of modulation. Since the transmitter modulated the original message so that it could travel over the channel, the receiver must undo the modulation to recover the intended message signal and deliver it to the user. Each modulation method has its own method of demodulation. The simplest demodulation scheme is the one used in AM. Since the message signal is represented as changes in amplitude within the carrier, we can recover it by using a special low-pass filter called an "envelope detector." This filter blocks the high frequency portion of the carrier signal while allowing the message to pass. The beauty of the AM demodulation approach is that its simple demodulator can be built using only a diode, resistor, and capacitor. An example of such a demodulation circuit is shown in Figure 12.12. This design makes AM receivers much simpler to construct and thus, more inexpensive than those for other modulation approaches.

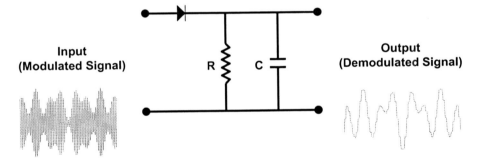

Figure 12.12 An example of a simple envelope detector circuit used in AM demodulation.

Thus, the receiver intercepts the signal that was placed on the channel by the transmitter and demodulates it to recover the message signal. This demodulated signal is then delivered to the final stage of our communication system, the destination.

12.7 The Destination

The destination in a communication system can be thought of as the final recipient of the information that we were trying to send. This could be another person, a piece of machinery that we are controlling, or a computer. The destination, regardless of whether it is human or machine, is the entity that makes use of the information that was sent. For example, if you were conveying an idea to another person, as the destination of your communication, they would comprehend your idea and respond appropriately. Likewise, if you were issuing commands to a remote space probe, the computer on the probe would perform the actions specified in the transmission.

So, in the preceding sections, we have discovered that a basic communication system consists of five parts. We have examined these five parts and learned a little about what is involved in the process of sending information from one point to another, as well as, some of the problems we may encounter while doing so. In the remainder of this chapter, we will look at an area of communication that is very important, particularly in digital communication systems. That area is the encoding, or representation of the information that we are trying to send.

12.8 Introduction to Coding

Whether we realize it or not, we work with encoded information all the time. At its basic level, both speech and writing are codes. When we speak to each other, we encode ideas into sounds. The person who hears these sounds must be familiar with our code (language) in order to understand what we are saying. Likewise, writing is a method of encoding ideas by representing the spoken language as a series of written characters (the alphabet). Numbers can also be thought of as a code. When we use a number, we are representing a physical quantity with a symbol. We saw in chapter one that we can use many different codes (called bases) to represent these quantities. We even learned that we could convert from one system of numeric encoding to another. Just as we use codes to represent numbers and ideas to make them easier to convey to others, communication systems use codes to aid them in their task of moving information from place to place. Codes are particularly prevalent in the world of digital communication where their use can help us overcome the effects of noise. In the remainder of this chapter, we will look at some common ways that digital systems encode both numbers and text. We will also see how these codes can be implemented to detect or even correct errors in our data. Let's begin our study by looking at some of the simple ways in which a digital system deals with numeric quantities.

12.9 Basic Numeric Coding Schemes

As we have learned, computers and digital communication systems cannot represent numbers in the same way that we do. A computer stores information as a series of "on" and "off" values. Likewise, digital communication systems transmit information as a series of ones and zeros. These values may be represented as high and low voltage levels on a connection or by using one of the digital modulation techniques, ASK, FSK, or PSK. Regardless of how the signal is transmitted, we must find a way to represent values in terms of zeros and ones if we ever want to send them digitally. When dealing with numbers, there are several ways this can be done. Two of the most common of these are straight binary coding (which we have already seen in chapter 1) and binary coded decimal. Let's see how these methods operate.

12.9.1 Straight Binary Coding

As we learned in chapter one, numbers can be represented in a variety of "bases." We found that a particularly useful base was base 2, also known as binary. This base proved useful because it allowed us to represent a number as a series of zeros and ones. Thus, it became easy to represent numbers in computers, digital logic circuits, and digital communication systems. So, the easiest way for us to represent a number in a digital communication system is to convert it to binary and transmit it as a series of binary bits. This method of encoding numbers is known as straight binary coding. Let's look at an example of how it works.

Example 12.2

___

Problem: You need to transmit the number 122 over a digital communication system. Using straight binary coding, prepare this number for transmission.

Solution: Here we are asked to encode the number 122 using straight binary coding. Since straight binary encoding only requires us to represent the number as its binary equivalent, this can be easily accomplished by using decimal to binary conversion (for a review of decimal to binary conversion see chapter 1). Thus, we simply divide our number repeatedly by two to yield:

$$
\begin{array}{r r}
 & \dfrac{0}{2\,\overline{)1}} \quad 1 \\[4pt]
 & 2\,\overline{)3} \quad 1 \\[4pt]
 & 2\,\overline{)7} \quad 1 \\[4pt]
 & 2\,\overline{)15} \quad 0 \\[4pt]
 & 2\,\overline{)30} \quad 1 \\[4pt]
 & 2\,\overline{)61} \quad 0 \\[4pt]
 & 2\,\overline{)122}
\end{array}
$$

So, our straight binary encoding of 122 is 1111010.

___

Example 12.3

___

Problem: You have just received a signal representing the binary sequence 1001011 on your digital communication system. Since you know that the communication system uses straight binary coding, decode the value to discover what decimal number was sent.

Solution: Here we have a decoding problem. Since we know that the sequence was encoded using the basic straight binary encoding method, we can simply view the sequence as the binary representation of a decimal number and perform binary to decimal conversion (see chapter 1) to convert it to its decimal equivalent. Thus, we get:

$$1(64)+0(32)+0(16)+1(8)+0(4)+1(2)+1(2) = 75$$

So, the sequence represents the decimal number 75.

___

Thus, the examples show us that straight binary coding works in exactly the same way as our decimal to binary conversions in chapter one. This is obviously the most straightforward approach to encoding numbers for digital communication.

12.9.2 Binary Coded Decimal

Another popular method of encoding numbers is called binary coded decimal, or BCD. In this approach we encode each digit of the decimal number independently. Since there are ten possible values for each digit in a decimal number, we must use four binary digits for each decimal place. Table 12.1 illustrates the association between each decimal digit and its binary representation.

Table 12.1 A listing of the binary values used to represent the decimal digits in a binary coded decimal system.

Binary Value	Decimal Digit	Binary Value	Decimal Digit
0000	0	1000	8
0001	1	1001	9
0010	2	1010	Not Used
0011	3	1011	Not Used
0100	4	1100	Not Used
0101	5	1101	Not Used
0110	6	1110	Not Used
0111	7	1111	Not Used

As the table illustrates, there are more binary combinations with four digits than there are possible values for a given decimal digit. This means that binary representations from 1010 to 1111 are not used to encode the decimal digits. While it may seem like a waste to simply not use six of the available values, they are, in a way, used. For example, we may want a value to separate numbers that we are transmitting so that all the digits do not just run together. In addition, we can use these extra values as special flags to represent the start and end of a transmission. The appearance of any of these unused values where they are not expected can indicate an error in the data stream. Thus, they can be used to a limited extent to detect data errors. So, how do we use all this to encode a binary number? Basically, we convert each digit to its matching four digit binary equivalent and then chain all of these four digit values together to form a BDC value. Let's look at an example to see how the encoding process works.

Example 12.4

Problem: Encode the decimal number 4379 using the binary coded decimal approach.

Solution: Here we are asked to encode a decimal value in BCD. To do this, we first find the four digit binary equivalent for each digit in the decimal number. This gives us:

$$4 - 0100$$

$$3 - 0011$$

$$7 - 0111$$

$$9 - 1001$$

Now, we simply chain them together to form the BCD value.

$$4379 \rightarrow (0100001101111001)_{BCD}$$

Thus, from the example, we see that the process of encoding a decimal number using binary coded decimal is a simple process.

Decoding a BCD value is also a simple task. We simply group the binary digits into groups of four and convert each of these four digit binary values back to its decimal equivalent. If we encounter any of the unused groups in our BCD number, we know that an error has occurred. Let's look at an example of the decoding process.

Example 12.5
_____

Problem: You have just received the following strings of binary digits on your digital communication system. Use binary coded decimal to decode the numbers into their original decimal representation.

    a. 0110100000111001
    b. 0111110000010101

Solution: Here we have been given two numbers encoded using the binary coded decimal approach. We now want to determine what their original decimal representations were. Let's begin by taking the first sequence and breaking it down into four digit segments. Thus, we have:

    0110 1000 0011 1001

Now, we convert each of these four digit groups back into its one digit decimal equivalent.

    0110 – 6

    1000 – 8

    0011 – 3

    1001 – 9

So, the original decimal number was 6839.

Now let's look at the second sequence. Breaking it down into groups of four we find:

    0111 1100 0001 0101

Now converting these groups back into their decimal equivalents we find:

    0111 – 7

    1100 – 12 (Error)

    0001 – 1

    0101 – 5

In this case, we have encountered one of the unused symbols. Because of this, it is not possible to convert this BCD value back to its decimal equivalent. An error has apparently occurred during the transmission process.

So, from the two previous examples, we see that the process for encoding and decoding decimal numbers using the BDC approach is a simple task.

12.10 Basic Alphanumeric Encoding Schemes

Up to now, we have looked at methods that can be used to represent numbers in a digital environment. What we have found is that, numbers tend to be easy to convert into a digital equivalent, thanks largely to binary numbers. Thus, it is a fairly simple task for numbers to be converted into a form that; can be manipulated, stored, and transmitted digitally. Numbers are not the only data that we wish to communicate, however. In many cases we want to be able to send text. This requires that alphabetic characters be encoded into a form that can be worked with in a digital environment. There has been a great deal of work done researching methods that, we can use to represent text as digital signals. Two of the simpler approaches are outlined below.

12.10.1 The ASCII Code

Since we have already seen how easy it is to convert numbers to work in a digital environment, the easiest way to send text would obviously be to associate each character that we might need to send with a specific number. Then we could transmit the associated numbers in their binary representations as a digital signal. As long as the receiver knows which numbers were associated with which characters, the message can be reconstructed by chaining together the characters represented by the numbers that it receives. In order to be sure that everyone uses the same numbers to represent the characters in their message, a standard mapping has been developed. This mapping is known as the American Standard Code for Information Interchange or ASCII for short.

The ASCII code uses seven binary bits to represent 128 different possible characters including upper and lower case letters, numbers, punctuation, symbols, and some special control characters. Thus, each character transmitted is encoded into its associated seven bit binary representation. Upon receipt, each group of seven bits is replaced by its corresponding symbol and the message can be read by the end user. Table 12.2 illustrates the characters that can be encoded using ASCII. As you can see, the first 32 characters are special control characters. After that the binary values correspond to actual characters. Let's use the contents of this table to solve some example problems.

Example 12.6

Problem: You are using a digital communication system and wish to send your friend the text message 'Hello'. Using the ASCII method and Table 12.2, determine what binary sequence should be transmitted.

Solution: In this problem we simply need to use Table 12.2 to find the seven digit binary numbers that represent each character in our message. So, finding the binary representation for the characters in our message, we get:

H – 1001000

e – 1100101

l – 1101100

l – 1101100

o – 1101111

Now, the message that we want to send is the combination of these seven digit binary codes.

(1001000110010111011001101100 1101111)$_{ASCII}$

Example 12.7

Problem: You have received the following binary sequence. Using ASCII and Table 12.2, decode the message into its appropriate text message.

100010111011101100111110100111011101100101110010111100101101001110111 0 1100111

Solution: In this case, we need to use Table 12.2 to decode a message encoded in ASCII. First, let's split the binary sequence into seven digit binary numbers. Doing this yields:

1000101 1101110 1100111 1101001 1101110 1100101 1100101 1110010
1101001 1101110 1100111

Now, we need to associate each of the seven digit codes with its appropriate character. Doing this we get:

1000101 – E

1101110 – n

1100111 – g

1101001 – i

1101110 – n

1100101 – e

1100101 – e

1110010 – r

1101001 – i

1101110 – n

1100111 – g

So, putting these together, we find that the message that we received was 'Engineering'.

Table 12.2 A listing of the binary values and their associated text characters as used by the ASCII coding method.

Decimal	Binary	Symbol	Decimal	Binary	Symbol	Decimal	Binary	Symbol
000	0000000	(Null)	043	0101011	+	086	1010110	V
001	0000001	SOH	044	0101100	,	087	1010111	W
002	0000010	STX	045	0101101	-	088	1011000	X
003	0000011	ETX	046	0101110	.	089	1011001	Y
004	0000100	EOT	047	0101111	/	090	1011010	Z
005	0000101	ENQ	048	0110000	0	091	1011011	[
006	0000110	ACK	049	0110001	1	092	1011100	\
007	0000111	BEL	050	0110010	2	093	1011101	]
008	0001000	BS	051	0110011	3	094	1011110	^
009	0001001	HT	052	0110100	4	095	1011111	_
010	0001010	LF	053	0110101	5	096	1100000	`
011	0001011	VT	054	0110110	6	097	1100001	a
012	0001100	FF	055	0110111	7	098	1100010	b
013	0001101	CR	056	0111000	8	099	1100011	c
014	0001110	SO	057	0111001	9	100	1100100	d
015	0001111	SI	058	0111010	:	101	1100101	e
016	0010000	DLE	059	0111011	;	102	1100110	f
017	0010001	DC1	060	0111100	<	103	1100111	g
018	0010010	DC2	061	0111101	=	104	1101000	h
019	0010011	DC3	062	0111110	>	105	1101001	i
020	0010100	DC4	063	0111111	?	106	1101010	j
021	0010101	NAK	064	1000000	@	107	1101011	k
022	0010110	SYN	065	1000001	A	108	1101100	l
023	0010111	ETB	066	1000010	B	109	1101101	m
024	0011000	CAN	067	1000011	C	110	1101110	n
025	0011001	EM	068	1000100	D	111	1101111	o
026	0011010	SUB	069	1000101	E	112	1110000	p
027	0011011	ESC	070	1000110	F	113	1110001	q
028	0011100	FS	071	1000111	G	114	1110010	r
029	0011101	GS	072	1001000	H	115	1110011	s
030	0011110	RS	073	1001001	I	116	1110100	t
031	0011111	US	074	1001010	J	117	1110101	u
032	0100000	(Space)	075	1001011	K	118	1110110	v
033	0100001	!	076	1001100	L	119	1110111	w
034	0100010	"	077	1001101	M	120	1111000	x
035	0100011	#	078	1001110	N	121	1111001	y
036	0100100	$	079	1001111	O	122	1111010	z
037	0100101	%	080	1010000	P	123	1111011	{
038	0100110	&	081	1010001	Q	124	1111100		
039	0100111	'	082	1010010	R	125	1111101	}
040	0101000	(	083	1010011	S	126	1111110	~
041	0101001	)	084	1010100	T	127	1111111	DEL
042	0101010	*	085	1010101	U			

As we have seen, working with ASCII is not all that difficult. This is probably why ASCII is a very widely used method of encoding text characters for storage and transmission on digital systems. In fact, computers make extensive use of the ASCII system. Each time a key is pressed on the keyboard, the ASCII code representing the character on that key is sent to the computer. Likewise, when the computer uses its COM or parallel ports, the data is transmitted in ASCII. Thus, we see that ASCII is an incredibly useful coding scheme.

12.10.2 Huffman Coding

As we just saw, we can easily represent characters by mapping each character to a specific number and then transmitting the binary representation of that number. While this is an effective approach, it requires that we send seven bits of data to represent each character. While this is not a bad thing, we could use our channel more efficiently if we could use fewer bits to represent our characters, especially the characters that we use extensively. A coding method that takes this approach is Huffman coding. The Huffman code is what is known as a variable length code, meaning that each character is not represented by the same number of digital bits. The Huffman code attempts to use the least number of bits to transmit characters that appear often in the data, like the letter 'E', and saves the longer representations for characters that we do not see as often, such as 'Z'. This is achieved by arranging all of the characters that we wish to transmit in what is known as a binary tree structure. A binary tree is basically an arrangement of data such that, from each point (or node) in the tree we can branch off in at most two directions, one represented by '0' and the other by '1'. Some of the nodes in the tree are known as junction nodes. These do not contain any data. Instead, they are intermediate points that connect to other junction nodes or leaf nodes. Leaf nodes are the points in the tree that contain data. They are located around the outer edges of the tree. The junction node at the top of the tree is known as the root node and it is from this node that you begin all operations when working with the tree.

So, how does the Huffman code make use of this structure? Basically, it begins by building the tree. The tree is constructed by looking at the frequency that characters occur in the data. The characters that occur most frequently are placed on the tree so that they can be reached via the shortest path from the root node. The less frequently a character appears, the longer the path will be to reach it. Let's take a look at how the tree is built. Table 12.3 displays the letters in the English language and a percentage indicating how often they appear in the English language (based on the type of material you are transmitting, these values may be somewhat different).

Table 12.3 The letters in the English alphabet and their percentage of usage.

Letter	%	Letter	%
E	12.32	P	2.58
T	8.37	G	2.77
A	8.23	U	2.43
I	8.08	F	2.28
R	6.67	Y	1.46
S	7.64	B	1.26
O	7.59	W	1.07
N	6.81	V	0.97
C	4.04	K	0.43
H	3.94	X	0.29
L	3.79	J	0.14
D	3.40	Q	0.14
M	3.06	Z	0.09

As the table shows, the letter 'E' is the most common letter and 'Z' is the least common. When a tree for Huffman encoding is built, it is built from the bottom up using the following steps:

1. Assign each character to a node. Label each node with the probability that that character will appear (in this case the percentage it is used)
2. Choose the two nodes that have the smallest probabilities (percentages) and link them to a junction node. Label the junction node with the sum of the probabilities of the two nodes that were connected to it.
3. Repeat step two until you arrive at a single junction node with all the characters below it, the root node.

Let's apply these steps to the English alphabet by using the percentages in Table 12.3.

We begin by locating the two letters with the least percentage of usage. In this case, it is the letter 'Z' and the letter 'Q'. Now, we attach these two to a junction node and label that node with the sum of their appearance percentages. Thus, we arrive at the following structure.

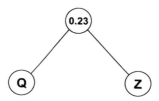

Now, we remove the percentages of 'Q' and 'Z' from our list and insert the sum contained in the junction node. Now we repeat the second step and find the next two lowest values. These would be 'J' with 0.14 percent and the node we just created with 0.23 percent. So, we join these two nodes under one junction node as follows:

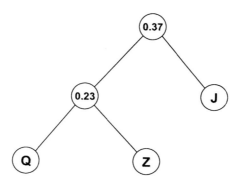

Repeating step two again, we find that our two nodes with minimum percentages are 'X' and our new node with 0.37 percent. Once again these are joined as follows:

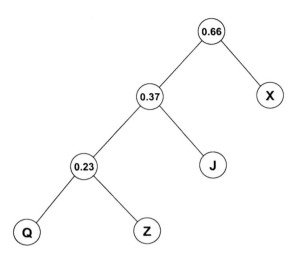

If we continue to combine nodes in this way, we will eventually combine all of the character nodes together to form a single tree structure. This would yield a tree like the one shown in Figure 12.13.

Now that we have a tree structure, we can begin to encode the characters in our message for transmission. This is done as follows:

1. Start at the root node

2. Trace through each junction node until you reach the leaf node containing the

3. character that is to be encoded.

4. At each of these junction nodes, record a zero if you take the left path and a one if you take the right path

5. The binary value resulting from the values you recorded as you moved through the tree is the code for that character.

So, if we simply trace through the tree and keep track of the paths that we take, we arrive at a code word for our character. The resulting code words generated for each letter in the English alphabet by tracing through the tree in Figure 12.13 are given in Table 12.4.

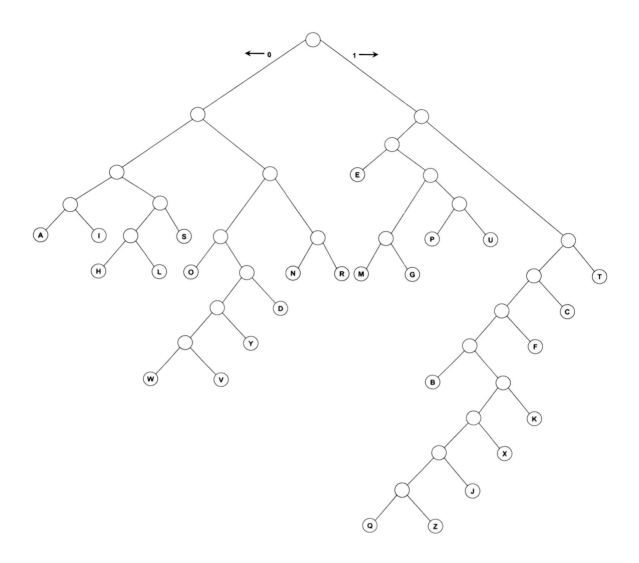

Figure 12.13 A binary tree of the English alphabet for use with the Huffman encoding method.

As the table shows, some of our characters can be transmitted with very few bits. For example, 'E' requires only three bits to represent. On the other hand, the characters that occur less often require more binary digits. The longest of these, 'Z', requires ten binary bits to transmit. Thus, if we are transmitting standard English text that follows the character percentages that we used to generate our binary tree, we should be able to send our messages with fewer characters than are required for the same message using ASCII. Let's look at an example of encoding using the Huffman code.

Table 12.4 The code words generated for each letter of the alphabet using the binary tree in Figure 12.13.

Letter	Code Word	Letter	Code Word
A	0000	N	0110
B	110000	O	0100
C	1101	P	10110
D	01011	Q	1100010000
E	100	R	0111
F	11001	S	0011
G	10101	T	111
H	00100	U	10111
I	0001	V	0101001
J	110001001	W	0101000
K	1100011	X	11000101
L	00101	Y	010101
M	10100	Z	1100010001

Example 12.8

Problem: You wish to transmit the word "ENGINEERING'" using your digital communication system. Using the Huffman code and the tree in Figure 12.13, encode the message.

Solution: Here we are asked to simply encode a message using the tree that we constructed in Figure 12.13. We can either trace each letter through the tree to find its code word or we can use the code words listed in Table 12.4. Either way, we find the following code words for each letter.

E – 100

N – 0110

G – 10101

I – 0001

N – 0110

E – 100

E – 100

R – 0111

I – 0001

N – 0110

G – 10101

Now, we obtain our encoded message by chaining these together to get:

10001101010100010110100100011100010110010101

Thus, we needed 43 binary digits to represent this message. The same message encoded in seven bit ASCII would require 77 binary digits. In this particular case, Huffman coding has saved us 34 bits. Obviously, if we were transmitting a large amount of text, small savings like this would add up to increased efficiency in channel usage. It should be noted that, since the Huffman code requires more than seven bits for some letters, it is possible that messages could arise that would be longer when encoded using the Huffman code than they would be in ASCII. On the average however, some savings is realized through the use of the Huffman code.

Now that we know how to encode text using the Huffman code, how do we decode it? Huffman codes are decoded by basically reversing the encoding process. It should be noted that, both, the transmitter and receiver must use the same binary tree to represent the data, or they will not be able to communicate properly. Once both the transmitter and receiver have the same tree, the incoming data is decoded as follows:

1. Start at the root node of the tree.

2. Look at the current binary bit. If it is a one, take the path to the right, if it is a zero, take the path to the left.

3. Repeat step two for each bit of incoming data until a leaf node is reached

4. When a leaf node is reached, output the associated character and return to step one to decode the next character.

So, decoding the signal is not too difficult. Let's take a look at an example.

Example 12.9

Problem: You have just received the following string of binary digits. Use the Huffman coding approach and the tree in Figure 12.13 to decode the message.

0010010000101001010100

Solution: In this problem, we need to decode a string of binary digits into text using the Huffman code. If we follow our tracing procedure starting with the first digit of the stream, we will get the following results:

Root Node → Go Left → Go Left → Go Right → Go Left → Go Left → **H**

Root Node → Go Right → Go Left → Go Left → **E**

Root Node → Go Left → Go Left → Go Right → Go Left → Go Right → **L**

Root Node → Go Left → Go Left → Go Right → Go Left → Go Right → **L**

Root Node → Go Left → Go Right → Go Left → Go Left → **O**

Thus, the message that we received was "HELLO."

So, we have now seen how to decode a message using the Huffman code. As you worked through the example, you may have thought of one of the major drawbacks to variable length codes like the Huffman code. Since you don't know how long each code word is until you trace through the tree and find the character, you do not know where the code for each character begins and ends. As a result one error somewhere in the sequence can make it impossible for us to decode any of the information after it. Thus, variable length codes must be used with caution and the benefits gained from the reduced code lengths must be weighed against the possibilities of corrupted data due to errors. You do not realize any transmission savings if you constantly have to retransmit large amounts of data due to the occurrence of small errors in the data stream.

12.11 Dealing with Errors

In earlier sections, we talked about the impact of noise on a signal. We noted that noise impacts a signal by distorting it in various ways. These distortions can cause errors in the data that we are trying to transmit. As a result, finding ways to detect these errors and possibly even correct them, is a very important area of study. In an analog system, correcting problems with the signal can be very difficult. In digital systems, however, we can utilize some special error detection and correction coding schemes to help us out. In the remainder of the chapter, we will look at some of these approaches. We will begin by looking at one of the simplest methods of detecting errors, the parity check.

12.11.1 Parity Error Detection

If you have ever configured a COM port on a computer, you have dealt with parity error checking. If you recall the parameters that you can set, one of them is called parity and has the option of even, odd, and none. So, what exactly is parity, how does it work, and what does even and odd parity mean? Parity checking is a very simple method of detecting errors in data that is sent over a communication channel. Basically, parity works as follows. For every seven bits of data, we add an eighth bit called a parity bit. This parity bit takes on a value of zero or one depending on two factors, first, the number of ones in the preceding seven bits of data and secondly whether we are using even or odd parity. So, to set the value of the parity bit, we count up the number of ones in the seven bits of data that we are adding the parity bit to. Then we set the parity bit so that the number of ones in the resulting eight bits of data will be even or odd depending on our parity setting. If we are using even parity, we want an overall even number of ones. Likewise, if we are using odd parity we want an odd number of ones. Let's take a look at how this works.

Example 12.10

Problem: Given the following seven bit data values, determine the value for the parity bit using the parity type that is specified.

    a. 1010110_ Even Parity
    b. 1001011_ Odd Parity
    c. 0110111_ Even Parity
    d. 1001010_ Odd Parity

Solution: In this problem we simply need to determine what the parity bit will be in each of the four cases specified. Let's start with the first value. We see that we want to use even parity, so let's count the number of ones in the data. A quick count tells us that 1010110 has four ones. This is an even number, so our parity bit will be zero to preserve the even count. Thus we can write:

a.   1010110_ Even Parity → 1010110<u>0</u>

Now, let's look at the second set of binary digits. In this case, we are using odd parity so we want an odd number of ones. Counting the number of ones we find that there are four. Thus we make the eighth bit a one to make the count odd.

b.   1001011_ Odd Parity → 1001011<u>1</u>

The third set requires even parity again. In this case, there are an odd number of ones. Thus our parity bit will be one to make the count even.

c.   0110111_ Even Parity → 0110111<u>1</u>

Finally, the last set of bits requires odd parity and has an odd number of ones. Thus, we set our parity bit to zero to preserve the odd count.

d.   1001010_ Odd Parity → 1001010<u>0</u>

Thus, the example shows us that the process of setting the parity bit is easy, but how do we make use of this bit to detect errors? Well, let's assume that one of the bits in our eight bit sequence is impacted by channel noise and is flipped to its opposite value. This will change the number of ones in the value. For example, if the bit that flipped was a zero, it becomes a one and the count of ones in the value increases. If the bit was a one and is flipped to a zero, however, the count will decrease. In either case, if the count was initially an even number, it will now be odd and if it was odd it will now be even. As a result, when the receiver checks to see if the count of ones fits the parity setting it will not agree and the error will be detected. Let's see how this works.

Example 12.11

Problem: Using even parity, check to see if the following eight bit binary values contain errors.

a.   10101010
b.   11011010

Solution: In this example, we are simply checking to make sure that each sequence of binary bits contains the appropriate number of ones. If it does, the value arrived without error, if it does not an error must have occurred.

a. Starting with the first binary sequence, we count the number of ones. We see that there are four ones in this value. Since we are using even parity and four is an even number, we can assume that the data arrived without error.

b. Now, looking at the second sequence, we count its ones and find that there are five. Since five is an odd number and we are using even parity, we know that an error occurred during transmission.

At this point, you may be wondering what happens if more than one bit is changed. This is one of the weaknesses of parity error detection. What happens for example, if two bits are changed to their opposite values? If this occurs, their changes cancel each other out and the count of ones in the series of binary digits will remain the same. Thus, when it is checked for even or odd parity, the data will be incorrect, but the problem will not be detected. So, we can see that parity error detection is easy to implement and can detect some errors that occur during transmission, but it is not capable of detecting errors in which an even number of bits are flipped.

12.11.2 Two-Dimensional Parity

To try to guard against the chances of an even number of bit flips causing an error to go undetected, parity error checking can be augmented into what is known as two-dimensional parity. In two-dimensional parity, we arrange our data in a seven bit by seven bit block and then generate a parity bit for each row and column of the data as well as a parity check bit that implements parity error checking on the parity values for the rows and columns. This arrangement is illustrated in Figure 12.14. As was the case with regular parity checking, the parity bit is set based upon the number of ones in each row or column and whether even or odd parity is being used. Let's look at an example of how a two-dimensional parity is encoded.

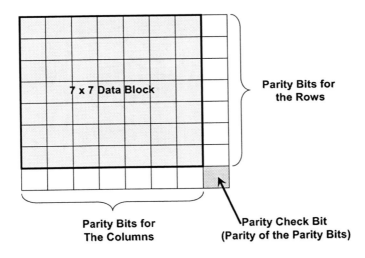

Figure 12.14 An illustration of the structure of two-dimensional parity.

Example 12.12

Problem: Given the following 7 x 7 block of data, generate the resulting two-dimensional parity structure using odd parity.

```
1 0 1 0 1 0 1
0 1 0 1 0 1 1
0 1 0 1 0 1 1
1 1 0 1 1 1 0
1 1 1 0 1 1 1
1 1 0 1 0 1 1
1 1 1 1 1 0 1
```

Solution: Here we have been given a 7 x 7 block of data and have been asked to generate the appropriate two-dimensional parity structure using odd parity. Let's begin by calculating the appropriate parity values for each of the rows by setting the parity bit to insure that there are an odd number of ones in each row.

```
1 0 1 0 1 0 1 1
0 1 0 1 0 1 1 1
0 1 0 1 0 1 1 1
1 1 0 1 1 1 0 0
1 1 1 0 1 1 1 1
1 1 0 1 0 1 1 0
1 1 1 1 1 0 1 1
```

Now, let's do the same thing for each of the columns of the data block.

```
1 0 1 0 1 0 1 1
0 1 0 1 0 1 1 1
0 1 0 1 0 1 1 1
1 1 0 1 1 1 0 0
1 1 1 0 1 1 1 1
1 1 0 1 0 1 1 0
1 1 1 1 1 0 1 1
0 1 0 0 1 0 1
```

Finally, we set the parity check bit so that the number of ones in the column and row of parity bits is an odd number.

```
1 0 1 0 1 0 1 1
0 1 0 1 0 1 1 1
0 1 0 1 0 1 1 1
1 1 0 1 1 1 0 0
1 1 1 0 1 1 1 1
1 1 0 1 0 1 1 0
1 1 1 1 1 0 1 1
0 1 0 0 1 0 1 0
```

Thus, we have completed the two-dimensional parity coding for the block that we were given.

Now that we see how two-dimensional parity is encoded, let's look at what benefits it gives us. First, we can immediately see that a much larger number of errors can be reliably detected using this approach. In fact, this implementation of parity checking is capable of detecting any error in the data as long as an even number of errors does not occur in both the same row and column at the same time. Thus, while there are some patterns of errors that cannot be detected, there is now a much greater chance of detecting errors than there was with regular parity error detection. Another important benefit is that, this approach allows us to correct a single error in the data. To understand how we would be able to do that, let's look at an example.

Example 12.13
_____

Problem: Determine if any errors occurred in the following transmission. The data has been encoded using even two-dimensional parity. If there are any errors, can they be corrected?

```
1 0 0 1 1 1 0 0
0 1 1 0 0 1 0 1
1 1 1 0 0 0 1 1
0 0 1 1 0 0 1 1
0 0 0 1 0 1 0 0
1 0 1 0 1 1 0 0
0 0 1 1 1 0 0 1
1 0 1 1 1 0 0 0
```

Solution: In this example, we have been given a block of data encoded using even two-dimensional parity. We want to find out if any errors occurred and if so, can be corrected. Let's begin by checking the parity for each row.

```
1 0 0 1 1 1 0 0   ← 4 Ones – Even – OK
0 1 1 0 0 1 0 1   ← 4 Ones – Even – OK
1 1 1 0 0 0 1 1   ← 5 Ones – Odd – Error
0 0 1 1 0 0 1 1   ← 4 Ones – Even – OK
0 0 0 1 0 1 0 0   ← 2 Ones – Even – OK
1 0 1 0 1 1 0 0   ← 4 Ones – Even – OK
0 0 1 1 1 0 0 1   ← 4 Ones – Even – OK
1 0 1 1 1 0 0 0   ← 4 Ones – Even – OK
```

So, we have found that there is an error in the third row. Let's continue by checking the parity for the columns.

```
1 0 0 1 1 1 0 0
0 1 1 0 0 1 0 1
1 1 1 0 0 0 1 1
0 0 1 1 0 0 1 1
0 0 0 1 0 1 0 0
1 0 1 0 1 1 0 0
0 0 1 1 1 0 0 1
1 0 1 1 1 0 0 0
4 2 6 5 4 4 2 4
```

Here we see that the fourth column has five ones. This tells us that there is an error in the fourth column of the data block.

So, now we know that an error did occur during the transmission of this data. In addition, since only one error occurred, we also know the exact row and column of the element that is incorrect. Thus, if we flip this bit, we can correct the error without having to request that the data be retransmitted.

Thus, the example has demonstrated that when a single error occurs in data encoded with two-dimensional parity, we can correct the error. This can be a major benefit and save a great deal of channel bandwidth by not requiring us to retransmit data that has been corrupted by only one error.

12.11.3 Cyclic Redundancy Check – CRC

The final error detection method that we will look at is known as cyclic redundancy checking, or CRC for short. It is an extremely powerful technique that is widely used in data communication. In fact, the CRC approach is what is used to check the data that is sent on the internet. The Cyclic redundancy check is based upon the concept of binary division. We take a predefined binary divisor that is known by both the sender and receiver. The sender then divides the message data by this divisor and arrives at some binary remainder value. This remainder is appended to the end of the message data much as we previously appended the parity bit. When the receiver receives the message, they divide the received information by the predefined binary divisor. If no errors occurred in the transmission the remainder of this division will be zero. Thus, if the receiver finds that there is a remainder after the division, an error has occurred and the data must be retransmitted. To see the power of the CRC approach, let's look at a simplified example of the encoding and decoding process.

Example 12.14

Problem: Encode the 10 bit message $M=1010001101$ using the six bit divisor pattern $P=110101$

Solution: In this example, we are asked to encode a binary sequence for transmission by using the cyclic redundancy check encoding approach. To do this, we must find the remainder of the division of the message by the divisor pattern. We should note that the remainder of this division will have one less binary digit than the divisor, so in this case, we will have a five digit remainder. To perform this encoding we will carry out the following steps.

1. Given the message signal M and divisor pattern, P. Find the length of the divisor pattern, L_P and from that length determine the length of the remainder ($L_R = L_P - 1$). In our case, this yields:

$$M = 1010001101$$

$$P = 110101$$

$$L_P = 6 \text{ bits}$$

$$L_R = L_P - 1 = 5 \text{ bits}$$

2. Add L_R zeros to the end of the message, M.

   Modified M = 101000110100000

3. Divide the modified message signal by the divisor pattern.

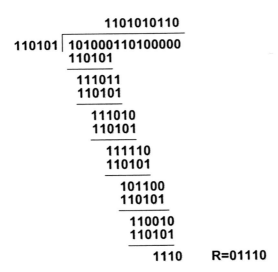

```
                      1101010110
              110101 │ 101000110100000
                       110101
                       ──────
                        111011
                        110101
                        ──────
                         111010
                         110101
                         ──────
                          111110
                          110101
                          ──────
                           101100
                           110101
                           ──────
                            110010
                            110101
                            ──────
                              1110        R=01110
```

4. Attach the 5 digit remainder, R, to the end of the original message to get the encoded message, T.

   T = 101000110101110

Thus, we have encoded the message using the cyclic redundancy check approach.

Once the message is encoded, it is transmitted to the receiver where it must be decoded to check for errors.

Example 12.15

Problem: Verify that dividing the message *T* encoded in the previous example by the divisor will result in a remainder of zero.

Solution: In this problem we need to verify that the message transmitted in example 12.14 will indeed yield a zero remainder when divided by the divisor as the cyclic redundancy check approach says it should.

So, let's begin our division.

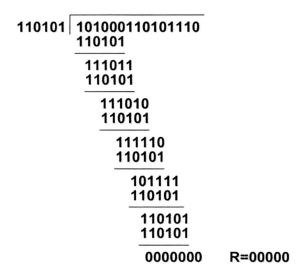

Thus, we see that we do indeed get a zero remainder when dividing the transmitted message by the divisor. If any of the bits had changed, however, we would have gotten a remainder.

From the previous two examples, we can see that this form of error detection is very powerful. In practice, however, the messages and divisors are much longer that the ones we used here. The divisors, for example are usually 16 or 32 bits long.

In the previous few sections, we have investigated some of the basic techniques that can be used to detect, and sometimes correct errors in the digital data that we transmit. Since no channel is immune to noise, this is an important area of study. A great deal of research is being done to improve and develop new approaches to this problem.

12.12 Chapter Summary

In this chapter we explored communication systems and some of the methods that are used to send data from place to place.

- We began by looking at the basic communication system. We learned that all communication systems have five basic parts: The source, transmitter, channel, receiver, and destination.

- Next, we looked at each of the five parts in more detail, beginning with the source. We looked at several different types of sources including audio signals, video signals, and computer data.

- The transmitter is responsible for converting the message generated by the source into a signal that can travel over the channel. This typically includes modulation and in some cases coding.

- Most transmitters use modulation to transmit the signal over the channel. We learned that this is done for two purposes. First to use the channel more effectively and secondly to decrease the size of antennas required to receive the signal in wireless applications.

- We looked at a few of the approaches to both analog and digital modulation including AM, FM, PM in the analog case and ASK, FSK, PSK in the digital case.

- Next, we investigated the channel. We found that the channel is simply the medium over which the signal is transmitted. Channels can be either wireless or wired and can consist of materials such as air, copper wire, and fiber optic cable.

- Both wired and wireless channels have their benefits and drawbacks. Wireless channels provide mobility at the expense of available bandwidth, poor security, and increased susceptibility to noise while wired channels provide increased bandwidth, improved security, and less susceptibility to noise at the expense of requiring a fixed connection between the transmitter and receiver.

- Noise can be a major problem on any channel. It can cause distortions in the signal as it travels to the receiver. We looked at some of the approaches that can be taken to reduce the impact of noise in a communication channel.

- The receiver takes the signal off the channel and converts it back into a form that can be understood by the end user. This usually involves demodulation and sometimes decoding.

- The final part of a communication system is the destination. It makes use of the signal that the source sent.

- In the last part of the chapter, we looked at several methods we can use to encode digital data. First, we looked at how numbers could be represented. The two methods we examined were straight binary coding and binary coded decimal.

- Next, we looked at two ways to encode text for digital transmission. We started by looking at ASCII which operates by creating a simple mapping between each character and a seven bit binary number. These numbers can then be transmitted to represent the appropriate characters.

- After ASCII, we looked at a method of encoding characters with a variable number of binary bits, Huffman coding. We learned that we could use a smaller number of bits to represent the most often used characters while reserving longer codes for the less often used characters.

- Finally, we examined some methods of checking transmitted data for errors. We looked at parity checking which encodes an extra bit for every seven bits of data. This bit is set so that the number of ones in the resulting eight bits will be even or odd depending on the type of parity used.

- Parity checking works well for finding single errors in data, but cannot detect an even number of errors.

- Two-dimensional parity is an extension of regular parity checking that creates parity checks on all the rows and columns of a seven by seven block of data. It can detect many more errors and even correct a single error in the block.

- Cyclic redundancy checking involves the binary division of the message with a predefined divisor. The message is divided to obtain a remainder that is appended to the message and sent to the receiver. When the data is received it is divided by the divisor once again. If the division results in a remainder of zero then no error occurred during transmission. If there is a remainder other than zero, an error occurred.

12.13 Review Exercises

Section 12.2 – Parts of a Communication System

1. List the five parts of a communication system. Explain the purpose of each part.

Section 12.4 – The Transmitter

2. What is modulation?

3. Why would anyone want to modulate a signal?

4. Calculate the antenna lengths required to receive the following signal frequencies.

   a. 50 KHz   b. 650 KHz   c. 87 MHz
   d. 180 MHz   e. 3 GHz   f. 50 Hz
   g. 20 GHz   h. 3 KHz   i. 99 MHz

5. Determine the minimum frequency that the following lengths of antenna can receive.

   a. 1 m   b. 5 cm   c. 10 m
   d. 1 cm   e. 50 m   f. 15 cm
   g. 1 Km   h. 1 mm   i. 25 cm

6. List the three types of analog modulation. Briefly describe the idea behind each approach.

7. List the three types of digital modulation. Briefly describe the idea behind each approach.

Section 12.5 – The Channel

8. List some common channel types. Provide an example use for each channel type.

9. What is bandwidth?

10. How can the available bandwidth be divided up and allocated?

11. Explain the difference between circuit switching and packet switching.

12. What is channel noise?

13. Explain what can be done to help reduce the impact of channel noise.

Section 12.9 – Basic Numeric Coding Schemes

14. What is straight binary coding?

15. Use straight binary coding to encode the following numeric values.

    a. 117   b. 12   c. 250
    d. 191   e. 28   f. 129
    g. 85   h. 188   i. 348

16. How does binary coded decimal coding differ from straight binary coding?

17. Use BCD to encode the following values.

    a. 928   b. 1234   c. 5294
    d. 9214   e. 132   f. 14936
    g. 223   h. 1892   i. 572

18. Encode your phone number using BCD.

19. Encode your zip code using BCD.

20. Decode the following BCD values. If the value contains an impossible value, note the error.

    a. $(1000100100110111)_{BCD}$ b. $(0111000000101000011)_{BCD}$

    c. $(0101001101111001)_{BCD}$ d. $(1000000110101001)_{BCD}$

    e. $(10010001001101110101)_{BCD}$ f. $(10000010100001110101)_{BCD}$

    g. $(10110001100001010001)_{BCD}$ h. $(01010001100010010101)_{BCD}$

Section 12.10 – Basic Alphanumeric Encoding Schemes

21. What does the abbreviation ASCII stand for?

22. Using the ACSII code, encode the following text strings.

    a. Electricity   b. Encoder   c. Student
    d. College   e. AbCdXyZ   f. Computer
    g. Internet   h. Keyboard   i. Driver

23. Use ASCII to encode your name.

24. Decode the following 7-bit ASCII encoded strings.

   a. $\left(1000101100111010000100\right)_{ASCII}$
   b. $\left(10100101010101011001110\right)_{ASCII}$
   c. $\left(10000101100001110000100\right)_{ASCII}$
   d. $\left(1010000101001010010011001110101000100\right)_{ASCII}$
   e. $\left(10000101100001111010011101001100101111001011111001\right)_{ASCII}$
   f. $\left(\begin{array}{l}100010010010011010011100001110011111\\10101101000101101001010101011001\end{array}\right)_{ASCII}$

25. Describe the basic reasoning behind the Huffman code.

26. Encode the following strings using Huffman coding. (Use the code tree in Figure 12.13)

   a. PAPER   b. SOFTWARE   c. ENCODE
   d. HARDWARE   e. HUFFMAN   f. RECORD
   g. DECODE   h. CALCULATOR   i. MODULATE

27. Decode the following Huffman encoded strings. (Use the code tree in Figure 12.13)

   a. $\left(0101001000000101000101011\right)_{HUFFMAN}$
   b. $\left(1111001100010111111000001000100110 0011\right)_{HUFFMAN}$
   c. $\left(10100010001100001111010 00111\right)_{HUFFMAN}$
   d. $\left(10011000101000010100\right)_{HUFFMAN}$
   e. $\left(1100000001011000000111010101\right)_{HUFFMAN}$
   f. $\left(0101001010000101 1110011\right)_{HUFFMAN}$
   g. $\left(00001010010110 1000111100\right)_{HUFFMAN}$
   h. $\left(1101000101111101101110001111\right)_{HUFFMAN}$

Section 12.11 – Dealing with Errors

28. Given the following 7-bit data values, determine the necessary parity bit based upon the type of parity requested.

   a. 0101101_ Even Parity   b. 1100110_ Odd Parity
   c. 0110111_ Even Parity   d. 1111111_ Even Parity
   e. 1100111_ Even Parity   f. 1110111_ Odd Parity
   g. 0000000_ Odd Parity   h. 1010101_ Odd Parity

29. Determine if there is an error in each of the following 8-bit values based upon the given parity type.

    a. 00101001 Odd Parity   b. 10101101 Even Parity
    c. 01110110 Odd Parity   d. 01110101 Even Parity
    e. 11101111 Odd Parity   f. 11101110 Even Parity
    g. 01011011 Odd Parity   h. 00000011 Even Parity

30. What is the weakness of parity error detection?

31. Generate the parity bits for the following two dimensional parity blocks. Use the parity type specified.

a. Even Parity

```
1  0  1  0  0  1  0
0  1  1  0  1  0  0
1  1  1  1  0  0  0
1  0  1  0  1  0  0
0  1  1  1  0  0  0
1  0  0  0  0  1  0
1  0  0  0  0  0  0
```

b. Odd Parity

```
1  1  1  1  1  1  1
0  0  1  1  0  0  0
0  0  0  0  0  1  1
0  0  0  0  0  0  0
1  1  1  0  1  0  1
0  0  0  0  1  1  1
1  1  1  1  0  0  0
```

c. Even Parity

```
1  1  0  1  1  0  1
1  0  0  0  1  0  1
0  1  1  1  0  1  0
1  1  1  1  0  1  0
1  1  1  1  1  0  1
1  1  1  0  0  0  1
0  0  0  1  1  1  1
```

d. Odd Parity

```
0  1  0  1  0  1  1
1  1  0  1  0  1  0
1  1  1  0  0  0  1
0  0  0  1  1  1  1
1  1  0  0  1  1  0
0  0  0  1  0  1  1
1  1  1  1  0  1  1
```

32. Determine if an error occurred in the following 2-D parity blocks. If possible, correct the error.

a. Even Parity

```
0  0  1  1  0  0  1  1
1  1  1  0  1  1  1  0
0  1  1  0  1  1  1  1
0  0  0  1  1  0  0  1
1  1  1  1  1  1  1  1
0  0  0  0  1  1  0  0
1  0  1  0  1  1  0  0
1  1  1  0  0  1  0  0
```

b. Odd Parity

```
0  1  1  1  0  1  0  1
1  1  1  1  0  0  0  0
1  1  0  1  1  1  0  0
1  1  1  0  0  0  1  1
1  1  1  1  1  1  1  0
0  0  1  0  1  1  1  1
1  1  1  1  0  0  1  0
0  1  1  0  1  1  1  0
```

c. Even Parity

```
1  1  1  0  1  1  1  0
0  0  0  1  0  0  0  1
0  0  0  0  1  0  0  0
1  1  1  1  1  0  1  0
0  0  1  1  1  0  1  0
0  1  1  1  0  1  0  0
1  0  1  0  1  0  0  1
1  1  1  1  1  0  1  0
```

d. Odd Parity

```
0  0  0  0  1  1  1  0
0  1  0  1  1  1  1  0
1  0  0  1  1  0  0  0
0  1  1  1  0  0  1  1
1  1  0  0  1  1  0  1
1  1  1  0  0  0  1  1
0  0  0  1  1  1  0  0
0  1  1  1  0  1  1  0
```

33. Using CRC, encode the following messages with the provided divisor pattern.

a. M = 1101011101   P = 110101
c. M = 0100100110   P = 101101

b. M = 101101101001   P = 110101
d. M = 11101101110101   P = 110010

34. Given the following received values use CRC to determine if any errors exist.

a. R = 110110010100000   P = 110101
b. R = 10110111011101010   P = 101110
c. R = 10001101110011010   P = 110101
d. R = 1111010110101111   P = 101101

UNIT 5

Laboratory Experiments and Design Problems

Introduction

This unit is divided into *modules* and *exercises*. Each module addresses a specific topic in engineering software and/or hardware. The overall goal of this unit is to provide the student with an introduction to several concepts and tools that they will be utilizing throughout their education in engineering. Furthermore, the lab exercises help to bridge the gap between lecture and application by providing the students with hands on experience. We feel it is very important to make the transition from theory to application early in the educational process, allowing the student to see how knowledge gained through lecture and reading is applied in the real world.

The exercises require knowledge of the subject material from one or more modules. Therefore, it is recommended that students complete the module(s) in order before completing a specific exercise. By following along, working examples included in the modules, and completing the exercises, the student will benefit by practicing the application of their new knowledge while developing some basic technical skills.

The first two modules require access to MATLAB and Multisim, formerly known as Electronics Workbench. To complete all of the modules and design exercises, you may need to either purchase or have access to the following list of components and equipment.

Equipment/Part	Number Recommended
Computer	1
Dual +/- 15 volt DC power supply	1
Digital Multimeter	1
Assortment 1/4 watt resistors	1
Assortment Capacitors	1
Bread Board	1
Hookup Wire Kit	1
General Purpose dual supply op amp	5
General Purpose PN diode	5
Multi-turn Potentiometer 5k	2
Multi-turn Potentiometer 10k	2
Analog switch or MUX (4 channel/CMOS)	2
Standard red LED	5
LED with fiber connector	2
Photodiode with fiber connector	2
1 meter plastic multimode fiber	1
Terminal Blocks (3 pin)	6
Null Modem cable (9 pin)	1
Serial cable (9 pin)	1
Oscilloscope	1
Signal Generator	1
Tool Set	1

Chapter 13. Module 1 – MATLAB

13.1 MATLAB Basic Math Functions and Matrices

Objectives

This section is designed to help you become familiar with MATLAB by introducing you to its basic features, commands, and functions. The exercise is intended to be very interactive. You should have MATLAB running and be entering commands as you work along with the examples.

Introduction

MATLAB can be used for solving all types of mathematical problems, ranging from the simplest calculation to complex engineering problems. MATLAB is distributed by **The Math Works, Inc**. The official Web site address is **http://www.mathworks.com**.

Starting MATLAB

To start a MATLAB session in the Windows operating system,
Click on the **Start menu > Programs > MATLAB**.

Go ahead and start the MATLAB program. When you see the command prompt in the command window, MATLAB is ready for you to enter commands.

The MATLAB workspace

The default MATLAB desktop contains three windows (**Workspace, Command History, and Command Window**) as well as tools for managing files, variables, and applications associated with MATLAB as shown in the Figure 13.1.1

You can change the layout of your workspace by selecting **Desktop > Desktop Layout >** and then selecting your preferences. Try it.

Getting Started

Follow along and complete all tasks, pay attention to how MATLAB responds to your commands.

1. Begin by creating a new folder on your Windows Desktop. Name it "mywork". The "mywork" folder will be your MATLAB working directory.

2. You can determine your current working directory, enter the following command at the command prompt:

>> **pwd**

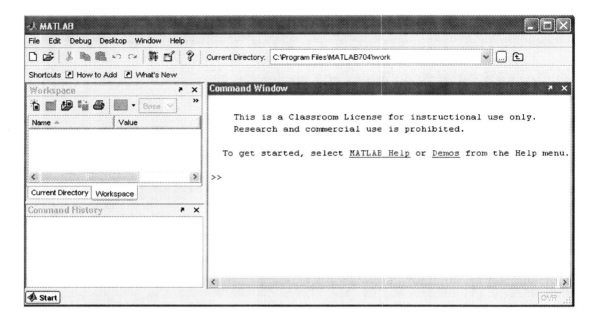

Figure 13.1.1 Command window, Workspace, and Command history

MATLAB's default working directory is named "work". Every time you start MATLAB your file path will point to the "work" directory. The *pwd* command should have produced an output similar to:

ans =

C:\...\MATLABXXX\work

Note: XXX identifies the version of MATLAB.

To list the contents of the working directory enter one of the following commands at the command prompt:

>ls

or

> dir

You should see a list of files or, ... , if the directory is empty.

Changing your working directory

To change the current working directory to the newly created folder, you can either type in the complete path in the Command Window or use the MATLAB toolbar. If you want to change the working directory using the command line the syntax is

cd('directory (full path)')

For example:

```
>> cd('C:\Documents and Settings\"username"\Desktop\mywork')
```

Replace **"username"** with your username:

Then enter

```
>> pwd
```

If your directory change was successful you should see

```
ans =

C:\Documents and Settings\"username"\Desktop\mywork
```

You can also change the working directory using the toolbar by clicking on the **Browse for Folder** icon and choosing a folder as depicted in Figure 13.1.2,

Figure 13.1.2 Toolbar

MATLAB Demos

One way to learn about MATLAB is to utilize its extensive demo library. To open the demo library type the command:

```
>> demo
```

Note: you can also access the demos from the **Start Menu Icon** located in the lower left corner of the MATLAB desktop.

To run the **Function Generator**, or any other demo of your choosing, start by expanding the file tree, in the **Help** window, to look like Figure 13.1.3

Select your demo and click **"Run This Demo"**. Go ahead and run a few, when you are finished close the **Help** and **Demo** windows.

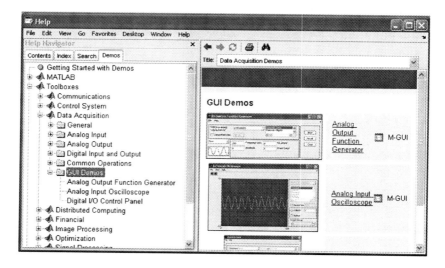

Figure 14.1.3 Demo Window

Basic arithmetic operations in MATLAB

MATLAB includes several commands for fundamental operations. A few are listed in Table 13.1.1

Table 14.1.1 Basic Arithmetic Operations

Type of Operation	MATLAB Syntax
Addition	x + y
Subtraction	x - y
Multiplication	2*x
Division	x/y
Exponentiation	x^2

Calculation Examples:

Follow along and enter the following expressions;

>> 3+4

>> 2*5

>> 50/5

>> (4*9+4)/(3+7)

  >> 4+5^2

Check and make sure your results make sense.

Defining and using Variables in MATLAB

Variables can be considered as storage units which allow the user to define frequently called values. The values stored in variables can then be recalled later and used in place of the values in MATLAB statements.

Syntax

Variable = Expression

Variable definition rules

1. Variables must begin with an alphabetic character.
2. MATLAB is case-sensitive (i.e., variable F ≠ variable f; variable A2 ≠ variable a2).
3. No spaces are allowed in your variable names.
4. It follows that your options are **a to z, A to Z**, numbers **0 to 9**, and the underscore character '_'.

Creating Your Own Variables

Follow along and enter the following commands;

>> x = 100

>> y = 101

>> x

>> y

>> average = (x + y) / 2

Check and make sure your results make sense.

Saving variables

Variables created during a MATALB session can be saved using the **save** function. Perform the following steps before quitting your MATLAB session:

1. To save all user-defined variables enter

```
>> save session_vars
```

All variables are saved into a file named session_vars.mat.

2. To save specific variables, say x and y, enter:

```
>> save session_vars x   y
```

Variables x and y are saved into session_vars.mat.

Follow along and save your variables x and y

enter:

```
>> save session_vars x y
```

All variables are saved in a ***.mat** file located in the current working directory. When you start MATLAB again you can use the stored variables after you load them using the

load *.mat command.

Related commands:

who	Gives the list of variables
whos	Get a more detailed listing of all the variables currently defined.
clear	To remove a variable and its value from memory.
clc	Clears the Command Window

At the command prompt, enter the following commands:

```
>> load session_vars
```

```
>> who
```

```
>> whos
```

```
>> clear x
```

The command **clear x** should have removed the variable x, check it.

```
>> who
```

```
>> clc
```

Suppressing output with a Semicolon (;)

If you do not wish to see MATLAB's output, you can eliminate it using a semicolon.

Enter the following and notice MATLAB's behavior

```
>> a = 50;
```

```
>> b = 145;
```

```
>> c = 35
```

```
>> d = 0
```

```
>> sum = (a + b + c + d)
```

Handling long command lines

Inserting 3 periods (...) followed by **Enter** at the end of any input allows you to use multiple line entries for a single command.

Enter the following at the command prompt

```
>> x = (5 + 6 + 7 - 8 + 9 + 10)/10 +...
12*(222-9)
```

Placing two or more commands on the same line

You can place two or more commands on the same command line by using commas (,) or semicolons (;) to separate them.

Enter the following at the command prompt

```
>> a = 4*1/2, b = a * 2; product = a*b
```

Notice the semicolon suppressed the output of **b**.

To terminate a MATLAB command in process

To terminate the program we have to use the following keys on the keyboard

`Ctrl+C`

As an example, MATLAB has a built in function called **lookfor**. The command allows the user to search for information on specific topics. For instance, if you need information on the syntax and use of the exponential function enter the following.

```
>> lookfor exponential
```

To stop the search, on the keyboard, press `Ctrl+C`. **For more information on exponentials use the help command.**

```
>> help exp
```

MATLAB Built-in Functions

MATLAB has several built-in functions. Here we will spend some time exploring them.

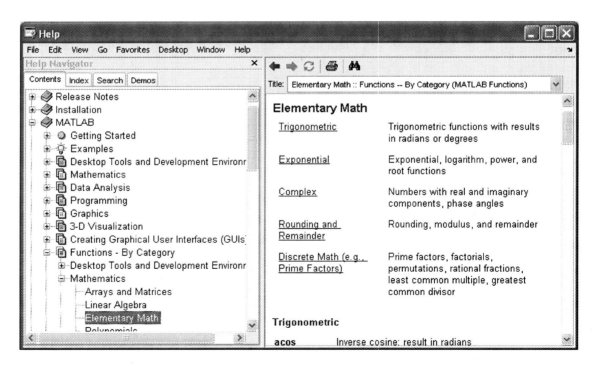

Figure 13.1.4 Help Window

At the command line type

```
>> helpwin
```

then expand the menu tree to look like Figure 13.1.4

Click on **Trigonometric** under **Elementary Math**, and then click on **cos** (Cosine). Notice that the section provides the **Syntax** and **Description**. This is where you can find basic information on using the **cos** function.

Follow along and enter the following commands

```
>> x = 2*pi;

>> y = cos(x)

>> b = sin(x/4+pi/2)

>> tan(x/3)-y
```

All MATLAB functions default to radian mode.   To convert from radians to degrees, simply multiply your answer by 180/pi or 360/2pi.

```
>>c = b*(180/pi)
```

Exponential function

In the **Help window** find and read the information on the **Exponential function,** then try the following commands

```
>> x = 2;

>> y = exp(x)
```

Natural logarithm function

In the **Help window** find and read the information on the **Natural Logarithm function** then try the following commands

```
>> x = 2;

>> log(x)

>> log2(x)
```

Square root function

In the **Help window** find and read the information on the **Square root** function, then try the following commands

`>> x = 16;`

`>> a = sqrt(x)`

`>> y = -16;`

`>> b = sqrt(y)`

Base Conversions in MATLAB

Syntax

D = base2dec('*string*',*base*)

This command converts a number '*string*' of the specified *base* into its decimal (base 10) equivalent. The **base** must be an integer between 2 and 36 and the **string** must be enclosed in single quotes.

Follow along using MATLAB and carry out the following conversions.

1. Binary to decimal conversion

`>> base2dec('0010110',2)`

2. Hexadecimal to decimal conversion

`>> base2dec('abcd',16)`

3. Octal to decimal conversion

`>> base2dec('151',8)`

MATLAB also has a built in function to convert from decimal to an arbitrary base

Syntax

N = dec2base(*int*,*base*)

This command converts a nonnegative integer **int** to the specified **base.** The **int** must be a nonnegative integer smaller than 2^{52}, and the **base** must be an integer between 2 and 36.

Follow along using MATLAB and carry out the following conversions

1. Decimal to binary

>> dec2base(22,2)

2. Decimal to Hexadecimal

>> dec2base(43981,16)

3. Decimal to octal

>> dec2base(105,8)

Using MATLAB with complex numbers

To input a complex number into MATLAB, you can use the **complex** command.

Syntax

Z=complex(a,b)

Here **a** represents the real coefficient and **b** represents the imaginary coefficient.

Follow along and enter the following complex numbers in rectangular form.

>> z = complex(3,4)

>> z1 = complex(0,2)

>> z2 = complex(2,0)

>> j

>> i

>> 4+i*4

As you can see, we can also enter a complex number into MATLAB as a sum between the real coefficient and the product of the imaginary coefficient and the built in constant **i** or **j**.

z = a + I * b

z = a + j * b

To extract the real and/or imaginary parts from a complex number use the following commands. Try it.

```
>> z = 4 +i*6
```

```
>> real(z)
```

```
>> imag(z)
```

Working with complex numbers in exponential polar form

A complex number can also be represented in polar form. The syntax is as follows

$$Z = Me^{j\theta}$$

Where M is the magnitude and 'θ' is the phase angle.

Follow along and enter the following commands

1. To find the magnitude use the command **abs**

```
>> z = 4 +i*4
```

```
>> abs(z)
```

2. To find the phase angle use the command **angle**

```
>> z = 4 +i*4
```

```
>> angle(z)
```

Don't forget, MATLAB provides the answer in radians.

Arrays, Vectors, and Matrices in MATLAB

An array can be entered in different ways in MATLAB, follow along using MATLAB

A. Syntax

```
A = initial value : step size : final value
```

where step size represents the difference between any two successive elements in the array.

For example, try entering

```
>> A = 1 : 3 : 28
```

```
>> B = 1 : 0.1 : 2
```

B. Syntax

A = initial value : final value

For this entry the **default increment** value is **1**.

Enter

>> A=1 : 7

C. You can also directly enter a set of values into an array as follows

Enter

>> A = [1 2 4 2 6 8 9]

Array operations

Element by element multiplication, division, and exponentiation between two matrices or vectors of the same size can be accomplished in MATLAB by preceding the arithmetic operator by a (.)

For example multiplication of the elements in two arrays A and B

If **A = [v1 v2 v3 v4]** and **B = [u1 u2 u3 u4]**

Then **C = A .* B**

Produces **C = [v1*u1, v2*u2, v3*u3, v4*u4]**

Similarly:

. *	Element -by-element multiplication
. /	Element -by-element left division
. ^	Element -by-element exponentiation

Follow along and enter the commands. Pay special attention to spaces between the numbers, the space separates the entries.

>> A = [2 4 1 2]

>> B = [1 3 4 2]

>> C =A .* B

```
>> C = A./B
```

```
>> C = A.^2
```

Matrices

Matrices can be entered into MATLAB using one of the following formats

A. Syntax

```
M = [first row; second row... ; last row]
```

Here the symbol " ; " separates the rows of the matrix

B. Syntax

```
M = [First row
Second row
.
.
.
Last row]
```

Follow along and enter the following

```
>> A = [3 4 7; 2 9 5; 9 1 0]
```

To extract the element located in the **second row-second column,** enter:

```
>> A(2,2)
```

Now try this one

```
>> A(2,3)
```

You can retrieve different blocks of a matrix (sub-matrices) using the following command

Syntax

```
M1 = M (from row : to row, form column : to column)
```

Where M is the original matrix and M1 is the desired sub-matrix.

Follow along and enter the commands

```
>> A = [2 3 1 6; 4 9 8 3; 9 8 3 0; 1 3 6 7]
```

```
>> A1=A(1:3,1:3)
```

```
>> A2=A(2:3,2:3)
```

You can also use arrays or vectors to generate a matrix, enter the commands:

```
>> A1 = [2 4 8 1]

>> A2 = [0 4 7 2]

>> A3 = [5 7 1 9]

>> A = [A1; A2; A3]
```

Basic matrix operations

If **A** and **B** are two (m x n) matrices then you can carry out the operations listed in Table 13.1.2

Table 13.1.2 Matrix Operations

Operation	Syntax	Description
Addition	C = A + B	Each element in A will be added to the corresponding element in B and will be stored in C
Subtraction	C = A - B	Each element in B will be subtracted from the corresponding element in A and will be stored in C
Multiplication	C = A * B	Standard matrix multiplication
Exponentiation	C = A.^n	Each element in A will be multiplied **n** times by itself and will be stored in C

Follow along and enter the commands

```
>> B= [3  6  2; 7  3  9; 0  8  2]

>> A = [2  5  8; 4  6  2; 9  0  2]

>> C = A+B

>> C=A-B

>> C=A*B
```

Check and make sure your answers make sense.

The Matrix Inverse

The inverse **A**$^{-1}$ of a matrix **A** is found using the command

```
B = inv(A)
```

Following along enter the commands

```
>> A = [2 3 1; 4 9 8; 9 8 3]
```

```
>> B=inv(A)
```

Solving Linear equations in MATLAB using the Matrix inverse method

We now demonstrate a method for solving a set of linear algebraic equations using MATLAB. The method is called the **"matrix inverse method"**.

Starting with the equations

$$2x + 4y = 12$$
$$3x = 2y + 2$$

Step 1. Rearrange the equations

Rearrange each equation so that all unknown quantities are on the left-hand side and all known quantities are on the right-hand side.

$$2x + 4y = 12$$
$$3x - 2y = 2$$

Step 2. Write the equations in matrix form

To write the equations in matrix form $A\overline{x} = \overline{b}$ you must arrange the unknowns in vector $\overline{x}$, the coefficients of the unknowns in matrix A, and the constants on the right hand side of the equations in vector $\overline{b}$.

It follows then that the coefficient matrix is

$$A = \begin{bmatrix} 2 & 4 \\ 3 & -2 \end{bmatrix}$$

Vector $\bar{x}$ will be

$$\bar{x} = \begin{bmatrix} x \\ y \end{bmatrix}$$

And the constant vector is

$$\bar{b} = \begin{bmatrix} 12 \\ 2 \end{bmatrix}$$

Thus, the final equation should have the form

$$A\bar{x} = \bar{b} \rightarrow \bar{x} = \frac{\bar{b}}{A} \rightarrow \begin{bmatrix} x \\ y \end{bmatrix} = A^{-1} \begin{bmatrix} 12 \\ 2 \end{bmatrix}$$

Here $\dfrac{\bar{b}}{A}$ means $\dfrac{1}{A}\bar{b} = A^{-1}\bar{b}$, thus $\bar{b}$ is multiplied by the inverse of A . This is not division!!!

Step 3: Solve the matrix equation in MATLAB

Example 1:

Follow along and solve the equations

```
>> A=[2   4; 3   -2]
>> b = [12; 2]
>>A1=inv(A)
>> x = A1*b
```

Your solution should be x = 2, y = 2, check your answers by substituting the x, y values in the original equations and verify.

Example 2: Starting with the equations;

$$x + 2y - 3z = 3$$
$$2x + y + 2z = 13$$
$$4x - 2y + z = 8$$

Step 1: Rearrange the equations

Here the equations are already in correct form, so there is no need to rearrange.

$$x + 2y - 3z = 3$$
$$2x + y + 2z = 13$$
$$4x - 2y + z = 8$$

Step 2: write the equations in the matrix form

Here $A = \begin{bmatrix} 1 & 2 & -3 \\ 2 & 1 & 2 \\ 4 & -2 & 1 \end{bmatrix}$

$$b = \begin{bmatrix} 3 \\ 13 \\ 8 \end{bmatrix}$$

$$x = \begin{bmatrix} x \\ y \\ z \end{bmatrix}$$

Step 3: Solve the matrix equation in MATLAB, follow along and enter the commands

```
>> A = [1 2 -3; 2 1 2; 4 -2 1]

>> b=[3;13;8]

>>A1=inv(A)

>> x = A1*b
```

The solution is x = 3; y = 3; z = 2. Check your answers by substituting x, y, and z values in the original equations and verify.

13.2 Plotting in MATLAB

Objectives

This section is designed to introduce you to basic plotting using MATLAB. The exercise is intended to be very interactive. You should have MATLAB running and be entering commands along with the examples.

Syntax 2-D Plots

```
plot (xvalue, yvalue, 'style-option')
```

Here **xvalue** and **yvalue** are arrays containing the **x** and **y** coordinates to be graphed. The two arrays must be the same length. Unequal length of the two arrays is the most common source of error when using the plot command.

The optional argument '**style-option**' can be used to specify options like color, line style (e.g. solid, dashed, dotted, etc.), and point-marker style (e.g. o, +, *, etc.)

Example: plotting y = sin x (-pi ≤ x ≤ pi), enter the following commands

```
>> x = -pi: pi/10: pi;
```

```
>> y = sin(x);
```

```
>> plot(x, y)
```

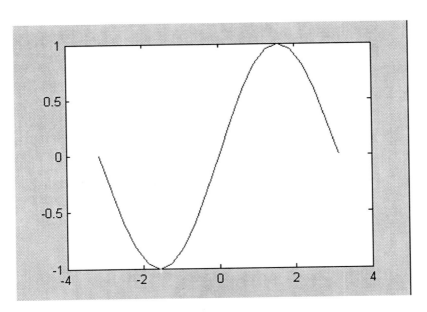

Plot 13.2.1: y = sin x (-pi ≤ x ≤ pi)

The plot function also works with a single array argument; in this case the elements of the array are plotted against its' row or column indices.

Syntax

```
plot(x)
```

Example: Plotting x = [1, 2, 3, 4, 6, 8, 10, 12], enter the following commands

```
>> x = [1, 2, 3, 4, 6, 8, 10, 12];

>> plot(x)
```

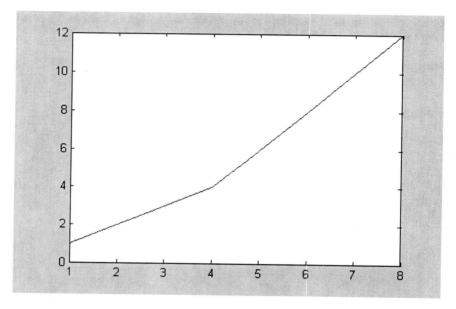

Plot 13.2.2: x = [1, 2, 3, 4, 6, 8, 10, 12]

Style options

MATLAB enables you to define several plot characteristics including: line style, line width, line color, marker type, and marker size. Some of the most common specifiers are listed below.

A. Line Style Specifiers

Specifiers	Line Style
-	Solid line (default)
--	Dashed line
:	Dotted line
-.	Dash-dot line

B. Marker Specifiers

Specifiers	Marker Type
+	Plus sign
o	Circle
*	Asterisk
.	Point
x	Cross
s	Square
d	Diamond
^	Upward pointing triangle
v	Downward pointing triangle
>	Right pointing triangle
<	Left pointing triangle
h	Six-pointed star (hexagram)
p	Five-pointed star (pentagram)

C. Color Specifiers

Specifiers	Color
r	Red
g	Green
b	Blue
c	Cyan
m	Magenta
y	Yellow
k	Black
w	White

For Example:

`plot(x, y, 'r')`   plots y vs. x with a red solid line.

`plot(x, y, ':')`   plots y vs. x with a dotted line.

`plot(x, y, 'm--')`   plots y vs. x with a magenta dashed line.

`plot(x, y, '+')`   plots y vs. x as unconnected points marked by +.

`plot(x, y, '-.or')`   plots y vs. x using a dash-dot line (-.), places circular markers (o) at the data points, and colors both line and marker red (r).

Example: plotting with line spec arguments, enter the following commands

```
>> x = -pi: pi/10: pi;

>> y = sin(x);

>> plot(x, y, 'r')
```

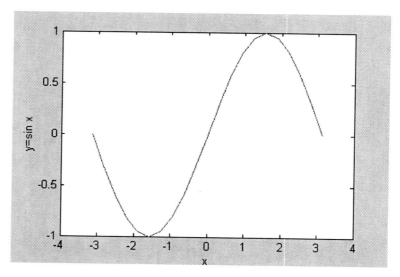

Plot 13.2.3: plot including Line spec arguments

```
>> plot(x, y, ':')
```

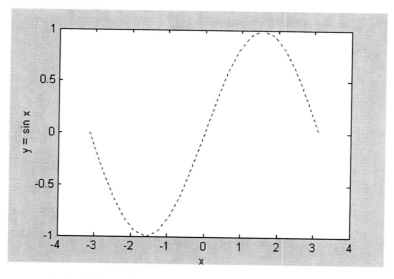

Plot 13.2.4: plot including Line spec arguments

```
>> plot(x, y, 'm--')
```

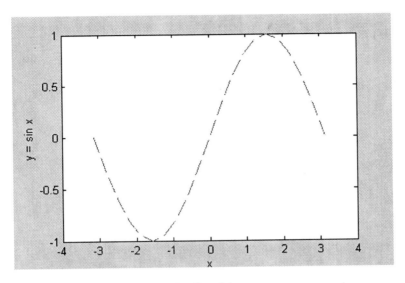

Plot 13.2.5: plot including Line spec arguments

Label and Title

Plots may be annotated with **xlabel**, **ylabel**, **title**, and **text** commands.

Syntax

```
xlabel('Time period')        labels the x-axis with "Time period"
ylabel('Amplitude')          labels the y-axis with "Amplitude"
title('sine wave')           titles the plot with "sine wave"
text(2,6, 'bob')             writes 'bob' at the location (2,6) in the plot coordinates
```

Enter the following commands

```
>> x = -pi: pi/10: pi;

>> y = sin(x);

>> plot(x,y)

>> xlabel('Time period');

>> ylabel('Amplitude');

>> title('Sine wave');

>> text(-3,0.7, 'sin(x)');
```

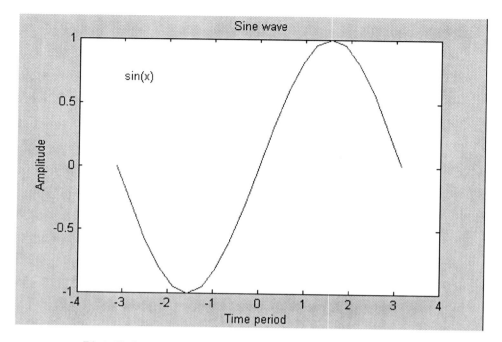

Plot 13.2.5: Annotated plot with Label, Title, and Text

Legend

The **legend** command produces a boxed legend on a plot.

Syntax

`legend(string1, string2, ..)`	Produces legend using the text in string1, string2, etc.
`legend(linestyle, string1, ..)`	Specifies the line style of each label.
`legend(.., pos)`	writes the legend outside the plot-frame if *pos* = - 1 and inside the frame if *pos* = 0.
`legend off`	deletes the legend from the plot.

Enter the following commands

```
>> x  = -pi: pi/10: pi;

>> y = sin(x);

>> plot(x,y)

>> xlabel('Time period');

>> ylabel('Amplitude');

>> title('Sine wave');

>> y = cos(x);

>> hold on;

>> plot(x,y,':r')

>> legend('sin(x)', 'cos(x)', -1);

>> legend('sin(x)', 'cos(x)', 0);

>> hold off
```

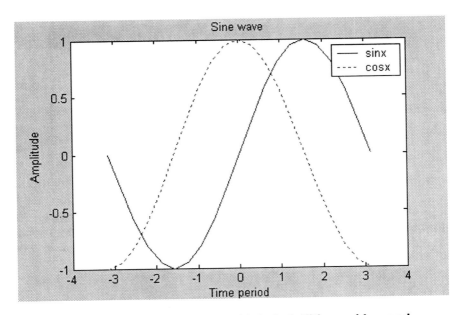

Plot 13.2.6: Annotated plot with Label, Title, and Legend

Axis control, zoom-in, and zoom-out

Once a plot is generated you can change the axes limits with the **axis** command.

Syntax

```
axis([xmin xmax ymin ymax])
```

The command changes the current axes limits to the specified new values **xmin** and **xmax** for the x-axis, and **ymin** and **ymax** for the y-axis.

For example, enter the following command

```
>>axis([0 pi -pi pi]);
```

Modifying plots using the plot editor

MATLAB has an interactive tool for modifying an existing plot. To use the GUI, go to the **Figure window**, click on the View>Plot Edit Toolbar shown in Figure 13.2.1, and click on the **left-leaning arrow (selection tool)** in the menu bar. Now you can select and double, or right, click on any object in the current plot to edit it.

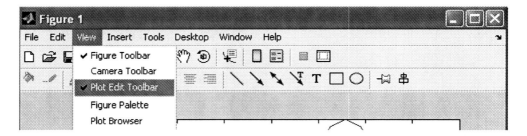

Figure 13.2.1: Figure window

Double clicking on the selected object brings up a **property editor window** where you can select and modify the current properties of the object. Other tools in the menu bar, for example, text (marked by A), arrow, and line, let you modify and annotate figures just like simple graphics packages do. Try it.

Overlay plots

There are three methods of generating overlay plots using MATLAB. The commands are **plot**, **hold**, and **line**.

A. Using the plot command.

Enter the following commands

```
>> x1 = -pi: pi/10: pi;
>> x2 = -pi: pi/10: pi;
>> y1 = sin(x1);
>> y2 = cos(x2);
>> plot(x1, y1, '-*', x2, y2, '-o')
```

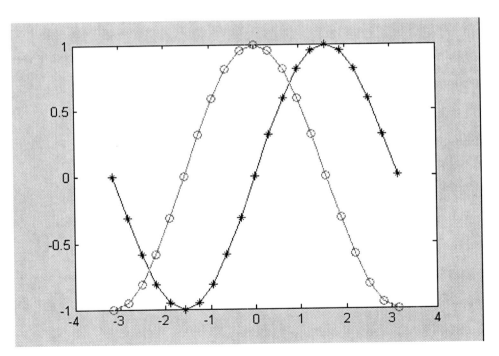

Plot 13.2.7: Plot for the equations y1 = sin(x1), y2 = cos(x2)

B. Using the hold command

Enter the following commands

```
>> x1 = -pi: pi/10: pi;
>> y1 = sin(x1);
>> plot(x1,y1, '-*');
>> hold on;
>> y2 = cos(x1);
>> plot(x1, y2, '-o')
>> hold off;
```

You should get the same plot as shown in Plot 13.2.7.

C. Using the line command

The **line** is a low-level graphics command, which is used by the **plot** command to generate lines. Once a plot exists in the graphics window, additional lines may be added by using the **line command** directly.

Syntax

line(xdata, ydata, ParameterName, ParameterValue)

Enter the following commands

```
>> x = -pi: pi/10: pi;
>> y1 = sin(x);
>> y2 = cos(x);
>> plot(x,y1, '-*');
>> line(x,y2,'linestyle', '--');
>> line(x,y2,'marker', 'o');
```

Once more, you should get the same plot as shown in Plot 13.2.7.

Sub-plots

The command **subplot** divides the current figure into rectangular panes that are numbered row-wise. Each pane contains a set of axes. Subsequent plots are output to the current pane at the designated position.

Syntax

```
subplot(m,n,p) = subplot(#rows, #columns, plot row position)
```

The command creates axes in the pth plane of a figure divided into an m-by-n matrix of rectangular planes.

Enter the following commands

```
>> x = [3.2 4.1 5.0 5.6];
```

```
>> y = [2.5 4.0 3.35 4.9];
```

```
>> subplot(2,1,1);
```

```
>> plot(x)
```

```
>> subplot(2,1,2);
```

```
>> plot(y)
```

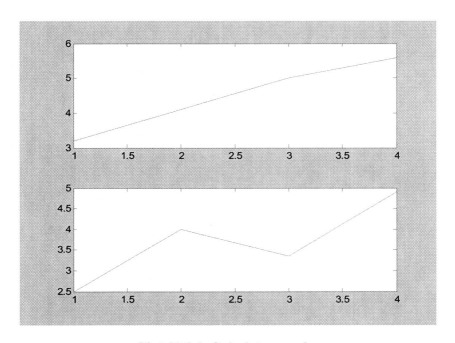

Plot 13.2.8: Subplots x and y

Enter the following commands

```
>> x = -pi: pi/20: pi;

>> y1 = sin(x);

>> y2 = cos(x);

>> y3 =2+sin(x);

>> y4 = 2+cos(x);

>> subplot(2,2,1)

>> plot(x,y1)

>> subplot(2,2,2)

>> plot(x,y2)

>> subplot(2,2,3)

>> plot(x,y3)

>> subplot(2,2,4)

>> plot(x,y4)
```

Pay close attention to the position of each plot, and the **subplot** entry for it.

(2,2,x)

1	2
3	4

Three-dimensional (3-D) plotting

Three-dimensional data can be displayed in several different styles using MATLAB. The line, surface, and wire mesh plots are examples.

Line plots

The **plot3** command in MATLAB is used to display data using connected lines.

Syntax

```
plot3(x, y, z)
```

Where **x**, **y**, and **z** are three arrays of the same size.

Close any open plot windows then enter the following commands

```
>> x = -12: 0.001: 12;

>> y = sin(x);

>> z = sin(x + pi/4);

>> plot3(x,y,z)

>> grid on;

>> title('Plot3'),xlabel('sin(x)'),ylabel('sin(x+pi/4)'),zlabel('x');
```

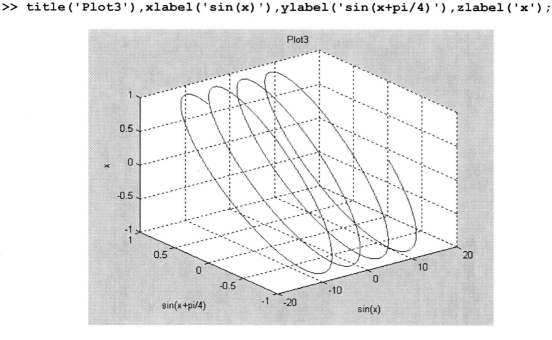

Plot 13.2.9: x = [-12...12], y = sin x, z = sin (x +pi/4)

Enter the following commands to produce different projections of Plot 13.2.9.

```
>>plot(x, y)
>>plot(z, y)
>>plot(x, z)
```

Mesh plots

Mesh plots are useful for plotting functions of two variables such as **z = f(x, y)**. The mesh surface is defined by the z-coordinates of points above a rectangular grid in the x-y plane. The **mesh** command forms a plot by joining adjacent points with straight lines.

There are four steps involved in generating mesh plots.

Step 1:   Define a desired range, plane size, for two variables, say x, y.

Step 2:   Use the **meshgrid** command to generate a mesh of points related to x and y

```
[x, y] = meshgrid(x, y).
```

Step 3:   Evaluate the function f(x, y), for example z = sqrt(x ^2 + y^2).

Step 4:   Use the **mesh** command to plot the mesh corresponding to z:

```
mesh(x, y, z)
```

Generating a mesh plot for $z = \sqrt{x^2 + y^2}$, enter the following commands

```
>> x = -10: 1: 10;

>> y = x;

>> [x, y] = meshgrid(x, y);

>> z = sqrt(x.^2 + y.^2);

>> mesh(x, y, z)

>> title('3D Plot'), xlabel('x'), ylabel('y'), zlabel('z');
```

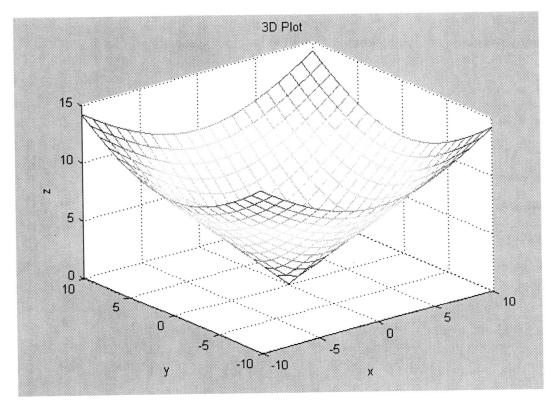

Plot 13.2.10: Mesh Plot x = [-10...10], y = x, $z = \sqrt{x^2 + y^2}$

Surface Plots

A surface plot is similar to a mesh plot except the patches (the spaces between the lines) are filled with a specified color. The **surf** command creates colored parametric surfaces specified by x, y, and z.

There are four steps involved in generating surface plots.

Step 1:   Define a desired range, plane size, for two variables, say x, and y.

Step 2:   Use the **meshgrid** command to generate a mesh of points related to x and y

        `[x, y] = meshgrid(x, y).`

Step 3:   Evaluate the function f(x, y), for example z = sqrt(x^2 + y^2).

Step 4:   Use the **surf** command to plot the mesh corresponding to z:

        `surf(x, y, z).`

Plotting a surface plot for $z = \sqrt{x^2 + y^2}$, enter the following commands

```
>> x = -10: 1: 10;

>> y = x;

>> [x, y] = meshgrid(x, y);

>> z = sqrt(x.^2 + y.^2);

>> surf(x, y, z)

>> title('3D Plot'), xlabel('x'), ylabel('y'), zlabel('z');
```

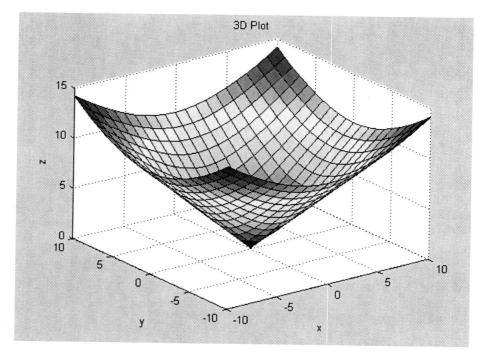

Plot 13.2.11: Surface Plot x = [-10...10], y = x, $z = \sqrt{x^2 + y^2}$

13.3 MATLAB M-Files

Objectives

This section is designed to introduce MATLAB M-Files. We cover opening, creating, saving, and the execution of M-Files. The exercise is intended to be very interactive. You should have MATLAB running and be entering commands along with the examples.

M-Files

Up to this point, when solving problems using MATLAB, you have entered a series of commands at the command prompt. For example, if you are interested in the power dissipated by a resistor, you would simply enter an expression to calculate power at the command line.

To find the power dissipated, assuming you know the current through and voltage across the resistor, you would enter the command:

```
>> I = 0.001;

>> V = 10:

>> P =I*V
```

P = 0.01

This is a relatively painless task. However, if you need to perform more than a few power calculations or carry out this calculation on a regular basis, there is a better way. You can create an M-File for calculating power. Then instead of typing your commands one by one in the command window, you type them only once and save them in an M-File. You can then use the M-File to compute the values whenever needed.

There are two types or forms of M-Files.

   1. Script form.
   2. Function form.

Scripts are suitable when you need to perform a long sequence of commands. Function files offer more flexibility as they allow the user to pass and return values, as you would with MATLAB's built in functions. A comparison is shown in Table 13.3.1.

Table 13.3.1: Comparison of Script and Function M-Files

Script M-Files	Function M-Files
Do not accept input arguments or return output arguments	Can accept input arguments and return output arguments
Operate on data in the current workspace	Internal variables are local to the function by default
Useful for automating a series of steps you need to perform many times	Useful for extending the MATLAB language for your application

Note: This module only covers script files since function files usually require some basic programming knowledge.

Creating and editing an M-File

M-Files are ordinary text files that can be created using a text editor. MATLAB provides a built-in editor to create and modify M-Files. Keep in mind that any text editor can be used, but you must use an ".m" extension on your file. This is how MATLAB identifies an M-File. To open MATLAB's editor, in the command window, go to the **File menu**, choose **New**, and then **M-File**. You can also open an M-File, for editing, from the Command window by typing the command,

>>edit *filename.m*

provided the file is located in your current working directory.

Saving a M-File

Once you have finished creating or editing an M-File you must save your work. To save your file, choose **save as** from the **File menu** of the text editor or click on the **disk icon** as shown in Figure 13.3.1.

Click here

Figure 13.3.1: Saving an M-file

Execution of an M-File

To execute an M-file, first you must open it. This brings up the question of locating both user written M-Files and built-in MATLAB functions. Consider the function (M-File) rot90.m, which rotates a matrix, you can use the function **which** to find its' location.

For example, to find the location of the function rot90.m, enter the command:

```
>> which rot90.m
```

The answer returned by MATLAB gives the location of the M-File rot90.m. To open the file you will need to include the path or change to the directory first.

Once you have created (or opened) an M-File, you can execute it by either; 1) typing the name of M-File at the command prompt or 2) by choosing **Save and Run** from the **Debug menu** in the command window as shown in Figure 13.3.2.

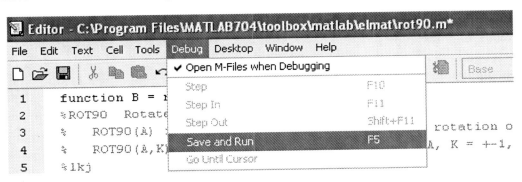

Figure 13.3.2: Running/Executing a M-File

Example: Rotating a matrix by 90 degrees with the function **rot90**, Follow along and enter the commands, but first change your working directory to the directory that contains **rot90.m**, which you can find using the **which** command.

```
>> which rot90.m

>> cd('directory path')

>> y = [2 5 3; 3 3 6; 4 5 6]

>> z = rot90(y)
```

From inspection did **rot90** perform the expected task?

NOTE: Before continuing, change back to your home working directory.

To create a script file, follow along and complete the task.

1.   Choose **New** from the **File menu** and select **M-file.**

2.   In the **editor window** enter the information shown in Figure 13.3.3.

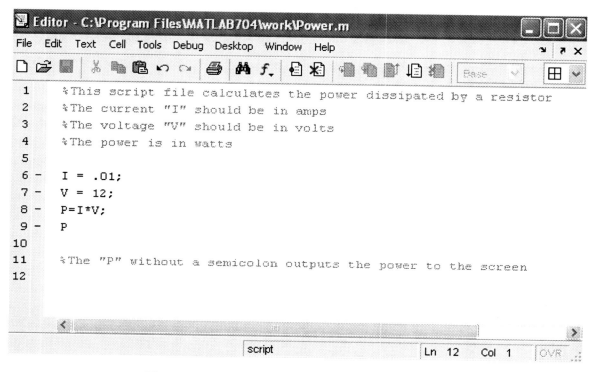

Figure 13.3.3: Resistor power dissipated M-File

3.   Save the file as **power.m** and run it. Your results will show up in the MATLAB command window.
4.   Change **I** and run it a few more times confirming the results (Don't forget to resave the file after you make changes).
5.   In the MATLAB **command window** enter the command:

>>**help power.**

The first four lines of the M-file should appear on the screen. The **%** sign in an M-File indicates a comment and allows you to create your own on-line help. The **help** command displays only those comments before the beginning of the actual code or body of the M-File. For example line 6 in Figure 13.3.3 is the beginning line of the code.

6. Make the following changes to the file power.m

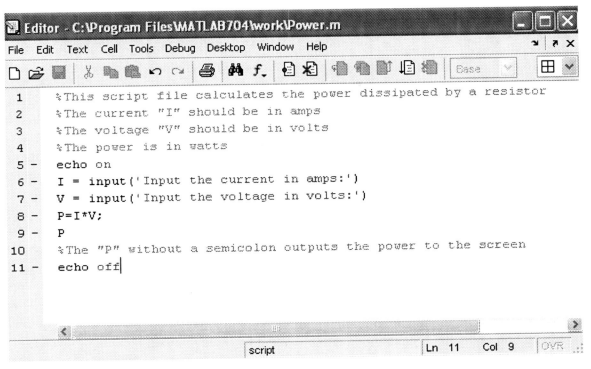

Figure 13.3.4: Computes power dissipated, user inputs I and V.

7. **Save** the edited M-File and **run** it again.

Notice the effects of your changes;

1. The **echo on** command prints the M-File to screen as it runs. If there is an error in the code, the user can see which command (or line) caused the problem. This is useful when debugging the M-File. Once your M-File executes correctly, you can remove the **echo on** and **echo off** commands.

2. The MATLAB variables **I** and **V** now require an input from the user. That is, the user is prompted to input a value for the current and voltage when the M-File is executed. This is accomplished by adding the lines:

```
I = input('Input the current in amps:')
V = input('Input the voltage in volts:')
```

We have presented you with an introduction to MATLAB mathematics, plotting, and M-File writing. Keep in mind, MATLAB is a very powerful computational tool and we have barely scratched the surface of its capabilities. The goal was to provide you a basic set of computational tools to get started with, which should prove useful in your engineering, as well as other, course work.

Chapter 14. Module 2 – Multisim

14.1 Multisim CAD Design and Electric Measurements

Objectives

This module is designed to introduce you to the CAD/Simulation software Multisim and a few basic electric measurement techniques. The module is intended to be very interactive. You should have Multisim running and perform the tasks along with the examples.

Introduction

Multisim, formerly Electronics Workbench, is an electronic CAD software package that allows the user to build; test and troubleshoot circuits on the computer.

Open Multisim by double clicking on your **desktop icon**, or by using the **Start-Programs menu**. You should see Figure 14.1.1.

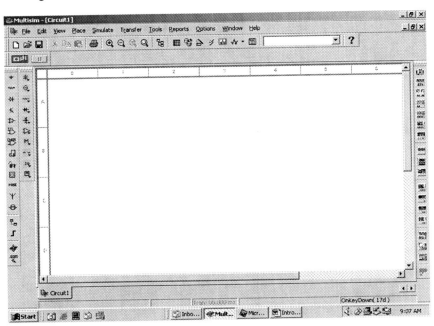

Figure 14.1.1: Multisim workspace

At the top left of the screen you will see several menus available to you. Click on each of these with your left mouse button and explore the menus.

On the left hand side of the workspace there are two **component (parts) tool bars**. Here you will find libraries of components. The user has two choices for picking parts, the left most toolbar allows for adding **specific parts** and the inner toolbar contains **virtual parts**. See Figure 14.1.2.

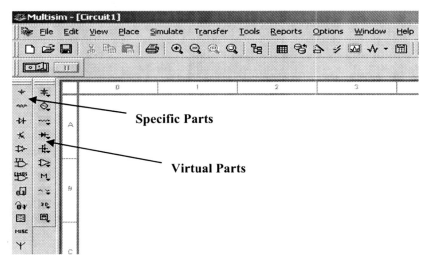

Figure 14.1.2: Component tool bars

The choice of which toolbar to use depends upon the circuit. For a fast build and simulate you may use the <u>virtual components</u> since they are set by default to <u>exhibit ideal component behavior</u>.

Click on the component button that is second from the top in the left-most toolbar. It is a squiggly symbol. A window should open as shown in Figure 14.1.3.

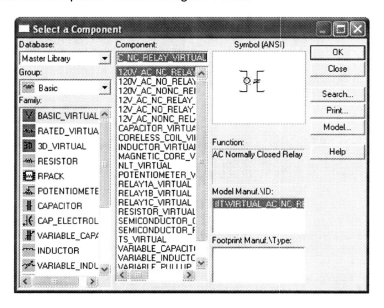

Figure 14.1.3: Basic components library

From this screen you can change groups by dropping down the **Group menu**. Then you can select other parts. Go ahead and change groups and browse the component libraries.

Close the **Select a Component** dialog box. Now click on the resistor symbol on the **Virtual Toolbar**, (the inner toolbar). It is the third icon down. You should see the virtual **Basic Components toolbar** as shown in Figure 14.1.4.

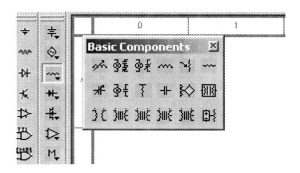

Figure 14.1.4: Basic Components

This toolbar contains some commonly used circuit components. You can see the name of the parts by placing your mouse cursor over the icon, try it.

On the right hand side of your workspace you will find the **Instrument Toolbar**, which is illustrated in Figure 14.1.5.

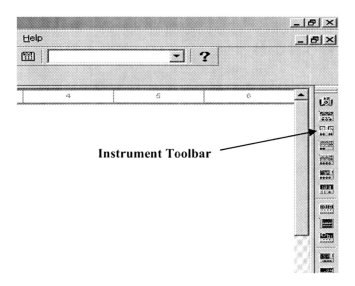

Figure 14.1.5: Instrument Toolbar

This is the location where Multisims' test instruments are kept. These instruments include, the multimeter, signal generator, oscilloscope, and so on. Placing your mouse cursor over the instrument buttons will reveal the instrument type, just as it did for the virtual components.

Building a Circuit

1. Click on the Resistor parts bin with your left mouse button. Left click on the resistor symbol and release. Move your mouse pointer to the workspace area. Then left click the mouse button. This places a resistor in your workspace. Your workspace should look similar to Figure 14.1.6.

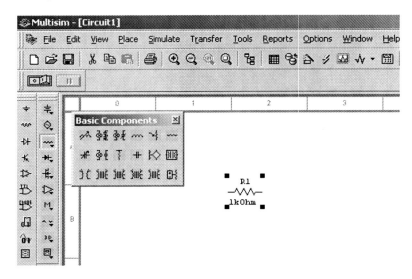

Figure 14.1.6: Placing a part

Note: to close the parts bin click on the x icon

2. Now make a copy of the resistor. First, right click on the resistor, scroll down and select copy. See Figure 14.1.7.

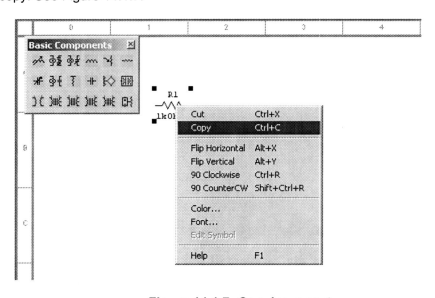

Figure 14.1.7: Copying a part

Now place the mouse pointer in an open space, in your work area, and right click. Next, scroll down select paste. Click once more to place the resistor. You should now have 2 resistors in your workspace.

3. Components can be rotated by right clicking on one of them, then scrolling down to **Rotate** and left clicking on it. Go ahead and rotate one resistor by 90^0.

Next the component values must be selected, or changed from their default values, to be utilized in the test circuit.

4. Double click on a resistor, and then select the **Value** tab. In the **Resistance** text box change the value from 1 to 47 kΩ. See Figure 14.1.8.

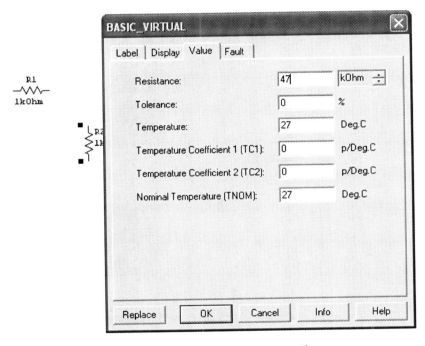

Figure 14.1.8: Component properties

Next click on **OK**, you have just changed the value of the selected resistor from 1 (default) to 47 kΩ.

5. Add a **Ground** to your workspace; click on the **Power Source Component icon**, then select **Place Ground,** and place it just as you did the resistors.

6. Connect the components with a wire. Using the mouse, point to the terminal of one component until a dot appears, then depress your left mouse button and release, drag the wire to the next component and again depress and release. Wire your circuit EXACTLY as shown in Figure 14.1.9.

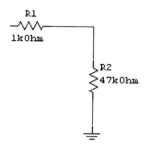

R1
1kOhm

R2
47kOhm

Figure 14.1.9: First circuit

7. In the same fashion that you added components to your workspace, add the following instruments; Multimeter and Oscilloscope. When you are finished your workspace should resemble Figure 14.1.10

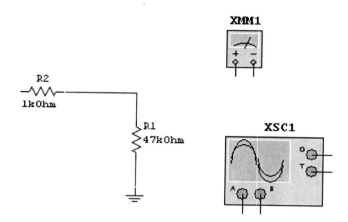

Figure 14.1.10: Multimeter and oscilloscope

Next, you will make a series of measurements. Keep in mind the techniques applied in the virtual environment also apply when you are making real physical measurements. However, if you make your connections wrong in Multisim nothing will explode or catch on fire, and no one will get a shock, as opposed to what could happen when making a real physical measurement.

A. Resistance Measurements:

   a. Single Resistor:

      i. Remove the wire connecting the two resistors in your circuit. Connect the multimeter to the 1 kΩ resistor. Double click on the multimeter and depress the ohm (Ω) symbol. See Figure 14.1.11.

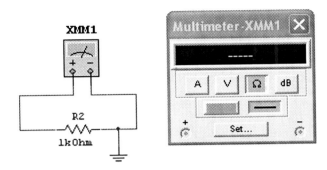

Figure 14.1.11: Resistance measurement

ii. Now hit the Run/stop simulation switch in the upper right corner of your Multisim window. When you see a reading on the multimeter stop the simulation.

This is an example of measuring the resistance of a single resistor. Does your reading make sense?

b. Resistors in Series:

i. Connect two resistors in series and reconnect the multimeter as shown in Figure 14.1.12. Run a simulation.

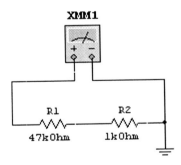

Figure 14.1.12: Resistors in series

This is an example of measuring the resistance of resistors in series which should obey the formula,

$$R_{Total} = R_1 + R_2 + ... + R_n$$

Do your readings agree?

c. Resistors in Parallel:

    i. Re-configure your workspace by placing the resistors and multimeter in a parallel configuration as shown in Figure 14.1.13. Now, run a simulation.

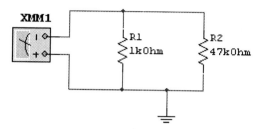

Figure 14.1.13: Resistors in parallel

This is an example of measuring the resistance of resistors in parallel which should obey the formula,

$$\frac{1}{R_{Total}} = \frac{1}{R_1} + \frac{1}{R_2} + ... + \frac{1}{R_n}$$

Do your readings agree?

B. DC Voltage Measurement:

a. Add a **DC voltage source**, located under the **Power Source icon**, and re-configure your workspace as shown in Figure 14.1.14. Be sure to depress the volt (V) button and the DC button on the multimeter, then run a simulation.

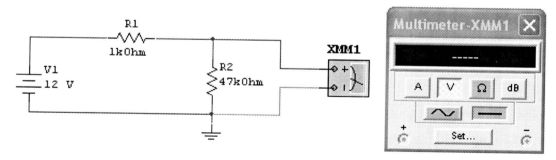

Figure 14.1.14: Voltage divider

This is an example of a direct current (dc) voltage measurement taken across resistor R2. According to the voltage divider equation, your reading on the multimeter should adhere to,

$$V_{R_2} = \left(\frac{R_2}{R_1 + R_2} \right) V_{source}$$

Do your readings agree?

C. AC Voltage Measurement:

a. Replace your **DC voltage source** with an **AC power source** and re-configure your workspace as shown in Figure 14.1.15. On the multimeter, be sure to depress the volt (V) button and the AC button. Now run a simulation.

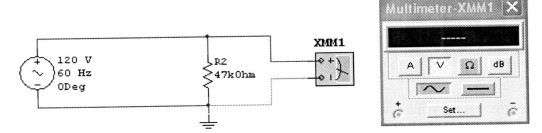

Figure 14.1.15: AC voltage measurement

This is an example of an alternating current (ac) voltage measurement taken across resistor R2. The multimeter is actually reporting an RMS voltage value, which is defined by; see Appendix B for details,

$$V_{RMS} = \sqrt{\frac{1}{T} \int_0^T v^2(t)\,dt}$$

For a sinusoidal function then,

$$V_{RMS} = \frac{V_{Peak}}{\sqrt{2}} = 0.707 V_{Peak} \,.$$

b. Hook up the oscilloscope, and set the **Channel A** and **Time base** settings as shown in Figure 14.1.16. Run the Simulation.

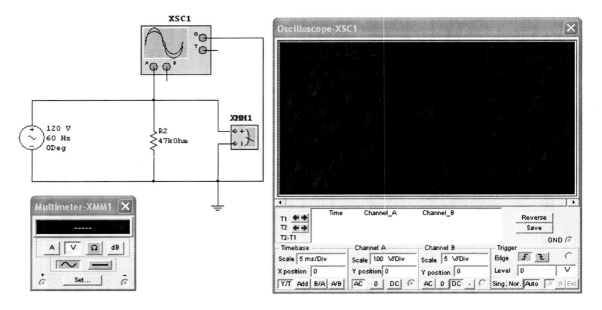

Figure 14.1.16: Peak voltage vs. RMS

Does the multimeter RMS reading match the expected value using the RMS definition and the peak voltage reading from the oscilloscope?

D. DC Current Measurement:

  a. Next you will be measuring a DC current. The main difference in instrument setup for current and voltage measurements is that for a **voltage** reading you **connect** the multimeter **across an element** and measure the difference between two points. Whereas, when performing a current measurement, you are measuring the charge flow through the circuit. Therefore, to make a **current** measurement you must **place the multimeter in the circuit** so that charge flows through it. Compare the two setups in Figure 14.1.17.

  b.

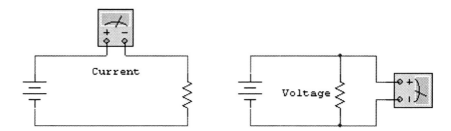

Figure 14.1.17: Current vs. Voltage measurement

  c. Set up the circuit shown in Figure 14.1.18. Be sure your **multimeter is inline** and it has the current (A) and DC buttons depressed. Run a simulation.

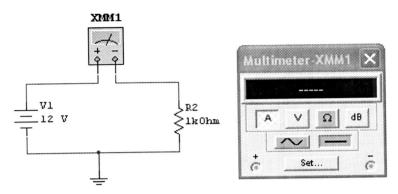

Figure 14.1.18: DC current measurement

From ohm's law your current should be equal to

$$I = \frac{V}{R}.$$

Does your reading from the multimeter agree?

E. AC Current Measurement:

a. Set up a circuit as shown in Figure 14.1.19. Be sure to depress the current (A) and AC button on the multimeter, then check the settings on the oscilloscope. Run a simulation.

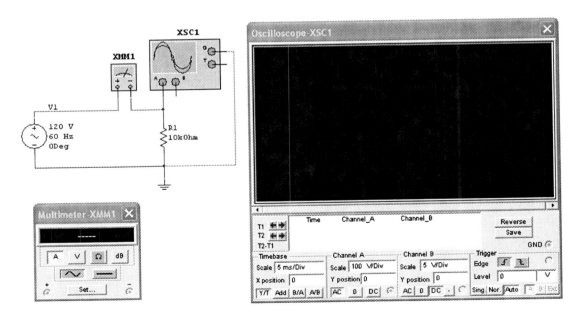

Figure 14.1.19: AC current with oscilloscope

This is an example of an alternating current (ac) measurement taken through resistor R1. The multimeter is actually reporting an RMS current value, which is defined by; see Appendix B for details,

$$I_{RMS} = \sqrt{\frac{1}{T}\int_0^T i^2(t)\,dt}$$

For a sinusoidal function we have,

$$I_{RMS} = \frac{I_{Peak}}{\sqrt{2}} = 0.707\,I_{Peak}\,.$$

Does the multimeter RMS reading match the expected value using the RMS definition for current and the peak voltage reading from the oscilloscope? **Don't forget to use Ohm's law to find** I_{Peak} **from** V_{Peak} **, the oscilloscope is not measuring current directly.**

F. *Power Measurement:*

    a. Set up a circuit as shown in Figure 14.1.20. The instrument shown is a Wattmeter which measures power by monitoring the current and voltage of a circuit. Recall that power is a function of current and voltage.

$$P = IV$$

Be sure you break the circuit for the current connection. The voltage is measured across R_L .

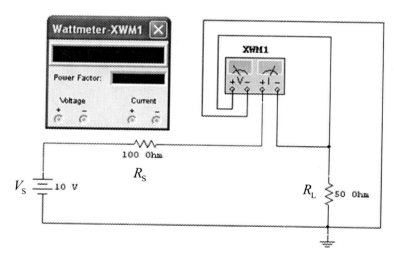

Figure 14.1.20: Power Measurement

b. Measure the power output at the load resistance R_L for the values;

R_L = 50, 60, 70, 80, 90, 100, 110, 120, 130, 140, and 150 Ohms. For each value of R_L record the power reading.

c. The configuration in Figure 14.1.20 is basically a power source consisting of a source voltage V_S with a source impedance R_S, that is delivering power to a device of load impedance R_L. In this configuration the maximum power transfer should occur when the source impedance matches the load impedance, or

$$R_S = R_L \, ,$$

Do your results confirm this? This is basically an optimization problem. For maximum power transfer the product of the current and voltage at R_L must be at its maximum value. A complete derivation can be found in Appendix C.

14.2 Digital Design

In addition to analog simulations and measurements, Multisim offers digital circuit layout and simulation tools. We will start with a simple circuit consisting of an AND gate and two SPST (single-pole single-throw) switches.

a. Build a circuit as illustrated in Figure 14.2.1. Switches can be found in the specific parts toolbar under the **basic icon.** The virtual voltmeter is located in the virtual toolbar under the **measurement components icon**; gates are under the **TTL icon.**

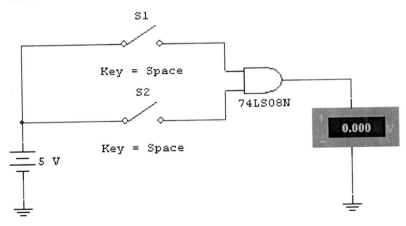

Figure 14.2.1: Simple logic circuit

This configuration allows you to toggle on and off the input switches while monitoring the associated outputs.

b.   Double click on one of the switches, go to the **Value** tab and set the **Key for Switch** to **A**. Do the same for the other switch, but set the **Key for Switch** to **B**.

c.   The **Key for Switch** allows you to control the switches, turn them on and off, with a unique key press. Try it and confirm the truth table in Table 15.2.1. Don't forget to start your simulation before you change the switch settings.

Table 15.2.1: AND gate truth table

A	B	F
0	0	0
0	1	0
1	0	0
1	1	1

Although this method works, it is a crude and limited method for determining a truth table. A better approach is to use Multisim's Logic Converter.

The Logic Converter:

The Logic Converter allows you to generate truth tables and Boolean expressions for a circuit. In addition, it can even be used to generate a circuit from a Boolean expression. Note there isn't a "real world" equivalent to the Logic Converter, it is a purely simulation tool.

Build the circuit as shown in Figure 14.2.2. Then double click on the Logic Converter and locate the Out pin. This is the last pin on the Logic Converter. Connect the circuit output to the Out pin on the Logic Converter.

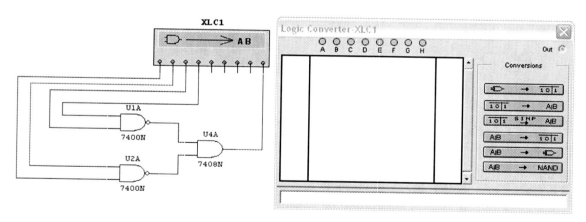

Figure 14.2.2: Multisim's Logic Converter

b. Click on the **circuit-to-truth table icon** to produce a truth table.

.

The Logic Converter will determine a Boolean expression for any truth table that you enter into it.

c. **Delete your entire circuit**, place a **new logic converter** in your workspace, and then click on the input radio buttons A - D, to activate, as shown in Figure 14.2.3.

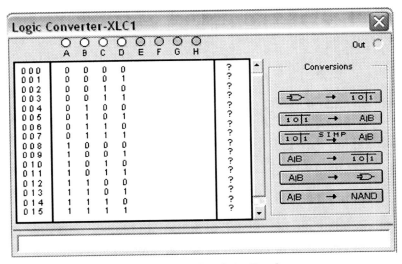

Figure 14.2.3: Truth to Boolean

d. Next, set the outputs by clicking on the **? marks**. Each time you click, the output will cycle through 0, 1, and X. You can choose some at random for this exercise.

e. Click the **truth table-to-Boolean icon** [1 0 1 → AIB]. Your expression will be displayed at the bottom of the Logic Converter window.

f. To simplify the expression click on the **SIMP icon** [1 0 1 SIMP AIB].

The Logic Converter also has a built in function to generate a truth table or circuit schematic from a Boolean expression.

g. Enter the expression AB+A'B in the bottom window of the Logic converter. Click the **Boolean-to-truth table icon** [AIB → 1 0 1], a truth table will be generated.

h. To generate a circuit schematic click on the **Boolean expression-to-circuit icon** AIB → , a schematic will appear similar to Figure 14.2.4. Be sure tp clear your workspace first.

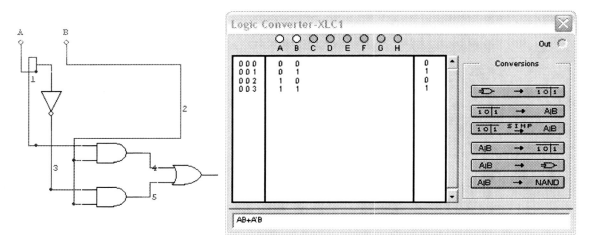

Figure 14.2.4: Boolean to circuit schematic

The Logic Analyzer

The Logic Analyzer is an instrument that allows simultaneous monitoring of logic inputs and outputs. This device does have a "real world" counterpart. Its' operation is best introduced by demonstration.

a. Set up a circuit as shown in Figure 14.2.5. Take note, the **GNDs** are digital grounds.

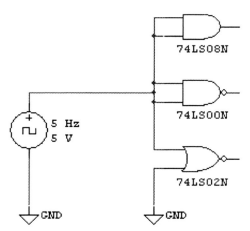

Figure 14.2.5: Digital test circuit

b. Add a Logic Analyzer to your workspace and make the connections shown in Figure 14.2.6.

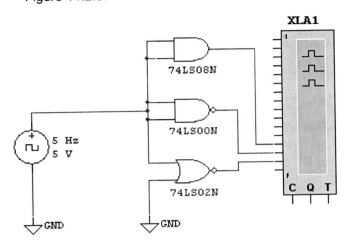

Figure 14.2.6: Multisim's Logic Analyzer

c. Open the Logic Analyzer display by double clicking on it, see Figure 14.2.7 and then click on the Clock **Set** button. Set the **Clock Source** to internal, **Clock Rate** to 10 Hz, and the **Threshold Volt. (V)** to 1 volt.

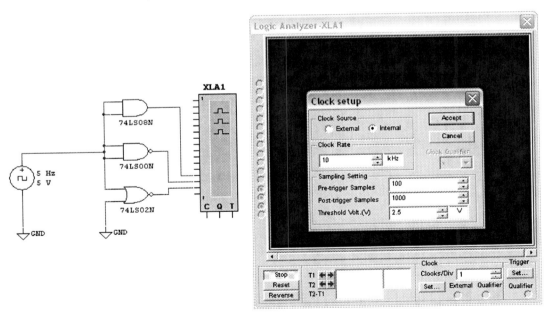

Figure 14.2.7: Logic Analyzer setup

d. Click **Accept** to close the Clock **setup** window.

e. Start a simulation. Let the simulation run for one full display length and then click stop.

f. Click the **Reverse** icon to change your display to a white back ground. You should be able to see the internal clock and all three of the outputs on the Logic Analyzer display as shown in Figure 14.2.8.

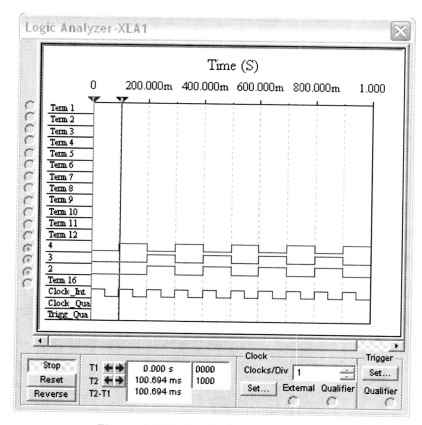

Figure 14.2.8: Logic Analyzer output

15.1 Definitions/RS232 Standards

Objectives

This module is designed to introduce you to serial communications and the RS232 standard. In addition, the Windows communication software HyperTerminal is introduced. You should follow along and enter the required commands in the HyperTerminal section.

Serial data communication

Data in a computer is represented as binary, two state, electrical signals usually referred to as bits. In order for two devices to exchange information (data) a connection must exist which allows data to be passed from one device to the other. For successful communication to take place between two, or more, devices the senders and the receivers must agree upon how the data will be represented and interpreted in advance. To achieve this objective several standard character codes have been established. One of the most prominent is the *American Standard Code for Information Interchange* or ASCII. ASCII assigns a unique binary representation for specific characters and symbols as shown in Table 15.1.1.

In the case of serial data communications the bits of a character are sent and received one at a time, hence the term serial communication. As an example, the character 'Y' coded using ASCII as 1011001 is shown in Figure 15.1.1. To transmit the character 'Y' the sender transmits the string of bits 1011001 as a serial sequence of high and low voltage levels, relative to some baseline reference, typically zero volts, and the receiver receives the bits one after the other.

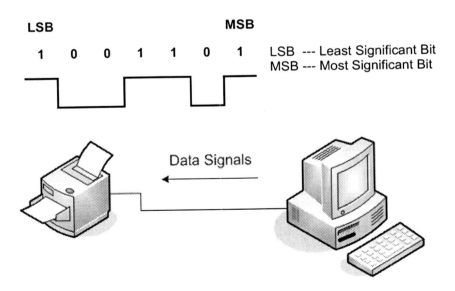

Figure 15.1.1: Serial communication

Table 15.1.1: Map of ASCII character set

Decimal	Binary	Symbol
000	0000000	(Null)
001	0000001	SOH
002	0000010	STX
003	0000011	ETX
004	0000100	EOT
005	0000101	ENQ
006	0000110	ACK
007	0000111	BEL
008	0001000	BS
009	0001001	HT
010	0001010	LF
011	0001011	VT
012	0001100	FF
013	0001101	CR
014	0001110	SO
015	0001111	SI
016	0010000	DLE
017	0010001	DC1
018	0010010	DC2
019	0010011	DC3
020	0010100	DC4
021	0010101	NAK
022	0010110	SYN
023	0010111	ETB
024	0011000	CAN
025	0011001	EM
026	0011010	SUB
027	0011011	ESC
028	0011100	FS
029	0011101	GS
030	0011110	RS
031	0011111	US
032	0100000	(Space)
033	0100001	!
034	0100010	"
035	0100011	#
036	0100100	$
037	0100101	%
038	0100110	&
039	0100111	'
040	0101000	(
041	0101001	)
042	0101010	*

Decimal	Binary	Symbol
043	0101011	+
044	0101100	,
045	0101101	-
046	0101110	.
047	0101111	/
048	0110000	0
049	0110001	1
050	0110010	2
051	0110011	3
052	0110100	4
053	0110101	5
054	0110110	6
055	0110111	7
056	0111000	8
057	0111001	9
058	0111010	:
059	0111011	;
060	0111100	<
061	0111101	=
062	0111110	>
063	0111111	?
064	1000000	@
065	1000001	A
066	1000010	B
067	1000011	C
068	1000100	D
069	1000101	E
070	1000110	F
071	1000111	G
072	1001000	H
073	1001001	I
074	1001010	J
075	1001011	K
076	1001100	L
077	1001101	M
078	1001110	N
079	1001111	O
080	1010000	P
081	1010001	Q
082	1010010	R
083	1010011	S
084	1010100	T
085	1010101	U

Decimal	Binary	Symbol
086	1010110	V
087	1010111	W
088	1011000	X
089	1011001	Y
090	1011010	Z
091	1011011	[
092	1011100	\
093	1011101	]
094	1011110	^
095	1011111	_
096	1100000	`
097	1100001	a
098	1100010	b
099	1100011	c
100	1100100	d
101	1100101	e
102	1100110	f
103	1100111	g
104	1101000	h
105	1101001	i
106	1101010	j
107	1101011	k
108	1101100	l
109	1101101	m
110	1101110	n
111	1101111	o
112	1110000	p
113	1110001	q
114	1110010	r
115	1110011	s
116	1110100	t
117	1110101	u
118	1110110	v
119	1110111	w
120	1111000	x
121	1111001	y
122	1111010	z
123	1111011	{
124	1111100		
125	1111101	}
126	1111110	~
127	1111111	DEL

RS232 is the recommended standard for connecting computers and peripherals via serial interfaces for the purpose of communication. The standard defines a set of rules governing the connections and communication between data terminal equipment (DTE) and data communication equipment (DCE) employing serial binary data interchange. An example of a DTE device would be a computer or computer terminal, and DCE devices are typically modems, or printers. The RS232 standard dictates that the serial data signal amplitudes occur between ± 15 volts. The standard defines a negative voltage level between -15 and -3 volts as a *binary bit 1* and positive voltage levels between +3 and +15 volts as a *binary bit 0*. The voltage level between ± 3 volts is *undefined*. Therefore, the actual signal representing the character 'Y' when sent serially using the RS232 standard is depicted in Figure 15.1.2.

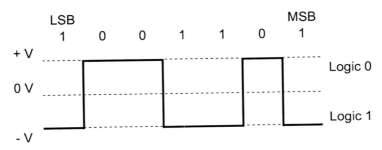

Figure 15.1.2: RS232 representation of the character 'Y' using the ASCII code

Start and Stop bits

Serial transmissions include a start bit and a stop bit that encapsulate the actual data bits. The start bit lets the receiver know that data bits will follow. The stop bit indicates the end of the transmission of a particular character or symbol. The ASCII character 'Y' with the start and stop bits included is shown in Figure 15.1.3. It is important to note that the Data line signal is always held in the logic-1 state when transmissions are idle.

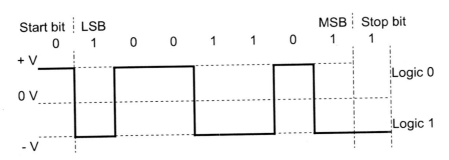

Figure 15.1.3: RS232 representation of the character 'Y' including start and stop bits

Parity bits

Data signals are susceptible to the influence of noise and electrical interference, during their transmission over a cable. This can lead to some bits of the data signal being corrupted, for example unintentionally being flipped from 0 to 1 or from 1 to 0. The unintentionally flipped bit is referred to as an error bit. There is a long history of techniques for detecting and correcting bit errors. One of the simplest techniques for detecting bit errors is the parity check method.   In the parity check scheme the sender makes the number of 1s or 0s, representing an ASCII character or symbol, odd or even by appending a redundant bit at the end of each character or symbol transmission. Of courses this scheme must be defined and agreed upon by both the sender and receiver prior to communication.   An example of the Odd-parity checking, using logic-1 bits, is depicted in Figure 15.1.4.

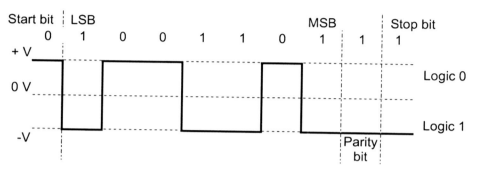

Figure 15.1.4: RS232 representation of the character 'Y' including, logic-1, using odd parity checking

In this example, the receiver checks whether the number of 1s is odd or not. If it is odd, the receiver assumes no errors occurred during the transmission. If it is determined the number of 1s is even, the receiver would assume an error had occurred during transmission and requests the sender to resend the information. Obviously, the parity check scheme is only useful when a single bit error occurs.   When more errors are likely to occur, it would be wise to use other error detecting/correcting codes.

RS232 Connections

Serial RS232 communication requires a cable connection using a 25-pin (DB25) or a 9-pin (DB9) connector.   The DB9 connector is the more commonly used connector. You can find it on the back of most personal computers as pictured in Figure 15.1.5. The fixed connector is also referred to as a *serial* or *COM port* on the device. The detailed pin numbers are shown in Figure 15.1.6, and their functions are listed Table 15.1.2.

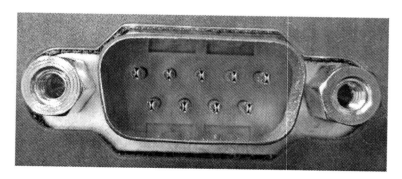

Figure 15.1.5: Fixed male DB 9 serial connector

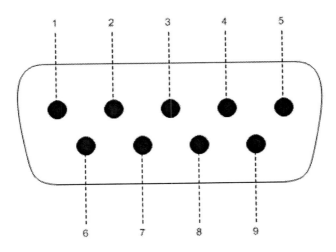

Figure 15.1.6: Male DB 9 serial connector pin numbers, typically found on the back of a personal computer

Table 15.1.2: RS232 Pin Assignments (DB9 connector)

Pin	Signal	Pin	Signal
1	Data Carrier Detect (DCD)	6	Data Set Ready (DSR)
2	Received Data (RD)	7	Request To Send (RTS)
3	Transmitted Data (TD)	8	Clear To Send (CTS)
4	Data Terminal Ready (DTR)	9	Ring Indicator (RI)
5	Signal Ground (SG)		

DTE to DCE Communication

An RS232 serial communication link can be established between a computer (DTE) and a modem (DCE) simply by connecting a standard RS232 serial cable between them. The connection is depicted in Figure 15.1.7.

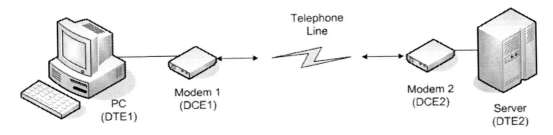

Figure 15.1.7: DTE to DCE serial communication link

The DTE-DCE communication flow can be summarized in chronological order as follows:

1. DTE1 raises a *Request to Send* to start a data transmission.
2. DCE1 checks if DTE2 has its *Request to Send line* in use (high).
3. If the *Request to Send line* on DCE2 is high, DCE1 does not give *Clear To Send* to DTE1, and DTE1 drops *Request to Send* and returns to step 1.
4. If the *Request to Send line* on DCE2 is not high, DCE1 gives a *Clear To Send* signal to DTE1. At this point, DTE1 has control of the serial link for data transmission.
5. DTE1 then sends data on the *Transmitted data line*, and DCE1 relays this to DCE2.
6. DCE2 then retransmits the data to DTE2.
7. DTE1 will continue to hold its *Request to Send line* high until all data is transmitted. Then it drops its *Request to Send line*, causing the line to be idle again.
8. Now, either DTE1 or DTE2 can raise a *Request to Send line* to obtain control of the serial link for data transmission.

This DTE-DCE communication flow represents *half-duplex communication*, where data is transmitted in one direction at a time. Since separate cable lines for *Transmitted Data* (TD) and *Received Data* (RD) are available, two-way simultaneous data communication, called *full-duplex communication*, is possible if both DTE and DCE devices support full-duplex mode. In full-duplex mode, both DTE1 and DTE2 have their RTS line held high and data transmission occurs simultaneously in both directions.

DTE to DTE Communication

Another example of serial data communication connection is a direct link between two computers, DTE to DTE, as pictured in Figure 15.1.8.

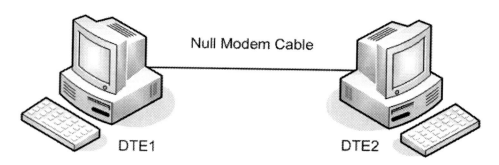

Figure 15.1.8: DTE to DTE serial communication link

In the previous DTE-DCE serial example, each DB9 connector pin is connected to the corresponding pin on the two devices. That is to say pin 1 of DTE1 is connected to the pin 1 of DCE1 and so on for all 9 pins in use. The cable used for the DTE-DCE connection is often called a *Straight-Through* serial cable. On the other hand for DTE to DTE serial connections several of the pin connections must be changed, the cable is no longer a straight through connection. The DTE-DTE pin connections are defined in Table 15.1.3. A cable that has been manufactured with these pin assignments is known as *Null-Modem* cable.

Table 15.1.3: Null-Modem Pin Assignments for serial DB9 cable

Pin Name	DTE 1 DB9 pin number	DTE 2 DB9 pin number	Pin Name
Received Data (RD)	2	3	Transmitted Data (TD)
Transmitted Data (TD)	3	2	Received Data (RD)
Data Terminal Ready (DTR)	4	1, 6	Data Carrier Detect (DCD) Data Set Ready (DSR)
Signal Ground (SG)	5	5	Signal Ground (SG)
Data Carrier Detect (DCD) Data Set Ready (DSR)	1, 6	4	Data Terminal Ready (DTR)
Request To Send (RTS)	7	8	Clear To Send (CTS)
Clear To Send (CTS)	8	7	Request To Send (RTS)
Ring Indicator (RI)	9	-	-

Once the Null-Modem cable is connected, all other communication takes place just as in the DTE to DCE serial communications example. It is important to note, that serial communication can be established using only three wires; *Transmitted Data* (TD), *Receive Data* (RD) and *Signal Ground* (SG). In this scenario TD and RD may or may not be crossed dependant on the type of devices, DTE or DCE. The remaining pins are attached locally as 1 to 4 to 6, and 7 to 8 to 9. A summary is presented in Figure 15.1.9.

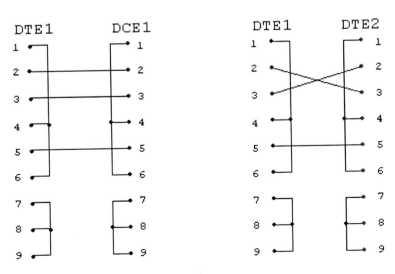

Figure 15.1.9: 3 wire configuration for DTE to DCE and DTE to DTE serial communication

15.2 HyperTerminal

HyperTerminal is a basic terminal communications program included with Microsoft Windows. One function of HyperTerminal is that it allows for a simple serial communications interface. By design, it guides the user through the COM port settings needed to establish communications. In this section you will be introduced to the HyperTerminal program and make a physical serial communication connection between two computers.

Initial Setup

1. Since computer to computer communication is *DTE to DTE*, use a null modem cable to connect two computers directly through their RS232 serial interface.

Figure 15.2.1: Null Modem cable

Simply connect the male connectors of both the computers to the female connecters of the null modem cable.

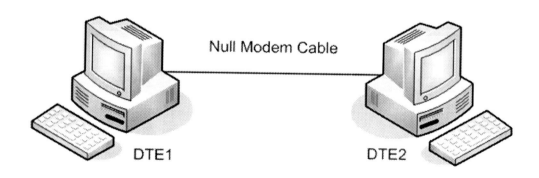

2. Once the connection is made, from the windows start menu, select
    Programs → Accessories → Communications and click on *HyperTerminal*.

3. You should now see the HyperTerminal window, as shown in Figure 15.2.2, with a dialog box, which prompts you for a new connection name.

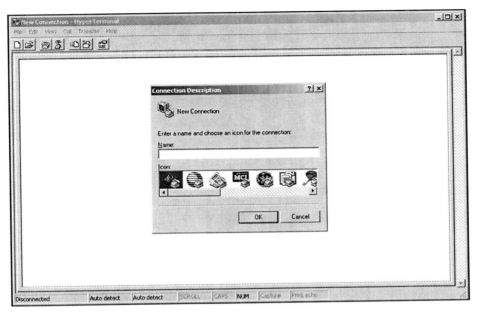

Figure 15.2.2: HyperTerminal program

4. Enter a connection name (e.g., Serial-Test) and choose an icon. After you click on *OK*, another dialog box appears as shown in Figure 15.2.3. Select *COM1* and then click *OK*.

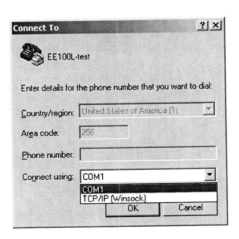

Figure 15.2.3: COM port selection

Note: If your computer has two serial ports you will also see COM2 in the pull-down menu. In this case you may need to select COM2. The other listing TCP/IP (Winsock) stands for Windows sockets, which is an Application Program Interface (API) for developing windows programs.

5. Next the COM1 Properties dialog box should appear. Select the properties as shown in Figure 15.2.4 and click on *OK*. A summary of COM port properties settings is given at the end of this section.

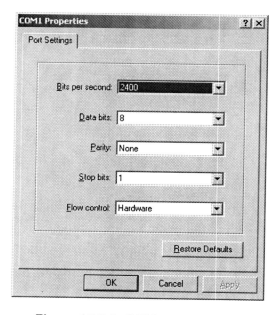

Figure 15.2.4: COM port settings

6. Now, if successful, HyperTerminal should be running as shown in Figure 15.2.5. Check and make sure the clock, circled by a dashed-line in Figure 15.2.5, is ticking.

Figure 15.2.5: Successful HyperTerminal connection

7. On the remote computer, the other pc you have connected to the Null modem cable, repeat procedures 2 through 6 again. The computers should now be ready to communicate with each other. When you type a character on one of the computers, the character will be sent to the other computer and appear on its monitor. Note: you will not be able to view the typed characters on the local computer unless you have *echo local characters typed* enabled.

Summary of COM Port Properties Setting

Bits per second: *110 ~ 921600 bps*
This is a list of transmission speeds in *bits per second*. You can choose one of the transmission speeds from the pull-down menu.

Data bits: *5, 6, 7, 8 bits*
Defines the number of data bits that can be used to represent ASCII characters and symbols. You can change the number of bits from the pull-down menu, dependant on the DCE's data format specifications.

Parity: *Even, Odd, None, Mark, Space*
This specifies the type of parity checking that should be used. One of the parity settings: Even, Odd, or None, can be selected from the pull-down menu. Usually, Mark and Space parities are not used because the parity bit is always set to one with the Mark parity and zero with the Space parity.

Stop bits: *1, 1.5, 2 bit(s)*
Here the length of the stop bit can be chosen from the pull-down menu.

Flow control: *Xon/Xoff, Hardware, None*
Serial data communication can be implemented with three separate cable lines for *Transmitted Data* (TD), *Received Data* (RD), and *Signal Ground* (SG). However, when the processing speeds of devices are different, the receiving device may not be able to handle all of the incoming data, resulting in data loss. Thus, a data flow control scheme in general, needs to be implemented in software or hardware.   In the case of software implementation, the receiver will send "Xoff" character to pause the transmission and "Xon" character to resume the transmission. For three wire serial connections typically the setting *None* should be chosen.

Chapter 16. Module 4 – Operational Amplifiers (Op amps)

16.1 Operational Amplifier Theory

Objectives

This module is designed to introduce you to the theory needed to utilize operational amplifiers (op amps) in circuits. The module is designed to supply you with enough information to use op amps as building blocks in the circuits you will be constructing in the exercises that follow.

Introduction

An operational amplifier is an integrated circuit (IC) consisting of transistors, resistors, diodes, and other components packaged on a silicon chip. It gets its name from the fact that it can be used to perform mathematical operations such as addition, subtraction, integration and also amplification of current and voltage, hence operational amplifier. In the lab, the op amps that you will typically use for prototyping are packaged in a dual inline package, also called a DIP for short. A typical pinout for a standard op amp is shown in Figure 16.1.1, and Table 16.1.1 (*National Semiconductor LM741*).

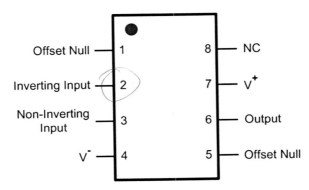

Figure 16.1.1: Typical op amp pinout (LM741)

Table 16.1.1: List of typical op amp pinouts (LM741)

Pins 1 and 5	Offset nulls	Can be biased to help null voltage offsets
Pin 2	Inverting input	When used as an input, it produces an inverted signal at the op amp output. (or 180^0 out of phase with input)
Pin 3	Non-inverting input	When used as an input, it produces a non-inverted signal at the op amp output. (or in phase with input)
Pin 4	V^-	Connection for the negative power supply, this typically requires (-12) to (-15) volts
Pin 6	Output	Signal output
Pin 7	V^+	Connection for the positive power supply, this typically requiresr (12) to (15) volts
Pin 8	NC	Not used

The op amp is a useful, inexpensive device, which can be configured for multiple tasks by simply biasing it properly with a few resistors and capacitors. It is the basic building block for many more complex circuits. All op amps require a dc power source to function. The dc source supplies the energy, or power, that is needed for signal amplification. Typically op amps require dual dc sources one positive and one negative. It is important to note that the amount of gain that an op amp can produce is limited by its power source. For example; an op amp powered by a dual ± 12 volt supply cannot amplify a signal to exceed the ± 12 volt limits. Any output larger than the ± 12 volt limit would be clipped, or experience amplifier saturation.

The internals of an op amp are designed in such a way that its output voltage v_{out} will be equal to the difference between its input voltages v_1 and v_2, at the inverting and non-inverting input terminals, multiplied by a constant term A. Mathematically we can express this relationship as;

$$v_{out} = A(v_2 - v_1)$$

Here v_{out}, v_2, and v_1 are defined in Figure 16.1.2. The constant 'A' is classified as the open loop gain (i.e. with no feedback network). The open-loop gain is typically on the order 10^5 or greater in magnitude, so often it is approximated to be an infinite gain.

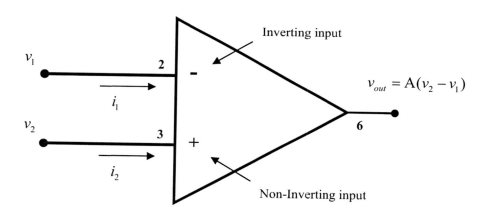

Figure 16.1.2: *Ideal* Op amp

For the *Ideal* Op Amp, with an infinite open-loop gain, four assumptions are made that greatly simplify the analysis of op amp circuits:

1. The current i_1 = 0 at the inverting input terminal.
2. The current i_2 = 0 at the non-inverting input terminal.

- Another way to express assumptions 1 and 2 is to say that the input impedance of an ideal op amp is infinite.

3. The output voltage of the op amp is <u>independent</u> of the current drawn by a load.

    •   Assumption 3 is also a statement that the output impedance of an ideal op amp is zero.

4. The differential input is governed by $(v_2 - v_1) \cong 0$

    •   To justify assumption 4, start with the definition of the operational amplifier's open-loop gain, and rearrange it as:

$$v_2 - v_1 = \frac{v_{out}}{A}$$

Since the open-loop gain is assumed infinite for an ideal amplifier, the following must be true.

$$v_2 - v_1 \cong 0$$

From this statement notice then it must all so be true that

$$v_2 \cong v_1.$$

To summarize, for an *Ideal* op amp; where the open loop-gain " A " is assumed to be infinite;

$$i_1 = i_2 = 0$$

$$v_2 - v_1 \cong 0$$

$$v_2 \cong v_1.$$

It is recommended that you commit these *Ideal* op amp assumptions to memory, most op amp circuits you will encounter can be analyzed using them.

16.2 Operational Amplifier Circuits

Objectives

This module is designed to present you with the configuration details and gain equations governing some commonly encountered operational amplifier circuits. Pay close attention to the biasing of the inverting and non-inverting terminals, and under what circumstances the *ideal* op amp assumptions are applied.

A. Inverting Amplifier

An inverting amplifier configuration is pictured in Figure 16.2.1, Recall inverting means the output signal will be 180^0 out of phase with respect to the input signal or of opposite polarity, dependant on the input signal being either an AC or DC type.

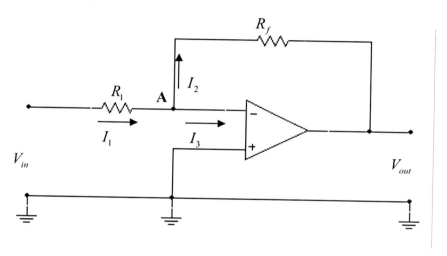

Figure 16.2.1: Inverting Op Amp Configuration

The circuit behavior can be determined by applying KCL and applying the *ideal* op amp assumptions at node "**A**".

$$I_1 = I_2 + I_3$$

From the assumptions the current I_3 = 0 for an *Ideal* op amp, then using ohm's law,

$$I_1 = \frac{V_{in} - V_A}{R_1}, I_2 = \frac{V_A - V_{out}}{R_f}$$

Note: V_A is the voltage at node "**A**" with respect to ground.

it follows

$$\frac{V_{in} - V_A}{R_1} = \frac{V_A - V_{out}}{R_f}.$$

Notice the non-inverting input is grounded, it follows V_A and must also be equal to zero which leaves,

$$\frac{V_{in}}{R_1} = \frac{-V_{out}}{R_f}$$

$$V_{out} = -\frac{R_f}{R_1} V_{in}.$$

Then the closed-loop gain, call it " G ", for the inverting op amp is simply

$$G = -\frac{R_f}{R_1}.$$

B. Non-Inverting Amplifier

A non-inverting amplifier configuration is pictured in Figure 16.2.2. Recall that non-inverting means the output signal will be in phase with respect to the input signal or of same polarity.

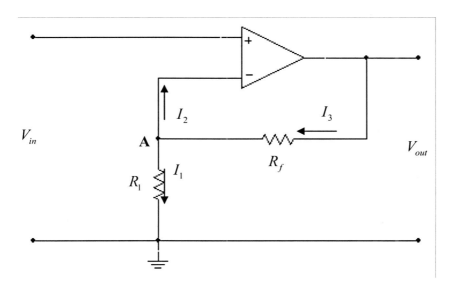

Figure 16.2.2: Non-inverting Op Amp Configuration

As with the inverting configuration, begin by applying KCL and the op amp assumptions at node **"A"**.

$$I_1 = I_2 + I_3$$

Once again $I_2 = 0$, then recasting I_1 and I_3 using ohm's law gives,

$$I_1 = \frac{V_A}{R_1}, \; I_3 = \frac{V_{out} - V_A}{R_f}$$

and,

$$\frac{V_A}{R_1} = \frac{V_{out} - V_A}{R_f}.$$

For the *Ideal* op amp, recall the terminal voltages are assumed equal, so that $V_A = V_{in}$, it then follows,

$$\frac{V_{in}}{R_1} = \frac{V_{out} - V_{in}}{R_f}$$

$$\frac{R_f}{R_1} = \frac{V_{out} - V_{in}}{V_{in}} = \frac{V_{out}}{V_{in}} - 1$$

$$V_{out} = \left(1 + \frac{R_f}{R_1}\right) V_{in}$$

The closed-loop gain, "G", for the non-inverting op amp is

$$G = 1 + \frac{R_f}{R_1}.$$

C. *Voltage-follower Op Amp Configuration*

A voltage-follower, or buffer amplifier, configuration is pictured in Figure 16.2.3.

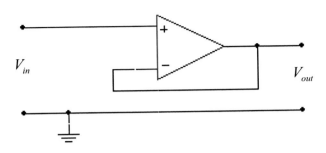

Figure 16.2.3: Voltage-follower Op Amp Configuration

Returning to the standard analysis techniques and assumptions, the voltage at the inverting input terminal is V_{out}, which leads to,

$$V_{in} = V_{out}$$

$$G = 1.$$

Another way of viewing the circuit is to notice that the voltage follower is nothing more than a non-inverting op amp configuration with the added requirement,

$$R_f = 0.$$

It then follows that

$$V_{out} = \left(1 + \frac{0}{R_1}\right) V_{in}$$

$$V_{out} = V_{in}$$

$$G = 1$$

This configuration is often used as a one-to-one buffer in circuits, especially where impedance matching or isolation between two components is necessary. It is also the building block of a simple signal attenuator.

D. Non-inverting Op Amp Attenuator

A non-inverting attenuator configuration is pictured in Figure 16.2.4.

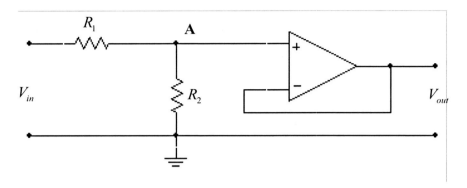

Figure 16.2.4: Non-inverting Op Amp Attenuator

From inspection, notice this configuration is simply a standard voltage divider connected to the non-inverting input of a voltage follower amplifier. The voltage V_A is expressed as

$$V_A = \left(\frac{R_2}{R_1 + R_2} \right) V_{in}.$$

It follows, from the op amp assumptions, that the voltage at the inverting input is equal to V_A or,

$$V_{out} = V_A$$

$$V_{out} = \left(\frac{R_2}{R_1 + R_2} \right) V_{in}$$

$$G = \frac{R_2}{R_1 + R_2}.$$

Therefore, by carefully choosing the appropriate values for R_1 and R_2 the circuit acts to attenuate the input.

E. Op Amp Current to Voltage Amplifier

An op amp current to voltage amplifier configuration is pictured in Figure 16.2.5.

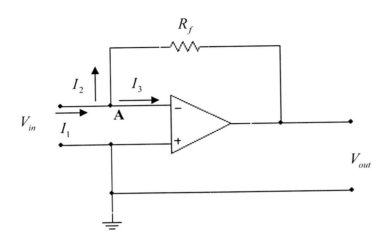

Figure 16.2.5: Current to Voltage Amplifier

From inspection, at node "**A**", KCL can be used to write

$$I_1 = I_2 + I_3.$$

But again from the op amp assumptions I_3 = 0, it then follows $I_1 = I_2$.
It then follows that

$$I_1 = \frac{V_A - V_{out}}{R_f}$$

Applying the op amp assumptions once more, V_A = 0, produces

$$I_1 = -\frac{V_{out}}{R_f}$$

$$V_{out} = -I_1 R_f.$$

In conclusion, this configuration produces a output voltage that is proportional to the product of an input current I_1 and the feedback resistance R_f.

F. The Adder or Summing Amplifier

An op amp summing amplifier configuration is pictured in Figure 16.2.6.

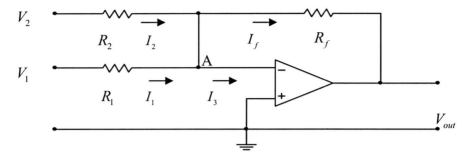

Figure 16.2.6: Adder or Summing Amplifier

From inspection, at node **"A"**, KCL can be used to write

$$I_1 + I_2 - I_f = I_3$$
$$I_1 + I_2 = I_3 + I_f$$

From the op amp assumptions I_3 = 0, it then follows that

$$I_1 + I_2 = I_f.$$

Using Ohm's law the expression can be recast as

$$I_1 = \frac{V_1 - V_A}{R_1}, \; I_2 = \frac{V_2 - V_A}{R_2}, \; I_f = \frac{V_A - V_{out}}{R_f}.$$

Note: V_A = 0, as a result of the non-inverting input being grounded.

$$-\frac{V_{out}}{R_f} = \frac{V_1}{R_1} + \frac{V_2}{R_2} \;\; \text{Or} \;\; V_{out} = -\frac{R_f}{R_1}V_1 - \frac{R_f}{R_2}V_2.$$

This configuration can be extended as,

$$V_{out} = -\frac{R_f}{R_1}V_1 - \frac{R_f}{R_2}V_2 - \text{.........} - \frac{R_f}{R_n}V_n.$$

Notice if $R_1 = R_2 = ...R_n = R_f$, then

$$V_{out} = -(V_1 + V_2 + \text{.........} + V_n).$$

Chapter 17.   Module 5 – Light Emitting and Photodiodes

17.1 Diodes

Objectives

This module introduces diodes (pn junctions) including the general purpose silicon, light emitting, and photodiode. In section 17.2 (optional) the characterization of semiconductor materials and a more detailed explanation of the pn junction is provided. The material in section 17.2 is for the student wishing to have a clearer understanding of the diode.

A. Diode (pn junction)

A diode is a device that has two possible operational states. In state one it acts as an open switch. It will not support a current and has no charge flow. In state two, it acts as a closed switch. It will support a current, and charge is allowed to flow. Formally, a diode is a device that supports a current in one direction but not the other. The diodes behavior is determined by its connection and how it is biased in a circuit, which will be discussed shortly. Figure 17.1.1 shows the schematic symbol and pictures of some general purpose diodes.

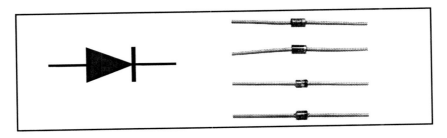

Figure 17.1.1: Common general purpose diode

The terminal on the left side of the schematic symbol, (arrowhead side), is referred to as the anode, or positive side. The right side, (vertical line side), is defined as the cathode or negative side.

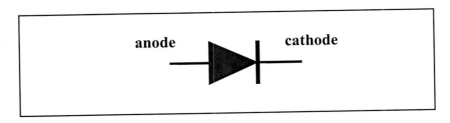

Figure 17.1.2: diode anode and cathode

A diode usually includes a band that identifies the cathode, as seen in Figure 17.1.1.

Diodes are made from special materials called semiconductors, which are unique in that they can behave as both an insulator and conductor. Typically, the semiconductors used to make diodes are classified as either being a p-type or n-type material. A p-type material is fashioned such that it has excess positive charge carriers, or in other words an abundance of positive charge that can be used for conduction. An n-type material, on the other hand, has an excess of negative charge carriers, or an abundance of negative charge available for conduction. A diode is created by joining a p-type and n-type material together as depicted in Figure 17.1.3. Note that the leads on the left and right side allow for connections to the external world.

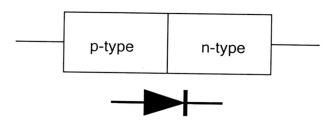

Figure 17.1.3: Simple diode from combining an n-type and p-type material

When the materials are joined, a special region, called the depletion region, sets up between them. As a result of the formation, a small *built-in voltage* V_{bi} sets up across the junction as depicted in Figure 17.1.4.

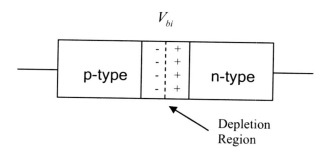

Figure 17.1.4: pn junction depicting depletion region and resulting built-in voltage V_{bi}

The depletion region forms and the built-in voltage manifests because of charge migration. Specifically, on the p-side there is a build up of *fixed negative charge*, and on the n-side a build up of *fixed positive charge* close to the joining junction. By fixed, it is implied that once the charges are in place they are not free to move, thus establishing a permanent built-in voltage.

The magnitude of V_{bi} is a function of the material used to make the diode. Usually silicon is used for a general purpose diode and the built-in voltage is approximately equal to 0.6 volts. This is also known as the turn-on voltage, because the diode will not conduct, in a forward bias configuration, until the applied voltage reaches or exceeds V_{bi}.

Diode Biasing

We mentioned a diode's ability to conduct is a function of how it is biased. We now address the connections and apply two different biasing schemes to the diode and characterize the resulting behavior.

1. Forward bias: A forward bias configuration is accomplished by connecting a positive source to the p-side (anode) of the pn junction and a negative source to the n-side (cathode). This scheme results in an applied voltage V_A across the depletion region that opposes the built-in voltage V_{bi}, notice the polarity of V_{bi} is opposite the applied voltage polarity, as shown in Figure 17.1.5.

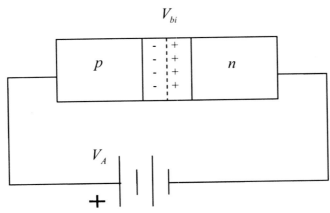

Figure 17.1.5: forward biased pn junction

Due to the built-in voltage, charge will not flow, until the condition $V_A - V_{bi} > 0$ is met. This means charge will not move from the p-side to the n-side, as a result of V_A, until the applied voltage is larger than the internal built-in voltage. This is why the built-in voltage is commonly called the turn-on voltage. Once the condition is met the diode is said to be on and current will pass. The schematic representation is shown in Figure 17.1.6.

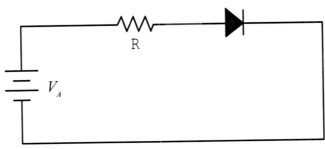

Figure 17.1.6: Forward biased diode, assuming V_A is larger than the turn-on voltage, current will pass. Notice the current is approximately V_A/R.

2. **Reversed bias:** The opposite scheme or *reversed bias* configuration is shown in Figure 17.1.7.

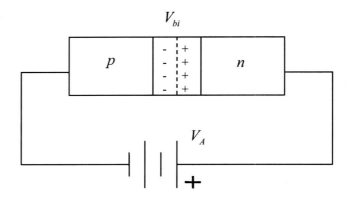

Figure 17.1.7: reversed biased pn junction

In the reverse bias configuration the applied voltage V_A acts to re-enforce the built-in voltage V_{bi}. Recall that the built-in voltage V_{bi} is the voltage that opposes charge flow in the junction. The net result is that the turn-on voltage is increased and no current is created resulting from the biasing. In this configuration the diode is said to be off, allowing no net current to pass. The schematic representation is shown in Figure 17.1.8.

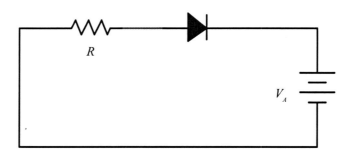

Figure 17.1.8: Reversed biased diode configuration, notice the current in this case is V_A/R, and always equal to zero as a result of the biasing.

B. Light Emitting Diode

A Light Emitting Diode (LED) is nothing more than a simple pn junction diode designed to produce visible light when it is forward biased. Several LEDs are pictured in Figure 17.1.9, along with their schematic symbol.

Figure 17.1.9: Light Emitting Diodes and schematic symbol

To explain the operation of an LED one must consider the atomic structure of a semiconductor material. A semiconductor has two energy bands defined as the conduction band and non-conduction (valance) band. The conduction band occupancy, or lack thereof, determines the insulating or conducting behavior of the material. In addition, an energy gap, more commonly referred to as a band gap, exists between the conduction and valance bands. This gap must be crossed by charge moving between bands. The bands are identified in Figure 17.1.10.

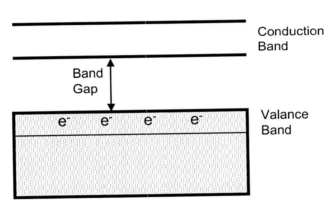

Figure 17.1.10: Energy band model representing the valance and conduction bands, and identifying the energy separation, band gap, between the two.

The valance band is where the material's electrons reside when they are not available for conduction. A semiconductor with all of it's electrons in this band acts as an insulator. The conduction band is where the electrons reside when they are available for conduction. Proper forward-biasing of a pn junction allows electrons to move from the valance band to the conduction band; in effect the electron has been supplied energy equal to the energy needed to make it across the band gap into the conduction band. Over time the electrons that originally left the valance band will return. When this occurs the electron moves from a higher energy state to a lower state. This means the electron must give up the energy that it gained transversing the band gap in the first place. The release of energy is accomplished via an emission of electromagnetic radiation. This process occurs in all pn junctions and is formally referred to as the Generation-Recombination process. For an LED the band gap is chosen so that the emission of electromagnetic radiation is in the visible light, or infrared, spectrum. This is accomplished by choosing the right material and doping agents that make up the semiconductor. Of course the light must also have a clear path to escape, so the pn junctions are packaged in a transparent material.

The emission wavelength λ can be determined from the equation,

$$\lambda = \frac{1.24}{E_{Gap}}$$

where λ is in μm and the Gap Energy is in electron volts eV . Energies are typically reported in the units of electron volts, $1\ eV = 1.6 \times 10^{-19}\ Joules$. Some common values are listed in Table 17.1.1. Note the materials can be combined in different concentrations to achieve other transmission wavelengths.

Table 17.1.1: Semiconductor materials

Material	E_{Gap} (eV)	λ (μm)	λ (nm)
Silicon (Si)	1.1	1.127	1127
Gallium phosphate (GaP)	2.2	0.563	563
Gallium arsenide (GaAs)	1.4	0.886	866
Gallium-aluminum phosphate	0.8-2.0	1.55-0.620	1550-620

    A simple LED circuit is shown in Figure 17.1.11. The resistor is used to limit the current to prevent burn out of the device. The LED's current and voltage limitations are provided by the manufacturer. A combination of voltage and resistance values are used to set the diodes operating current, typically referred to as the diodes forward current I_f . It is important to note the LED can also be driven by an alternating source, such as a square wave. This makes it a useful device in communication systems.

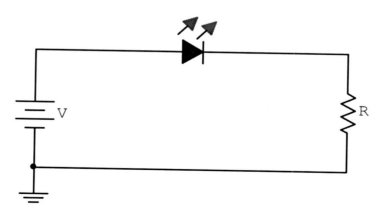

Figure 17.1.11: A simple DC Light Emitting Diode circuit

C. Photodiode

A photodiode can be viewed as a Light Emitting Diode working in reverse. Technically a photodiode is a device that converts incident light energy into an electrical current. Several photodiodes are pictured in Figure 17.1.12, along with their schematic symbol.

Figure 17.1.12: Photodiodes and schematic symbol

The operation of a photodiode relies on the bias, or built-in voltage, in the depletion region. Some photodiodes allow you to apply an additional reverse bias voltage which creates an even larger voltage, or E-field, to exist across the depletion region. Recall in a reverse bias configuration the applied voltage acts to reinforce the pn junction built-in voltage. One characteristic of the depletion region is that, if we drop an electron into or near it, the electron will be swept to the n-side of the device. This occurs because the internal built-in voltage produces an electric field (E-field) across the depletion region due to the fixed charge distribution. All electrical fields have a force field associated with them. This force is directly proportional to the magnitude of the voltage across the junction. It follows that the larger the applied reverse biased voltage, the larger the force across the depletion region will be. If you drop a few free electrons into this region a current will manifest. These free electrons in the case of the photodiode are produced by the absorption of incident light energy. The photodiode by design allows incoming radiation to strike the pn material. When the light, or photons, strike the material, they can transfer their energy by being absorbed by the materials electrons. Note, this energy transfer can only occur if the semiconductors have been designed to have a band gap that matches the energy the photon carries, which happens to be unique for different wavelengths in the electromagnetic spectrum. When energy is transferred, electrons sent into the conduction band and those near, or that migrate near, the depletion region are acted upon by the electric field. The end result is a measurable current that arises in the material which is proportional to the light (number of photons) absorbed.

Since photodiodes produce current, a useful circuit configuration must include a component capable of converting current to voltage. One method to accomplish this is to implement an op amp current to voltage amplifier. A simple photodiode detection circuit using a current to voltage op amp configuration is pictured in Figure 17.1.13.

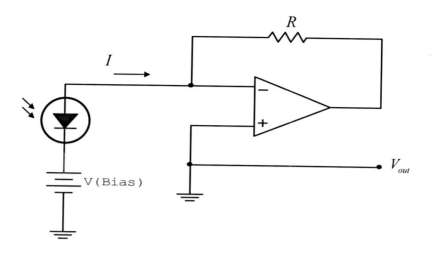

Figure 17.1.13: Biased photodiode detection circuit, note the bias voltage can be zero, then the cathode is connected directly to ground and the set up is referred to as a non-biased photodiode detection circuit.

17.2 PN Junctions (Optional)

A. Semiconductor model

We begin by introducing some basic properties of semiconductor materials, which are the building block of the pn junction. A semiconductor is unique, in that it demonstrates characteristics of both conductors and insulators. A conductor has electrons available for conduction under the influence of an electrical field, and an insulator does not conduct under the influence of an electrical field. A semiconductor is capable of functioning as both. It acts as a conductor, when an electrical field is applied, if certain criteria are met. If these criteria are not met, a semiconductor behaves as an insulator. This behavior is a direct result of a semiconductors atomic structure.

Examples of semiconductor materials are silicon and germanium. Of the two, silicon is by far the most popular. Examining silicon's molecular structure reveals that it has 14 electrons, 10 of which are strongly bounded inner, or core, electrons and 4 weakly bound outer, or valance, electrons. These weakly bound electrons are the electrons available for conduction when certain conditions are met. A simple energy band model of silicon identifying the valance and conduction bands is shown in Figure 17.2.1.

In Figure 17.2.1, the valance and conduction energy levels are depicted by $E_{valance}$ and $E_{conduction}$. Below the valance energy level is where the electrons, represented by e⁻, spend most of their time under equilibrium conditions. In this state silicon acts as an insulator. However, if enough energy, thermal, electric, or some other form, is imparted to the electrons, they can be excited to the conduction band. Once an electron resides in the conduction band it is classified as unbounded, free to move, and available for conduction. In this state, silicon acts as a conductor. The specific energy needed to move an electron from the valance to the conduction band is equal to ($E_{conduction}$ - $E_{valance}$), commonly defined as the material's band gap or energy gap E_{gap}, which is a tabulated quantity for such materials. Of course a sufficient external electrical field must be applied for conduction to occur, once free electrons reside in the conduction band.

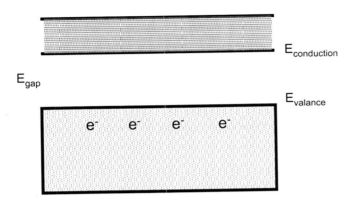

Figure 18.2.1: Energy band model representing the valance and conduction energy levels, and identifying the energy separation E$_{gap}$ between the two.

B. Semiconductor charge concentrations

A pure semiconductor, free of impurities, is classified as an intrinsic material. In an intrinsic semiconductor charge carriers, negative and positive charge, exist in pairs due to charge neutrality. It follows if we define "n" to represent the number of electrons and "p" to represent the number of positive charge we can write

$$n_i = p = n .$$

Here n_i represents the intrinsic electron concentration which is a tabulated quantity for materials.

A semiconductor that has had impurities added is classified as an extrinsic material. The process of adding impurities to semiconductor material is termed doping the material.

In an extrinsic semiconductor the impurities are added to produce a charge imbalance or an excess of a particular charge. The resulting charge distribution is described by,

$$n \neq p \neq n_i .$$

A semiconductor doped with excess negative charge is referred to as an n-type material. A p-type material is a semiconductor with excess positive charge. The process of creating an n-type material produces a new energy level in the energy band diagram called the Fermi Energy level, as pictured in Figure 17.2.2.

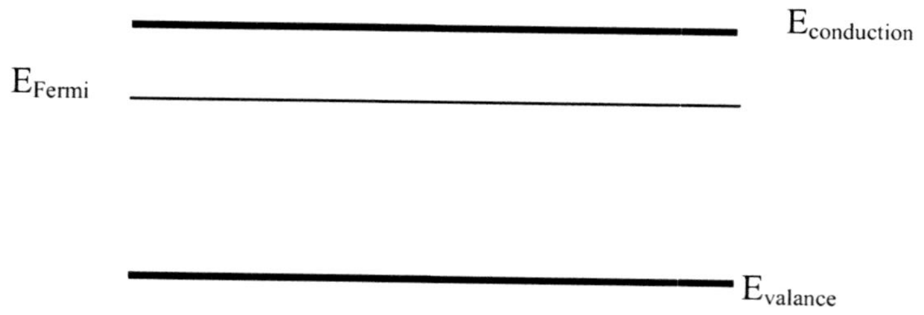

$$E_{conduction}$$

$$E_{Fermi}$$

$$E_{valance}$$

Figure 17.2.2: n-type material and Fermi Energy level.

The Fermi level conceptually represents a reduction in the energy gap, as seen by the electrons, which must be overcome to reach the conduction band. The result is a larger number of charge carriers will be available for conduction if, the required energy is provided.

C. Semiconductor charge transport mechanisms

There are four basic types of charge transport taking place in a semiconductor. They are; Thermal motion, Drift, Diffusion, and Generation-Recombination. Note: The drift and diffusion currents are typically the major contributors to any measurable net current in a material.

Thermal motion: is due to the random processes associated with collisions and general movement of particles inside of the material.

Drift currents: arise from charged particle motion that occurs as a direct result of an applied electrical field.

Diffusion currents: arise from localized charge concentrations. For example, put a lot of negative or positive charge together in a small space and they will tend to move apart.

Generation-Recombination: is usually related to thermal or photo generation. When an electron has been provided with a sufficient energy to leave its energy state, a free electron and stationary hole pair are created. Generation has occurred. At some later time, the electron will return to its equilibrium state, thus, the electron and hole cease to exist as separate entities and Recombination has occurred.

Summarizing: The most common semiconductor material is silicon. An intrinsic semiconductor is a pure semiconductor with no impurities. An extrinsic semiconductor is a semiconductor that has been doped. Dopants are impurities added to increase negative and/or positive charge concentrations in a material. An extrinsic semiconductor's properties are determined by its' excess carriers, electrons, for an n-type material, or holes for a p-type material. As a result of doping, the energy gap that charge must overcome to get into the conduction band is reduced, which is represented by a new energy level termed the Fermi level. The net result of doping is that higher charge concentrations can be made available for conduction. Thermal motion, Drift, Diffusion, and Generation-Recombination are the four basic types of charge transport taking place in a semiconductor.

D. The pn junction

Now that you have a basic understanding of semiconductor material we proceed to the simple pn junction. A pn junction is created by bringing together a p-type and n-type semiconductor material. Recall that when we refer to a p-type or n-type, it is implied that the material has been doped to have either excess positive or negative carriers. Before looking at the details of a pn junction, it is important to define a few simplifying assumptions. First, we assume equilibrium conditions meaning; no applied voltages, no light is illuminating the device, and there are no thermal gradients (implying uniform temperature). Second, we assume no external electrical or magnetic fields are being applied. With the assumptions in place, we can represent the pn junction as in Figure 17.2.3.

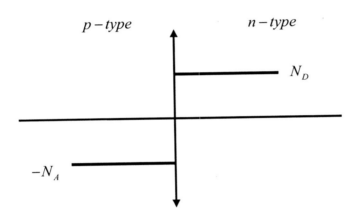

Figure 17.2.3: pn junction and charge concentrations

In Figure 17.2.3 N_A and N_D signify the excess charge concentrations that occurred from the doping process. Here $-N_A$ represents the excess positive charge, commonly referred to as the acceptor concentration and N_D represents the excess negative charge, or the donor concentration of the junction material. Under equilibrium conditions these are equivalent to the hole, positive charge p, and electron n concentrations. In the ideal case, all positive charge would reside on the p-side and all electrons on the n-side. However, in reality, once the junction is formed a diffusion current appears. As a result, some electrons migrate to the p-side and some holes migrate to the n-side. Figure 17.2.4 represents the redistribution of charge once the $p-type$ and $n-type$ materials are joined.

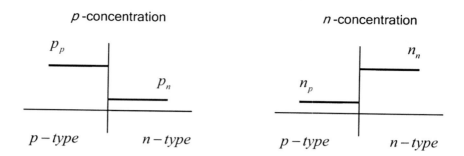

p-concentration   n-concentration

Figure 17.2.4: charge concentrations

The subscripts denote which side of the junction the concentration is on. For example p_p stands for "holes on the p-side" while p_n stands for "holes on the n-side". Due to the migration of charges across the junction, a non-uniform distribution is set up in the central region of the junction. This region is referred to as the depletion region. This region is graphically depicted in Figure 17.2.5.

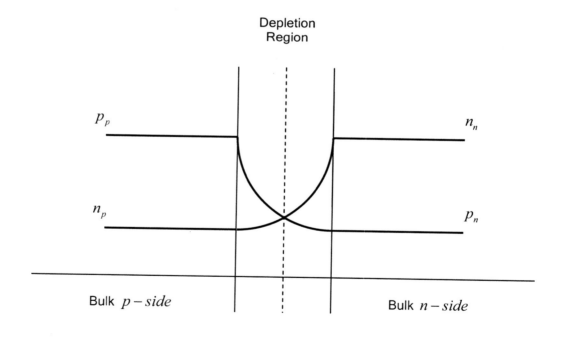

Figure 17.2.5: illustration of depletion region occurring due to charge redistribution

As one may suspect, the diffusion will not continue until bulk charge neutrality is reached. The reason being, the excess charges produced by doping are free charges, free from molecular bonds, to move about in the atomic lattice. However, for each hole that leaves the p-side, left behind is a fixed ionized acceptor site. In addition, each electron that leaves the n-side leaves behind a fixed donor site. The acceptor and donor sites are fixed in the atomic lattice, not free to move. The end result is a negative charge distribution builds up on the p-side and a positive charge distribution builds up on the n-side. A resulting localized electric field is realized in the depletion region as pictured in Figure 17.2.6.

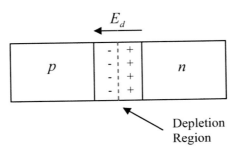

Figure 17.2.6: pn junction depicting depletion region and resulting electric field E_d

The electric field E_d gives rise to a so called, "built-in voltage" across the depletion region " V_{bi} ". The equation used to determine the numerical value for the built-in voltage is,

$$V_{bi} = \frac{KT}{q} \ln \left[\frac{N_D N_A}{n_i^2} \right].$$

Here K is the Boltzmann constant, T is the temperature in Kelvin, and q is the fundamental charge, at room temperature $KT/q \cong 0.026$ Volts , N_D and N_A are the number of donor and acceptor sites, and n_i is the intrinsic electron concentration of the material. For a more detailed treatment and derivation of V_{bi} , see Appendix D.

For silicon, at room temperature, $N_D = N_A = 10^{15}/cm^3$, $n_i = 10^{10}/cm^3$, calculating the V_{bi} gives,

$$V_{bi} = 0.026 \ln \left[10^{10} \right] \text{ volts}$$
$$= 0.599 \text{ volts}$$

If you are not familiar with this number you soon will be, it is the characteristic turn-on voltage of a silicon diode.

In summary: A simple pn junction can be formed by bringing together a $p-type$ and $n-type$ semiconductor. When the pn junction is created the initial difference in positive and negative charge concentrations leads to carrier diffusion. This diffusion does not continue until charge balance is reached, due to the fact the donor and acceptor sites left behind are fixed in the atomic lattice. This leads to a non uniform charge distribution across the adjoining junction where a localized internal electric field appears. This region is termed the depletion region. The depletion region has a so called "built-in voltage" resulting from the internal electric field which is a function of the junction material and doping concentrations.

Chapter 18. Module 6 – Fiber Optic Communications Link

18.1 Light Reflection and Refraction

Objectives

This module is designed to introduce a fiber optics communications link. The approach is to first introduce and characterize light reflections, refractions, and then the concept of total internal reflection. This information provides the foundation for understanding how light propagates in a fiber optic cable.

Introduction

Light is a part of the electromagnetic spectrum. The portion of the spectrum that the human eye is sensitive to ranges from approximately 400 to 700 nm. As shown in Figure 18.1.1.

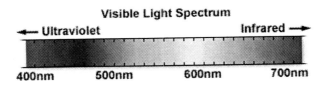

Figure 18.1.1: Visible light spectrum

Light is an interesting phenomena, it exhibits properties of both waves and particles. The tools used to study and characterize light depend on what particular aspect of its behavior you are interested in. For example, are you investigating position, image formation, light propagation, interference, diffraction, or polarization? For characterizing light transmissions in fiber, as a first approximation, we rely on the tools of Ray Optics. Ray Optics is the simplest model of light. This model allows for the treatment of light geometrically and is useful for describing the location and direction of light rays. In fiber optic communication systems, LEDs and lasers are typically used to transmit information. By design these devices transmit in the visible and infrared portions of the spectrum. The transmissions are guided from source to receiver via fiber optic cables. The fiber optic cable acts as a waveguide, much the same as the coaxial cable does for electrical transmissions. To understand how optical waveguides work, we must first discuss the reflection and refraction of light, and the role they play in the optical waveguide.

A. Reflection of Light

To begin we define how light behaves when it is reflected from a smooth surface. The Law of Reflection states: when a ray of light is incident on an interface, of two uniform media, all reflected rays will be reflected at an angle equal to the incident light ray angle.

The angles are measured relative to a defined normal line which is taken perpendicular to the reflecting surface. Mathematically the law of reflection it is expressed as,

$$\theta_{in} = \theta_{reflected} \, .$$

The angles θ_{in} and $\theta_{reflected}$ and the surface normal are defined in Figure 18.1.2.

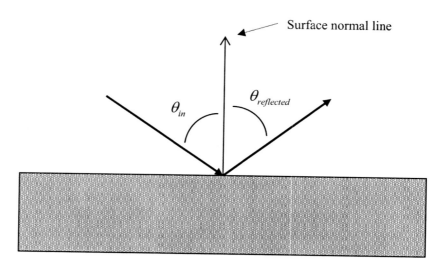

Figure 18.1.2: Law of Reflection

B. Refraction of Light

Next, consider what happens when light passes through an interface. The law of refraction, Snell's Law, states: when a ray of light is incident on an interface of two uniform media, the transmitted ray will be refracted; the amount of refraction is related to the incident ray by the equation,

$$n_1 \sin\left(\theta_{in}\right) = n_2 \sin\left(\theta_{refracted}\right).$$

Again the angles are with respect to the normal line of the interface. In the equation, n_1 and n_2 are tabulated values, also known as a material's *index of refraction*. A material's index of refraction gives the ratio of the speed of light in a vacuum, $c \approx 3.0x10^8$ m/s, to the speed of light in the material v, which happens to be wavelength dependant. This relationship is mathematically expressed as,

$$n_i = \frac{c}{v_i} \, .$$

For more information on the index of refraction and some common values see Appendix G.

Visually the refraction of a light ray is depicted in Figure 18.1.3; (here we are ignoring any reflected rays for convenience).

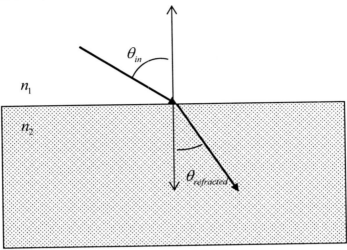

Figure 18.1.3: Snell's Law, for $n_1 < n_2$.

C. Total Internal Reflection

Notice in Figure 18.1.3, that the incident angle was larger than the refracted angle. This is always the case when light passes from a media of lower index of refraction to a medium of a higher index. This behavior can be explained using Snell's law,

$$n_1 \sin\left(\theta_{in}\right) = n_2 \sin\left(\theta_{refracted}\right).$$

When $n_1 < n_2$,

$$\frac{n_1}{n_2} \propto \frac{\theta_{refracted}}{\theta_{in}} < 1.$$

It follows that the refracted angle is less than the incident angle. On the other hand, when $n_1 > n_2$ the refracted angle is larger than the incident angle,

$$\frac{n_1}{n_2} \propto \frac{\theta_{refracted}}{\theta_{in}} > 1$$

The latter situation is pictured in Figure 18.1.4; again we are ignoring any reflected rays for convenience.

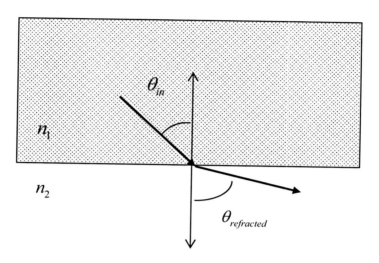

Figure 18.1.4: Snell's law, for $n_1 > n_2$.

When $n_1 > n_2$, it is possible to increase the incident angle to the point that all the light rays are reflected. That is to say, total internal reflection occurs at some angle θ_{in}. This angle has a special name, it is denoted as the *Critical angle* or θ_c. Simply stated there is a critical angle, once surpassed by the incident ray, which will cause all the light to be totally internally reflected, provided the light is passing from a higher index to a lower index material. The critical angle is found by setting the refracted angle, in Snell's law, equal to 90°. This is the angle at which the refracted ray is parallel to, or along, the surface of the interface.

$$n_1 \sin(\theta_{in}) = n_2 \sin(\theta_{refracted})$$

$$set\ \theta_{refracted} = 90°$$

$$and\ \theta_{in} = \theta_c$$

$$n_1 \sin(\theta_c) = n_2$$

$$\theta_c = \sin^{-1}\left(\frac{n_2}{n_1}\right)$$

It follows for a given ration of $n_2/n_1 < 1$ an angle can be defined so that all light will be internally reflected and none will pass into the material of the lower index of refraction. This is exactly the behavior one would expect a waveguide to exhibit.

18.2 Fiber Optic Cable and Connectors

A. Fiber Optic Cable

There are two basic types of optical fiber, multimode and single-mode. Fiber is available as both glass and plastic cable. The materials used to manufacture fiber are constructed to exploit the phenomena of total internal reflection. Fiber is generally cylindrical in shape and contains an inner core and outer cladding as shown in Figure 18.2.1. Typically the fiber also includes an outside protective coating or jacket. The fiber jacket protects it from external elements and plays no role in the fibers light guiding properties.

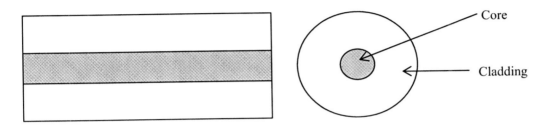

Figure 18.2.1: Optical fiber

By design, the cables core refractive index is made larger than the refractive index of the cladding. Therefore, sending a light signal into the fiber at an entrance angle less than or equal to the critical angle, smaller as a result of how θ_c is defined in Figure 18.2.2, results in the light rays experiencing total internal reflection. Theoretically, the entire signal entering the fiber will be guided to the fiber exit as pictured in Figure 18.2.2.

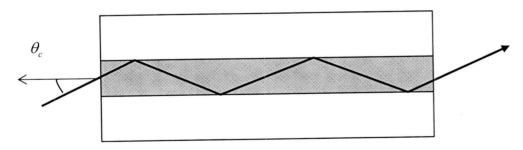

Figure 18.2.2: Total internal reflection in fiber.

The fiber's *acceptance angle* is defined as a cone around the fiber entrance that includes all angles at which total internal reflection will occur, commonly referred to in literature as the fibers numerical aperture denoted by N_A. The numerical aperture and formula are defined in Figure 18.2.3. Notice the numerical aperture is calculated using θ_c which is equal to half the full cone angle.

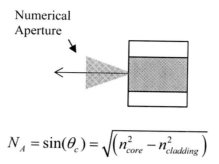

Numerical
Aperture

$$N_A = \sin(\theta_c) = \sqrt{\left(n^2_{core} - n^2_{cladding}\right)}$$

Figure 18.2.3: Fiber acceptance cone or numerical aperture

The numerical aperture is important because it provides a numeric measure of the fibers coupling requirements to external devices such as fiber transmitters and receivers. Recall that any light entering the fiber outside of the numerical aperture will be lost, passing through the cladding.

Multimode fiber typically has a core/cladding size of $62.5/125\mu m$ or $50/125\mu m$. Multimode is the least expensive type of fiber and is generally used for applications requiring less than 2 kilometers transmission length. Single-mode fiber typically has a core/cladding size of $8.3/125\mu m$ or $7.1/125\mu m$, it is more robust, and costly, than multimode fiber. Single-mode fiber provides data integrity over longer distances, and typical applications would be cable television, telephone and large area network point-to-point connections. The two basic fiber types can be further characterized by application specific requirements, but the selection process can be very in-depth and requires more knowledge than we have provided here. The goal of this section was to clarify the mechanism used to transmit light in a fiber.

B. Fiber Optic Cable Connectors

Cable can be purchased in bulk or in precut and assembled lengths known as patch cables or pigtails. The two most common connectors are the ST-type and SC-type. Patch cables come ready to use and are typically packaged for simplex (single cable) or duplex (twin pair) connections. Bulk cable is available in single and multi-strand. The two types are identified in Figure 18.2.4.

There are several tools available on the market designed for cutting fiber and attaching connectors. These tools can range in cost from as little as $10 to several thousand. The fiber application dictates how well the cable must be cut, polished, and aligned since fiber connections are the leading contributor to signal loss

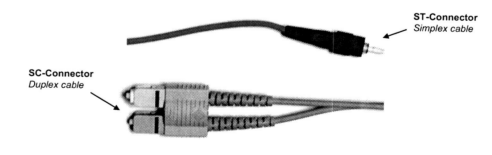

ST-Connector
Simplex cable

SC-Connector
Duplex cable

Figure 18.2.4: Common Fiber connectors, with simplex and duplex patch cable

Chapter 19.   Module 7 – Printed Circuit Board Design

19.1 Introduction to ExpressPCB Software

Objectives

This module will introduce you to basic printed circuit board (PCB) design using the software ExpressPCB. The module is intended to be very interactive. You should have ExpressPCB running and be entering commands along with the examples.

Introduction

ExpressPCB is a free CAD program used for designing and laying out Printed Circuit Boards (PCBs). The program includes two Windows applications: ExpressSCH for drawing and simulating circuit schematics and ExpressPCB for designing printed circuit boards. The two applications can be used together or as stand-alone programs. For example, ExpressSCH can be used to generate and simulate a circuit schematic that can later be imported by ExpressPCB to produce a PCB layout. If you already have a schematic or know the circuit details you wish to design a PCB for, you can enter it directly into ExpressPCB and bypass using the ExpressSCH program.

This module will focus on printed circuit board design and discuss only the use of the software component ExpressPCB. To download ExpressPCB or for information on the software, including tutorials on using ExpressSCH, visit www.expresspcb.com.

ExpressPCB Design Window

The design window pictorially represents the PCB itself. The available design tools are located on the side and top toolbars as pictured in Figure 19.1.1

Summary Design Procedure using ExpressPCB

At this point you should have the ExpressPCB program running. Follow along and try the following.

The first step to a new design is to accurately define the size and shape of the board you wish to create. There are two methods to accomplish this;

1. Drag the corners of the yellow rectangle, which represents the board perimeter, to match the shape and the size of the board you wish to design. Note that the upper left corner is fixed at the coordinates (0, 0). Try it.
2. Enter the coordinates of the points by double-clicking at the corners on the yellow rectangle. Additional corners can be added by clicking on **Insert corner on board perimeter** (*icon 12*) and then clicking on the yellow line where you wish to place the new corner. Try it.

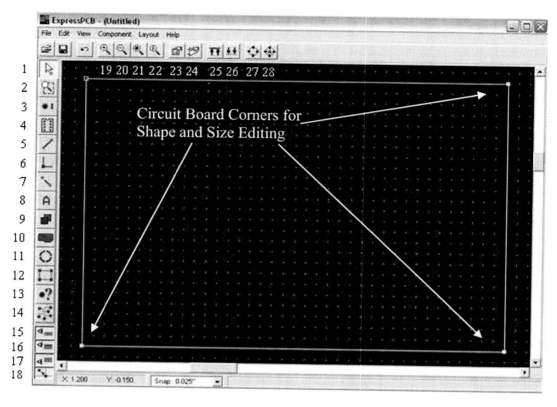

The Toolbar Icon Functions

1. Select, drag or set properties of an item in the layout
2. Zoom to view a selected area
3. Place a pad (either a through hole or surface mount pad)
4. Place a component
5. Place a trace
6. Insert a corner into a trace
7. Disconnect a trace
8. Place text in the layout
9. Place a rectangle
10. Place a circle or an arc
11. Place a filled power or ground plane
12. Insert an edge segment in the board perimeter
13. Display information about a pad
14. Highlight net connections

15. Toggle display of silkscreen layer
16. Toggle display of top copper layer
17. Toggle display of bottom copper layer
18. Toggle Snap-to-grid
19. Zoom in
20. Zoom out
21. Zoom to see all
22. Zoom to previous size
23. Display "options" dialog box
24. Display the component manager
25. Top layer
26. Bottom layer
27. Rotate 90° excluding text
28. Rotate 90° including text

Figure 19.1.1: Express PCB Design Window

Step two requires defining the board properties. On the menu select **Layout** then **Board properties**. The **Board properties** window should open. Note the default settings

- Number of copper layers = 2
- Default via = 0.059" round via with 0.029" hole
- Default clearance around traces (inches) = 0.012
- Default clearance around pads (inches) = 0.012

Click on the **Help setting these values...** button. Here you will find the information for each setting.

At this point, it would be a good time to name your file and save the settings. After closing the **Board properties** window go ahead and save your project using the **Save As** command in the **File** menu.

Step three is to select and place the circuit components (chips, resistors, capacitors...). Components can be selected from a list using the **Component Manager** (*icon 24*) or by right clicking in your workspace and selecting **Component Manager**.

Once you have opened the Component manager, select (click on) let's say the *Cap-0.1 uf axial ceramic, decoupling..*, and then click on **Insert component into PCB** button. Highlight the component and place it within the PCB perimeter. Once the component is in place, assign the component a **Part ID** by right clicking on the component, then selecting **Set component properties**. In the **Part ID** field enter C1 and click **OK**. Repeat the procedure but place a resistor on the board, assign the resistor a Part ID of R1.

Once all required components have been placed and assigned part IDs the final step is to layout and add the copper traces to your board.

To place a trace, first you must choose the board layer you want to place the trace on. The board layer views are selected using icons *15 (silkscreen layer)*, *16 (top copper layer)* or *17 (bottom copper layer)*. Depress icons 15 and 16, but unselect icon 17. This allows you to see the components and the top layer of the PCB. The silkscreen layer shows the components placed. It can be turned off for a better view of the board layers and through holes.

Select **Place a trace** (*icon 5*), then select the board layer you wish to place the trace on by selecting **Top copper layer** (*icon 25*) or the **Bottom copper layer** (*icon 26*). Now mouse click on one end point of R1, then click on one end point of C1, press the Spacebar or right click to complete the place trace operation. The default trace width is set to 0.010" suitable for most digital and analog signals. Double clicking on a trace will open, **Set properties of the selected trace** window. Go ahead and open the window and note the setting options. Read the **Help setting these values...** for setting information.

To create a corner in a trace requires the use of **Insert a corner in a trace** (*icon 6*). Select *icon 6* then click on the trace you wish to add the corner to. Place the corner and then drag it to a new position of your choice.

19.2 Simple Board Design

Your task in this section is to design a printed circuit board for the circuit pictured in Figure 19.2.1. The circuit establishes a rhythm at a specific tempo using a 555 timer IC and potentiometer.

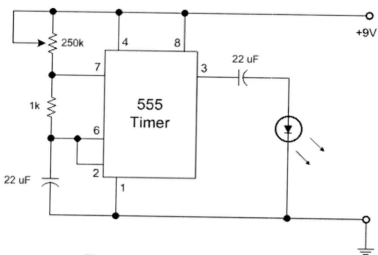

Figure 19.2.1: Metronome Circuit

Figure 19.2.2 shows the detailed pin out for the 555 timer IC. When placing the IC, pay attention to the pin configuration as compared to the schematic. The pins on the schematic typically do not reflect the actual position on the IC, they are placed for component to IC hook up clarity.

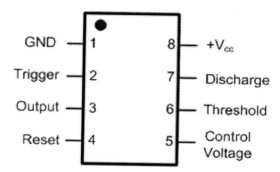

Figure 19.2.2: LM555 Timer

Begin by selecting a 2 layer board (Top and Bottom) with the default via. Don't worry about adjusting the board size at this point. The size of the board will be adjusted later since the size is not a design constraint in this example. Place a 555 timer socket in the middle of the board by inserting **Dip-8 pin** from the library components window. Recall that to access the components you right click in your workspace and select **Component manager**.

Now that the 555 timer socket is in place, design the circuit board following Pin by Pin connections. First connect Pin 2 to Pin 6 across the timer: select **Place a trace** *icon 5*, select the bottom layer (Bottom copper layer) *icon 26*. This places the trace connection on the bottom layer of the PCB. Click on the starting and ending Pins to add a trace, right click to end placing the trace. Double click on the trace and change it to a 0.030" type, (make **all traces for this PCB 0.030" type**). Next work your way out from Pin 3 to the ground: Insert a **LED (T1)** and a **capacitor (Radial Electrolytic 2mm)** and wire them up, using **Place trace** following the schematic in Figure 19.2.1. Select **Place a pad** (*icon 3*) to insert the ground. You should end up with an arrangement similar to Figure 19.2.3.

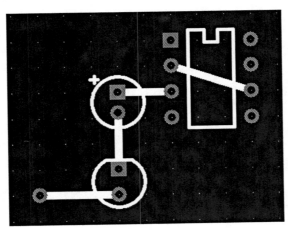

Figure 19.2.3: Board Design - Step 1

Next, place a **capacitor (Radial Electrolytic 2mm)** between Pin 6 and Pin 1. The challenge here is to be able to reproduce the schematic without having traces intersecting with each other. For this purpose notice that Pin 6 and Pin 2 are connected to each other and therefore connecting the capacitor straight to the Pin 6 or the Pin 2 will result in the same circuit. Also, insert a **resistor (0.125W)** between Pin 6 and Pin 7. Now your arrangement should resemble Figure 19.2.4.

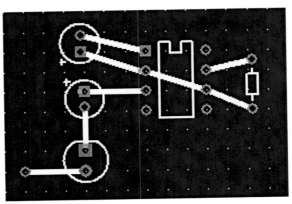

Figure 19.2.4: Board Design - Step 2

The final steps are to insert the **variable resistor (Potentiometer – Bourns 3386F)** between Pin 7 and Pin 4 and to add the **power trace** to Pin 4 and Pin 8. Complete the final connections;

refer to Figure 19.2.1. Now experiment a bit by adding trace corners and rearranging the board trying to create a more compact design. Be careful not to create intersections of traces that are not supposed to cross. Finally, adjust the size and shape of the board to fit your design by selecting and dragging the corners of the yellow perimeter. The arrangement shown in Figure 19.2.5 reflects one possible outcome.

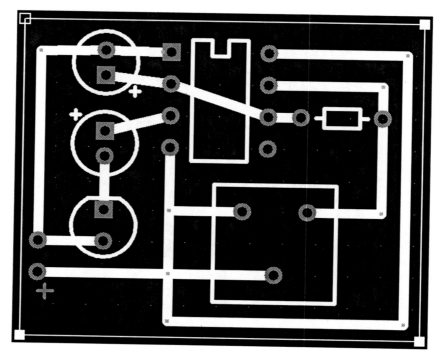

Figure 19.2.5: Metronome Board Design

Chapter 20 Exercises

20.1 MATLAB Basics

1. Evaluate the function $y = \dfrac{x^2 - 6x}{3}$, for x = -5, -2, -1, 0, 1, 2, 5

2. Find the average of the set {42, 84, 85 , 85, 90, 93, 93, 100}

3. Find the following using the arrays $a = \begin{bmatrix} 2 & 5 & 8 \end{bmatrix}$, $b = \begin{bmatrix} 5 & 3 & -1 \end{bmatrix}$

   a. $c = a + b$

   b. $c = a * b$

   c. $c = a.* b$

4. Convert the binary numbers to a decimal.

   a. 10111011

   b. 10101110

5. Convert the decimal numbers to binary.

   a. 1000

   b. 1024

6. Use the listed complex numbers to find the following:

   $$z_1 = 2 + j7 \qquad z_2 = e^{j(45^\circ)}$$

   a. Compute $z_1 + z_2$ c. Compute $z_1 z_2$.   d. Compute $\dfrac{z_2}{z_1}$.

7. Use the matrix inverse method to solve the system of equations.

a.
$$2x+3y+z = 3$$
$$x+3y-z = 6$$
$$2x+2y = 7$$

b.
$$4x-y+z = 10$$
$$y = 3x+z$$
$$2x-8y = 0$$

c.
$$x+y = 3$$
$$2x-3y = 7$$

d.
$$4x-2y = 15$$
$$3y-7z = 0$$
$$5x+4y-z = 3$$

20.2 MATLAB Plotting and M-Files

1. Plot the following on the interval $0 \le x \le 2\pi$, using 5, 20, and 50 points.

a. $y = \cos(x)$

b. $y = \sin(x)$

2. Create a 3-D line plot for $x(t) = t^2$, $y(t) = t^5$, $z(t) = t^8$ on the interval: $0 \le t \le 3$.

3. Plot $y = \cos(x)$, and $z = 1 - \dfrac{x^2}{2} + \dfrac{x^4}{24}$, for $0 \le x \le 2\pi$, on the same figure using:

a. The overlay plots command

b. The subplot command

4. Write a script file that calculates the voltage drop across a resistor, and the power dissipated by the resistor, assume you are given the current through and resistance of the resistor. Make sure the returned value, the program output, includes units.

5. Complete the following:

a. Write a script file that automates the procedure for creating the surface plot, using the code below.

```
x = -10:1:10;
y = x;
[x, y] = meshgrid(x, y);
z = sqrt(x.^2 + y.^2);
surf (x, y, z)
title('3D plot'), xlabel('X'), ylabel('Y'), zlabel('Z');
```

b. Change the script file so that the user is prompted to input the **function z**, and **title**. Your M-File should plot the inputted function and output the inputted plot title. Note the title must be a string; look up the command **input** in MATLAB help to determine the correct syntax.

Section 20.3 Multisim Design

1. A simple voltage divider can be assembled using two resistors:

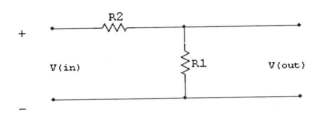

Where

$$V(out) = \left(\frac{R1}{R1 + R2} \right) V(in)$$

Design a voltage divider that

a) Produces a V(out) that is 0.5 the value of V(in).
b) Produces a V(out) that is 0.1 the value of V(in).

For both cases verify your results using Multisim and print out your schematics and instruments for turn in.

2. An RC low-pass filter is a device that passes low frequency signals and blocks high frequency signals. The configuration is:

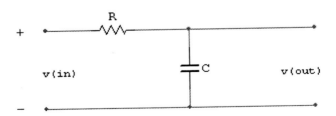

The cut off frequency (f_c) of the filter can be defined as

$$f_c = \frac{1}{2\pi RC}$$

At the cut off frequency, the amplitude of the output signal to the amplitude of the input signal is defined by

$$\frac{v(out)}{v(in)} = 0.707$$

Using this information, design a low-pass filter with a cut off frequency of 10 kHz. Use Multisim to simulate your design using a five volt peak to peak sine wave as your input signal. View both the input and output signals on the oscilloscope and find the frequency where the ratio of the output is approximately 0.707 the magnitude of the input. Print out your circuit and the oscilloscope display. Determine and report the minimal input frequency that causes the output amplitude to go to zero.

3. An RC high-pass filter is a device that passes high frequency signals and blocks low frequency signals. The configuration is:

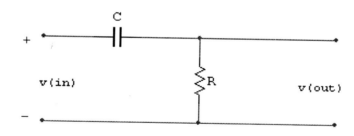

The cut off frequency and ratio of output to input signal magnitudes obey the same equations as the RC low-pass filter. Repeat exercise 2 for the RC high-pass filter, but make your design with a cut off frequency equal to 1 kHz. Print your results and report the maximum input frequency that maintains the output amplitude equal to zero.

4. Design a digital circuit that has one input and two outputs. The circuit should reproduce the input signal and its complement at the outputs. At the input apply a 5 volt 10 Hz clock signal. Use the Logic analyzer to simultaneously view the input and outputs. Turn in a print out of your circuit and the input/outputs displayed on the Logic Analyzer.

5. Using Multisim's Logic Converter.

a. Generate and print the truth table for,

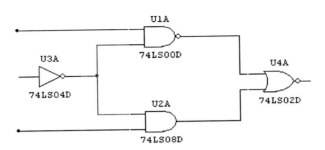

b. Generate and print a Boolean expression for the following truth table, note F is the output.

A	B	C	F
0	0	0	1
0	0	1	0
0	1	0	0
0	1	1	1
1	0	0	1
1	0	1	0
1	1	0	1
1	1	1	0

20.4 RS232 Point-to-Point Communications

1. Using a 9-pin null modem cable connect two computers together using their serial (COM) ports. Note the bare minimum pin configuration needed to establish a connection are pins 2 to 3, 3 to 2, and a signal ground; pin 5 to 5.

   a. Open HyperTerminal in Windows and establish a connection (use HyperTerminal's default settings). Once HyperTerminal is open, on the menu bar, click on **File** then **Properties**. Click on the **Settings tab**, and then open the **ASCII Setup**. Put a check mark in the boxes next to **Send line ends with line feeds and Echo typed characters locally**. Each time you open a new connection in HyperTerminal perform these steps, they allow you to see what you are typing locally and send line feeds to the remote connection. Now send some text.

   b. Set the computers up for different baud rates and try sending some text. You will have to close and reopen the connection to set this. What is the result of mismatched baud rates?

2.   Using a serial cable create a loop back connection for the serial port.

     The loop back pinout is;

     Connect pins 1, 4, and 6 together
     Connect pins 2 and 3 together
     Connect pins 7, 8, and 9 together

     You may need to cut a serial cable and use an ohm meter or a continuity tester to
     figure out which pin on the cable is connected to which wire. The female side of
     your serial cable connects to your computer. Cut the serial cable close to the
     male end. The motivation being when you connect the
     cable to the computer you have as much cable as possible to work with.

     Check your connection by opening HyperTerminal and enter the settings;

     Bits per second: 2400
     Data bits: 7
     Parity: None
     Stop bits: 1
     Flow control: None

     You should see the characters you type echoed back to the screen. Once you are
     sure the loop back is working move on to the next step.

3.   Connect an oscilloscope to monitor the serial data. Using your loop back connector,
     connect the positive side of the oscilloscope probe to pins 2 and 3 and negative side of
     the probe to pin 5 (Gnd). Send several different characters, for example depress and hold
     down the letter "c", while monitoring the signal outputs on your scope.

     a.   What are the voltage levels of the RS232 signal on the oscilloscope?

     b.   What does each level represent, for example which level represents a logic one and a
          logic zero?

     c.   How many bits are used for each character? How many data bits did you set when
          you opened HyperTerminal?

     d.   Sketch the scope output for a lower case "v". On your sketch identify the bits that
          represent Start, Stop, and the Character "v", along with the *least significant bit (LSB)*
          and the *most significant bit (MSB)*.

4. Close your connection and open a new connection, with your loop back still in place, with the settings:

>   Bits per second: 2400
>   Data bits: 8
>   Parity: None
>   Stop bits: 1
>   Flow control: None

a. Repeat part d from step 3.

b. What effect did changing your Data bits settings from 7 to 8 bits produce?

Section 20.5 Op amp Circuits

For all circuits: Use the appropriate data sheets for your op amps to determine pin connections and power supply requirements. Try to use a general purpose 741 amplifier, if available and set the power supplies to $\pm$ 15 volts for powering the op amp. For several of the exercises you will need a variable 5 volt dc supply, as an alternate use a fixed 5 volt dc supply along with the adjustable voltage divider from Appendix D.

1. Design and build an inverting amplifier with the gains, 10, 20, and 100 using resistors in the $k\Omega$ range.

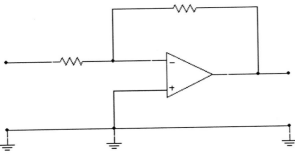

a. For each configuration apply a DC input voltage starting at 100 mV. Increase your input to 1 Volt incrementing by 100 mV steps while monitoring your output.

b. For each value of gain, at what input voltage does the amplifier saturate (the output stops increasing as the input voltage is increased)? Report your $\pm$ op amp supply voltages, along with the input voltage value that caused saturation.

2. Design and build a non-inverting amplifier with the gains, 11, 21, and 101 using resistors in the kΩ range.

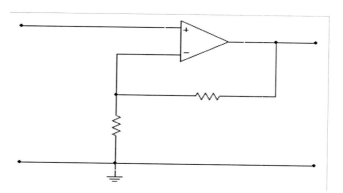

c. For each configuration apply a DC input voltage starting at 100 mV. Increase your input to 1 Volt incrementing by 100 mV steps while monitoring your output.

d. For each value of gain, at what input voltage does the amplifier saturate (the output stops increasing as the input voltage is increased)? Report your ± op amp supply voltages, along with the input voltage value that caused saturation?

3. Design and build an op amp attenuator, as depicted below. Use a variable and fixed resistor combination that allows for an adjustable attenuation from 0% down to 10% of the applied input voltage. Try to choose the fixed resistance between 50 and 100 Ω. Use a 5 volt dc input for your input source.

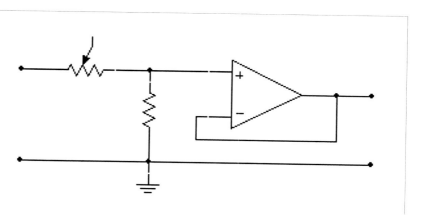

a. What is the minimum and maximum voltage you can produce at the op amp output?

b. Can you produce a zero output, if not what is causing the residual voltage?

c. Switch the position of the adjustable and fixed resistors and determine the maximum and minimum voltage outputs.

4. Design and build a current to voltage amplifier that has a gain of $1V/mA$, as pictured below, with R = to 5 k. Ω Note the left half of the figure will supply a $1\ mA$ current.

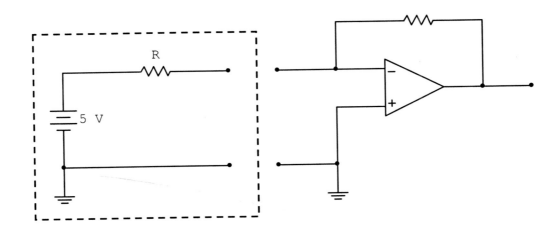

Vary your input current, by varying your 5 volt supply, from 0 to 5 volts. How does the output respond, is the output linear with respect to the input? What is the polarity of the output?

20.6 Diodes: LEDs and Photodiodes

1. Build a circuit, as pictured, using a general purpose silicon pn diode. Use a resistance $R = 1$ kΩ, and apply 5 volts dc to your circuit. This is the diode's forward bias configuration.

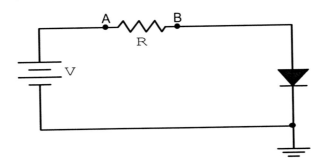

a. Measure the voltage across your resistor, between nodes A and B. Is this voltage equal to 5 volts? Measure the voltage drop across the diode, between node B and ground.

b. Insert a current meter in your circuit between node B and the diode. Then measure and record the current. Replace your resistor with a 10 kΩ resistor and measure the current again. What function does the resistor serve in the circuit?

c. Replace the 10 kΩ resistor with the original 1 kΩ resistor and keep the current meter connected in the circuit. Using an adjustable dc power supply, start with zero volts and slowly increase the voltage until you see a current arise. Be sure you have the current meter set on its most sensitive scale. Measure your applied voltage. Is it equal to or greater than the reported diode turn-on voltage? Is your voltage reading, required to produce a current, the same as your neighbors?

2. Build a LED driver circuit as shown below. Assume your input voltage, to the op amp, has a range equal to 1 to 2 volts DC. Use the non-inverting op amp to amplify the input signal so that the LED will be on for both 1 and 2 volt inputs. The voltage and current specifications will be listed on the LED datasheet typically as a maximum diode forward current $I_{f_{MAX}}$. You must adjust the resistor R so that the circuit current I_f does not exceed the maximum current limit of the LED which is dependant on the output voltage V of the op amp and the resistance R.

$$I_f \cong \frac{V}{R} < I_{f_{MAX}}$$

The equation does not address any forward voltage produced by the diode which would subtract from V. Typically you should choose an operating current 50 to 75% lower than $I_{f_{MAX}}$, to protect against transients or current surges. For example, if $I_{f_{MAX}}$ is reported as 35 mA for a LED you might choose a biasing voltage of 5 volts and a resistor value equal to 250 Ω. This produces an operating current equal to 20 mA.

Vout =

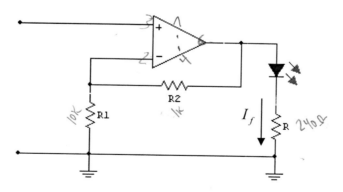

a. Using a standard LED, test your circuit using 1 and 2 volt inputs at the op amp. You should see a reasonably bright output for both inputs. If not, adjust resistor R to increase I_f until a reasonable brightness is obtained. Turn in a schematic with all component values included. Do not disassemble your circuit; you will need it for exercise 3.

3. Assemble the photodiode detector circuit as shown below using a photodiode housed in a fiber optic connector. This is a current to voltage op amp configuration. Start with a 50 kΩ resistor for R.

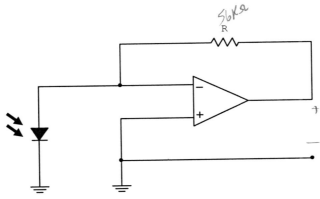

a. Using your LED driver circuit, from exercise 2, switch the general purpose LED out for an LED with a fiber optic connector housing. Connect a fiber between your LED and photodiode and generate a signal using a 2 volt input at the driver circuit (LED) op amp.

b. Increase the gain of your detector circuit, increase R, until your photodiode output is approximately 5 volts in magnitude. Turn in a schematic with all component values included.

20.7 Design Exercise RS232-Fiber Point-to-Point Communications Link

Engineering Process/Example

In this exercise you are required to design, test, and implement an RS232-fiber communications link. Your finished link will be similar to the diagram in Figure 20.7.1.

Figure 20.7.1: Block diagram of RS232-Fiber Communication System

Since this is your first project you will be guided through a predefined design while applying the "Engineering Design Process". The intent is to introduce you to a structured approach and thought process that is useful when confronted with the need to solve design problems. In parts of this exercise, we identify, discuss, and complete the steps for you. In other parts it is up you to fill in the blanks and complete the process. The skills you acquire should better prepare you to solve the other practical design problems in this section. We begin by defining the "Engineering Design Process" in a general fashion, and then apply it directly to the RS232-Fiber Point-to-Point Communication Link. This is intended to be a Hands-On exercise so you should treat it as such.

The Engineering Design Process can be informally viewed as a set of steps and rules that, when applied properly, provide a directed approach, with built-in checks and balances, to solving a design problem. Ultimately solving the design problem will lead to the creation of a new product or the enhancement of an existing product. There is an abundance of information available on the subject of the "Engineering Design Process". The information can be as little as a single page flow chart, or as complex as an entire book covering the topic in great detail. We will limit ourselves to the following six steps;

1. Clearly understand the problem
2. Gather the information necessary to facilitate a solution to the problem
3. Create a set of solutions that solve the problem
4. Choose the solution(s) to pursue
5. Implement, test, and evaluate the solution(s)
6. Final implementation

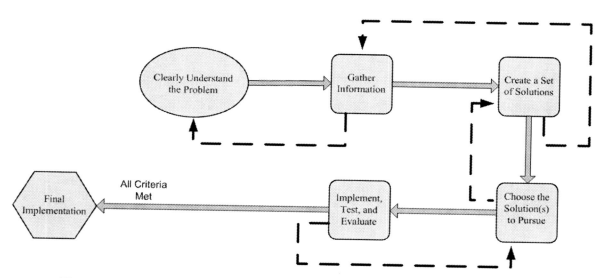

Figure 20.7.2: Engineering Design Process Flow Chart, note that any step in the process may lead back to a previous step.

1. Clearly understand the problem: At the beginning it is crucial the problem be clearly defined and understood. The problem may be associated with an existing design which calls for a modification or it could require the creation of a new product that fulfills a current or future societal need. In this step, you must clearly define the problem, perhaps with a concise problem statement, and all constraints/criteria associated with solving the problem.

2. Gather the information necessary to facilitate a solution to the problem: At this point you should gather information and locate resources required to generate solutions for the problem. You should make use of your past experience; existing solutions/methods if any, do research, and consult other experts in your field, as well as other fields of expertise as the problem dictates.

3. Create a set of solutions that solve the problem: Once you have a clear understanding of the problem and have gathered relevant information you should generate as many solutions as possible that show promise of solving the problem. This step, which is sometimes referred to as the solution brainstorming phase, can be accomplished through individual or group effort. Here you must apply creative thinking and gathered knowledge, from step two, to generate ideas.

4. Choose the solution(s) to pursue: Now that you have a set of plausible solutions, you should analyze and evaluate the pros and cons of each. This step requires thoughtful consideration of such things as; how well does the solution meet the initial constraints, what are each solutions chances of success, difficulty of final product manufacturability, available resources, cost of production, and evaluation of current, as well as future, market demand. You should narrow down the solutions to "one best" however there may be occasions when you wish to pursue multiple solutions simultaneously. For example if the speed at which you reach the implementation stage out weights the initial cost and resources, then it would make more sense to be working on multiple solutions in case one or more do not reach completion or experience unpredicted setbacks.

5. Implement, test, and evaluate the solution(s): Finally you are ready to implement, test, and evaluate your solution(s). For example if your solution involved an electrical circuit, now is the time the circuit should be prototyped and tested. During the testing phase evaluation must take place to determine how well your solution meets the initial constraints and criteria. In the case of the circuit, some questions to be answered may include; 1) does it do what it is supposed to do, 2) are the voltages, currents, and EMI characteristics within tolerances.

6. Final implementation: If all expectations are realized in step five you are now ready to make the final implementation of your solution. For example have your product put into production and marketed. If there are problems encountered you will move back to step one and repeat the process until all constraints are satisfied. More often then not you will find yourself returning to previous steps and refining throughout the entire design process.

Applying the process to the RS232-Fiber Point-to-Point Communications Link

1. **Clearly understand the problem:**

The Problem: Two computer systems at a local company communicate regularly using serial communications. Every summer the data between the two systems becomes corrupted. For no apparent reason at random times during the day the communication link fails due the excessive data errors.

After further investigation by the company's tech support team, requested by a staff engineer, the following information is found.

- The computer systems are checked out with loop-back connecters and no local problems are found with the computer hardware

- The communication software is tested and found to be working properly on both systems

- The serial cable connecting the two systems is replaced with new cable and it is found that the problem still exists

- During the cable replacement it is noted that the cable runs very close to the buildings Air Conditioning unit which is suspected as a possible interference source

- Further investigation shows that the data errors occur only when the Air Conditioning unit in running

- After careful consideration it is decided that connecting the two systems via a fiber optic serial link should fix the problem.

The information above is part of the process leading to clearly understanding the problem. Note after identifying the problem several steps were necessary to come to the conclusion that the buildings Air Conditioning system is the most probable cause of the problem.

The following information identifies the solutions constraints/criteria. In this case the list would more than likely be developed by the Engineer tasked with solving the problem and the company's tech support team.

The design/system constraints

1. Windows HyperTerminal will be used as the communications software.
2. The 9-pin serial port will be the links interface to the local and remote computers.
3. The over all design should accomplish the following tasks:

   a. Convert the local computer transmitted serial data into an optical signal, of appropriate signal levels, that will be transmitted to a remote computer over a multimode fiber cable for processing.

   b. Convert optical singles received from the remote computer, via a separate fiber cable, to RS232 signal levels that are to be relayed to the local computer for processing.

   c. The system should have an acceptable bit error rate, comparable to standard RS232 serial communications.

4. Transmitted data will be converted from RS232 and sent over the fiber link using a fiber housed LED.
5. Received data will be detected with a standard fiber housed photodiode and converted to an RS232 signal so it can be processed by the local computer.
6. When complete, the device should connect to a local computer via a serial cable and to a remote computer via a pair of 25 meter fiber optic cables.
7. The link will only use three wires of the serial cable; pin 2 (Receive Data), pin 3 (Transmitted Data), and pin 5 (Common, or Signal Ground).

2. Gather the information necessary to facilitate a solution to the problem:

The following is a list of design questions you should address. Writing down a list can help you stay focused and provides a systematic approach when completing this step of the "Design Process". By answering the questions you will be gathering necessary information and knowledge to generate solutions.

Transmitter

A. What does the serial data look like coming from the RS232 port? What are the signal levels, which level represents a logic-1, and which represents a logic-0?

B. What type of signal inputs will the fiber housed LED accept, transmit?

C. What signal levels generated by the LED will represent a logic-1 and logic-0?

D. What circuitry will be needed to bridge the RS232 signal to the LED?

Receiver

A. What does the data look like arriving at, and leaving, the fiber housed photodiode, what are the signal levels, which level represents a logic-1, and which represents a logic-0?

B. How do I handle noise or imperfect signals at the receiver?

C. What signal inputs will the RS232 port recognize?

D. What circuitry will be needed to bridge the data from the photodiode to the RS232 port?

3. Create a set of solutions that solve the problem:
4. Choose the solution(s) to pursue:

At this point you would generate as many solutions as possible that show promise of solving the problem. That is choosing different configurations and/or components to create a RS232-Fiber Point-to-Point Communications Link that will satisfy the requirements from step one. Here you must apply creative thinking and the knowledge gathered in step two to generate ideas. After generating possible solutions, in step four you would narrow down your solutions to "one best solution". Since this is a predefined exercise we will dictate that you use the solution that is schematically represented below. You are encouraged to see how many different solutions you can generate on your own, there are several.

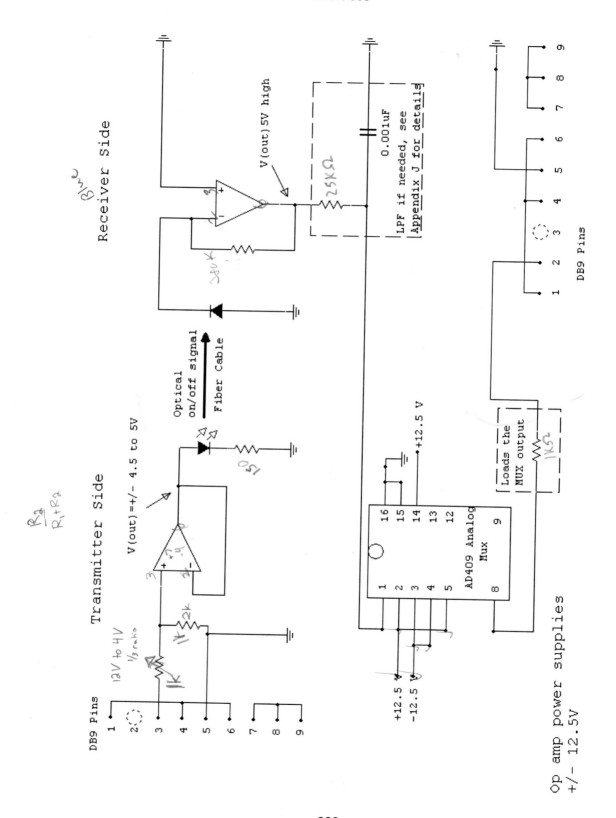

5. Implement, test, and evaluate the solution(s):

Presented below is an outlined step by step procedure that will guide you through building, testing, and evaluating the RS232-Fiber Point-to-Point Communications Link. This might be a similar outline you would create to fulfill the requirements in step five of the "Engineering Design process".

Step	Procedure
	Transmitter Design
1	Identify and pinout the RS232 cable wires by connecting them to a breadboard using terminal blocks. Be systematic; note the shield wire has no connection required. After attaching all pins, recheck your work for correct configuration. Use a Sharpie, or other method, to identify pin 1 at the terminal blocks.
2	Using the breadboard connectors, wire your terminal blocks for a loop back connection. Connect the serial cable to a computer and open HyperTerminal to test your wiring configuration. The pin outs for the loop back are 1-4-6, 7-8-9, and 2-3 with 5 as *Signal ground*.
3	Design and build an attenuator circuit that will lower the RS232 signal levels to ± 4 to 5 volts. Use a simple voltage divider and buffer op amp for this task. Be sure to check your RS232 signal levels before calculating the attenuator gain. Once complete, wire in the *Transmitted data* wire (pin 3) and *Signal ground* (pin 5) to the attenuator and verify the attenuated output levels, at the op amp output, using an oscilloscope. At this point leave *Receive data* (pin 2) connected to *Transmitted data* (pin 3) so you can see the character you are sending echoed in HyperTerminal.
4	Connect a standard LED, visible, with a current limiting resistor at the attenuator op amp output. Choose your resistor by factoring in the output voltage and the operating current you wish to set for the LED, typically the operating current will be 30 to 35 mA . Once in place retest your configuration by sending data from HyperTerminal; you should see the LED blinking. Once the LED is working properly, the final step to complete the transmitter section is to replace the visible LED with a fiber housed LED.
	Receiver Design
5	Design and assemble an optical receiver using a fiber housed photodiode and an op amp current-to-voltage amplifier. At this point do not bias the photodiode, connect it's anode to ground and the cathode to the input on the current-to-voltage amplifier (produces neg. current or positive output at op amp). Start out with a 100 kOhm feedback resistor for the current-to-voltage amplifier. Once assembled, connect an optical fiber between the transmitter LED and photodiode. Connect an oscilloscope to the output of the current-to-voltage amplifier. Using HyperTerminal send a continuous string of characters and observe the output on the oscilloscope. You should have a signal that resembles a non symmetric square wave with amplitude ranging from a small negative voltage to a positive 4.5 to 5 volts. If necessary change your feedback resistor to achieve the appropriate signal levels.
6	Add a low-pass RC filter at the output of the current-to-voltage amplifier as detailed on your circuit schematic. Recheck the signal; you should have a cleaner, better defined, output. The details on the low-pass RC filter can be found in Appendix I.
7	Next, you will need to convert the signal to RS232 levels. There are multiple options for accomplishing this task. A very simple approach is to use an analog MUX, which is basically a selector switch that allows for the selection of multiple output signals determined by an addressable input. For this design CH 1 of the MUX is supplied with -12.5V and CH 2 with +12.5V. The *Analog Devices* AD409 is the device used in your circuit schematic, look through its data sheet for the details of its operation.

8	Wire in the AD409 as depicted in the schematic, be sure to include the 1 kΩ resistor between the output of the MUX and the *Receive data* (pin 2) and remove your loop-back connection, the jumper wire connecting pins 2-3.

6. Final implementation:

You should now be able to view the characters you send in HyperTerminal just as you would with a standard loop-back connection. Next, remove the fiber connecting your LED and photodiode and use 2 pieces of fiber to connect your transceiver to your neighbor's transceiver. The two computers should be able to communicate, try sending some data to confirm this.

If all goes well you are now ready to implement your final product. If there are malfunctions you should return to step one and apply the steps to the problem at hand.

Appendix A

Resistor Color Codes

Color		Numeric 1	Numeric 2	Numeric 3	Multiplier	Tolerance
Black		0	0	0	1	
Brown		1	1	1	10	±1%
Red		2	2	2	100	±2%
Orange		3	3	3	1K	
Yellow		4	4	4	10K	
Green		5	5	5	100K	±0.5%
Blue		6	6	6	1M	±0.25%
Violet		7	7	7	10M	±0.10
Gray		8	8	8		±0.05%
White		9	9	9		
Gold					0.1	5%
Silver					0.01	10%

Four band Resistor

1. The first two bands indicate numeric values.
2. The third band is the multiplier (power of ten).
3. The fourth band is the tolerance.

Example Calculation:

Red	Green	Yellow	Silver
2	5	10k	10%

Resistor value = 25 x 10k = 250 kΩ ± 10%

Five band Resistors (High precision)

1. The first three bands are numerical values.
2. The fourth band is the multiplier.
3. The fifth band is the tolerance.

Example Calculation:

Blue	Gray	Red	Blue	Brown
6	8	2	1M	1%

Resistor value = 682 x 1M = 682 MΩ ± 1%

Appendix B

Average (DC) and Root-Mean-Square (RMS) Values

1. Average Value

The average or the mean value of a function, $f(t)$, over the interval (t_1, t_2) is defined as;

$$F_{avg} = \frac{1}{(t_1 - t_2)} \int_{t_1}^{t_2} f(t)dt \tag{0.1}$$

For periodic functions with period T, the average value also known as the dc value is calculated over one period. That is,

$$F_{avg} = F_{dc} = \frac{1}{T} \int_{t_0}^{t_0+T} f(t)dt \tag{0.2}$$

Where t_0 is an arbitrary time.

Typically AC voltages and currents are represented with a sinusoidal function of the form,

$$f(t) = A\sin(2\pi ft + \phi), \text{ or} \tag{0.3}$$

$$f(t) = A\sin(\frac{2\pi}{T}t + \phi), \text{ or} \tag{0.4}$$

$$f(t) = A\sin(\omega t + \phi). \tag{0.5}$$

Where A = amplitude, f = frequency in Hz (Hertz), $T = \frac{1}{f}$ = period, in second, $\omega = 2\pi f$ =angular frequency, in radians, and ϕ = relative phase.

For a pure sinusoidal signal, $v(t) = V_0 \sin(2\pi ft + \phi)$, Equation 1.2 takes the form of

$$V_{avg} = \frac{1}{T} \int_{t_0}^{t_0+T} V_0 \sin(2\pi ft + \phi)dt. \tag{0.6}$$

.

For a sinusoidal voltage that includes a dc component, V_{dc}, Equation (1.3) becomes,

$$V_{avg} = \frac{1}{T} \int_{t_0}^{t_0 + T} V_{dc} + V_0 \sin(2\pi ft + \phi) dt \qquad (0.7)$$

2. Root-Mean-Square (RMS) Values

Equation (1.1) calculates the mean value of a function $f(t)$. Using (1.1), we can calculate the mean value of $f^2(t)$, the square of $f(t)$, that is:

$$\text{Average Value of } f^2 = \frac{1}{(t_2 - t_1)} \int_{t_1}^{t_2} f^2(t) dt \qquad (0.8)$$

The Average value of $f^2(t)$, equation (1.5), is also called "mean square" value of $f(t)$. Taking the square root of the equation (1,5), we obtain the root mean square, "RMS" value of f(t)., that is

:

$$F_{RMS} = \sqrt{\frac{1}{(t_2 - t_1)} \int_{t_1}^{t_2} f^2(t) dt} \qquad (0.9)$$

To describe AC voltages and currents, RMS values are used due to the fact that; the power that is dissipated by a device or load, supplied by a DC source, will require an equivalent RMS AC voltage equal to the magnitude of a DC voltage.

The RMS value for periodic functions, $f(t)$, is given by:

$$F_{RMS} = \sqrt{\frac{1}{T} \int_0^T f^2(t) dt} \qquad (0.10)$$

Here we assumed 0 for t_0.

Setting the phase in Equation 1.2 equal to zero, and combining with Equation 1.1 yields,

$$F_{RMS} = \sqrt{\frac{A^2}{T} \int_0^T \sin^2\left(\frac{2\pi}{T} t\right) dt} \qquad (0.11)$$

Using a standard integration table Equation 1.3 evaluates to,

$$F_{RMS} = \frac{A}{\sqrt{2}} = 0.7071A \tag{0.12}$$

Equation 1.4 can be used to RMS value of a sinusoidal voltage or current by replacing;

$$F_{RMS} \text{ with } V_{RMS} \text{ or } I_{RMS}$$

$$\tag{0.13}$$

$$A \text{ with } V_{peak} \text{ or } I_{peak}$$

Referenced: *Introductory Electronics for Scientists and Engineers 2ed*, Robert E. Simpson, Prentice Hall, 1987.

Appendix C

Optimization of Power Transfer

To determine optimal power transfer, consider the circuit in Figure C.1.

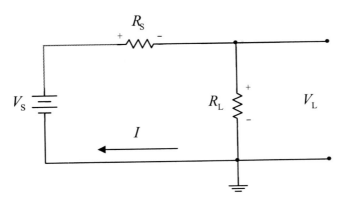

Figure C.1: Optimizing for power transfer

From Figure C.1 we can write;

$$I = \left(\frac{V_S}{R_S + R_L} \right), \quad V_L = \left(\frac{V_S R_L}{R_S + R_L} \right). \tag{1.1}$$

The power expression is,

$$P = IV_L = \left(\frac{V_S^2 R_L}{R_L^2 + 2 R_S R_L + R_S^2} \right). \tag{1.2}$$

At this point define the values for V_S and R_S as constants then,

$$P(R_L) = \left(\frac{V_S^2 R_L}{R_L^2 + 2 R_S R_L + R_S^2} \right) \tag{1.3}$$

Now the power has been reduced to a function of the load impedance R_L, to find the maximum power transfer we must find the extrema by evaluating the expression

$$\frac{dP(R_L)}{dR_L} = 0 \tag{1.4}$$

Recall that the derivative can be found using

$$\frac{d}{dx}\left(\frac{u}{v}\right) = \frac{v(du/dx) - u(dv/dx)}{v^2}$$

(1.5)

To this end

$$P(R_L) = \left(\frac{V_S^2 R_L}{R_L^2 + 2R_S R_L + R_S^2}\right)$$

(1.6)

$$u = V_S^2 R_L, \quad v = R_L^2 + 2R_S R_L + R_S^2, \quad v^2 = \left(R_L^2 + 2R_S R_L + R_S^2\right)^2$$

$$\frac{d}{dR_L}\left(V_S^2 R_L\right) = V_S^2, \quad \frac{d}{dR_L}\left(R_L^2 + 2R_S R_L + R_S^2\right) = 2R_L + 2R_S$$

(1.7)

It follows then;

$$\frac{dP(R_L)}{dR_L} = \left(\frac{\left(V_S^2 R_L^2 + 2V_S^2 R_S R_L + V_S^2 R_S^2\right) - \left(2V_S^2 R_L^2 + 2V_S^2 R_S R_L\right)}{\left(R_L^2 + 2R_S R_L + R_S^2\right)^2}\right)$$

(1.8)

After simplifying,

$$\frac{dP(R_L)}{dR_L} = \frac{V_S^2 R_S^2 - V_S^2 R_L^2}{\left(R_L^2 + 2R_S R_L + R_S^2\right)^2}$$

(1.9)

Finally we set Equation 1.9 equal to zero and solve for the roots.

$$\frac{V_S^2 R_S^2 - V_S^2 R_L^2}{\left(R_L^2 + 2R_S R_L + R_S^2\right)^2} = 0$$

$$V_S^2 R_S^2 - V_S^2 R_L^2 = 0$$

$$R_L^2 = R_S^2$$

$$R_L = R_S$$

Therefore for the circuit model in Figure C.1, maximum power transfer always occurs when $R_L = R_S$ and is independent of the source voltage.

Appendix D

Adjustable Voltage Divider

Consider the following circuit in Figure D.1.

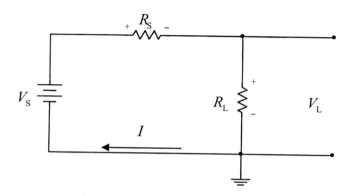

Figure D.1: Fixed voltage divider

This is an example of a simple voltage divider consisting of two resistors and is described by the equation

$$V_L = \left(\frac{R_L}{R_L + R_S} \right) V_S .$$

Here R_L is the resistor V_L is taken across.

For fixed values of R_S and R_L the ratio of the output voltage to the source voltage is fixed. However, if R_S is replaced with a variable resistor it is possible to change the values of the output voltage for a fixed source voltage.

This arrangement is depicted in Figure D.2.

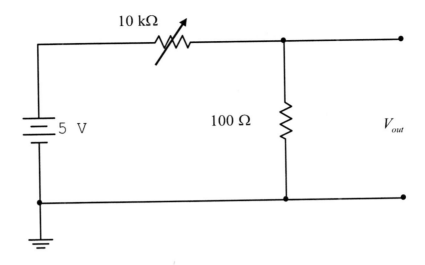

Figure D.2: Adjustable voltage divider.

In this configuration V_{out} is a function of the adjustable 10 kΩ resistor. When the adjustable resistance is set to zero, V_{out} is 5 volts. When the adjustable resistance is set to 10 kΩ, V_{out} is approximately 50 mV. It is recommended to use a Multi-turn Potentiometer for finer output voltage control.

Appendix E

Viewing RS232 Signals with an Oscilloscope

Oscilloscopes are well suited for viewing periodic waveforms by design. The serial data from the RS232 port is transmitted a bit at a time and not periodic. However, it is possible to create a periodic data stream by simply sending the same ASCII character repetitively. For example, in HyperTerminal depressing the letter "v" on the keyboard will send "v" over and over until you release the key. You can view the serial output on an oscilloscope, using a loop-back connection, by attaching a test probe to pin 3 (Transmitted data) and pin 5 (Ground). The loop-back connection is realized by connecting pins 1-4-6 together, pins 7-8-9 together, and finally connecting pins 2 and 3 together. If you are using a digital storage oscilloscope you can record the data being sent, if not you must continue to send the character while viewing. Figure D.1 shows the detailed output resulting from the ASCII character "v" on an oscilloscope.

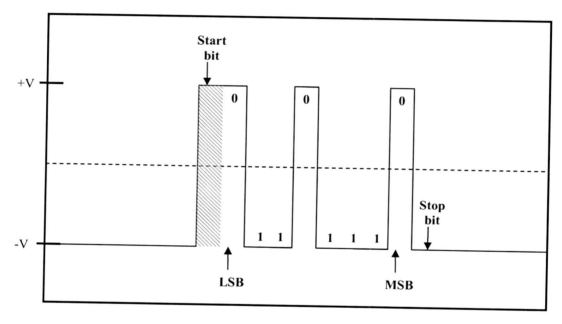

Figure D.1: ASCII character "v", Data bits: 8, Parity: none, and Stop bit: 1, on the oscilloscope display.

For RS232 a logic-0 is represented by a positive signal level, typically 8 to 12 volts, and a logic-1 is represented by a negative signal level, typically -8 to -12 volts. From Figure D.1 notice that the signal also includes a *start bit* and a *stop bit*. It is important to note the first bit sent after the *start bit* is the *least significant bit (LSB)* and the *most significant bit (MSB)* is sent last. So from Figure D.1 the ASCII character "v" is

0	1	1	1	0	1	1	0

here, the *start* and *stop bits* have been truncated.

Appendix F

PN Junction Bias Voltage

Under perfect equilibrium conditions there is no net current inside a pn device, so that the current densities, represented by J_i, can be expressed as,

$$J_n = J_p = J_{NET} = 0.$$ (1.1)

Choosing to work with only one of the current terms from Equation 1.1 and the standard textbook definitions for J_{n_Drift} and J_{n_Diff} we can expand as,

$$J_n = J_{n\_Drift} + J_{n\_Diff}$$

$$J_n = q\mu_n nE + qD_n \frac{dn}{dx}$$

$$q\mu_n nE + qD_n \frac{dn}{dx} = 0$$

$$E = -\frac{D_n}{\mu_n} \frac{1}{n} \frac{dn}{dx}$$

$$\frac{D_n}{\mu_n} = \frac{KT}{q}$$

Here the expression $D_n/\mu_n = KT/q$ is the Einstein relationship for electrons. After substitution,

$$E = -\frac{KT}{q} \frac{1}{n} \frac{dn}{dx}.$$ (1.2)

Writing the definition of the electric field and substituting V_{bi} for V leads to;

$$V_{bi}(x) = -\int_{-\infty}^{\infty} E(x)dx$$ (1.3)

Using Equation 1.2 in Equation 1.3 produces

$$V_{bi} = -\frac{KT}{q} \int_{-\infty}^{\infty} \frac{1}{n} \frac{dn}{dx} dx \qquad (1.4)$$

$$= -\frac{KT}{q} \int_{-\infty}^{\infty} \frac{dn}{n},$$

$$V_{bi} = \left(\frac{KT}{q}\right) [\ln(n)]_{-\infty}^{+\infty}. \qquad (1.5)$$

It is common practice to define negative infinity to lie on the p -side of the semiconductor where the " n " concentration will be " n_p ", approximately zero, and positive infinity on the n -side where " n " is " n_n ".

Then assuming equilibrium, that is $p = N_A$ and $n = N_D$, and using the np -product, $n_i^2 = np$, we can write

$$n(-\infty) = n_p = \frac{n_i^2}{N_A}$$

$$n(\infty) = n_n = \frac{n_i^2}{N_D}.$$

Substitution into Equation 1.5 yields

$$V_{bi} = \frac{KT}{q} \left[\ln(n_p) - \ln(n_n) \right]$$

$$= \frac{KT}{q} \ln \left[\frac{N_D N_A}{n_i^2} \right]$$

$$V_{bi} = \frac{KT}{q} \ln \left[\frac{N_D N_A}{n_i^2} \right]. \qquad (1.6)$$

Referenced: Semiconductor Fundamentals, Volume 1, 2ed, Robert F. Pierret, Addison-Wesley, 1988

Appendix G

Indices of Refraction

Vacuum	1.000	Ethyl Alcohol	1.362
Air	1.00028	Ice	1.31
Water	1.333	Crown Glass	1.50-1.62
Gallium arsenide	3.6	Silicon	3.4

Frequency dependence of the indices:

The frequency dependence of the index of refraction for a given material is known as the Chromatic dispersion of the material. Basically, light of different wavelengths travels at different speeds though the same material; if this were not true white light would not separate into its constituents when passing through a prism. This is important for fiber optics systems since all LED and Laser sources transmit multiple wavelengths, although laser sources typically have a much narrower bandwidth then an LED. The different speeds at which different wavelengths of light travel in the fiber causes overlap of the signals and reduces the length of fiber over which a discernable signal can be recovered. Typically, for longer wavelengths of light the index of a material decreases. For optical fiber typically, a graph or table will accompany the product, which defines the refractive index for different source wavelengths.

Appendix H

American Standard Code for Information Interchange (ASCII) Table

The ASCII code uses seven binary bits to represent 128 different possible characters including upper and lower case letters, numbers, punctuation, and symbols.

Decimal	Binary	Symbol	Decimal	Binary	Symbol	Decimal	Binary	Symbol
000	0000000	(Null)	043	0101011	+	086	1010110	V
001	0000001	SOH	044	0101100	,	087	1010111	W
002	0000010	STX	045	0101101	-	088	1011000	X
003	0000011	ETX	046	0101110	.	089	1011001	Y
004	0000100	EOT	047	0101111	/	090	1011010	Z
005	0000101	ENQ	048	0110000	0	091	1011011	[
006	0000110	ACK	049	0110001	1	092	1011100	\
007	0000111	BEL	050	0110010	2	093	1011101	]
008	0001000	BS	051	0110011	3	094	1011110	^
009	0001001	HT	052	0110100	4	095	1011111	_
010	0001010	LF	053	0110101	5	096	1100000	`
011	0001011	VT	054	0110110	6	097	1100001	a
012	0001100	FF	055	0110111	7	098	1100010	b
013	0001101	CR	056	0111000	8	099	1100011	c
014	0001110	SO	057	0111001	9	100	1100100	d
015	0001111	SI	058	0111010	:	101	1100101	e
016	0010000	DLE	059	0111011	;	102	1100110	f
017	0010001	DC1	060	0111100	<	103	1100111	g
018	0010010	DC2	061	0111101	=	104	1101000	h
019	0010011	DC3	062	0111110	>	105	1101001	i
020	0010100	DC4	063	0111111	?	106	1101010	j
021	0010101	NAK	064	1000000	@	107	1101011	k
022	0010110	SYN	065	1000001	A	108	1101100	l
023	0010111	ETB	066	1000010	B	109	1101101	m
024	0011000	CAN	067	1000011	C	110	1101110	n
025	0011001	EM	068	1000100	D	111	1101111	o
026	0011010	SUB	069	1000101	E	112	1110000	p
027	0011011	ESC	070	1000110	F	113	1110001	q
028	0011100	FS	071	1000111	G	114	1110010	r
029	0011101	GS	072	1001000	H	115	1110011	s
030	0011110	RS	073	1001001	I	116	1110100	t
031	0011111	US	074	1001010	J	117	1110101	u
032	0100000	(Space)	075	1001011	K	118	1110110	v
033	0100001	!	076	1001100	L	119	1110111	w
034	0100010	"	077	1001101	M	120	1111000	x
035	0100011	#	078	1001110	N	121	1111001	y
036	0100100	$	079	1001111	O	122	1111010	z
037	0100101	%	080	1010000	P	123	1111011	{
038	0100110	&	081	1010001	Q	124	1111100		
039	0100111	'	082	1010010	R	125	1111101	}
040	0101000	(	083	1010011	S	126	1111110	~
041	0101001	)	084	1010100	T	127	1111111	DEL
042	0101010	*	085	1010101	U			

Appendix I

RC Low-Pass Filter

A simple RC low-pass filter is an inexpensive device used for high frequency noise attenuation.

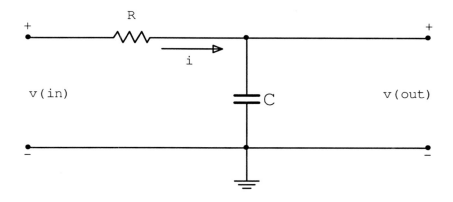

The filter design includes a capacitor so we need to define capacitive reactance (X_c), which can be thought of as a frequency dependant resistor.

$$X_c = \frac{1}{j\omega C}$$

ω = angular frequency, or $2\pi f$

C = capacitance

Writing expressions for the input and output voltages, v(in) and v(out);

$$v(in) = iR + iX_C$$

$$v(out) = iX_C$$

Notice we can create a gain (G) expression if we write the ratio,

$$\frac{v(out)}{v(in)} \geq 1, \; G \geq 1, \qquad \frac{v(out)}{v(in)} < 1, \; G < 1$$

Substituting in the expressions yields,

$$\frac{v(out)}{v(in)} = \frac{iX_C}{iR + iX_C} = \frac{X_C}{R + X_C} = \frac{1}{1 + R/X_C}$$

$$\text{recall}: X_C = \frac{1}{j\omega C}, \text{ then}$$

$$\frac{v(out)}{v(in)} = \frac{1}{1 + j\omega RC}.$$

We are interested in the ratio of amplitudes, which is found by taking the magnitude of the expression.

$$\left|\frac{v(out)}{v(in)}\right| = \left[\left(\frac{1}{1 + j\omega RC}\right)\left(\frac{1}{1 - j\omega RC}\right)\right]^{1/2}$$

$$\left|\frac{v(out)}{v(in)}\right| = \frac{1}{\sqrt{1 + \omega^2 R^2 C^2}}$$

So the final expression for the low-pass filter is:

$$\left|\frac{v(out)}{v(in)}\right| = \frac{1}{\sqrt{1 + \omega^2 R^2 C^2}}$$

From the equation, it should be clear that for a fixed value R and C, the filter passes low frequency signals and blocks high frequency signals. It is characterized by a defined cutoff frequency;

$$f_c = \frac{1}{2\pi RC}$$

The output amplitude with respect to the input at the cutoff frequency is expressed as

$$\frac{v(out)}{v(in)} = 0.707$$

Plot of frequency response for the LPF

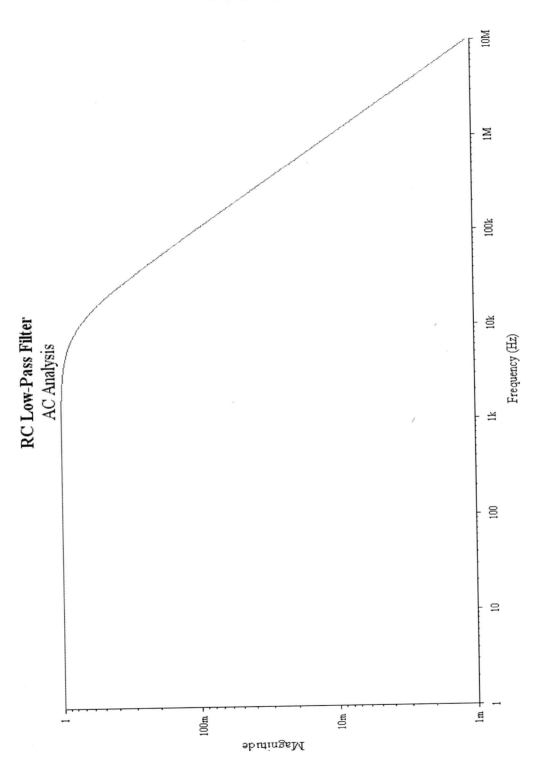

REFERENCES

The following publications were referred to during the creation of this text. Since the goal of this text was to present an overview of the subject matter, you are encouraged to consult these sources should you require additional or more in-depth information on any of the topics that were presented.

- Fraleigh Beauregard, *Linear Algebra*, New York: Addison-Wesley Publishing Company, 1995.

- Alan Clements, *Microprocessor Systems Design*, Boston, MA: PWS Publishing Company, 1997.

- Thomas L. Floyd, *Electric Circuits Fundamentals*, Upper Saddle River, NJ: Prentice Hall, 2004.

- Rafael C. Gonzalez and Richard E. Woods, *Digital Image Processing*, Upper Saddle River, NJ: Prentice Hall, 2002.

- Charles A. Gross, *Power System Analysis*, New York: John Wiley & Sons, 1986

- Simon Haykin, *Communication Systems*, New York: John Wiley & Sons, 1994.

- Roland E. Larson, Robery P. Hostetler, and Bruce H. Edwards, *Calculus with Analytic Geometry*, Lexington, MA: D. C. Heath and Company, 1994.

- M. Morris Mano, *Digital Design*, Upper Saddle River, NJ: Prentice Hall, 2001.

- Carl D. Meyer, *Matrix Analysis and Applied Linear Algebra*, Philadelphia, PA: Siam, 2000.

- Sophocles J. Orfanidis, *Introduction to Signal Processing*, Upper Saddle River, NJ: Prentice Hall, 1996.

- Behrooz Parhami, *Computer Architecture: From Microprocessors to Supercomputers*, New York: Oxford University Press, 2005.

- Samir S. Soliman and Mandyam D. Srinath, *Continuous and Discrete Signals and Systems*, Upper Saddle River, NJ: Prentice Hall, 1998.

- Andrew S. Tanenbaum, *Computer Networks*, Upper Saddle River, NJ: Prentice Hall, 1996.

- Paul A. Tipler, *Physics for Scientists and Engineers*, New York: Worth, 1991.

- Rodger E. Ziemer, William H. Tranter, and D. Ronald Fannin, *Signals & Systems: Continuous and Discrete*, Upper Saddle River, NJ: Prentice Hall, 1998.

References

- William J. Palm III, *Introduction to MATLAB 6 for Engineers,* New York: McGraw Hill, 2001.
- Douglas E. Comer, *Computer Networks and Internets 4th Edition,* Upper Saddle River, NJ: Prentice Hall, 2004.
- Martin D. Seyer, *RS-232 MADE EASY 2nd Edition,* Upper Saddle River, NJ: Prentice Hall, NJ, 1991
- ARC Electronics, *RS-232 Data Interface,* http://www.arcelect.com/rs232.htm
- Harold Kolimbiris, *Digital Communications Systems,* Upper Saddle River, NJ: Prentice Hall, 2000.
- Robert E. Simpson, *Introductory Electronics for Scientist and Engineers 2nd Edition,* Upper Saddle River, NJ: Prentice Hall, 1987.
- Robert F. Pierret, *Semiconductor Fundamentals 2nd Edition,* New York: Addision-Wesley Publishing Company, 1988.
- Gerold W. Neudeck, *The PN Junction Diode 2nd Edition,* New York: Addision-Wesley Publishing Company, 1989.
- Bahaa E. A. Saleh and Malvin Carl Teich, *Fundamentals of Photonics,* New York: John Wiley & Sons, Inc, 1991.
- Adel S. Sedra and Kenneth C. Smith, *Microelectronic Circuits 5th Edition,* New York: Oxford University Press, 2004.
- Seyyed Diablo Khandani, *Engineering Design Process Education Transfer Plan,* Valley College, Pleasant Hill California, 2005.
- G. C. Orsak, S. L. Wood, S. C. Douglas, D. C. Munson Jr., J. R. Treichler, R. Athale, M. A. Yoder, *Engineering Our Digital Future,* Upper Saddle River, NJ: Prentice Hall, New Jersey, 2004.

ACKNOWLEDGEMENTS

The authors would like to gratefully acknowledge the contribution of two UAH students, Yoshito Kanamori and Damien Galzi, who contributed to the development of chapters 15 (Module 3 – Serial Communications) and 19 (Module 7 – Printed Circuit Board Design).

INDEX

Index

Printed in the United States
152192LV00004B/2/A

9 781581 129717